Glacial Geology

Glacial Geology

Edited by **Brandon Holland**

SYRAWOOD
PUBLISHING HOUSE
New York

Published by Syrawood Publishing House,
750 Third Avenue, 9th Floor,
New York, NY 10017, USA
www.syrawoodpublishinghouse.com

Glacial Geology
Edited by Brandon Holland

International Standard Book Number: 978-1-64740-116-0 (Hardback)

Cataloging-in-Publication Data

Glacial geology / edited by Brandon Holland.
p. cm.
Includes bibliographical references and index.
ISBN 978-1-64740-116-0
1. Glaciers. 2. Glacial landforms. 3. Glacial epoch. 4. Geology, Stratigraphic. I. Holland, Brandon.
GB2405 .G53 2022
551.312--dc23

TABLE OF CONTENTS

PREFACE

Glaciers are the massive bodies of ice that are formed from the accumulation of snow over a long period of time. Glacial geology is a scientific study that deals with the natural process that involves glaciers and the natural phenomena that are related to ice. It is an interdisciplinary field that assimilates physical geography, geophysics, geology, climatology, meteorology, etc. There are two types of glaciation, alpine and continental glaciation. The rate of movement of ice depends upon factors such as temperature of the ice, gradient of the slope, thickness of the glacier and sub glacial water dynamics. There are three zones of glaciers, namely, accumulation, ablation and snow line. Study within glaciology includes two major areas namely glacial history and the reconstruction of past glaciation. This field also deals with the study of glacial deposits and glacial erosive features on the landscape. This book unravels the recent studies in the field of glacial geology. It presents researches and studies performed by experts across the globe. Those in search of information to further their knowledge will be greatly assisted by this book.

This book is the end result of constructive efforts and intensive research done by experts in this field. The aim of this book is to enlighten the readers with recent information in this area of research. The information provided in this profound book would serve as a valuable reference to students and researchers in this field.

At the end, I would like to thank all the authors for devoting their precious time and providing their valuable contribution to this book. I would also like to express my gratitude to my fellow colleagues who encouraged me throughout the process.

Editor

Ice crystal *c*-axis orientation and mean grain size measurements from the Dome Summit South ice core, Law Dome, East Antarctica

Adam Treverrow[1], Li Jun[2], and Tim H. Jacka[1]

[1]Antarctic Climate and Ecosystems Cooperative Research Centre, University of Tasmania, Hobart 7004, Australia

[2]SGT Inc., NASA Goddard Space Flight Center, Greenbelt, MD, USA

Correspondence to: Adam Treverrow (adam.treverrow@utas.edu.au)

Abstract. We present measurements of crystal *c*-axis orientations and mean grain area from the Dome Summit South (DSS) ice core drilled on Law Dome, East Antarctica. All measurements were made on location at the borehole site during drilling operations. The data are from 185 individual thin sections obtained between a depth of 117 m below the surface and the bottom of the DSS core at a depth of 1196 m. The median number of *c*-axis orientations recorded in each thin section was 100, with values ranging from 5 through to 111 orientations. The data from all 185 thin sections are provided in a single comma-separated value (csv) formatted file which contains the *c*-axis orientations in polar coordinates, depth information for each core section from which the data were obtained, the mean grain area calculated for each thin section and other data related to the drilling site. The data set is also available as a MATLAB™ structure array. Additionally, the *c*-axis orientation data from each of the 185 thin sections are summarized graphically in figures containing a Schmidt diagram, histogram of *c*-axis colatitudes and rose plot of *c*-axis azimuths.

1 Introduction

The greatest source of uncertainty in forecasts of sea level rise during the 21st century is the contribution from the Antarctic and Greenland ice sheets (e.g. Willis and Church, 2012; Gregory et al., 2013). The poor constraints on predictions of grounded ice discharge are related to the currently inadequate description of ice dynamic processes and boundary conditions in numerical models used to simulate ice sheet evolution (e.g. Alley and Joughin, 2012; Gregory et al., 2013; Vaughan et al., 2013; Carson et al., 2014). One of the key components of all numerical ice sheet models is the relationship which governs ice flow rates as a function of temperature and the stresses driving the deformation. The primary consideration in the development of a numerical flow relation for ice sheet modelling is to provide a realistic description of

ice rheology that does not significantly decrease the computational efficiency of the model.

Typically, ice is modelled as a very slow flowing, nonlinear viscoplastic fluid where the effective viscosity is highly temperature dependent, varying by ~ 3 orders of magnitude over the in situ temperature range. Despite being modelled as a high-viscosity fluid, the polar ice sheets are massive polycrystalline aggregates of solid ice, where the largest linear dimension of individual grains is in the order of several millimetres to centimetres (e.g. Faria et al., 2014a; Ng and Jacka, 2014). Furthermore, ice flow rates are significantly influenced by the microstructural evolution which occurs during deformation (e.g. Budd and Jacka, 1989; Cuffey and Paterson, 2010).

Under terrestrial conditions ice exists in the hexagonal Ih phase and individual crystals possess a high level of plastic anisotropy because their dominant mode of deformation is slip on crystallographic basal planes (e.g. Duval et al., 1983; Schulson and Duval, 2009; Cuffey and Paterson, 2010). Within a polycrystalline aggregate, the spatial orientation of the basal planes of an individual crystal is defined by the orientation of its crystallographic c-axis, which is normal to the basal planes and is the axis of hexagonal symmetry in an ice crystal. During the high-strain deformation, typical of the polar ice sheets, patterns of preferred crystal c-axis orientations (often referred to as crystal orientation fabrics) evolve to accommodate crystallographic slip on basal planes. This leads to the development of polycrystalline anisotropy and an associated reduction in the ice viscosity (e.g. Budd and Jacka, 1989; Cuffey and Paterson, 2010). A physically accurate parameterization of these effects is fundamental to improving the predictive capability of ice sheet models, in order to (i) more accurately quantify ice sheet contributions to global sea level and (ii) reduce uncertainty in the depth–age relationships used to constrain ice core palaeoclimate records.

Various numerical flow relations for ice, where the effect of c-axis orientations on the flow properties is explicitly included as a rheological variable, have been proposed (e.g. Lile, 1978; van der Veen and Whillans, 1994; Azuma and Goto-Azuma, 1996; Svendsen and Hutter, 1996; Thorsteinsson, 2002; Gillet-Chaulet et al., 2005; Placidi et al., 2010). In such flow relations the anisotropic viscosity of a polycrystalline aggregate is obtained from the orientation relationship between the grain c-axes and the stress configuration, or some parameterization of these orientation effects. In general such flow relations are not suited to implementation within models used to simulate the large-scale evolution of the polar ice sheets, being either too numerically complex or lacking the ability to accurately describe anisotropic flow effects (e.g. Treverrow et al., 2015). The importance of a realistic description of anisotropic ice rheology to accurate modelling of ice sheet dynamics has been demonstrated in regional-scale ice sheet models (e.g. Seddik et al., 2011; Wang et al., 2012; Zwinger et al., 2014), where the task of determining the three dimensional distribution of stresses within an ice mass is computationally tractable.

Accordingly, the ongoing value of physically motivated, yet complex flow relations lies in their role as tools to understand intra- and intercrystalline microdeformation, recovery and recrystallization processes (e.g. Montagnat et al., 2014b; Faria et al., 2014b). In turn, these models can inform the development of parameterizations of key rheological variables, which are necessary for the specification of numerically simpler ice flow relations. The continued development and validation of physically accurate flow relations for ice sheet modelling requires observations of ice microstructures from deep drilled ice cores, including the patterns of preferred crystal c-axis orientations within an ice mass. Such data can be obtained from the analysis of thin section samples. These measurements are made routinely in ice core and laboratory deformation studies investigating the links be-

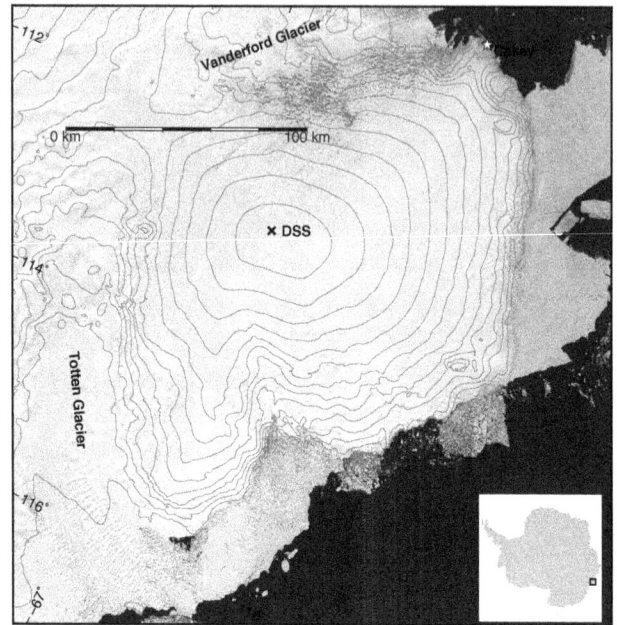

Figure 1. Location of the drilling site for the 1196 m Dome Summit South (DSS) ice core on Law Dome, East Antarctica. The drill site is 4.7 km SSW from the dome summit. The background image is from the Landsat Image Mosaic of Antarctica (Bindschadler et al., 2008) and 100 m elevation contours from Bamber et al. (2009).

tween the microstructure of ice and the large-scale dynamics of ice sheets (e.g. Gow and Williamson, 1976; Russell-Head and Budd, 1979; Gao and Jacka, 1987; Pimienta et al., 1987; Budd and Jacka, 1989; Tison et al., 1994; Li et al., 2000; Durand et al., 2007; Gow and Meese, 2007a; Treverrow et al., 2012; Montagnat et al., 2014a).

Here we present ice crystallographic c-axis orientation and grain size data from the Dome Summit South (DSS) ice core drilled 4.7 km SSW of the summit of Law Dome, East Antarctica (66.770° S, 112.807° E; Table 1; Li, 1995; Morgan et al., 1997; Li et al., 1998; Morgan et al., 1998). This 1195.9 m ice core was drilled by the Australian Antarctic Division during the austral summers of 1987–1988 to 1992–1993. Law Dome (Fig. 1) is a coastal ice cap ~ 200 km in diameter with a summit elevation of 1370 m. Grounded ice at Law Dome extends to lower latitudes than any other region of the Antarctic Ice Sheet, with the exception of the northern Antarctic Peninsula. As such, the local climate has a strong maritime influence and the summit region of Law Dome experiences a high rate of annual snow accumulation and relatively low wind speeds (Bromwich, 1988; Curran et al., 1998). The ice dynamics of Law Dome are relatively independent of the adjoining East Antarctic Ice Sheet as it is isolated from the Aurora subglacial basin by the Vanderford trench (Pfitzner, 1980; Roberts et al., 2011).

A primary consideration in the selection of the DSS drilling site was the identification of a location where the

Table 1. DSS borehole site and ice core information (Morgan et al., 1997, 1998).

Latitude	66.770° S
Longitude	112.807° E
Surface elevation	1370 m
Ice thickness	1220 ± 20 m
Borehole depth	1195.9 m
Borehole ice-equivalent depth	1174.07 m
Mean annual surface temperature	−21.8 °C
Bottom of borehole temperature	−6.9 °C
Annual ice accumulation	0.69 m a^{-1}
Ice surface velocity and bearing	(2.04 ± 0.11) m a^{-1} at 225° ± 3°
Surface GPS strain rate (parallel to surface flow direction)	(3.22 ± 0.016) × 10^{-4} a^{-1}
Surface GPS strain rate (normal to surface flow direction)	(4.50 ± 0.027) × 10^{-4} a^{-1}

rates of ice deformation, and those upstream, were insufficient to significantly disturb the chronology of annually accumulated ice layers. Ice-penetrating radar was used to assist site selection by identifying locations where the regional bedrock topography was least likely to have created flow-induced folding or other discontinuities in the ice laminae (Hamley et al., 1986; Etheridge, 1990; Morgan et al., 1997). The well-preserved layering at the DSS site, combined with the maritime influence on the orographically driven high accumulation rates on Law Dome has made the DSS ice core a valuable resource for generating mid- and high-latitude palaeoclimate proxy records from the Holocene and Last Glacial Maximum (LGM; van Ommen and Morgan, 2010; Plummer et al., 2012; Vance et al., 2013; Roberts et al., 2015; Vance et al., 2015).

2 Methods

The upper 96 m of the DSS core was obtained with a 270 mm diameter thermal drill during the 1987–1988 austral summer and the borehole was cased to a depth of 82 m. In the following season a drill shelter was constructed over the borehole and thermal drilling continued to 117 m depth using a smaller 120 mm drill. Over the 1989–1990 and 1990–1991 field seasons the electromechanical drill used to recover the main DSS ice core was assembled and commissioned (Morgan et al., 1997). This drill was based on a modified Danish ISTUK design (Gundestrup et al., 1984) and produced 100 mm diameter core sections. Coring to a depth of 553.9 m was completed during the 1991–1992 field season and in the following season drilling continued to the final depth of 1195.6 m. Coring was halted when the cutting head of the drill was damaged through contact with a rock, which was not recovered in the core. This final core section was found to contain other, small rock fragments. The estimated total ice thickness of 1220 ± 20 m at the DSS site was determined

using ice-penetrating radar (Morgan et al., 1997). Based on this value, the bedrock is expected to be within 5 to 45 m of the bottom of the borehole. The DSS core was recovered in 1261 sections with a mean length of 0.949 m and standard deviation $\sigma = 0.092$ m.

2.1 Thin section preparation

Ice crystal c-axis orientation and grain size measurements were made on horizontal thin sections obtained from 185 of the 1261 individual ice core sections that make up the DSS core. These samples were obtained at intervals of ∼ 5 to 6 m between the depths of 117.1 and 1195.9 m. The thin sections were prepared during drilling operations in order to minimize the time between core recovery and analyses. Typically, the interval between drilling and thin section preparation was < 1 to 16 h, with a mean of ∼ 3 h (Li et al., 1998). This procedure reduced the potential for any post-drilling microstructural evolution to adversely influence the analyses. Furthermore, all sample preparation and analyses were conducted in a sub-surface laboratory, excavated adjacent to the drilling pit. Since the temperature in the laboratory was approximately equivalent to the mean annual surface temperature of −21.8 °C all analyses were conducted at temperatures below the in situ values. Working at these low temperatures reduced the potential for post-drilling microstructural modification. Conducting the microstructural observations at the drilling site also eliminated the risk of damaging the samples during long-term storage and/or transportation to laboratories outside Antarctica.

The ∼ 10 mm thick horizontal sections were cut perpendicular to the long axis of the ice core. These samples were sanded to a smooth finish using wet-and-dry type abrasive paper mounted on a flat board before being thermally bonded to glass slides. Following mounting, the thickness of the samples was reduced to between 0.4 and 0.7 mm using a microtome. Thicknesses towards the lower end of this range were necessary to clearly resolve individual grains in those samples with a smaller mean grain size. Two different microtomes were used to reduce the sample thickness. An electromechanical microtome, based on a modified electric woodworking plane, was better suited to rapidly reducing the section thickness while a manually operated sledge microtome provided superior control when making the final fine adjustments to the section thickness. The sledge microtome was also found to be less likely to induce fracturing when working with the extremely brittle ice encountered at depths from 552 to 1190 m.

2.2 Crystallographic c-axis measurements

The crystal c-axis orientations were measured manually using a universal (Rigsby) stage following the standard techniques described by Langway Jr. (1958). The instrument used was modified to provide digital orientation data for each grain examined (Morgan et al., 1984). These output data are corrected for differences in the refractive indices of ice and air using the values of Langway Jr. (1958).

As outlined by Langway Jr. (1958), several potential sources of error can influence the orientations determined using a universal stage. These include (i) uncertainty in precisely locating the extinction position for high-angle orientations, (ii) systematic operator bias due to not correctly aligning the viewing direction (line of sight) with the crystal c-axis being examined and (iii) instrument backlash due to inherent and finite non-zero tolerances that influence the reproducibility of measurements. Langway Jr. (1958) estimates that the latter may contribute to 1 to $2°$ of measurement error and that the combined maximum error from all sources will typically be $< 5°$, particularly when measurements are made with consistently high levels of care. Because the universal stage used to measure the DSS ice core crystal orientations employs sensors to determine the position of the instrument axes, incorrect reading of the axis scales is eliminated as a source of error in these data. An upper limit of the possible angular error in azimuth or colatitude data is $\pm 5°$.

Due to the time-consuming procedure required to manually determine c-axis orientations using the universal stage in a challenging work environment, a maximum of ~ 100 orientations were recorded from each thin section, including those fine-grained samples where the total number of grains, N, within the thin section was $\gg 100$. Sampling bias, including the preferential selection of larger grains was avoided by (i) tracking those grains that had been measured on a polaroid image of the thin section and (ii) selecting a continuous (i.e. gap-free) set of neighbouring grains from within an enclosed region so that each grain had at least one neighbour within the region of interest. Analyses were completed at the rate of ~ 100 c-axis orientations per 3 h. For those thin sections containing larger grains with $N \ll 100$, c-axis orientations were recorded for all identifiable grains. For each of the 185 DSS thin sections, the number of measured c-axis orientations was $5 < N < 111$, with a median of $N = 100$.

During the 1990s when the DSS ice core was drilled, the measurement of ice crystal c-axis orientations using a universal stage, as used in this study, was considered state of the art. In the time since these measurements were made considerable technological advances in the instruments available to measure crystal orientations have occurred (e.g. Yun and Azuma, 1999; Wilen et al., 2003; Wilson et al., 2007; Wilson and Peternell, 2011). Modern instruments provide several benefits over the universal stage including (i) higher resolution, enabling small-scale differences in crystallographic orientations within individual grains to be detected, (ii) the ability to spatially map c-axis orientations, (iii) automated operation and a significantly higher speed of determining orientations, and (iv) lower levels of uncertainty in both c-axis azimuth and colatitude. Since analytical techniques for both microstructural and chemical analyses of ice cores are destructive, only a small proportion of the original ice core cross section remains over the full length of the core. This remainder is insufficient to allow a detailed reanalysis of c-axis orientations and grain size using a modern instrument. Additionally, the remaining core material is prioritized for chemical reanalyses, should any be required.

2.3 Crystal size measurements

The mean grain size was determined from polaroid photographs of the thin sections. A purpose-built stand was used to position the thin sections between orthogonal plane polarizing filters. This allowed individual grains to be visually distinguished by their orientation-dependent birefringence colours. A polaroid camera, mounted directly to the stand, captured high-quality colour images on a 1 : 1 scale. A transparent cover plate, etched with a 10 mm square grid, was placed over the samples to superimpose a grid of the same dimensions onto the polaroid images.

The mean grain area was calculated from the number of grains contained within a specified area of the thin section. The region of interest was determined by placing a sheet of low-opacity tracing paper over the polaroid image; its transparency allowed an irregularly shaped region to be traced along grain boundaries and the number of grains within this region to be counted. A digitizing tablet was used to accurately determine the area of the irregularly shaped region marked on the tracing paper. Typically these areas for grain size analysis varied between 880 and 2400 mm^2 (Li, 1995). The area, in conjunction with the total number of grains counted within the traced region, was used to calculate an arithmetic mean estimate of cross sectional grain area. Figures incorporating the depth profile of mean grain size at the DSS site have been presented previously (e.g. Morgan et al., 1997; Li et al., 1998), but until now the data have not been easily accessible.

Uncertainty in the mean grain area estimates may originate from errors in counting the number of grains within the traced region or when determining the area of the traced region using a digitizing tablet. For digitizing tablets, instrument-related planimetric position errors are typically $\ll 1$ mm and are negligible in comparison with any operator error. Consequently, the uncertainty in the calculated mean grain areas is strongly related to both operator skill and consistency.

To determine an upper limit on the uncertainty in the calculated mean area we assume the maximum error in the position of any section of the traced boundary enclosing the counted grains to be $\sim \pm 1$ mm. Applying this level of uncertainty to the entire perimeter of the traced region results in an area uncertainty of $\sim \pm 10 \%$. In most cases this upper limit on uncertainty will be an overestimate of the true uncertainty; however, precise estimates of uncertainty were not made at the time of measurement, so we are unable to calculate the specific error for the individual mean grain areas.

Some further general comments on the methods used to specify the grain size of polycrystalline materials are warranted. While thin section analysis is the only practical means of routinely estimating grain area, it only provides a two dimensional estimate of a volumetric object. Furthermore, all methods used to determine grain areas from thin sections underestimate the actual grain dimensions because the sectioning plane almost never intersects each grain across its plane of maximum cross-sectional area (e.g. Feltham, 1957; Gow, 1969). There are a variety of techniques for determining the grain size of polycrystalline materials from thin sections that also take the sectioning bias into account.

In the methods described by Krumbein (1935), Pickering (1976), Baker (1982) and Jones and Chew (1983), grain size is estimated from the mean length of the maximum linear intercept through a grain. This requires assumptions to be made regarding the grain shape and distribution of grain sizes. Stephenson (1967) and Gow (1969) calculated mean grain areas based on the measurement of minimum and maximum axial dimensions of individual grains. To further reduce the effect of sectioning on underestimating the three-dimensional distribution of grain size, Gow (1969) restricted measurement to the 50 largest grains identified within a thin section; however, the results obtained using this technique are sensitive to the number of grains within the section.

Methods of grain size analysis that require measurement of linear intercepts through grains (e.g. Krumbein, 1935; Stephenson, 1967; Gow, 1969; Baker, 1982) are time consuming and were considered incompatible with the program of obtaining crystal c-axis orientation and size measurements during drilling and field analysis of the DSS ice core. As noted by Jacka (1984), the mean linear intercept methods of grain size estimation, and in particular those assumptions that attempt to correct for sectioning effects, may be inappropriate when the actual distributions of grain size and shape are either unknown or variable. In such cases Jacka (1984) suggests that the arithmetic mean grain area, based on the crystal count per unit area, is a superior estimate of grain size. For polycrystalline metals, Feltham (1957) has demonstrated that the three-dimensional distribution of grain size can be represented by the planar distribution obtained from thin sections provided that grain shape is not correlated with the crystallographic c-axis orientation. This result is directly applicable to other polycrystalline materials, including ice, for which the grain size does not vary with the orientation of the sectioning plane. In this case the mean grain area (count per unit area) can be considered representative of the corresponding volumetric grain size. While the mean grain areas presented here will underestimate the corresponding three-dimensional grain size due to the aforementioned sectioning effects, the data remain valuable as they are indicative of the mean grain size at a specific depth and also describe the evolution in grain size as a function of depth.

The single mean grain size measurements per thin section included in this data set provide a coarse representation of grain size in comparison to the data that can be obtained using modern instruments (e.g. Durand et al., 2006). Since the original thin sections used for c-axis and mean grain size measurements no longer exist, higher-resolution analyses using such an instrument are not possible. For those interested in extracting additional microstructural information, such as the distribution of grain size and/or shape, digital analysis of the original thin section images is a possibility. A full set of polaroid photographs of the DSS ice core thin sections remain in existence and these can be accessed by contacting the authors via the Australian Antarctic Data Centre (http://data.antarctica.gov.au).

3 Data

Measurements of c-axis orientations and mean grain area were made on 185 horizontal thin sections obtained from the upper end of the selected core sections. The reported depth for each thin section is that from the ice sheet surface to the top of the core section. In the upper part of an ice sheet the observed ice density increases with depth as snow undergoes a transition to firn and then glacial ice. The very slow flow of glaciers and ice sheets is the result of gravity-induced deformation that is dependent upon several factors, including the ice sheet geometry (ice thickness, surface elevation, bedrock topography) and the ice density (e.g. Cuffey and Paterson, 2010).

In the numerical models used to simulate ice sheet dynamics, it is more practical to specify a constant ice density when calculating the magnitude of the stresses driving ice flow. The assumption of a constant density is reasonable for ice sheets where ice thicknesses may be hundreds or thousands of metres; however, it is necessary to convert any data sets used for calibration or validation of the models to an ice-equivalent depth scale. Similarly, the interpretation and application of ice core chemistry in palaeoclimate studies is simplified by conversion to an ice-equivalent depth scale. This is particularly convenient when synchronizing multiple ice core records from different geographical locations. The crystal c-axis orientation and mean grain area data are reported on both actual and ice-equivalent depth scales. At an actual depth χ, the ice-equivalent depth

$$d(\chi) = \int_0^{\chi} \frac{\rho(\chi')}{\rho_{ice}} d\chi', \tag{1}$$

where $\rho(\chi')$ is the firn and ice density profile and $\rho_{ice} = 917 \, kg \, m^{-3}$ is the maximum ice density (Roberts et al., 2015). The depth-varying $\rho(\chi')$ is obtained from an empirical fit to density measurements from the DSS core (van Ommen et al., 1999). The firn to ice transition marks the depth where the interconnected pore space between individual grains is closed off, forming discrete bubbles and $\rho_{ice} \sim 830 \, kg \, m^{-3}$. At DSS this transition

Table 2. Description of the data fields in each row of the csv formatted DSS ice core c-axis orientation and mean grain area data file, `DSS_fabric_data.csv`

Name	Identifier of the DSS ice core section
Longitude	DSS ice core site longitude (decimal degrees)
Latitude	DSS ice core site latitude (decimal degrees)
Number_of_c_axes	the number of c-axes in the section identified by "name"
Depth_actual	the actual depth measured downwards from the ice sheet surface to the top of the core section (metres)
Depth_ice_equivalent	"depth_actual" converted to an ice-equivalent depth (metres)
Section_orientation	thin section orientation ("vertical" or "horizontal")
Mean_horz_grain_area	mean grain area measured from horizontal thin sections (perpendicular to the vertical ice core axis; mm^2)
Grain_index	the identifier of the data point within the thin section
Colatitude	c-axis orientation colatitude (degrees, lower hemisphere projection)
Azimuth	c-axis orientation azimuth (degrees, lower hemisphere projection)

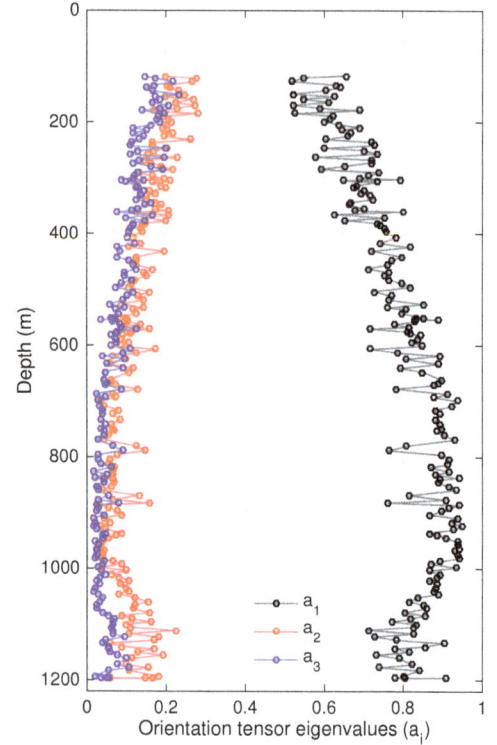

Figure 2. Variation in the eigenvalues, a_i, of the second-order orientation tensor, Λ, as a function of actual depth. See text for details of a_i.

occurs at ~ 40 m depth. Below this depth density variations are negligible and the ice-equivalent depth offset from the actual depth has a constant value of -21.53 m (van Ommen et al., 2004).

3.1 Crystal orientation and mean grain size data products

The combined crystal c-axis orientation and mean grain size data are provided in two formats.

1. A single comma-separated value (csv) file containing c-axis orientation information for all grains in each of the 185 thin sections analysed. Each row in the file contains c-axis orientation data for an individual grain expressed in polar coordinates (azimuth and colatitude), the mean grain size calculated for the parent thin section and other information relevant to the measurements. See Table 2 for a full description of the fields in the data set.

2. A MATLAB™ structure array containing c-axis vectors expressed in both polar and Cartesian coordinates (MATLAB and Statistics Toolbox Release 2015b, The MathWorks Inc., Natick, Massachusetts, United States). Other data included in the structure array are described in Table 3. Some derived quantities are also presented in the structure array. These include the eigenvalues a_i ($i = 1, 2, 3$) of the second-order c-axis orientation tensor, Λ, and its corresponding eigenvectors $\boldsymbol{v}$.

The crystallographic c axis and mean crystal area data are summarized as a function of actual depth in Figs. 2 and 3 respectively. The pattern of c-axis orientations can be expressed by the eigenvalues, a_i, of the c-axis orientation tensor, Λ (Fig. 2, Woodcock, 1977). Calculation of the orientation tensor requires transformation of the c-axis data from polar to Cartesian coordinates. For a population of N c-axis orientation vectors the normalized form of Λ is defined in terms of their N outer products:

$$\Lambda = \frac{\sum_{i=1}^{N} \hat{c}_i \otimes \hat{c}_i}{N}. \tag{2}$$

The eigenvalues are related as $a_1 + a_2 + a_3 = 1$, where by convention $0 \leq a_3 \leq a_2 \leq a_1 \leq 1$. The eigenvalues provide a statistical representation of the pattern of c-axis orientations as each a_i defines the degree of clustering about its corresponding eigenvector, $\hat{v}_i$. The distribution of orientations is minimized about the eigenvector $\hat{v}_1$ of the maximum eigenvalue, a_1, whilst $\hat{v}_3$ is the direction about which the distribution is largest. The distribution of orientations becomes smaller, i.e. fabrics become stronger or more clustered as $a_1 \rightarrow 1$, whilst for an isotropic (random) distribution of orientations $a_1 = a_2 = a_3 = \frac{1}{3}$. As the area of individual grain orientations was not recorded, volume weighting of the ori-

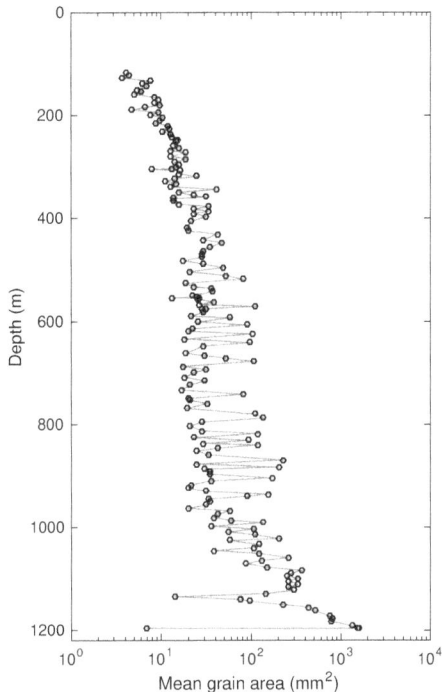

Figure 3. Variation in the DSS ice core mean grain area with actual depth. See text for details of the mean grain area calculation. All values were determined from horizontal thin sections (after Li et al., 1998).

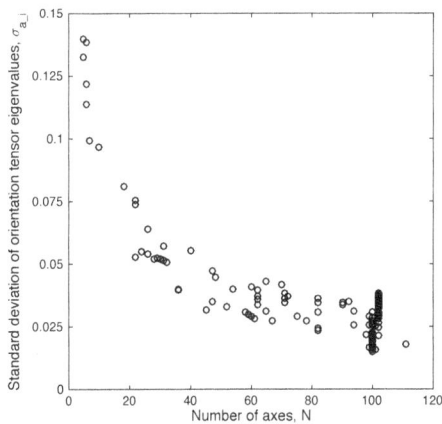

Figure 4. Variation of the standard deviation, $\sigma_{a_i}^{p}$, of the eigenvalues of the second-order orientation tensor with N, the number of c-axis orientations measured in each of the 185 thin sections. Values of $\sigma_{a_i}^{p}$ are calculated according to Eq. (3). The median value of $N = 100$ and there are 50 measurements at $N = 100$ and 70 at $N = 102$.

entation data (e.g. Durand et al., 2006) was not possible and all orientations contribute equally to $\mathbf{\Lambda}$ (Fig. 2).

Allowances can be made for the distribution of grain sizes encountered in polycrystalline materials by weighting the contribution of individual c-axis orientations accord-

Table 3. Format of the MATLAB[TM] R2015b structure array, DSS_fabric_data.mat, which contains DSS ice core crystallographic c-axis orientation and mean grain area data.

Name	DSS ice core section name
C_cartesian	[$N \times 3$] array of Cartesian c-axis unit vectors; N is the number of c-axes in the section
C_polar	[$N \times 2$] array of polar c-axis vectors; colatitude and azimuth (degree)
Depth_actual	the actual depth measured downwards from the ice sheet surface to the top of the core section (metres)
Depth_ice_equivalent	"depth_actual" converted to an ice-equivalent depth (metres)
Mean_horz_grain_area	mean grain area measured from horizontal thin sections (perpendicular to the vertical ice core axis; mm^2)
Orientation_tensor_a	eigenvalues, $\mathbf{a}$, of the 2nd order orientation tensor, $\mathbf{\Lambda}$
Orientation_tensor_V	eigenvectors, $\mathbf{v}$, of the 2nd order orientation tensor, $\mathbf{\Lambda}$

ing to their area in quantitative descriptions of fabric, such as $\mathbf{\Lambda}$. The pixel-scale orientation data provided by modern automated fabric analysers makes area weighting of c-axis orientations a routine aspect of microstructural analysis. In turn, this allows for an improved representation of microstructures extracted from thin sections. For example, Gagliardini et al. (2004) note that with area weighting of orientations, the mean error in second-order orientation tensor-based representations of fabric (e.g. Woodcock, 1977; Durand et al., 2006) may be up to ~ 2.5 times less than that for equally weighted orientations. Notwithstanding the restriction of this data set to equal weighting of orientations, it represents a valuable resource for the quantitative assessment of ice flow relations and microstructural evolution.

For a population of N c-axis orientations, the contribution of the population size to the standard deviation of the orientation tensor eigenvalues, $\sigma_{a_i}^{p}$, can be estimated from,

$$\sigma_{a_i}^{p} = \left[-1.64 \times (a_1)^2 + 1.86 \times a_1 - 0.14 \right] \times \frac{1}{N^{1/2}}. \quad (3)$$

Equation (3) was derived by Durand et al. (2006) from the statistical analysis of multiple subsamples of $100 < N < 1000$ orientations from a parent population of 10^4 orientations. Values of $\sigma_{a_i}^{p}$ calculated using Eq. (3) for each of the 185 thin sections in the DSS data set are presented in Fig. 4 and clearly demonstrate the influence of N on the variability in fabric statistics; $\sigma_{a_i}^{p}$ ranges from 0.0148 up to a maximum of 0.140 for the lowest values of N. As expected from Eq. (3), $\sigma_{a_i}^{p}$ decreases with larger N, and for the majority of the DSS data set, where $N > 40$ the estimated variability in $\sigma_{a_i}^{p}$ is rela-

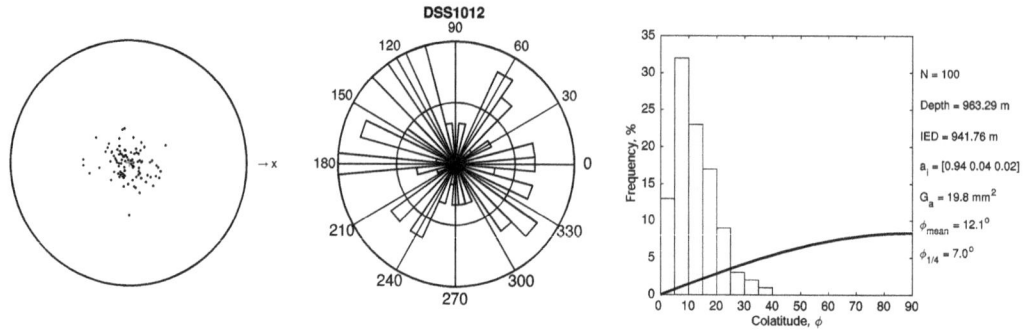

Figure 5. Sample of the thin-section figures included with the data set. From left to right: (i) lower hemisphere c-axis Schmidt plot, (ii) a rose plot of the c-axis azimuth distribution and (iii) a histogram of the c-axis colatitude, ϕ. The colatitude histograms are annotated with the number of grains, N, in the distribution; the actual ice depth; ice-equivalent depth (IED); eigenvalues, a_i, of the orientation tensor Λ; mean grain area, G_a; mean colatitude, ϕ_{mean}; and the cone angle containing the first quartile of c-axis colatitudes, $\phi_{\frac{1}{4}}$.

tively low, as indicated by mean and median values of 0.0355 and 0.0337 respectively.

A component of the observed variability in the DSS fabric and mean grain size data (Figs. 2 and 3) over small vertical distances (e.g. $\sim$ 5–20 m) is due to the influence of ice chemistry, impurities and the dynamic conditions at the drilling site on microstructural evolution. This variability is therefore an inherent feature of the data and similar effects have been reported for other Antarctic and Greenland ice cores (e.g. Durand et al., 2007; Gow and Meese, 2007b).

Estimating the magnitude of impurity effects on the variability in derived microstructural parameters is not possible with the DSS data set; however, observations from Morgan et al. (1997, Fig. 8) indicate that localized variability in the fabric strength that is superimposed on the large-scale pattern of fabric evolution with depth corresponds to changes in the mean grain size. In particular, thin sections with stronger fabrics (lower mean c-axis colatitude) tended to have smaller mean grain sizes than adjacent thin sections with weaker fabrics and correspondingly larger mean grain sizes. Based on analysis of the insoluble impurity content in the DSS ice core, Li et al. (1998, Fig. 2) suggest that locally high levels of microparticles are associated with a refinement of the mean grain size and the preservation of stronger crystal orientation fabrics due to a retardation of recrystallization processes.

The data set made available for download at the Australian Antarctic Data Centre also includes graphical representations of the c-axis orientations. For each thin section a figure containing three subfigures is provided (see Fig. 5). These include a lower-hemisphere Schmidt plot of the c-axes; a rose plot of the distribution of c-axis azimuth, θ; and a histogram of the c-axis colatitude, ϕ. The colatitude histograms are annotated with the number of grains, N, in the distribution; the actual depth; ice-equivalent depth (IED); eigenvalues, a_i, of the orientation tensor Λ; mean grain area, G_a; mean colatitude, ϕ_{mean}; and the cone angle containing the first quartile of c-axis colatitudes, $\phi_{\frac{1}{4}}$.

4 Summary

Observations of ice microstructures from deep drilled polar ice cores play a vital role in the development and validation of ice flow relations for numerical ice sheet modelling. In particular, measurements of the patterns of ice crystal c-axis orientations and grain size from thin sections, obtained at regular depth intervals along an ice core, provide detailed information on the relationships between large-scale ice sheet dynamics and microstructural evolution.

Measurements of crystal c-axis orientations and grain size were obtained from the Dome Summit South (DSS) ice core, drilled at Law Dome, East Antarctica. The data were obtained from 185 horizontal thin sections taken at intervals of approximately 5 to 6 m between the depths of 117 and 1196 m. All c-axis orientation measurements were made using a universal (Rigsby) stage. The number of c-axis orientations recorded in each thin section varied according to the grain size and ranged from a minimum of 5 up to 111, with a median value of 100. For each thin section the arithmetic mean grain area was also determined. These data are made available in two formats: (i) as a single csv formatted file containing all c-axis orientations (in polar coordinates) and mean grain area for each of the 185 thin sections, plus other data relevant to the DSS drilling site, and (ii) as a MATLAB™ structure array containing the c-axis and grain size data for each thin section and derived quantities, including the eigenvalues and eigenvectors of the second-order orientation tensor. These data are available free of charge from the Australian Antarctic Data Centre (http://data.antarctica.gov.au) and are referenced by doi:10.4225/15/5669050CC1B3B.

Acknowledgements. The Australian Antarctic Division provided funding and logistical support for drilling the DSS ice core and subsequent data analysis through projects ASAC 15, AAS 757 and AAS 4289. The authors gratefully acknowledge the contribution of all participants in the Australian National Antarctic

Research Expeditions associated with retrieval of the DSS ice core. Preparation of the data for archiving was supported by the Australian Government Cooperative Research Centres Programme through the Antarctic Climate and Ecosystems Cooperative Research Centre (ACE CRC). Discussions with J. L. Roberts assisted with data management and manuscript preparation. B. Raymond assisted with data control and hosting. We are thankful for comments from Maurine Montagnat and an anonymous reviewer, who assisted in improving the manuscript. Adam Treverrow thanks R. C. Warner for stressing the importance of making these data widely available to the glaciological community.

Edited by: O. Eisen

References

Alley, R. and Joughin, I.: Modeling ice-sheet flow, Science, 336, 551–552, doi:10.1126/science.1220530, 2012.

Azuma, N. and Goto-Azuma, K.: An anisotropic flow law for ice sheet ice and its implications, Ann. Glaciol., 23, 202–208, 1996.

Baker, R.: A flow equation for anisotropic ice, Cold Reg. Sci. Technol., 141–148, 1982.

Bamber, J. L., Gomez-Dans, J. L., and Griggs, J. A.: A new 1 km digital elevation model of the Antarctic derived from combined satellite radar and laser data – Part 1: Data and methods, The Cryosphere, 3, 101–111, doi:10.5194/tc-3-101-2009, 2009.

Bindschadler, R., Vornberger, P., Fleming, A., Fox, A., Mullins, J., Binnie, D., Paulsen, S., Granneman, B., and Gorodetzky, D.: The Landsat Image Mosaic of Antarctica, Remote Sens. Environ., 112, 4214–4226, doi:10.1016/j.rse.2008.07.006, 2008.

Bromwich, D. H.: Snowfall in High Southern Latitudes, Rev. Geophys., 26, 149–168, 1988.

Budd, W. and Jacka, T.: A review of ice rheology for ice sheet modelling, Cold Reg. Sci. Technol., 16, 107–144, doi:10.1016/0165-232X(89)90014-1, 1989.

Carson, C. J., McLaren, S., Roberts, J. L., Boger, S. D., and Blankenship, D. D.: Hot rocks in a cold place: high subglacial heat flow in East Antarctica, J. Geol. Soc., 171, 9–12, doi:10.1144/jgs2013-030, 2014.

Cuffey, K. and Paterson, W.: The Physics of Glaciers, Elsevier, 4th Edn., Burlington, Massachusetts, 2010.

Curran, M. A. J., Ommen, T. D. V., and Morgan, V.: Seasonal characteristics of the major ions in the high-accumulation Dome Summit South ice core, Law Dome, Antarctica, Ann. Glaciol., 27, 385–390, 1998.

Durand, G., Gagliardini, O., Thorsteinsson, T., Svensson, A., Kipfstuhl, S., and Dahl-Jensen, D.: Ice microstructure and fabric: an up-to-date approach for measuring textures, J. Glaciol., 52, 619–630, doi:10.3189/172756506781828377, 2006.

Durand, G., Gillet-Chaulet, F., Svensson, A., Gagliardini, O., Kipfstuhl, S., Meyssonnier, J., Parrenin, F., Duval, P., and Dahl-Jensen, D.: Change in ice rheology during climate variations – implications for ice flow modelling and dating of the EPICA Dome C core, Clim. Past, 3, 155–167, doi:10.5194/cp-3-155-2007, 2007.

Duval, P., Ashby, M., and Anderman, I.: Rate-controlling processes in the creep of polycrystalline ice, J. Phys. Chem., 87, 4066–4074, 1983.

Etheridge, D.: Scientific plan for deep ice drilling on Law Dome, ANARE Research Notes, Kingston, Australia, 1990.

Faria, S. H., Weikusat, I., and Azuma, N.: The microstructure of polar ice. Part I: Highlights from ice core research, J. Struct. Geol., 61, 2–20, doi:10.1016/j.jsg.2013.09.010, 2014a.

Faria, S. H., Weikusat, I., and Azuma, N.: The microstructure of polar ice. Part II: State of the art, J. Struct. Geol., 61, 21–49, doi:10.1016/j.jsg.2013.11.003, 2014b.

Feltham, P.: Grain growth in metals, Acta Metall., 5, 97–105, 1957.

Gagliardini, O., Durand, G., and Wang, Y.: Grain area as a statistical weight for polycrystal constituents, J. Glaciol., 50, 87–95, doi:10.3189/172756504781830349, 2004.

Gao, X. and Jacka, T.: The approach to similar tertiary creep rates for Antarctic core ice and laboratory prepared ice, J. Phys., 48, 289–296, 1987.

Gillet-Chaulet, F., Gagliardini, O., Meyssonnier, J., Montagnat, M., and Castelnau, O.: A user-friendly anisotropic flow law for ice-sheet modelling, J. Glaciol., 51, 3–14, doi:10.3189/172756505781829584, 2005.

Gow, A.: On the rates of growth of grains and crystals in south polar firn, J. Glaciol., 53, 241–252, 1969.

Gow, A. and Meese, D.: Physical properties, crystalline textures and c-axis fabrics of the Siple Dome (Antarctica) ice core, J. Glaciol., 183, 573–584, doi:10.3189/002214307784409252, 2007a.

Gow, A. and Meese, D.: The distribution and timing of tephra deposition at Siple Dome, Antarctica: possible climatic and rheologic implications, J. Glaciol., 53, 585–596, 2007b.

Gow, A. and Williamson, T.: Rheological implications of the internal structure and crystal fabrics of the West Antarctic ice sheet as revealed by deep ice core drilling at Byrd station, Geol. Soc. Am. Bull., 87, 1665–1677, 1976.

Gregory, J., White, N., Church, J., Bierkens, M., Box, J., van den Broeke, M., Cogley, J., Fettweis, X., Hanna, E., Huybrechts, P., Konikow, L., Leclercq, P., Marzeion, B., Oerlemans, J., Tamisiea, M., Wada, Y., Wake, L., and van de Wal, R.: Twentieth-century global-mean sea-level rise: is the whole greater than the sum of the parts?, J. Climate, 26, 4476–4499, doi:10.1175/JCLI-D-12-00319.1, 2013.

Gundestrup, N. S., Johnsen, S. J., and Reeh, N.: ISTUK: A deep ice core drill system, in: Proceedings of the Second International Workshop/Symposium on Ice Drilling Technology, edited by: Holdsworth, G., Kuivinen, K. C., and Rand, J. H., CRREL Special Report 84-34, 7–19, 1984.

Hamley, T., Morgan, V., Thwaites, R., and Gao, X.: An ice-core drilling site at Law Dome summit, Wilkes Land, Antarctica, Tech. Rep. 37, ANARE Research Notes, 1986.

Jacka, T.: Laboratory studies on relationships between ice crystal size and flow rate, Cold Reg. Sci. Technol., 10, 31–42, 1984.

Jones, S. and Chew, H.: Effect of sample and grain size on the compressive strength of ice, Ann. Glaciol., 129–132, 1983.

Krumbein, W. C.: Thin-section mechanical analysis of indurated sediments, J. Geol., 43, 482–496, 1935.

Langway Jr., C.: Ice fabrics and the universal stage, Tech. Rep. 62, US Army Snow Ice and Permafrost Research Establishment, 1958.

Li, J.: Interrelation between the flow properties and crystal structure of snow and ice, PhD thesis, School of Earth Sciences, University of Melbourne, 1995.

Li, J., Jacka, T., and Morgan, V.: Crystal-size and microparticle record in the ice core from Dome Summit South, Law Dome, East Antarctica, Ann. Glaciol., 27, 343–348, 1998.

Li, J., Jacka, T., and Budd, W.: Strong single-maximum crystal fabrics developed in ice undergoing shear with unconstrained normal deformation, Ann. Glaciol., 30, 88–92, 2000.

Lile, R.: The effect of anisotropy on the creep of polycrystalline ice, J. Glaciol., 21, 475–483, 1978.

Montagnat, M., Azuma, N., Dahl-Jensen, D., Eichler, J., Fujita, S., Gillet-Chaulet, F., Kipfstuhl, S., Samyn, D., Svensson, A., and Weikusat, I.: Fabric along the NEEM ice core, Greenland, and its comparison with GRIP and NGRIP ice cores, The Cryosphere, 8, 1129–1138, doi:10.5194/tc-8-1129-2014, 2014a.

Montagnat, M., Castelnau, O., Bons, P., Faria, S., Gagliardini, O., Gillet-Chaulet, F., Grennerat, F., Griera, A., Lebensohn, R., Moulinec, H., Roessiger, J., and Suquet, P.: Multiscale modeling of ice deformation behavior, J. Struct. Geol., 61, 78–108, doi:10.1016/j.jsg.2013.05.002, 2014b.

Morgan, V., Davis, E., and Wehrle, E.: A Rigsby stage with remote computer compatible output, Cold Reg. Sci. Technol., 10, 89–92, 1984.

Morgan, V., Wookey, C., Li, J., van Ommen, T., Skinner, W., and Fitzpatrick, M.: Site information and initial results from deep ice drilling on Law Dome, Antarctica, J. Glaciol., 43, 3–10, 1997.

Morgan, V., van Ommen, T., Elcheikh, A., and Li, J.: Variations in shear deformation rate with depth at Dome Summit South, Law Dome, East Antarctica, Ann. Glaciol., 135–139, 1998.

Ng, F. and Jacka, T.: A model of crystal-size evolution in polar ice masses, J. Glaciol., 60, 463–477, doi:10.3189/2014JoG13J173, 2014.

Pfitzner, M. L.: The Wilkes Ice Cap Project, 1966, Series A (4) Glaciology 127, ANARE Scientific Reports, Melbourne, Australia, 1980.

Pickering, F. B.: The basis of quantitative metallography, Metals and Metallurgy Trust for the Institute of Metallurgical Technicians, London, 1976.

Pimienta, P., Duval, P., and Lipenkov, V. Y.: Mechanical behaviour of anisotropic polar ice, in: The Physical Basis of Ice Sheet Modelling, 57–65, IAHS Publ. 170, 1987.

Placidi, L., Greve, R., Seddik, H., and Faria, S.: Continuum-mechanical, Anisotropic Flow model, for polar ice masses, based on an anisotropic Flow Enhancement factor, Continuum Mech. Thermodyn., 22, 221–237, doi:10.1007/s00161-009-0126-0, 2010.

Plummer, C. T., Curran, M. A. J., van Ommen, T D., Rasmussen, S. O., Moy, A. D., Vance, T. R., Clausen, H. B., Vinther, B. M., and Mayewski, P. A.: An independently dated 2000-yr volcanic record from Law Dome, East Antarctica, including a new perspective on the dating of the 1450s CE eruption of Kuwae, Vanuatu, Clim. Past, 8, 1929–1940, doi:10.5194/cp-8-1929-2012, 2012.

Roberts, J., Plummer, C., Vance, T., van Ommen, T., Moy, A., Poynter, S., Treverrow, A., Curran, M., and George, S.: A 2000-year annual record of snow accumulation rates for Law Dome, East Antarctica, Clim. Past, 11, 697–707, doi:10.5194/cp-11-697-2015, 2015.

Roberts, J. L., Warner, R. C., Young, D., Wright, A., van Ommen, T. D., Blankenship, D. D., Siegert, M., Young, N. W., Tabacco, I. E., Forieri, A., Passerini, A., Zirizzotti, A., and Frezzotti, M.: Refined broad-scale sub-glacial morphology of Aurora Subglacial Basin, East Antarctica derived by an ice-dynamics-based inter-polation scheme, The Cryosphere, 5, 551–560, doi:10.5194/tc-5-551-2011, 2011.

Russell-Head, D. and Budd, W.: Ice-sheet flow properties derived from bore-hole shear measurements combined with ice-core studies, J. Glaciol., 24, 117–130, 1979.

Schulson, E. and Duval, P.: Creep and Fracture of Ice, Cambridge University Press, 2009.

Seddik, H., Greve, R., Zwinger, T., and Placidi, L.: A full Stokes ice flow model for the vicinity of Dome Fuji, Antarctica, with induced anisotropy and fabric evolution, The Cryosphere, 5, 495–508, doi:10.5194/tc-5-495-2011, 2011.

Stephenson, P.: Some considerations of snow metamorphism in the Antarctic Ice Sheet in the light of crystal studies, in: The Physics of Snow and Ice, Proceedings of the International Conference on Low Temperature Science, 1966, edited by: Ôura, H., 725–740, Institute of Low Temperature Science, Hokkaido University, Sapporo, Japan, 1967.

Svendsen, B. and Hutter, K.: A continuum approach for modelling induced anisotropy in glaciers and ice sheets, Ann. Glaciol., 262–269, 1996.

Thorsteinsson, T.: Fabric development with nearest-neighbour interaction and dynamic recrystallization, J. Geophys. Res.-Sol. Ea., 107, ECV 3-1–ECV 3-13, doi:10.1029/2001JB000244, 2002.

Tison, J.-L., Thorsteinsson, T., Lorrain, R., and Kipfstuhl, J.: Origin and development of textures and fabrics in basal ice as Summit, Central Greenland, Earth Planet. Sc. Lett., 125, 421–437, doi:10.1016/0012-821X(94)90230-5, 1994.

Treverrow, A., Budd, W., Jacka, T., and Warner, R.: The tertiary creep of polycrystalline ice: experimental evidence for stress-dependent levels of strain-rate enhancement, J. Glaciol., 58, 301–314, doi:10.3189/2012JoG11J149, 2012.

Treverrow, A., Warner, R., Budd, W., Jacka, T., and Roberts, J. L.: Modelled stress distributions at the Dome Summit South borehole, Law Dome, East Antarctica: a comparison of anisotropic ice flow relations, J. Glaciol., 61, 987–1004, doi:10.3189/2015JoG14J198, 2015.

van der Veen, C. and Whillans, I.: Development of fabric in ice, Cold Reg. Sci. Technol., 22, 171–195, doi:10.1016/0165-232X(94)90027-2, 1994.

van Ommen, T. and Morgan, V.: Snowfall increase in coastal East Antarctica linked with southwest Western Australian drought, Nat. Geosci., 3, 267–272, 2010.

van Ommen, T., Morgan, V., Jacka, T., Woon, S., and Elcheikh, A.: Near surface temperatures in the Dome Summit South (Law Dome, East Antarctica) borehole, Ann. Glaciol., 29, 141–144, doi:10.3189/172756499781821382, 1999.

van Ommen, T. D., Morgan, V., and Curran, M. A. J.: Deglacial and Holocene changes in accumulation at Law Dome, East Antarctica, Ann. Glaciol., 39, 359–365, doi:10.3189/172756404781814221, 2004.

Vance, T. R., van Ommen, T. D., Curran, M. A. J., Plummer, C. T., and Moy, A. D.: A Millennial Proxy Record of ENSO and Eastern Australian Rainfall from the Law Dome Ice Core, East Antarctica, J. Climate, 26, 710–725, doi:10.1175/JCLI-D-12-00003.1, 2013.

Vance, T. R., Roberts, J. L., Plummer, C. T., Kiem, A. S., and van Ommen, T. D.: Interdecadal Pacific variability and eastern Australian mega-droughts over the last millenium, Geophys. Res. Lett., 42, 129–137, doi:10.1002/2014GL062447, 2015.

Vaughan, D., Comiso, J., Allison, I., Carrasco, J., Kaser, G., Kwok, R., Mote, P., Murray, T., Paul, F., Ren, J., Rignot, E., Solomina, O., Steffen, K., and Zhang, T.: Climate Change 2013: The Physical Science Basis. Contribution of Working Group I to the Fifth Assessment Report of the Intergovernmental Panel on Climate Change, chap. 2013, Observations: Cryosphere, Cambridge University Press, Cambridge, United Kingdom and New York, NY, USA, 2013.

Wang, W., Li, J., and Zwally, H.: Dynamic inland propagation of thinning due to ice loss at the margins of the Greenland ice sheet, J. Glaciol., 58, 734–640, 2012.

Wilen, L., Diprinzio, C., Alley, R., and Azuma, N.: Development, principles, and applications of automated ice fabric analyzers, Microsc. Res. Techniq., 62, 2–18, 2003.

Willis, J. and Church, J.: Regional sea-level projection, Science, 336, 550–551, doi:10.1126/science.1220366, 2012.

Wilson, C. and Peternell, M.: Evaluating ice fabrics using fabric analyser techniques in Sørsdal Glacier, East Antarctica, J. Glaciol., 57, 881–894, 2011.

Wilson, C., Russell-Head, D., Kunze, K., and Viola, G.: The analysis of quartz c-axis fabrics using a modified optical microscope, J. Microscopy, 227, 30–41, 2007.

Woodcock, N.: Specification of fabric shapes using an eigenvalue method, Geol. Soc. Am. Bull., 88, 1231–1236, doi:10.1130/0016-7606(1977)88<1231:SOFSUA>2.0.CO;2, 1977.

Yun, W. and Azuma, N.: A new automatic ice-fabric analyzer which uses image-analysis techniques, Ann. Glaciol., 155–162, 1999.

Zwinger, T., Schäfer, M., Martín, C., and Moore, J. C.: Influence of anisotropy on velocity and age distribution at Scharffenbergbotnen blue ice area, The Cryosphere, 8, 607–621, doi:10.5194/tc-8-607-2014, 2014.

2

Global sea-level budget 1993–present

WCRP Global Sea Level Budget Group

A full list of authors and their affiliations appears at the end of the paper.

Correspondence: Anny Cazenave (anny.cazenave@legos.obs-mip.fr)

Abstract. Global mean sea level is an integral of changes occurring in the climate system in response to unforced climate variability as well as natural and anthropogenic forcing factors. Its temporal evolution allows changes (e.g., acceleration) to be detected in one or more components. Study of the sea-level budget provides constraints on missing or poorly known contributions, such as the unsurveyed deep ocean or the still uncertain land water component. In the context of the World Climate Research Programme Grand Challenge entitled "Regional Sea Level and Coastal Impacts", an international effort involving the sea-level community worldwide has been recently initiated with the objective of assessing the various datasets used to estimate components of the sea-level budget during the altimetry era (1993 to present). These datasets are based on the combination of a broad range of space-based and in situ observations, model estimates, and algorithms. Evaluating their quality, quantifying uncertainties and identifying sources of discrepancies between component estimates is extremely useful for various applications in climate research. This effort involves several tens of scientists from about 50 research teams/institutions worldwide (www.wcrp-climate.org/grand-challenges/gc-sea-level, last access: 22 August 2018). The results presented in this paper are a synthesis of the first assessment performed during 2017–2018. We present estimates of the altimetry-based global mean sea level (average rate of 3.1 ± 0.3 mm yr^{-1} and acceleration of 0.1 mm yr^{-2} over 1993–present), as well as of the different components of the sea-level budget (http://doi.org/10.17882/54854, last access: 22 August 2018). We further examine closure of the sea-level budget, comparing the observed global mean sea level with the sum of components. Ocean thermal expansion, glaciers, Greenland and Antarctica contribute 42 %, 21 %, 15 % and 8 % to the global mean sea level over the 1993–present period. We also study the sea-level budget over 2005–present, using GRACE-based ocean mass estimates instead of the sum of individual mass components. Our results demonstrate that the global mean sea level can be closed to within 0.3 mm yr^{-1} (1σ). Substantial uncertainty remains for the land water storage component, as shown when examining individual mass contributions to sea level.

1 Introduction

Global warming has already several visible consequences, in particular an increase in the Earth's mean surface temperature and ocean heat content (Rhein et al., 2013; IPCC, 2013), melting of sea ice, loss of mass of glaciers (Gardner et al., 2013), and ice mass loss from the Greenland and Antarctica ice sheets (Rignot et al., 2011a; Shepherd et al., 2012). On average over the last 50 years, about 93 % of heat excess accumulated in the climate system because of greenhouse gas emissions has been stored in the ocean, and the remaining 7 % has been warming the atmosphere and continents, and melting sea and land ice (von Schuckmann et al., 2016). Because of ocean warming and land ice mass loss, sea level rises. Since the end of the last deglaciation about 3000 years ago, sea level remained nearly constant (e.g., Lambeck, 2002; Lambeck et al., 2010; Kemp et al., 2011). However, direct observations from in situ tide gauges available since the mid-to-late 19th century show that the 20th century global mean sea level has started to rise again at a rate of 1.2 to 1.9 mm yr^{-1} (Church and White, 2011; Jevrejeva et al., 2014; Hay et al., 2015; Dangendorf et al., 2017). Since the early 1990s sea-level rise (SLR) is measured by high-precision altimeter satellites and the rate has increased to $\sim 3\,\mathrm{mm\,yr^{-1}}$ on average (Legeais et al., 2018; Nerem et al., 2018).

Accurate assessment of present-day global mean sea-level variations and its components (ocean thermal expansion, ice sheet mass loss, glaciers mass change, changes in land water storage, etc.) is important for many reasons. The global mean sea level is an integral of changes occurring in the Earth's climate system in response to unforced climate variability as well as natural and anthropogenic forcing factors, e.g., net contribution of ocean warming, land ice mass loss and changes in water storage in continental river basins. Temporal changes in the components are directly reflected in the global mean sea-level curve. If accurate enough, study of the sea-level budget provides constraints on missing or poorly known contributions, e.g., the deep ocean undersampled by current observing systems, or still uncertain changes in water storage on land due to human activities (e.g., groundwater depletion in aquifers). Global mean sea level corrected for ocean mass change in principle allows one to independently estimate temporal changes in total ocean heat content, from which the Earth's energy imbalance can be deduced (von Schuckmann et al., 2016). The sea level and/or ocean mass budget approach can also be used to constrain models of glacial isostatic adjustment (GIA). The GIA phenomenon has a significant impact on the interpretation of GRACE-based space gravimetry data over the oceans (for ocean mass change) and over Antarctica (for ice sheet mass balance). However, there is still no complete consensus on best estimates, a result of uncertainties in deglaciation models and mantle viscosity structure. Finally, observed changes in the global mean sea level and its components are fundamental for validating climate models used for projections.

In the context of the Grand Challenge entitled "Regional Sea Level and Coastal Impacts" of the World Climate Research Programme (WCRP), an international effort involving the sea-level community worldwide has been recently initiated with the objective of assessing the sea-level budget during the altimetry era (1993 to present). To estimate the different components of the sea-level budget, different datasets are used. These are based on the combination of a broad range of space-based and in situ observations. Evaluating their quality, quantifying their uncertainties and identifying the sources of discrepancies between component estimates, including the altimetry-based sea-level time series, are extremely useful for various applications in climate research.

Several previous studies have addressed the sea-level budget over different time spans and using different datasets. For example, Munk (2002) found that the 20th century sea-level rise could not be closed with the data available at that time and showed that if the missing contribution were due to polar ice melt, this would be in conflict with external astronomical constraints. The enigma has been resolved in two ways. Firstly, an improved theory of rotational stability of the Earth (Mitrovica et al., 2006) effectively removed the constraints proposed by Munk (2002) and allows a polar ice sheet contribution to 20th century sea-level rise of as much as $\sim 1.1\,\mathrm{mm\,yr^{-1}}$, with about $0.8\,\mathrm{mm\,yr^{-1}}$ beginning in the 20th century. In addition, more recent studies by Gregory et al. (2013) and Slangen et al. (2017), combining observations with model estimates, showed that it was possible to effectively close the 20th century sea-level budget within uncertainties, particularly over the altimetry era (e.g., Cazenave et al., 2009; Leuliette and Willis, 2011; Church and White, 2011; Llovel et al., 2014; Chambers et al., 2017; Dieng et al., 2017; X. Chen et al., 2017; Nerem et al., 2018). Assessments of the published literature have also been performed in past IPCC (Intergovernmental Panel on Climate Change) reports (e.g., Church et al., 2013). Building on these previous works, here we intend to provide a collective update of the global mean sea-level budget, involving the many groups worldwide interested in present-day sea-level rise and its components. We focus on observations rather than model-based estimates and consider the high-precision altimetry era starting in 1993. This era includes the period since the mid-2000s in which new observing systems, like the Argo float project (Roemmich et al., 2012) and the GRACE space gravimetry mission (Tapley et al., 2004a, b), provide improved datasets of high value for such a study. Only the global mean budget is considered here. Regional budget will be the focus of a future assessment.

Section 2 describes for each component of the sea-level budget equation the different datasets used to estimate the corresponding contribution to sea level, discusses associated errors and provides trend estimates for the two periods. Section 3 addresses the mass and sea-level budgets over the study periods. A discussion is provided in Sect. 4, followed by a conclusion.

2 Methods and data

In this section, we briefly present the global mean sea-level budget (Sect. 2.1) and then provide, for each term of the budget equation, an assessment of the most up-to-date published results. Multiple organizations and research groups routinely generate the basic measurements as well as the derived datasets and products used to study the sea-level budget. Sections 2.2 to 2.7 summarize the measurements and methodologies used to derive observed sea level, as well as steric and mass components. In most cases, we focus on observations but in some instances (e.g., for GIA corrections applied to the data), model-based estimates are the only available information.

2.1 Sea-level budget equation

Global mean sea level (GMSL) change as a function of time t is usually expressed by the sea-level budget equation:

$$\mathrm{GMSL}(t) = \mathrm{GMSL}(t)_{\mathrm{steric}} + \mathrm{GMSL}(t)_{\mathrm{ocean\,mass}}, \quad (1)$$

where $\mathrm{GMSL}(t)_{\mathrm{steric}}$ refers to the contributions of ocean thermal expansion and salinity to sea-level change, and $\mathrm{GMSL}(t)_{\mathrm{oceanmass}}$ refers to the change in mass of the oceans. Due to water conservation in the climate system, the ocean mass term (also noted as $\mathrm{M}(t)_{\mathrm{ocean}}$) can further be expressed as follows:

$$\begin{aligned}\mathrm{M}(t)_{\mathrm{ocean}} &+ \mathrm{M}(t)_{\mathrm{glaciers}} + \mathrm{M}(t)_{\mathrm{Greenland}} + \mathrm{M}(t)_{\mathrm{Antarctica}} \\ &+ \mathrm{M}(t)_{\mathrm{TWS}} + \mathrm{M}(t)_{\mathrm{WV}} + \mathrm{M}(t)_{\mathrm{Snow}} \\ &+ \mathrm{uncertainty} = 0, \quad (2)\end{aligned}$$

where $\mathrm{M(t)}_{\mathrm{glaciers}}$, $\mathrm{M(t)}_{\mathrm{Greenland}}$, $\mathrm{M(t)}_{\mathrm{Antarctica}}$, $\mathrm{M(t)}_{\mathrm{TWS}}$, $\mathrm{M(t)}_{\mathrm{WV}}$ and $\mathrm{M(t)}_{\mathrm{Snow}}$ represent temporal changes in mass of glaciers, Greenland and Antarctica ice sheets, terrestrial water storage (TWS), atmospheric water vapor (WV), and snow mass changes. The uncertainty is a result of uncertainties in all of the estimates. For the altimetry era, many studies have investigated closure of the sea-level budget and potentially missing mass terms, for example, permafrost melting.

From Eq. (2), we deduce the following:

$$\begin{aligned}\mathrm{GMSL}(t)_{\mathrm{ocean\,mass}} = &-[\mathrm{M}(t)_{\mathrm{glaciers}} + \mathrm{M}(t)_{\mathrm{Greenland}} \\ &+ \mathrm{M}(t)_{\mathrm{Antarctica}} + \mathrm{M}(t)_{\mathrm{TWS}} + \mathrm{M}(t)_{\mathrm{WV}} + \mathrm{M}(t)_{\mathrm{Snow}} \\ &+ \mathrm{missing\,mass\,terms}] \quad (3)\end{aligned}$$

In the next subsections, we successively discuss the different terms of the budget (Eqs. 1 and 2) and how they are estimated from observations. We do not consider the atmospheric water vapor and snow components, assumed to be small. Two periods are considered: (1) 1993–present (i.e., the entire altimetry era) and (2) 2005–present (i.e., the period covered by both Argo and GRACE).

2.2 Altimetry-based global mean sea level over 1993–present

The launch of the TOPEX/Poseidon (T/P) altimeter satellite in 1992 led to a new paradigm for measuring sea level from space, providing for the first time precise and globally distributed sea-level measurements at 10-day intervals. At the time of the launch of T/P, the measurements were not expected to have sufficient accuracy for measuring GMSL changes. However, as the radial orbit error decreased from $\sim 10\,\mathrm{cm}$ at launch to $\sim 1\,\mathrm{cm}$ presently, and other instrumental and geophysical corrections applied to altimetry system improved (e.g., Stammer and Cazenave, 2018), several groups regularly provided an altimetry-based GMSL time series (e.g., Nerem et al., 2010; Church et al., 2011; Ablain et al., 2015; Legeais et al., 2018). The initial T/P GMSL time

series was extended with the launch of Jason-1 (2001), Jason-2 (2008) and Jason-3 (2016). By design, each of these missions has an overlap period with the previous one in order to intercompare the sea-level measurements and estimate instrument biases (e.g., Nerem et al., 2010; Ablain et al., 2015). This has allowed the construction of an uninterrupted GMSL time series that is currently 25 years long.

2.2.1 Global mean sea-level datasets

Six groups (AVISO/CNES, SL_cci/ESA, University of Colorado, CSIRO, NASA/GSFC, NOAA) provide altimetry-based GMSL time series. All of them use 1 Hz altimetry measurements derived from T/P, Jason-1, Jason-2 and Jason-3 as reference missions. These missions provide the most accurate long-term stability at global and regional scales (Ablain et al., 2009, 2017a), and are all on the same historical T/P ground track. This allows computation of a long-term record of the GMSL from 1993 to present. In addition, complementary missions (ERS-1, ERS-2, Envisat, Geosat Follow-on, CryoSat-2, SARAL/AltiKa and Sentinel-3A) provide increased spatial resolution and coverage of high-latitude ocean areas, pole-ward of 66° N–S latitude (e.g., the European Space Agency/ESA Climate Change Initiative/CCI sea-level dataset; Legeais et al., 2018).

The above groups adopt different approaches when processing satellite altimetry data. The most important differences concern the geophysical corrections needed to account for various physical phenomena such as atmospheric propagation delays, sea state bias, ocean tides, and the ocean response to atmospheric wind and pressure forcing. Other differences come from data editing, methods to spatially average individual measurements during orbital cycles and links between successive missions (Masters et al., 2012; Henry et al., 2014).

Overall, the quality of the different GMSL time series is similar. Long-term trends agree well to within 6 % of the signal, approximately $0.2\,\mathrm{mm\,yr^{-1}}$ (see Fig. 1) within the GMSL trend uncertainty range ($\sim 0.3\,\mathrm{mm\,yr^{-1}}$; see next section). The largest differences are observed at interannual timescales and during the first years (before 1999; see below). Here we use an ensemble mean GMSL based on averaging all individual GMSL time series.

2.2.2 Global mean sea-level uncertainties and TOPEX-A drift

Based on an assessment of all sources or uncertainties affecting satellite altimetry (Ablain et al., 2017a), the GMSL trend uncertainty (90 % confidence interval) is estimated as 0.3 to $0.4\,\mathrm{mm\,yr^{-1}}$ over the whole altimetry era (1993–2017). The main contribution to the uncertainty is the wet tropospheric correction with a drift uncertainty in the range of 0.2–$0.3\,\mathrm{mm\,yr^{-1}}$ (Legeais et al., 2018) over a 10-year period. To a lesser extent, the orbit error (Couhert et al., 2015; Escudier

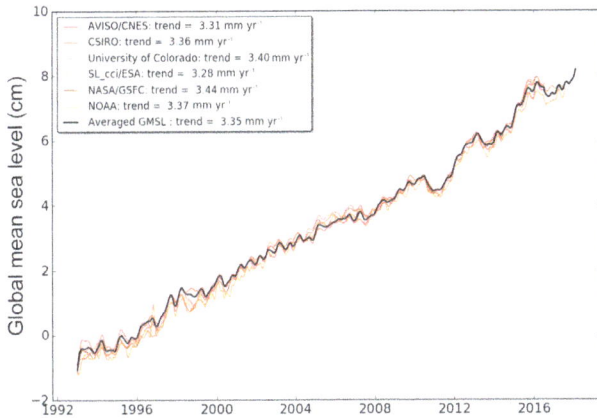

Figure 1. Evolution of GMSL time series from six different groups' (AVISO/CNES, SL_cci/ESA, University of Colorado, CSIRO, NASA/GSFC, NOAA) products. Annual signals are removed and 6-month smoothing applied. All GMSL time series are centered in 1993 with zero mean. A GIA correction of $-0.3\,\mathrm{mm\,yr^{-1}}$ has been subtracted from each dataset.

et al., 2018) and the altimeter parameters' (range, $\sigma 0$ and significant wave height – SWH) instability (Ablain et al., 2012) also contribute to the GMSL trend uncertainty, at the level of $0.1\,\mathrm{mm\,yr^{-1}}$. Furthermore, imperfect links between successive altimetry missions lead to another trend uncertainty of about $0.15\,\mathrm{mm\,yr^{-1}}$ over the 1993–2017 period (Zawadzki and Ablain, 2016).

Uncertainties are higher during the first decade (1993–2002), when T/P measurements display larger errors at climatic scales. For instance, the orbit solutions are much more uncertain due to gravity field solutions calculated without GRACE data. Furthermore, the switch from TOPEX-A to TOPEX-B in February 1999 (with no overlap between the two instrumental observations) leads to an error of $\sim 3\,\mathrm{mm}$ in the GMSL time series (Escudier et al., 2018).

However, the most significant error that affects the first 6 years (January 1993 to February 1999) of the T/P GMSL measurements is due to an instrumental drift of the TOPEX-A altimeter, not included in the formal uncertainty estimates discussed above. This effect on the GMSL time series was recently highlighted via comparisons with tide gauges (Valladeau et al., 2012; Watson et al., 2015; X. Chen et al., 2017; Ablain et al., 2017b), via a sea-level budget approach (i.e., comparison with the sum of mass and steric components; Dieng et al., 2017) and by comparing with Poseidon-1 measurements (Lionel Zawadsky, personal communication, 2017). In a recent study, Beckley et al. (2017) asserted that the corresponding error on the 1993–1998 GMSL resulted from incorrect onboard calibration parameters.

All approaches conclude that during the period January 1993 to February 1999, the altimetry-based GMSL was overestimated. TOPEX-A drift correction was estimated to be close to $1.5\,\mathrm{mm\,yr^{-1}}$ (in terms of sea-level trend) with an

uncertainty of ± 0.5 to $\pm 1.0\,\mathrm{mm\,yr^{-1}}$ (Watson et al., 2015; X. Chen et al., 2017; Dieng et al., 2017). Beckley et al. (2017) proposed to not apply the suspect onboard calibration correction on TOPEX-A measurements. The impact of this approach is similar to the TOPEX-A drift correction estimated by Dieng et al. (2017) and Ablain et al. (2017b). In the latter study, accurate comparison between TOPEX-A-based GMSL and tide gauge measurements leads to a drift correction of about $-1.0\,\mathrm{mm\,yr^{-1}}$ between January 1993 and July 1995, and $+3.0\,\mathrm{mm\,yr^{-1}}$ between August 1995 and February 1999, with an uncertainty of $1.0\,\mathrm{mm\,yr^{-1}}$ (with a 68 % confidence level, see Table 1).

2.2.3 Global mean sea-level variations

The ensemble mean GMSL rate after correcting for the TOPEX-A drift (for all of the proposed corrections) amounts to $3.1\,\mathrm{mm\,yr^{-1}}$ over 1993–2017 (Fig. 2). This corresponds to a mean sea-level rise of about $7.5\,\mathrm{cm}$ over the whole altimetry period. More importantly, the GMSL curve shows a net acceleration, estimated to be at $0.08\,\mathrm{mm\,yr^{-2}}$ (X. Chen et al., 2017; Dieng et al., 2017) and $0.084 \pm 0.025\,\mathrm{mm\,yr^{-2}}$ (Nerem et al., 2018) (note Watson et al., 2015 found a smaller acceleration after correcting for the instrumental bias over a shorter period up to the end of 2014.). GMSL trends calculated over 10-year moving windows illustrate this acceleration (Fig. 3). GMSL trends are close to $2.5\,\mathrm{mm\,yr^{-1}}$ over 1993–2002 and $3.0\,\mathrm{mm\,yr^{-1}}$ over 1996–2005. After a slightly smaller trend over 2002–2011, the 2008–2017 trend reaches $4.2\,\mathrm{mm\,yr^{-1}}$. Uncertainties (90 % confidence interval) associated with these 10-year trends regularly decrease through time from $1.3\,\mathrm{mm\,yr^{-1}}$ over 1993–2002 (corresponding to T/P data) to $0.65\,\mathrm{mm\,yr^{-1}}$ for 2008–2017 (corresponding to Jason-2 and Jason-3 data).

Removing the trend from the GMSL time series highlights interannual variations (not shown). Their magnitudes depend on the period ($+3\,\mathrm{mm}$ in 1998–1999, $-5\,\mathrm{mm}$ in 2011–2012 and $+10\,\mathrm{mm}$ in 2015–2016) and are well correlated in time with El Niño and La Niña events (Nerem et al., 2010, 2018; Cazenave et al., 2014). However, substantial differences (of 1–3 mm) exist between the six detrended GMSL time series. This issue needs further investigation.

For the sea-level budget assessment (Sect. 3), we will use the ensemble mean GMSL time series corrected for the TOPEX-A drift using the Ablain et al. (2017b) correction.

2.2.4 Comparison with tide gauges

Prior to 1992, global sea-level rise estimates relied on the tide gauge measurements, and it is worth mentioning past attempts to produce global sea-level reconstructions utilizing these measurements (e.g., Gornitz et al., 1982; Bartnett, 1984; Douglas, 1991, 1997, 2001). Here we focus on global sea-level reconstructions that overlap with satellite altimetry data over a substantial common time span. Some of

Table 1. TOPEX-A GMSL drift corrections proposed by different studies.

TOPEX-A drift correction	to be subtracted from the first 6 years (Jan 1993 to Feb 1999) of the uncorrected GMSL record
Watson et al. (2015)	1.5 ± 0.5 mm yr^{-1} over Jan 1993–Feb 1999
X. Chen et al. (2017), Dieng et al. (2017)	1.5 ± 0.5 mm yr^{-1} over Jan 1993–Feb 1999
Beckley et al. (2017)	No onboard calibration applied
Ablain et al. (2017b)	-1.0 ± 1.0 mm yr^{-1} over Jan 1993–Jul 1995 $+3.0 \pm 1.0$ mm yr^{-1} over Aug 1995–Feb 1999

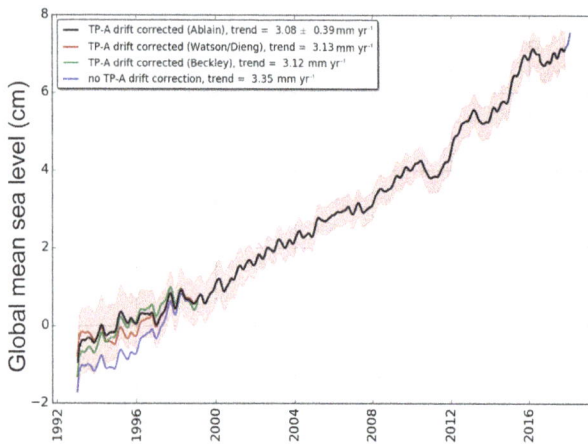

Figure 2. Evolution of ensemble mean GMSL time series (average of the six GMSL products from AVISO/CNES, SL_cci/ESA, University of Colorado, CSIRO, NASA/GSFC and NOAA). On the black, red and green curves, the TOPEX-A drift correction is applied respectively based on Ablain et al. (2017b), Watson et al. (2015) and Dieng et al. (2017), and Beckley et al. (2017). Annual signal removed and 6-month smoothing applied; GIA correction also applied. Uncertainties (90 % confidence interval) of correlated errors over a 1-year period are superimposed for each individual measurement (shaded area).

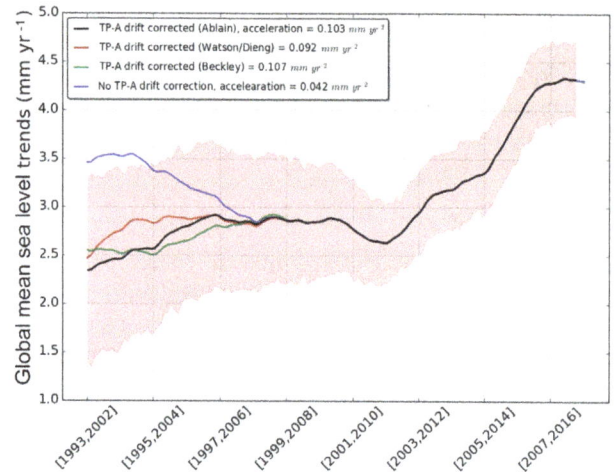

Figure 3. Ensemble mean GMSL trends calculated over 10-year moving windows. On the black, red and green curves, the TOPEX-A drift correction is applied respectively based on Ablain et al. (2017b), Watson et al. (2015) and Dieng et al. (2017), and Beckley et al. (2017). Uncorrected GMSL trends are shown by the blue curve. The shaded area represents trend uncertainty over 10-year periods (90 % confidence interval).

these reconstructions rely on tide gauge data only (Jevrejeva et al., 2006, 2014; Merrifield et al., 2009; Wenzel and Schroter, 2010; Ray and Douglas, 2011; Hamlington et al., 2011; Spada and Galassi, 2012; Thompson and Merrifield, 2014; Dangendorf et al., 2017; Frederikse et al., 2017). In addition, there are reconstructions that jointly use satellite altimetry, tide gauge records (Church and White, 2006, 2011) and reconstructions, which combine tide gauge records with ocean models (Meyssignac et al., 2011) or physics-based and model-derived geometries of the contributing processes (Hay et al., 2015).

For the period since 1993, with most of the world coastlines densely sampled, the rates of sea-level rise from all tide-gauge-based reconstructions and estimates from satellite altimetry agree within their specific uncertainties,

e.g., rates of 3.0 ± 0.7 mm yr^{-1} (Hay et al. 2015), 2.8 ± 0.5 mm yr^{-1} (Church and White, 2011; Rhein et al., 2013), 3.1 ± 0.6 mm yr^{-1} (Jevrejeva et al., 2014), 3.1 ± 1.4 mm yr^{-1} (Dangendorf et al., 2017) and the estimate from satellite altimetry 3.2 ± 0.4 mm yr^{-1} (Nerem et al., 2010; Rhein et al., 2013). However, classical tide-gauge-based reconstructions still tend to overestimate the interannual to decadal variability of global mean sea level (e.g., Calafat et al., 2014; Dangendorf et al., 2015; Natarov et al., 2017) compared to global mean sea level from satellite altimetry, due to limited and uneven spatial sampling of the global ocean afforded by the tide gauge network. Sea-level rise being non uniform, spatial variability of sea-level measured at tide gauges is evidenced by 2-D reconstruction methods. The most widely used approach is the use of empirical orthogonal functions (EOFs) calibrated with the satellite altimetry data (e.g., Church and

White, 2006). Alternatively, Choblet et al. (2014) implemented a Bayesian inference method based on a Voronoi tessellation of the Earth's surface to reconstruct sea level during the 20th century. Considerable uncertainties remain, however, in long-term assessments due to poorly sampled ocean basins such as the South Atlantic, or regions which are significantly influenced by open-ocean circulation (e.g., subtropical North Atlantic) (Frederikse et al., 2017). Uncertainties involved in specifying vertical land motion corrections at tide gauges also impact tide gauge reconstructions (Jevrejeva et al., 2014; Wöppelmann and Marcos, 2016; Hamlington et al., 2016). Frederikse et al. (2017) also recently demonstrated that both global mean sea level reconstructed from tide gauges and the sum of steric and mass contributors show a good agreement with altimetry estimates for the overlapping period 1993–2014.

2.3 Steric sea level

Steric sea-level variations result from temperature- (T) and salinity- (S) related density changes in sea water associated with volume expansion and contraction. These are referred to as thermosteric and halosteric components. Despite clear detection of regional salinity changes and the dominance of the salinity effect on density changes at high latitudes (Rhein et al., 2013), the halosteric contribution to present-day global mean steric sea-level rise is negligible, as the ocean's total salt content is essentially constant over multidecadal timescales (Gregory and Lowe, 2000). Hence, in this study, we essentially consider the thermosteric sea-level component.

Averaged over the 20th century, ocean thermal expansion associated with ocean warming has been the largest contribution to global mean sea-level rise (Church et al., 2013). This remains true for the altimetry period starting in the year 1993 (e.g., X. Chen et al., 2017; Dieng et al., 2017; Nerem et al., 2018). But total land ice mass loss (from glaciers, Greenland and Antarctica) during this period now dominates the sea-level budget (see Sect. 3).

Until the mid-2000s, the majority of ocean temperature data were retrieved from shipboard measurements. These include vertical temperature profiles along research cruise tracks from the surface sometimes all the way down to the bottom layer (e.g., Purkey and Johnson, 2010) and upper-ocean broad-scale measurements from ships of opportunity (Abraham et al., 2013). These upper-ocean in situ temperature measurements, however, are limited to the upper 700 m depth due to common use of expandable bathythermographs (XBTs). Although the coverage has been improved through time, large regions characterized by difficult meteorological conditions remained under-sampled, in particular the southern hemispheric oceans and the Arctic area.

2.3.1 Thermosteric datasets

Over the altimetry era, several research groups have produced gridded time series of temperature data for different depth levels, based on XBTs (with additional data from mechanical bathythermographs – MBTs – and conductivity–temperature–depth – CTD – devices and moorings) and Argo float measurements. The temperature data have further been used to provide thermosteric sea-level products. These differ because of different strategies adopted for data editing, temporal and spatial data gaps filling, mapping methods, baseline climatology, and instrument bias corrections (in particular the time-to-depth correction for XBT data, Boyer et al., 2016).

The global ocean in situ observing system has been dramatically improved through the implementation of the international Argo program of autonomous floats, delivering a unique insight into the interior ocean from the surface down to 2000 m depth of the ice-free global ocean (Roemmich et al., 2012; Riser et al., 2016). More than 80 % of initially planned full deployment of Argo float program was achieved during the year 2005, with quasi global coverage of the ice-free ocean by the start of 2006. At present, more than 3800 floats provide systematic T and S data, with quasi (60° S–60° N latitude) global coverage down to 2000 m depth. A full overview on in situ ocean temperature measurements is given for example in Abraham et al. (2013).

In this section, we consider a set of 11 direct (in situ) estimates, publicly available over the entire altimetry era, to review global mean thermosteric sea-level rise and, ultimately, to construct an ensemble mean time series. These datasets are as follows:

1. CORA = Coriolis Ocean database for ReAnalysis, Copernicus Service, France (marine.copernicus.eu/), product name: INSITU_GLO_ TS_OA_ REP_OBSERVATIONS_013_002_b;

2. CSIRO (RSOI) = Commonwealth Scientific and Industrial Research Organisation/Reduced-Space Optimal Interpolation, Australia;

3. ACECRC/IMAS-UTAS = Antarctic Climate and Ecosystem Cooperative Research Centre/Institute for Marine and Antarctic Studies-University of Tasmania, Australia (http://www.cmar.csiro.au/sealevel/thermal_expansion_ocean_heat_timeseries.html);

4. ICCES = International Center for Climate and Environment Sciences, Institute of Atmospheric Physics, China (http://ddl.escience.cn/f/PKFR);

5. ICDC = Integrated Climate Data Center, University of Hamburg, Germany;

6. IPRC = International Pacific Research Center, University of Hawaii, USA (http://apdrc.soest.hawaii.

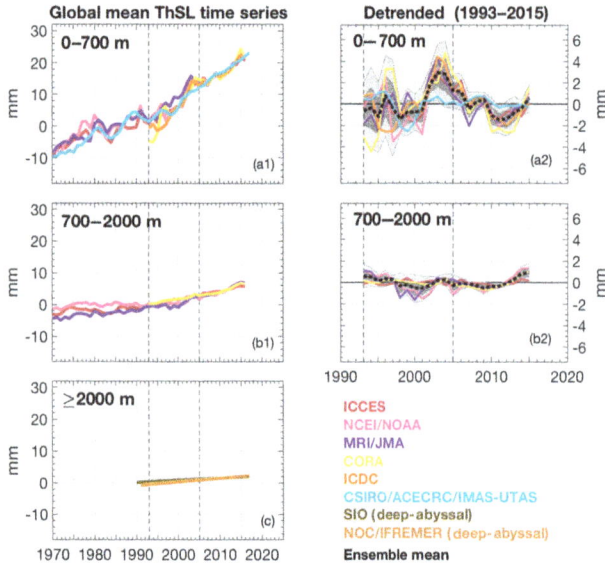

Figure 4. Left panels: annual mean global mean thermosteric anomaly time series since 1970, from various research groups (color) and for three depth integrations: 0–700 m (top), 700–2000 m (middle) and below 2000 m (bottom). Vertical dashed lines are plotted along 1993 and 2005. For comparison, all time series were offset arbitrarily. Right panels: respective linearly detrended time series for 1993–2015. Black bold dashed line is the ensemble mean and gray shadow bar the ensemble spread (1 standard deviation). Units are millimeters.

edu/projects/Argo/data/gridded/On_standard_levels/index-1.html);

7. JAMSTEC = Japan Agency for Marine-Earth Science and Technology, Japan (ftp://ftp2.jamstec.go.jp/pub/argo/MOAA_GPV/Glb_PRS/OI/);

8. MRI/JMA = Meteorological Research Institute/Japan Meteorological Agency, Japan (https://climate.mri-jma.go.jp/~ishii/.wcrp/);

9. NCEI/NOAA = National Centers for Environmental Information/National Oceanic and Atmospheric Administration, USA;

10. SIO = Scripps Institution of Oceanography, USA; Deep–abyssal: https://cchdo.ucsd.edu/;

11. SIO = Scripps Institution of Oceanography, USA; Deep–abyssal: https://cchdo.ucsd.edu/ (for the abyssal ocean).

Their characteristics are presented in Table 2.

2.3.2 Individual estimates

All in situ estimates compiled in this study show a steady rise in global mean thermosteric sea level, independent of depth integration and decadal or multidecadal periods (Figs. 4 and

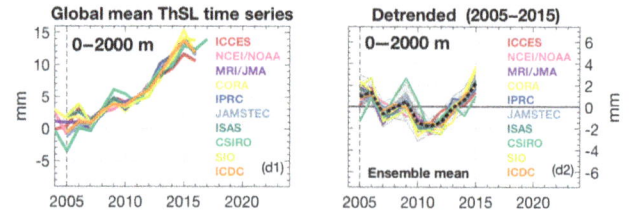

Figure 5. Left panel: annual mean global mean thermosteric anomaly time series since 2004, from various research groups (color) in the upper 2000 m. A vertical dashed line is plotted along 2005. For comparison, all time series were offset arbitrarily. Right panel: respective linearly detrended time series for 2005–2015. Black bold dashed line is the ensemble mean and gray shadow bar the ensemble spread (1 standard deviation). Units are millimeters.

5, left panels). As the deep–abyssal ocean estimate only illustrates the updated version of the linear trend from Purkey and Johnson (2010) for 1990–2010 extrapolated to 2016, it does not have any variability superimposed.

Interannual to decadal variability during the altimeter era (since 1993) is similar for both 0–700 and 700–2000 m, with larger amplitude in the upper ocean (Figs. 4 and 5, right panels). For the 0–700 m, there is an apparent change in amplitude before and after the Argo era (since 2005), mostly due to a maximum (2–4 mm) around 2001–2004, except for one estimate. Higher amplitude and larger spread in variability between estimates before the Argo era is a symptom of the much sparser in situ coverage of the global ocean. Interannual variability over the Argo era (Figs. 4 and 5, right panels) is mainly modulated by El Niño–Southern Oscillation (ENSO) phases in the upper 500 m of the ocean, particularly for the Pacific, the largest ocean basin (Roemmich and Gilson, 2011; Roemmich et al., 2016; Johnson and Birnbaum, 2017).

In terms of depth contribution, on average, the upper 300 m explains the same percentage (almost 70 %) of the 0–700 m linear rate over both altimetry and Argo eras, but the contribution from the 0–700 to 0–2000 m varies: about 75 % for 1993–2016 and 65 % for 2005–2016. Thus, the 700–2000 m contribution increases by 10 % during the Argo decade, when the number of observations within 700–2000 m has significantly increased.

2.3.3 Ensemble mean thermosteric sea level

Given that the global mean thermosteric sea-level anomaly estimates compiled for this study are not necessarily referenced to the same baseline climatology, they cannot be directly averaged together to create an ensemble mean. To circumvent this limitation, we created an ensemble mean in three steps, as explained below.

Firstly, we detrended the individual time series by removing a linear trend for 1993–2016 and averaged together to obtain an "ensemble mean variability time series". Sec-

Table 2. Compilation of available in situ datasets from different originators and/or contributors. The table indicates the time span covered by the data, the depth of integration, as well as the temporal resolution and latitude coverage.

	Product/institution	Period	Depth integration (m)				Temporal resolution/latitudinal range	Reference
			0–700	700–2000	0–2000	≥ 2000		
1	CORA	1993–2016	Y	Y	Y	–	Monthly 60° S–60° N	http://marine.copernicus.eu/services-portfolio/access-to-products/
2	CSIRO (RSOI)	2004–2017	Y/E (0–300)	Y/E	Y/E	–	Monthly 65° S–65° N	Roemmich et al. (2015), Wijffels et al. (2016)
3	CSIRO/ ACECRC/ IMAS-UTAS	1970–2017	Y/E (0–300)	–	–	–	Yearly (3-year running mean) 65° S–65° N	Domingues et al. (2008), Church et al. (2011)
4	ICCES	1970–2016	Y/E (0–300)	Y/E	Y/E	–	Yearly 89° S–89° N	Cheng et al. (2017)
5	ICDC	1993–2016	Y (1993)	–	Y (2005)	–	Monthly	Gouretski and Koltermann (2007)
6	IPRC	2005–2016	–	–	Y	–	Monthly	http://apdrc.soest.hawaii.edu/projects/argo (last access: 22 August 2018)
7	JAMSTEC	2005–2016	–	–	Y	–	Monthly	Hosoda et al. (2008)
8	MRI/JMA	1970–2016 (rel. to 1961–1990 averages)	Y/E (0–300)	Y/E	Y/E	–	Yearly 89° S–89° N	Ishii et al. (2009, 2017)
9	NCEI/NOAA	1970–2016	Y/E	Y/E	Y/E	–	Yearly 89° S–89° N	Levitus et al. (2012)
10	SIO	2005–2016	–	–	Y	–	Monthly	Roemmich and Gilson (2009)
11	SIO (Deep–abyssal)	1990–2010 (as of Jan 2018)	–	–	–	Y/E	Linear trend 89° S–89° N, as an aggregation of 32 deep ocean basins	Purkey and Johnson (2010)

ondly, we averaged together the corresponding linear trends of the individual estimates to obtain an "ensemble mean linear rate". Thirdly, we combined this "ensemble mean linear rate" with the "ensemble mean variability time series" to obtain the final ensemble mean time series. We applied the same steps for the Argo era (2005–2016).

To maximize the number of individual estimates used in the final full-depth ensemble mean time series, the three steps above were actually divided into depth integrations and then summed. For the Argo era, we summed 0–2000 m (nine estimates) and ≥ 2000 m (one estimate). For the altimetry era, we summed 0–700 m (six estimates), 700–2000 (four esti-

mates) and ≥ 2000 m (one estimate), although there is no statistical difference if the calculation was only based on the sum of 0–2000 m (4 estimates) and ≥ 2000 m (1 estimate). There is also no statistical difference between the full-depth ensemble mean time series created for the Altimeter and Argo eras during their overlapping years (since 2005).

Figure 6 shows the full-depth ensemble mean time series over 1993–2015 and 2005–2015. It reveals a global mean thermosteric sea-level rise of about 30 mm over 1993–2016 (24 years) or about 18 mm over 2005–2016 (12 years), with a record high in 2015. These thermosteric changes are equiv-

Global mean ThSL time series ensemble mean

(a)

(b)

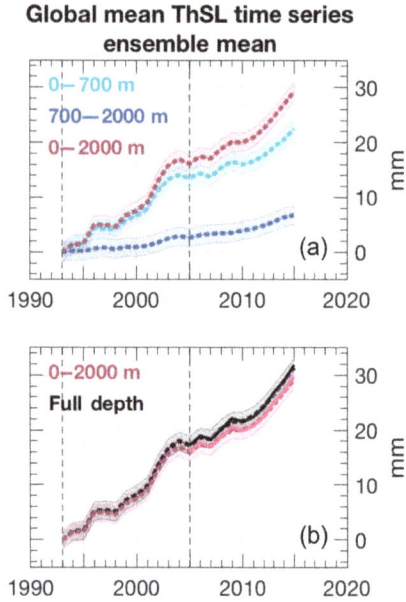

Figure 6. Ensemble mean time series for global mean thermosteric anomaly, for three depth integrations (**a**) and for 0–2000 m and full depth (**b**). In the bottom panel, dashed lines are for the 1993–2015 period whereas solid lines are for 2005–2015. Error bars represent the ensemble spread (standard deviation). Units are millimeters.

Figure 7. Linear rates of global mean thermosteric sea level for depth integrations (x axis), individual estimates and ensemble means, over 1993–2015 (**a**) and 2005–2015 (**b**). Ensemble mean rates with a black circle were used in the estimation of the time series described in Sect. 2.3.4. Error bars are standard deviation due to spread of the estimates except for ≥ 2000 m. Units are millimeters per year.

alent to a linear rate of 1.32 ± 0.4 and $1.31 \pm 0.4 \, \mathrm{mm \, yr^{-1}}$ respectively.

Figure 7 shows thermosteric sea-level trends for each of the datasets used over the 1993–2015 (a) and 2005–2015 (b) time spans and different depth ranges (including full depth), as well as associated ensemble mean trends. The full depth ensemble mean trend amounts to $1.3 \pm 0.4 \, \mathrm{mm \, yr^{-1}}$ over 2005–2015. It is similar to the 1993–2015 ensemble mean trend, suggesting negligible acceleration of the thermosteric component over the altimetry era.

2.4 Glaciers

Glaciers have strongly contributed to sea-level rise during the 20th century – around 40 % – and will continue to be an important part of the projected sea-level change during the 21st century – around 30 % (Kaser et al., 2006; Church et al., 2013; Gardner et al., 2013; Marzeion et al., 2014; Zemp et al., 2015; Huss and Hock, 2015). Because glaciers are time-integrated dynamic systems, a response lag of at least 10 years to a few hundred years is observed between changes in climate forcing and glacier shape, mainly depending on glacier length and slope (Johannesson et al., 1989; Bahr et al., 1998). Today, glaciers are globally (a notable exception is the Karakoram–Kunlun Shan region, e.g., Brun et al., 2017) in a strong disequilibrium with the current climate and are losing mass, due essentially to the global warming in the second half of the 20th century (Marzeion et al., 2018).

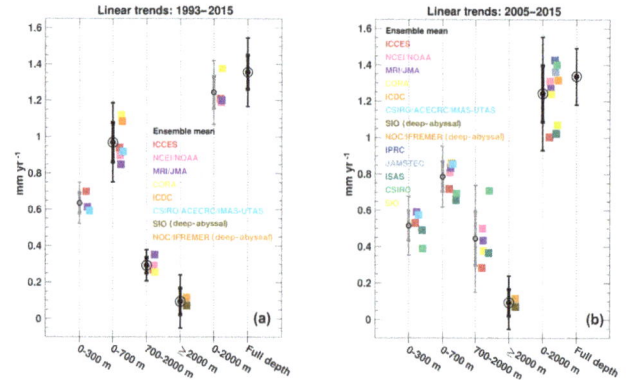

Global glacier mass changes are derived from in situ measurements of glacier mass changes or glacier length changes. Remote sensing methods measure elevation changes over entire glaciers based on differencing digital elevation models (DEMs) from satellite imagery between two epochs (or at points from repeat altimetry), surface flow velocities for determination of mass fluxes and glacier mass changes from space-based gravimetry. Mass balance modeling driven by climate observations is also used (Marzeion et al., 2017, provide a review of these different methods).

Glacier contribution to sea level is primarily the result of their surface mass balance and dynamic adjustment, plus iceberg discharge and frontal ablation (below sea level) in the case of marine-terminating glaciers. The sum of worldwide glacier mass balances does not correspond to the total glacier contribution to sea-level change for the following reasons:

– Glacier ice below sea level does not contribute to sea-level change, apart from a small lowering when replacing ice with seawater of a higher density. Total volume of glacier ice below sea level is estimated to be 10–60 mm sea-level equivalent (SLE, Huss and Farinotti, 2012; Haeberli and Linsbauer, 2013; Huss and Hock, 2015).

– There is incomplete transfer of melting ice from glaciers to the ocean: meltwater stored in lakes or wetlands, meltwater intercepted by natural processes and human activities (e.g., drainage to lakes and aquifers in endorheic basins, impoundment in reservoirs, agriculture use of freshwater, Loriaux and Casassa, 2013; Kääb et al., 2015).

Despite considerable progress in observing methods and spatial coverage (Marzeion et al., 2017), estimating glacier contribution to sea-level change remains challenging due to the following reasons:

- The number of regularly observed glaciers (in the field) remains very low (0.25 % of the 200 000 glaciers of the world have at least one observation and only 37 glaciers have multidecade-long observations, Zemp et al., 2015).

- Uncertainty of the total glacier ice mass remains high (Fig. 8, Grinsted, 2013; Pfeffer et al., 2014; Farinotti et al., 2017; Frey et al. 2014).

- Uncertainties in glacier inventories and DEMs are not negligible. Sources of uncertainties include debris-covered glaciers, disappearance of small glaciers, positional uncertainties, wrongly mapped seasonal snow, rock glaciers, voids and artifacts in DEMs (Paul et al., 2004; Bahr and Radić, 2012).

- Uncertainties of satellite retrieval algorithms from space-based gravimetry and regional DEM differencing are still high, especially for global estimates (Gardner et al., 2013; Marzeion et al., 2017; Chambers et al., 2017).

- Uncertainties of global glacier modeling (e.g., initial conditions, model assumptions and simplifications, local climate conditions; Marzeion et al., 2012).

- Knowledge about some processes governing mass balance (e.g., wind redistribution and metamorphism, sublimation, refreezing, basal melting) and dynamic processes (e.g., basal hydrology, fracking, surging) remains limited (Farinotti et al., 2017).

An annual assessment of glacier contribution to sea-level change is difficult to perform from ground-based or space-based observations apart from space-based gravimetry, due to the sparse and irregular observation of glaciers, and the difficulty of accurately assessing the annual mass balance variability. Global annual averages are highly uncertain because of the sparse coverage, but successive annual balances are uncorrelated and therefore averages over several years are known with greater confidence.

2.4.1 Glacier datasets

The following datasets are considered, with a focus on the trends of annual mass changes:

1. update of Gardner et al. (2013) (Reager et al., 2016), from satellite gravimetry and altimetry, and glaciological records, called G16;

2. update of Marzeion et al. (2012) (Marzeion et al., 2017), from global glacier modeling and mass balance observations, called M17;

3. update of Cogley (2009) (Marzeion et al., 2017), from geodetic and direct mass-balance measurements, called C17;

4. update of Leclercq et al. (2011) (Marzeion et al., 2017), from glacier length changes, called L17;

5. average of GRACE-based estimates of Marzeion et al. (2017), from spatial gravimetry measurements, called M17-G.

In general it is not possible to align measurements of glacier mass balance with the calendar. Most in situ measurements are for glaciological years that extend between successive annual minima of the glacier mass at the end of the summer melt season. Geodetic measurements have start and end dates several years apart and are distributed irregularly through the calendar year; some are corrected to align with annual mass minima but most are not. Consequently, measurements discussed here for 1993–2016 (the altimetry era) and 2005–2016 (the GRACE and Argo era) are offset by up to a few months from the nominal calendar years.

Peripheral glaciers around the Greenland and Antarctic ice sheets are not treated in detail in this section (see Sects. 2.5 and 2.6 for mass-change estimates that combine the peripheral glaciers with the Greenland ice sheet and Antarctic ice sheet respectively). This is primarily because of the lack of observations (especially ground-based measurements) and also because of the high spatial variability of mass balance in those regions, and the slightly different climate (e.g., precipitation regime) and processes (e.g., refreezing). In the past, these regions have often been neglected. However, Radić and Hock (2010) estimated the total ice mass of peripheral glaciers around Greenland and Antarctica as 191 ± 70 mm SLE, with an actual contribution to sea-level rise of around 0.23 ± 0.04 mm yr^{-1} (Radić and Hock, 2011). Gardner et al. (2013) found a contribution from Greenland and Antarctic peripheral glaciers equal to 0.12 ± 0.05 mm yr^{-1}.

Note that some new or updated datasets for peripheral glaciers surrounding polar ice sheets are under development and will hopefully be available in coming years in order to incorporate Greenland and Antarctic peripheral glaciers in the estimates of global glacier mass changes.

2.4.2 Methods

No globally complete observational dataset exists for glacier mass changes (except GRACE estimates; see below). Any calculation of the global glacier contribution to sea-level change has to rely on spatial interpolation or extrapolation or both, or to consider limited knowledge of responses to climate change (due to the heterogeneous spatial distribution of glaciers around the world). Consequently, most observational methods to derive glacier sea-level contribution must extend local observations (in situ or satellite) to a larger region. Thanks to the recent global glacier outline inventory (Randolph Glacier Inventory – RGI – first release in 2012) as well as global climate observations, glacier modeling can now also be used to estimate the contribution of glaciers to

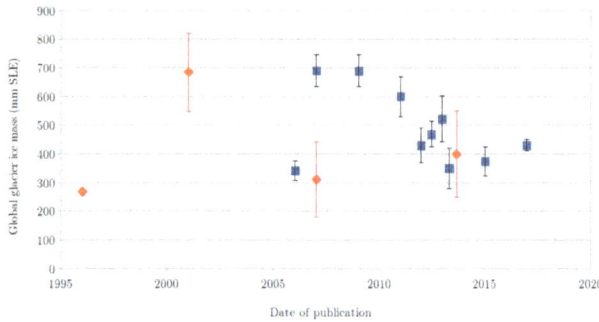

Figure 8. Evolution of global glacier ice mass estimates from different studies published over the past 2 decades, based on different observations and methods. The red marks correspond to IPCC reports. We clearly see the most recent publications lead to less scattered results. Note that Antarctica and Greenland peripheral glaciers are taken into account in this figure.

Table 3. Glacier contribution to sea level; all data are in millimeters per year of SLE.

	1993–2016 mm yr^{-1} SLE	2005–2016 mm yr^{-1} SLE
G16		0.70 ± 0.070[a]
M17	0.68 ± 0.032	0.80 ± 0.048
C17	0.63 ± 0.070	0.75 ± 0.070[b]
L17		0.84 ± 0.640[c]
M17-G		0.61 ± 0.070[d]

[a] The time period of G16 is 2002–2014. [b] The time period of C17 is 2003–2009. [c] The time period of L17 is 2003–2009. [d] The time period of M17-G is 2002/2005–2013/2015 because this value is an average of different estimates.

sea level (Marzeion et al., 2012; Huss and Hock, 2015; Maussion et al., 2018). Still, those global modeling methods need to globalize local observations and glacier processes which require fundamental assumptions and simplifications. Only GRACE-based gravimetric estimates are global but they suffer from large uncertainties in retrieval algorithms (signal leakage from hydrology, GIA correction) and coarse spatial resolution, not resolving smaller glaciated mountain ranges or those peripheral to the Greenland ice sheet.

The DEM differencing method is not yet global, but regional, and can hopefully in the near future be applied globally. This method needs also to convert elevation changes to mass changes (using assumptions on snow and ice densities). In contrast, very detailed glacier surface mass balance and glacier dynamic models are today far from being applicable globally, mainly due to the lack of crucial observations (e.g., meteorological data, glacier surface velocity and thickness) and of computational power for the more demanding theoretical models. However, somewhat simplified approaches are currently being developed to make the best use of the steadily increasing datasets. Modeling-based estimates suffer also from the large spread in estimates of the actual global glacier ice mass (Fig. 8). The mean value is 469 ± 146 mm SLE, with recent studies converging towards a range of values between 400 and 500 mm SLE global glacier ice mass. But as mentioned above, a part of this ice mass will not contribute to sea level.

2.4.3 Results (trends)

Table 3 presents most recent estimates of trends in global glacier mass balances.

The ensemble mean contribution of glaciers to sea-level rise for the time period 1993–2016 is 0.65 ± 0.051 mm yr^{-1} SLE and 0.74 ± 0.18 mm yr^{-1} for the time period 2005–2016 (uncertainties are averaged). Different studies refer to different time periods. However, because of the probable low variability of global annual glacier changes, compared to other components of the sea-level budget, averaging trends for slightly different time periods is appropriate.

The main source of uncertainty is that the vast majority of glaciers are unmeasured, which makes interpolation or extrapolation necessary, whether for in situ or satellite measurements, as well as for glacier modeling. Other main contributions to uncertainty in the ensemble mean stem from methodological differences, such as the downscaling of atmospheric forcing required for glacier modeling, the separation of glacier mass change to other mass change in the spatial gravimetry signal and the derivation of observational estimates of mass change from different raw measurements (e.g., length and volume changes, mass balance measurements, and geodetic methods), all with their specific uncertainties.

2.5 Greenland

Ice sheets are the largest potential source of future sea-level rise and represent the largest uncertainty in projections of future sea level. Almost all land ice ($\sim 99.5\%$) is locked in the ice sheets, with a volume in sea-level equivalent (SLE) terms of 7.4 m for Greenland and 58.3 m for Antarctica. It has been estimated that approximately 25 % to 30 % of the total land ice contribution to sea-level rise over the last decade came from the Greenland ice sheet (e.g., Dieng et al., 2017; Box and Colgan, 2017).

There are three main methods that can be used to estimate the mass balance of the Greenland ice sheet: (1) measurement of changes in elevation of the ice surface over time (dh/dt) either from imagery or altimetry; (2) the mass budget or input–output method (IOM), which involves estimating the difference between the surface mass balance and ice discharge; and (3) consideration of the redistribution of mass via gravity anomaly measurements, which only became viable with the launch of GRACE in 2002. Uncertainties due to the GIA correction are small in Greenland compared

Table 4. Datasets considered in the Greenland mass balance assessment, as well as covered time span and type of observations.

Reference	Time period	Method
Update from Barletta et al. (2013)	2003–2016	GRACE
Groh and Horwath (2016)	2003–2015	GRACE
Update from Luthcke et al. (2013)	2003–2015	GRACE
Update from Sasgen et al. (2012)	2003–2016	GRACE
Update from Schrama et al. (2014)	2003–2016	GRACE
Update from van den Broeke et al. (2016)	1993–2016	Input–output method (IOM)
Wiese et al. (2016a, b)	2003–2016	GRACE
Update from Wouters et al. (2008)	2003–2016	GRACE

to Antarctica: on the order of $\pm 20\,\mathrm{Gt\,yr^{-1}}$ mass equivalent (Khan et al., 2016). Prior to 2003, mass trends are reliant on IOM and altimetry. Both techniques have limited sampling in time and/or space for parts of the satellite era (1992–2002) and errors for this earlier period are, therefore, higher (van den Broeke et al., 2016; Hurkmans et al., 2014).

The consistency between the three methods mentioned above was demonstrated for Greenland by Sasgen et al. (2012) for the period 2003–2009. Ice-sheet-wide estimates showed excellent agreement although there was less consistency at a basin scale. We have, therefore, high confidence and relatively low uncertainties in the mass rates for the Greenland ice sheet in the satellite era (see also Bamber et al., 2018).

2.5.1 Datasets considered for the assessment

This assessment of sea-level budget contribution from the Greenland ice sheet considers the datasets shown in Table 4.

2.5.2 Methods and analyses

All but one of these datasets are based on GRACE data and therefore provide annual time series from ~ 2002 onwards. The one exception uses IOM (van Den Broeke et al., 2016) to give an annual mass time series for a longer time period (1993 onwards).

Notwithstanding this, each group has chosen their own approach to estimate mass balance from GRACE observations. As the aim of this global sea-level budget assessment is to compile existing results (rather than undertake new analyses), we have not imposed a specific methodology. Instead, we asked for the contributed datasets to reflect each group's 'best estimate' of annual trends for Greenland using the method(s) they have published.

Greenland contains glaciers and ice caps (GIC) around the margins of the main ice sheet, often referred to as peripheral GIC (PGIC), which are a significant proportion of the total mass imbalance (circa 15–20 %) (Bolch et al., 2013). Some studies consider the mass balance of the ice sheets and the PGIC separately but there has been, in general, no consistency in the treatment of PGIC and many studies do not

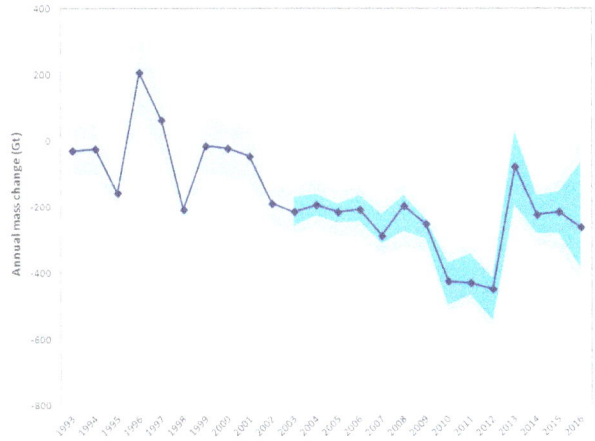

Figure 9. Greenland annual mass change from 1993 to 2016. The medium blue region shows the range of estimates from the datasets listed in Table 1. The lighter blue region shows the range of estimates when stated errors are included, to provide upper and lower bounds. The dark blue line shows the mean mass trend.

specify if they are included or excluded from the total. The GRACE satellites have an approximate spatial resolution of 300 km and the large number of studies that use GRACE, by default, include all land ice within the domain of interest. For this reason, the results below for Greenland mass trends all include PGIC.

From these datasets, for each year from 1993 to 2015 (and 2016 where available), we have calculated an average change in mass (calculated as the weighted mean based on the stated error value for each year) and an error term. Prior to 2003, the results are based on just one dataset (van den Broeke et al., 2016).

2.5.3 Results

There is generally a good level of agreement between the datasets (Fig. 9), and taken together they provide an average estimate of $171\,\mathrm{Gt\,yr^{-1}}$ of ice mass loss (or sea-level budget contribution) from Greenland for the period 1993 to 2016, increasing to $272\,\mathrm{Gt\,yr^{-1}}$ for the period 2005 to 2016 (Table 5).

All the datasets illustrate the previously documented accelerating mass loss up to 2012 (Rignot et al., 2011a; Velicogna, 2009) . In 2012, the ice sheet experienced exceptional surface melting reaching as far as the summit (Nghiem et al., 2012) and a record mass loss, since at least 1958, of over 400 Gt (van Den Broeke et al., 2016). The following years, however, show a reduced loss (not more than 270 Gt in any year). Inclusion of the years since 2012 in the 2005–2016 trend estimate reduces the overall rate of mass loss acceleration and its statistical significance. There is greater divergence in the GRACE time series for 2016. We associate this with the degradation of the satellites as they came to-

Table 5. Annual time series of Greenland mass change (GT yr^{-1}, negative values mean decreasing mass). The Δ mass is calculated as the weighted mean based on the stated error value for each year. The error for each year is calculated as the mean of all stated 1σ errors divided by sqrt(N) where N is the number of datasets available for that year, assuming that the errors are uncorrelated. The standard deviation (σ) is also given to illustrate the level of agreement between datasets for each year when multiple datasets are available (2003 onwards).

Year	Δ mass (Gt yr^{-1})	Error (Gt yr^{-1})	σ (Gt)
1993	−30	76	
1994	−25	77	
1995	−159	51	
1996	205	123	
1997	61	97	
1998	−209	45	
1999	−16	85	
2000	−24	85	
2001	−48	83	
2002	−192	58	
2003	−216	13	28
2004	−196	12	24
2005	−218	13	21
2006	−210	12	29
2007	−289	10	31
2008	−199	11	39
2009	−253	11	21
2010	−426	9	42
2011	−431	9	47
2012	−450	10	41
2013	−80	13	76
2014	−225	13	38
2015	−217	13	48
2016	−263	23	123
Average estimate 1993–2015	−167	54	
Average estimate 1993–2016	−171	53	
Average estimate 2005–2015	−272	11	
Average estimate 2005–2016	−272	13	

wards the end of their mission. For 2005–2012, it might be inferred that there is a secular trend towards greater mass loss and from 2010 to 2012 the value is relatively constant. Interannual variability in mass balance of the ice sheet is driven, primarily, by the surface mass balance (i.e., atmospheric weather) and it is apparent that the magnitude of this year-to-year variability can be large: it exceeded 360 Gt (or 1 mm sea-level equivalent) between 2012 and 2013. Caution is required, therefore, in extrapolating trends from a short record such as this.

2.6 Antarctica

The annual turnover of mass of Antarctica is about 2200 Gt yr^{-1} (over 6 mm yr^{-1} of SLE), 5 times larger than in Greenland (Wessem et al., 2017). In contrast to Greenland, ice and snow melt have a negligible influence on Antarctica's mass balance, which is therefore completely controlled by the balance between snowfall accumulation in the drainage basins and ice discharge along the periphery. The continent is also 7 times larger than Greenland, which makes satellite techniques absolutely essential to survey the continent. Interannual variations in accumulation are large in Antarctica, showing decadal to multidecadal variability, so that many years of data are required to extract trends, and missions limited to only a few years may produce misleading results (e.g., Rignot et al., 2011a, b).

As in Greenland, the estimation of the mass balance has employed a variety of techniques, including (1) the gravity method with GRACE since April 2002 until the end of the mission in late 2016; (2) the IOM method using a series of Landsat and synthetic-aperture radar (SAR) satellites for measuring ice motion along the periphery (Rignot et al., 2011a, b), ice thickness from airborne depth radar sounders such as Operation IceBridge (Leuschen, 2014a), and reconstructions of surface mass balance using regional atmospheric climate models constrained by re-analysis data (RACMO, MAR and others); and (3) a radar or laser altimetry method which mixes various satellite altimeters and correct ice elevation changes with density changes from firn models. The largest uncertainty in the GRACE estimate in Antarctica is the GIA, which is larger than in Greenland, and a large fraction of the observed signal. The IOM method compares two large numbers with large uncertainties to estimate the mass balance as the difference. In order to detect an imbalance at the 10 % level, surface mass balance and ice discharge need to be estimated with a precision typically of 5 to 7 %. The altimetry method is limited to areas of shallow slope; hence, it is difficult to use in the Antarctic Peninsula and in the deep interior of the Antarctic continent due to unknown variations of the penetration depth of the signal in snow and firn. The only method that expresses the partitioning of the mass balance between surface processes and dynamic processes is the IOM method (e.g., Rignot et al., 2011a). The gravity method is an integrand method which does not suffer from the limitations of surface mass balance models but is limited in spatial resolution (e.g., Velicogna et al., 2014). The altimetry method provides independent evidence of changes in ice dynamics, e.g., by revealing rapid ice thinning along the ice streams and glaciers revealed by ice motion maps, as opposed to large-scale variations reflecting a variability in surface mass balance (McMillan et al., 2014).

All these techniques have improved in quality over time and have accumulated a decade to several decades of observations, so that we are now able to assess the mass balance of the Antarctic continent using methods with reasonably low uncertainties and multiple lines of evidence as the methods are largely independent, which increases confidence in the results (see recent publication by the IMBIE Team, 2018). There is broad agreement in the mass loss from the Antarctic Peninsula and West Antarctica; most residual uncertainties are associated with East Antarctica as the signal is relatively small compared to the uncertainties, although most estimates tend to indicate a low contribution to sea level (e.g., Shepherd et al., 2012).

2.6.1 Datasets considered for the assessment

This assessment considers the datasets shown in Table 6.

In Table 6, the negative trend estimate by Zwally et al. (2016) is not added. It is worth noting that including it would only slightly reduce the ensemble mean trend.

2.6.2 Methods and analyses

The datasets used in this assessment are Antarctica mass balance time series generated using different approaches. Two estimates are a joint inversion of GRACE, altimetry and GPS data (Martín-Español et al., 2016) as well as GRACE and CryoSat data (Forsberg et al., 2017). Two methods are mascon solutions obtained from the GRACE intersatellite range-rate measurements over equal-area spherical caps covering the Earth's surface (Luthcke et al., 2013; Wiese et al., 2016b), three estimates use the GRACE spherical harmonics solutions (Velicogna et al., 2014; Wiese et al., 2016b; Wouters et al., 2013) and one uses gridded GRACE products (Sasgen et al., 2013).

All GRACE time series were provided as monthly time series (except for the one using the Martín-Español et al., 2016, method, which was provided as annual estimates). In addition, different groups use different GIA corrections, therefore the spread of the trend solutions also represents the error associated with the GIA correction which, in Antarctica, is the largest source of uncertainty. Sasgen et al. (2013) used their own GIA solution (Sasgen et al., 2017), as did Martín-Español et al. (2016); Luthcke et al. (2013), Velicogna et al. (2014), and Groh and Horwath (2016) used IJ05-R2 (Ivins et al., 2013). Wouter et al. (2013) used Whitehouse et al. (2012), and Wiese et al. (2016b) used A et al. (2013). In addition, Groh and Horwath (2016) did not include the peripheral glaciers and ice caps, while all other estimates do.

Table 6 shows the Antarctic contribution to sea level during 2005–2015 from the different GRACE solutions, and for the input and output method. There is a single IOM-based dataset that provides trends for the period 1993–2015 (update of Rignot et al., 2011a). For the period 2005–2015, we calculated the annual sea-level contribution from Antarctica using GRACE and IOM estimates (Table 7).

As we are interested in evaluating the long-term trend and interannual variability of the Antarctic contribution to sea level, for each GRACE dataset available in monthly time se-

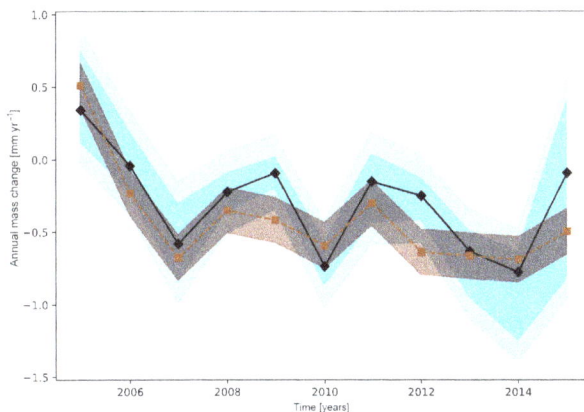

Figure 10. Antarctic annual sea-level contribution during 2005 to 2015. The black squares are the mean annual sea level calculated using the GRACE datasets listed in Table 6. The darker blue band shows the range of estimates from the datasets. The light blue band accounts for the error in the different GRACE estimates. The brown squares are the annual sea-level contribution calculated using the input–output method (updated from Rignot et al., 2011a); the light brown band is the associated error.

ries, we first removed the annual and subannual components of the signal by applying a 13-month averaging filter and we then used the smoothed time series to calculate annual mass change. Figure 10 shows the annual sea-level contribution from Antarctica calculated from the GRACE-derived estimates and for the input–output method. The GRACE mean annual estimates are calculated as the mean of the annual contributions from the different groups, and the associated error calculated as the sum of the spread of the annual estimates and the mean annual error.

2.6.3 Results

There is generally broad agreement between the GRACE datasets (Fig. 10), as most of the differences between GRACE estimates are caused by differences in the GIA correction. We find a reasonable agreement between GRACE and the IOM estimates although the IOM estimates indicate higher losses. Taken together, these estimates yield an average of $0.42 \, \mathrm{mm \, yr^{-1}}$ sea-level budget contribution from Antarctica for the period 2005 to 2015 (Table 7) and $0.25 \, \mathrm{mm \, yr^{-1}}$ sea level for the time period 1993–2005, where the latter value is based on IOM only.

All the datasets illustrate the previously documented accelerating mass loss of Antarctica (Rignot et al., 2011a, b; Velicogna, 2009). In 2005–2010, the ice sheet experienced ice mass loss driven by an increase in mass loss in the Amundsen Sea sector of West Antarctica (Mouginot et al., 2014). The following years showed a reduced increase in mass loss, as colder ocean conditions prevailed in the Amundsen Sea embayment sector of West Antarctica in 2012–2013 which reduced the melting of the ice shelves in

Table 6. Datasets considered in this assessment of the Antarctica mass change, and associated trends for the 2005–2015 and 1993–2015 expressed in millimeters per year of SLE. Positive values mean positive contribution to sea level (i.e., sea-level rise).

Reference	Method	2005–2015 trend (mm yr^{-1}) SLE	1993–2015 trend (mm yr^{-1}) SLE
Update from Martín-Español et al. (2016)	Joint inversion GRACE–altimetry–GPS	0.43 ± 0.07	–
Update from Forsberg et al. (2017)	Joint inversion GRACE–CryoSat	0.31 ± 0.02	–
Update from Groh and Horwath (2016)	GRACE	0.32 ± 0.11	–
Update from Luthcke et al. (2013)	GRACE	0.36 ± 0.06	–
Update from Sasgen et al. (2013)	GRACE	0.47 ± 0.07	–
Update from Velicogna et al. (2014)	GRACE	0.33 ± 0.08	–
Update from Wiese et al. (2016b)	GRACE	0.39 ± 0.02	–
Update from Wouters et al. (2013)	GRACE	0.41 ± 0.05	–
Update from Rignot et al. (2011b)	Input–output method (IOM)	0.46 ± 0.05	0.25 ± 0.1
Update from Schrama et al. (2014); version 1	GRACE ICE6G GIA model	0.47 ± 0.03	
Update from Schrama et al. (2014); version 2	GRACE Updated GIA models	0.33 ± 0.03	

Table 7. Annual sea-level contribution from Antarctica during 2005–2015 from GRACE and input–output method (IOM) calculated as described above and expressed in millimeters per year of SLE. Also shown is the mean of the estimate from the two methods; associated errors are the mean of the two estimated errors. Positive values mean positive contribution to sea level (i.e., sea-level rise).

Year	GRACE (mm yr^{-1}) SLE	IOM (mm yr^{-1}) SLE	Mean (mm yr^{-1}) SLE
2005	-0.34 ± 0.47	-0.51 ± 0.16	-0.42 ± 0.31
2006	0.04 ± 0.36	0.23 ± 0.16	0.14 ± 0.26
2007	0.58 ± 0.42	0.68 ± 0.16	0.63 ± 0.29
2008	0.22 ± 0.29	0.35 ± 0.16	0.29 ± 0.22
2009	0.09 ± 0.26	0.42 ± 0.16	0.26 ± 0.21
2010	0.74 ± 0.30	0.59 ± 0.16	0.67 ± 0.23
2011	0.15 ± 0.39	0.30 ± 0.16	0.23 ± 0.27
2012	0.25 ± 0.30	0.64 ± 0.16	0.44 ± 0.23
2013	0.63 ± 0.38	0.67 ± 0.16	0.65 ± 0.27
2014	0.78 ± 0.46	0.69 ± 0.16	0.73 ± 0.31
2015	0.09 ± 0.77	0.50 ± 0.16	0.29 ± 0.46
Average estimate 2005–2015	0.38 ± 0.06	0.46 ± 0.05	0.42 ± 0.06

time series in Antarctica, which have been obtained only by IOM and altimetry. The interannual variability in mass balance is driven almost entirely by surface mass balance processes. The mass loss of Antarctica, about 200 Gt yr^{-1} in recent years, is only about 10 % of its annual turnover of mass (2200 Gt yr^{-1}), in contrast with Greenland where the mass loss has been growing rapidly to nearly 100 % of the annual turnover of mass. This comparison illustrates the challenge of detecting mass balance changes in Antarctica, but at the same time, that satellite techniques and their interpretation have made tremendous progress over the last 10 years, producing realistic and consistent estimates of the mass using a number of independent methods (Bamber et al., 2018; the IMBIE Team, 2018).

2.7 Terrestrial water storage

Human transformations of the Earth's surface have impacted the terrestrial water balance, including continental patterns of river flow and water exchange between land, atmosphere and ocean, ultimately affecting global sea level. For instance, massive impoundment of water in man-made reservoirs has reduced the direct outflow of water to the sea through rivers, while groundwater abstractions, wetland and lake storage losses, deforestation, and other land use changes have caused changes to the terrestrial water balance, including changing evapotranspiration over land, leading to net changes in land–ocean exchanges (Chao et al., 2008; Wada et al., 2012a, b; Konikow, 2011; Church et al., 2013; Döll et al., 2014a, b). Overall, the combined effects of direct anthropogenic processes have reduced land water storage, increasing the rate of sea-level rise by 0.3–0.5 mm yr^{-1} during recent decades (Church et al., 2013; Gregory et al., 2013; Wada et al.,

front of the glaciers (Dutrieux et al., 2014). Divergence in the GRACE time series is observed after 2015 due to the degradation of the satellites towards the end of the mission.

The large interannual variability in mass balance in 2005–2015, characteristic of Antarctica, nearly masks out the trend in mass loss, which is more apparent in the longer time series than in short time series. The longer record highlights the pronounced decadal variability in ice sheet mass balance in Antarctica, demonstrating the need for multidecadal

2016). Additionally, recent work has shown that climate-driven changes in water stores can perturb the rate of sea-level change over interannual to decadal timescales, making global land mass budget closure sensitive to varying observational periods (Cazenave et al., 2014; Dieng et al., 2015a; Reager et al., 2016; Rietbroek et al., 2016). Here we discuss each of the major component contributions from land, with a summary in Table 8, and estimate the net terrestrial water storage contribution to sea level.

2.7.1 Direct anthropogenic changes in terrestrial water storage

Water impoundment behind dams

Wada et al. (2016) built on work by Chao et al. (2008) to combine multiple global reservoir storage datasets in pursuit of a quality-controlled global reservoir database. The result is a list of 48 064 reservoirs that have a combined total capacity of 7968 km^3. The time history of growth of the total global reservoir capacity reflects the history of the human activity in dam building. Applying assumptions from Chao et al. (2008), Wada et al. (2016) estimated that humans have impounded a total of 10 416 km^3 of water behind dams, accounting for a cumulative 29 mm drop in global mean sea level. From 1950 to 2000 when global dam-building activity was at its highest, impoundment contributed to the average rate of sea-level change at -0.51 mm yr^{-1}. This was an important process in comparison to other natural and anthropogenic sources of sea-level change over the past century, but has now largely slowed due to a global decrease in dam-building activity.

Global groundwater depletion

Groundwater currently represents the largest secular trend component to the land water storage budget. The rate of groundwater depletion (GWD) and its contribution to sea level has been subject to debate (Gregory et al., 2013; Taylor et al., 2013). In the IPCC AR4 (Solomon et al., 2007), the contribution of nonfrozen terrestrial waters (including GWD) to sea-level variation was not considered due to its perceived uncertainty (Wada et al., 2016). Observations from GRACE opened a path to monitor total water storage changes, including groundwater in data-scarce regions (Strassberg et al., 2007; Rodell et al., 2009; Tiwari et al., 2009; Jacob et al., 2012; Shamsudduha et al., 2012; Voss et al., 2013). Some studies have also applied global hydrological models in combination with the GRACE data (see Wada et al., 2016, for a review).

Earlier estimates of GWD contribution to sea level range from 0.075 to 0.30 mm yr^{-1} (Sahagian et al., 1994; Gornitz, 1995, 2001; Foster and Loucks, 2006). More recently, Wada et al. (2012b), using hydrological modeling, estimated that the contribution of GWD to global sea level increased from 0.035 (± 0.009) to 0.57 (± 0.09) mm yr^{-1} during the

20th century and projected that it would further increase to 0.82 (± 0.13) mm yr^{-1} by 2050. Döll et al. (2014b) used hydrological modeling, well observations and GRACE satellite gravity anomalies to estimate a 2000–2009 global GWD of 113 km^3 yr^{-1} (0.314 mm yr^{-1} SLE). This value represents the impact of human groundwater withdrawals only and does not consider the effect of climate variability on groundwater storage. A study by Konikow (2011) estimated global GWD to be 145 (± 39) km^3 yr^{-1} (0.41 $\pm$ 0.1 mm yr^{-1} SLE) during 1991–2008 based on measurements of changes in groundwater storage from in situ observations, calibrated groundwater modeling, GRACE satellite data and extrapolation to unobserved aquifers.

An assumption of most existing global estimates of GWD impacts on sea-level change is that nearly 100 % of the GWD ends up in the ocean. However, groundwater pumping can also perturb regional climate due to land–atmosphere interactions (Lo and Famiglietti, 2013). A recent study by Wada et al. (2016) used a coupled land–atmosphere model simulation to track the fate of water pumped from underground and found it more likely that ~ 80 % of the GWD ends up in the ocean over the long term, while 20 % re-infiltrates and remains in land storage. They estimated an updated contribution of GWD to global sea-level rise ranging from 0.02 (± 0.004) mm yr^{-1} in 1900 to 0.27 (± 0.04) mm yr^{-1} in 2000 (Fig. 11). This indicates that previous studies had likely overestimated the cumulative contribution of GWD to global SLR during the 20th century and early 21st century by 5–10 mm.

Land cover and land-use change

Humans have altered a large part of the land surface, replacing about 40 % of natural vegetation by anthropogenic land cover such as crop fields or pasture. Such land cover change can affect terrestrial hydrology by changing the infiltration-to-runoff ratio and can impact subsurface water dynamics by modifying recharge and increasing groundwater storage (Scanlon et al., 2007). The combined effects of anthropogenic land cover changes on land water storage can be quite complex. Using a combined hydrological and water resource model, Bosmans et al. (2017) estimated that land cover change between 1850 and 2000 has contributed to a discharge increase of 1058 km^3 yr^{-1}, on the same order of magnitude as the effect of human water use. These recent model results suggest that land-use change is an important topic for further investigation in the future. So far, this contribution remains highly uncertain.

Deforestation and afforestation

At present, large losses in tropical forests and moderate gains in temperate-boreal forests result in a net reduction of global forest cover (FAO, 2015; Keenan et al., 2015; MacDicken, 2015; Sloan and Sayer, 2015). Net deforestation releases carbon and water stored in both biotic tissues and soil, which leads to sea-level rise through three primary processes: deforestation-induced runoff increases (Gornitz et al., 1997), carbon loss-related decay and plant storage loss, and complex climate feedbacks (Butt et al., 2011; Chagnon and Bras, 2005; Nobre et al., 2009; Shukla et al., 1990; Spracklen et al., 2012). Due to these three causes, and if uncertainties from the land–atmospheric coupling are excluded, a summary by Wada et al. (2016) suggests that the current net global deforestation leads to an upper-bound contribution of $\sim 0.035\,\mathrm{mm\,yr^{-1}}$ SLE.

Wetland degradation

Wetland degradation contributes to sea level primarily through (i) direct water drainage or removal from standing inundation, soil moisture and plant storage, and (ii) water release from vegetation decay and peat combustion. Wada et al. (2016) consider a recent wetland loss rate of $0.565\,\%\,\mathrm{yr^{-1}}$ since 1990 (Davidson, 2014) and a present global wetland area of 371 mha averaged from three databases: Matthews natural wetlands (Matthews and Fung, 1987), ISLSCP (Darras, 1999), and DISCover (Belward et al., 1999; Lovel and Belward, 1997). They assume a uniform 1 m depth of water in wetlands (Milly et al., 2010), to estimate a contribution of recent global wetland drainage to sea level of $0.067\,\mathrm{mm\,yr^{-1}}$. Wada et al. (2016) apply a wetland area and loss rate as used for assessing wetland water drainage, to determine that the annual reduction of wetland carbon stock since 1990, if completely emitted, releases water equivalent to 0.003–$0.007\,\mathrm{mm\,yr^{-1}}$ SLE. Integrating the impacts of wetland drainage, oxidation and peat combustion, Wada et al. (2016) suggest that the recent global wetland degradation results in an upper bound of $0.074\,\mathrm{mm\,yr^{-1}}$ SLE.

Lake storage changes

Lakes store the greatest mass of liquid water on the terrestrial surface (Oki and Kanae, 2006), yet, because of their "dynamic" nature (Sheng et al., 2016; Wang et al., 2012), their overall contribution to sea level remains uncertain. In the past century, perhaps the greatest contributor in global lake storage was the Caspian Sea (Milly et al., 2010), where the water level exhibits substantial oscillations attributed to meteorological, geological, and anthropogenic factors (Ozyavas et al., 2010; Chen et al., 2017a). Assuming the lake level variation kept pace with groundwater changes (Sahagian et al., 1994), the overall contribution of the Caspian Sea, including both surface and groundwater storage variations through 2014, has been about $0.03\,\mathrm{mm\,yr^{-1}}$ SLE since 1900, 0.075 $(\pm 0.002)\,\mathrm{mm\,yr^{-1}}$ since 1995 and 0.109 $(\pm 0.004)\,\mathrm{mm\,yr^{-1}}$ since 2002. Additionally, between 1960 and 1990, the water storage in the Aral Sea Basin declined at a striking rate of $64\,\mathrm{km^3\,yr^{-1}}$, equivalent to $0.18\,\mathrm{mm\,yr^{-1}}$ SLE (Sahagian, 2000; Sahagian et al., 1994; Vörösmarty and Sahagian, 2000) due mostly to upstream water diversion for irrigation (Perera,

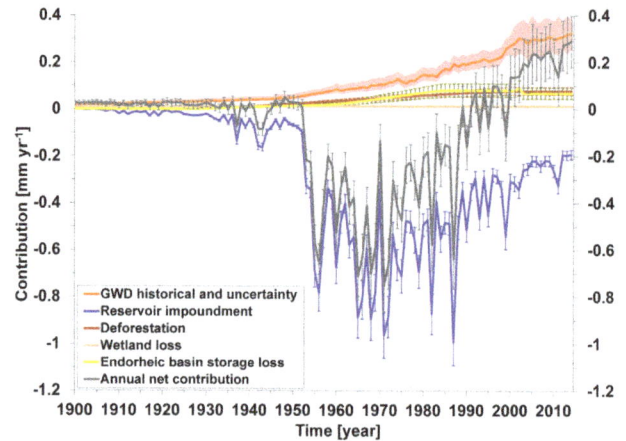

Figure 11. Time series of the estimated annual contribution of terrestrial water storage change to global sea level over the period 1900–2014 (rates in millimeters per year of SLE) (modified from Wada et al., 2016).

1993), which was modeled by Pokhrel et al. (2012) to be $\sim 500\,\mathrm{km^3}$ during 1951–2000, equivalent to $0.03\,\mathrm{mm\,yr^{-1}}$ SLE. Dramatic decline in the Aral Sea continued in recent decades, with an annual rate of 6.043 $(\pm 0.082)\,\mathrm{km^3\,yr^{-1}}$ measured from 2002 to 2014 (Schwatke et al., 2015). Assuming that groundwater drainage has kept pace with lake level reduction (Sahagian et al., 1994), the Aral Sea has contributed 0.0358 $(\pm 0.0003)\,\mathrm{mm\,yr^{-1}}$ to the recent sea-level rise.

Water cycle variability

Natural changes in the interannual to decadal cycling of water can have a large effect on the apparent rate of sea-level change over decadal and shorter time periods (Milly et al., 2003; Lettenmaier and Milly, 2009; Llovel et al., 2010). For instance, ENSO-driven modulations of the global water cycle can be important in decadal-scale sea-level budgets and can mask underlying secular trends in sea level (Fasullo et al., 2013; Cazenave et al., 2014; Nerem et al., 2018).

Sea-level variability due to climate-driven hydrology represents a super-imposed variability on the secular rates of global mean sea-level rise. While this term can be large and is important in the interpretation of the sea-level record, it is arguably the most difficult term in the land water budget to quantify.

2.7.2   Net terrestrial water storage

GRACE-based estimates

Measurements of non-ice-sheet continental land mass from GRACE satellite gravity have been presented in several recent studies (Jensen et al., 2013, Rietbroek et al., 2016; Reager et al., 2016; Scanlon et al., 2018) and can be used to

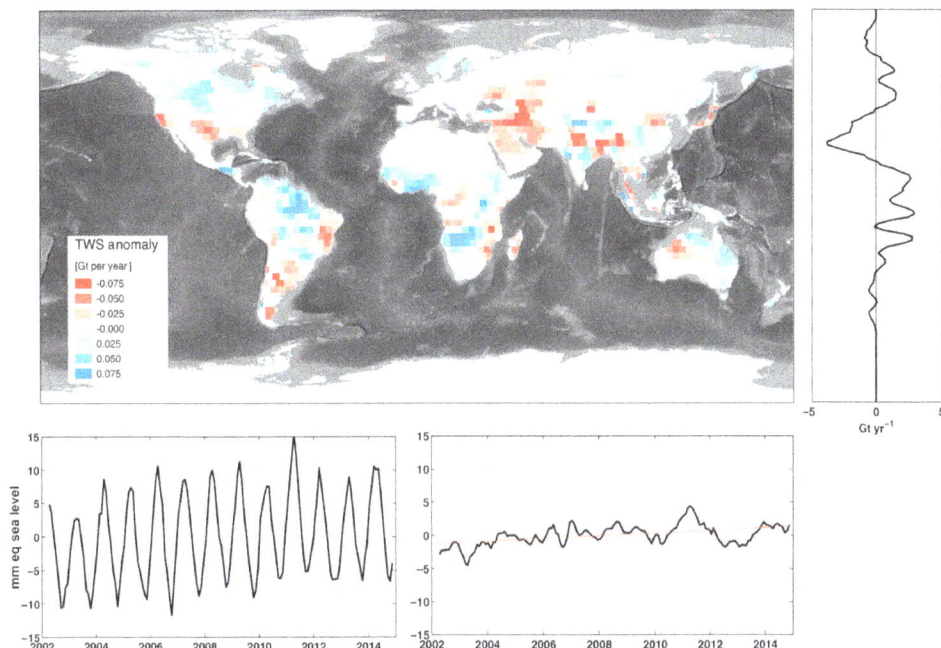

Figure 12. An example of trends in land water storage from GRACE observations, April 2002 to November 2014. Glaciers and ice sheets are excluded. Shown are the global map (gigatons per year), zonal trends and full time series of land water storage (in mm yr^{-1} SLE). Following methods detailed in Reager et al. (2016), GRACE shows a total gain in land water storage during the 2002–2014 period, corresponding to a sea-level trend of -0.33 ± 0.16 mm yr^{-1} SLE (modified from Reager et al., 2016). These trends include all human-driven and climate-driven processes in Table 1 and can be used to close the land water budget over the study period.

constrain a global land mass budget. Note that these "top-down" estimates contain both climate-driven and direct anthropogenic driven effects, which makes them most useful in assessing the total impact of land water storage changes and closing the budget of all contributing terms. GRACE observations, when averaged over the whole land domain following Reager et al. (2016), indicate a total TWS change (including glaciers) over the 2002–2014 study period of approximately $+0.32 \pm 0.13$ mm yr^{-1} SLE (i.e., ocean gaining mass). Global mountain glaciers have been estimated to lose mass at a rate of 0.65 ± 0.09 mm yr^{-1} (e.g., Gardner et al., 2013; Reager et al., 2016) during that period, such that a mass balance indicates that global glacier-free land gained water at a rate of -0.33 ± 0.16 mm yr^{-1} SLE (i.e., ocean losing mass; Fig. 12). A roughly similar estimate was found from GRACE using glacier-free river basins globally (-0.21 ± 0.09 mm yr^{-1}) (Scanlon et al., 2018). Thus, the GRACE-based net TWS estimates suggest a negative sea-level contribution from land over the GRACE period (Table 8). However, mass change estimates from GRACE incorporate uncertainty from all potential error sources that arise in processing and postprocessing of the data, including from the GIA model, and from the geocenter and mean pole corrections.

Estimates based on global hydrological models

Global land water storage can also be estimated from global hydrological models (GHMs) and global land surface models. These compute water, or water and energy balances, at the Earth's surface, yielding time variations of water storage in response to prescribed atmospheric data (temperature, humidity and wind) and the incident water and energy fluxes from the atmosphere (precipitation and radiation). Meteorological forcing is usually based on atmospheric model reanalysis. Model uncertainties result from several factors. Recent work has underlined the large differences among different state-of-the-art precipitation datasets (Beck et al., 2017), with large impacts on model results at seasonal (Schellekens et al., 2017) and longer timescales (Felfelani et al., 2017). Another source of uncertainty is the treatment of subsurface storage in soils and aquifers, as well as dynamic changes in storage capacity due to representation of frozen soils and permafrost, the complex effects of dynamic vegetation, atmospheric vapor pressure deficit estimation and an insufficiently deep soil column. A recent study by Scanlon et al. (2018) compared water storage trends from five global land surface models and two global hydrological models to GRACE storage trends and found that models estimated the opposite trend in net land water storage to GRACE over the 2002–2014 period. These authors attributed this discrepancy to model deficiencies, in particular soil depth limitations. These combined error sources are responsible for a range of storage trends across models of approximately 0.5 ± 0.2 mm yr^{-1} SLE. In terms of global land average, model differences can cause up to ~ 0.4 mm yr^{-1} SLE uncertainty.

2.7.3 Synthesis

Based on the different approaches to estimate the net land water storage contribution, we estimate that the corresponding sea-level rate ranges from -0.33 to 0.23 mm yr^{-1} during the period of 2002–2014/15 due to water storage changes (Table 8). According to GRACE, the net TWS change (i.e., not including glaciers) over the period 2002–2014 shows a negative contribution to sea level of -0.33 and -0.21 mm yr^{-1} by Reager et al. (2016) and Scanlon et al. (2018) respectively. Such a negative signal is not currently reproduced by hydrological models which estimate slightly positive trends over the same period (see Table 8). It is to be noted, however, that looking at trends only over periods on the order of a decade may not be appropriate due the strong interannual variability of TWS at basin and global scales. For example, Fig. 5 from Scanlon et al. (2018) (see also Fig. S9 from their Supplement), which compares GRACE TWS and model estimates over large river basins over 2002–2014, clearly shows that the discrepancies between GRACE and models occur at the end of the record for the majority of basins. This is particularly striking for the Amazon basin (the largest contributor to TWS), for which GRACE and models agree reasonably well until 2011, and then depart significantly, with GRACE TWS showing a strongly positive trend since then, unlike the models. Such a divergence at the end of the record is also noticed for several other large basins (see Scanlon et al., 2018, Fig. S9). No clear explanation can be provided yet, even though one may question the quality of the meteorological forcing used by hydrological models for the recent years. But this calls for some caution when comparing GRACE and other models on the basis of trends only because of the dominant interannual variability of the TWS component. Much more work is needed to understand differences among models, and between models and GRACE. Of all components entering in the sea-level budget, the TWS contribution currently appears to be the most uncertain one.

2.8 Glacial isostatic adjustment

The Earth's dynamic response to the waxing and waning of the late-Pleistocene ice sheets is still causing isostatic disequilibrium in various regions of the world. The accompanying slow process of GIA is responsible for regional and global fluctuations in relative and absolute sea level, 3-D crustal deformations, and changes in the Earth's gravity field (for a review, see Spada, 2017). To isolate the contribution of current climate change, geodetic observations must be corrected for the effects of GIA (King et al., 2010). These are obtained by solving the "sea-level equation" (Farrell and Clark, 1976; Mitrovica and Milne, 2003). The sea level can be expressed as $S = N - U$, where S is the rate of change in sea level relative to the solid Earth, N is the geocentric rate of sea-level change, and U is the vertical rate of displacement of the solid Earth. The sea-level equation accounts for solid Earth deformational, gravitational and rotational effects on sea level, which are sensitive to the Earth's mechanical properties and to the melting chronology of continental ice. Forward GIA modeling, based on the solution of the sea-level equation, provides predictions of unique spatial patterns (or *fingerprints*; see Plag and Juettner, 2001) of relative and geocentric sea-level change (e.g., Milne et al., 2009; Kopp et al., 2015). During recent decades, the two fundamental components of GIA modeling have been progressively constrained from the observed history of relative sea level during the Holocene (see, e.g., Lambeck and Chappell, 2001; Peltier, 2004). In the context of climate change, the importance of GIA has been recognized since the mid-1980s, when the awareness of global sea-level rise stimulated the evaluation of the isostatic contribution to tide gauge observations (see Table 1 in Spada and Galassi, 2012). Subsequently, GIA models have been applied to the study of the pattern of sea-level change from satellite altimetry (Tamisiea, 2011), and since 2002 to the study of the gravity field variations from GRACE. Our primary goal here is to analyze GIA model outputs that have been used to infer global mean sea-level change and ice sheet volume change from geodetic datasets during the altimetry era. These outputs are the sea-level variations detected by satellite altimetry across oceanic regions (n), the ocean mass change (w) and the modern ice sheets mass balance from GRACE. We also discuss the GIA correction that needs to be applied to GRACE-based land water storage changes. The GIA correction applied to tide-gauge-based sea-level observations at the coastlines is not discussed here. Since GIA evolves on timescales of millennia (e.g., Turcotte and Schubert, 2012), the rate of change of all isostatic signals can be considered constant on the timescale of interest.

2.8.1 GIA correction to altimetry-based sea level

Unlike tide gauges, altimeters directly sample the sea surface in a geocentric reference frame. Nevertheless, GIA contributes significantly to the rates of absolute sea-level change observed over the "altimetry era", which require a correction N_{gia} that is obtained by solving the SLE (e.g., Spada, 2017). As discussed in detail by Tamisiea (2011), N_{gia} is sensitive to the assumed rheological profile of the Earth and to the history of continental glacial ice sheets. The variance of N_{gia} over the surface of the oceans is much reduced, being primarily determined by the change in the Earth's gravity potential, apart from a spatially uniform shift. As discussed by Spada and Galassi (2016), the GIA contribution N_{gia} is strongly affected by variations in the centrifugal potential associated with Earth's rotation, whose fingerprint is

Table 8. Estimates of TWS components due to human intervention and net TWS based on hydrological models and GRACE.

Estimate terrestrial water storage contribution to sea level		2002–2014/15 $(\mathrm{mm\,yr}^{-1})$ SLE (positive values mean sea-level rise)
Human contributions by component		
Groundwater depletion	Wada et al. (2016)	0.30 ($\pm$0.1)
Reservoir impoundment	Wada et al. (2017)	-0.24 ($\pm$0.02)
Deforestation (after 2010)	Wada et al. (2017)	0.035
Wetland loss (after 1990)	Wada et al. (2017)	0.074
Endorheic basin storage loss		
Caspian Sea	Wada et al. (2017)	0.109 ($\pm$0.004)
Aral Sea	Wada et al. (2017)	0.036 ($\pm$0.0003)
Aggregated human intervention (sum of above)	Scanlon et al. (2018)	0.15 to 0.24
Hydrological model-based estimates		
WGHM model (natural variability plus human intervention) Döll et al. (2017)		0.15 $\pm$ 0.14
ISBA-TRIP model (natural variability only; Decharme et al., 2016) + human intervention from Wada et al. (2016) (from Dieng et al., 2017)		0.23 $\pm$ 0.10
GRACE-based estimates of total land water storage (not including glaciers) (Reager et al., 2016; Rietbroek et al., 2016; Scanlon et al., 2018)		-0.20 to -0.33 ($\pm$0.09–0.16)

dominated by a spherical harmonic contribution of degree $l = 2$ and order $m = \pm 1$. Since N_{gia} has a smooth spatial pattern, the global GIA correction to altimetry data can be obtained by simply subtracting its average $n = < N_{\mathrm{gia}} >$ over the ocean sampled by the altimetry missions. The computation of the GIA contribution N_{gia} has been the subject of various investigations, based on different GIA models. The estimate by Peltier (2001) of $n = -0.30\,\mathrm{mm\,yr}^{-1}$ is based on the ICE-4G (VM2) GIA model. Such a value has been adopted in the majority of studies estimating the GMSL rise from altimetry. Since n appears to be small compared to the global mean sea-level rise from altimetry ($\sim 3\,\mathrm{mm\,yr}^{-1}$), a more precise evaluation has not been of concern until recently. However, it is important to notice that n is of comparable magnitude as the GMSL trend uncertainty, currently estimated to be $\sim 0.3\,\mathrm{mm\,yr}^{-1}$ (see Sect. 2.2). In Table 9a, we summarize the values of n according to works in the literature where various GIA model models and averaging methods have been employed. Based on values in Table 9a for which a standard deviation is available, the average of n (weighted by the inverse of associated errors), assumed to represent the best estimate, is $n = (-0.29 \pm 0.02)\,\mathrm{mm\,yr}^{-1}$, where the uncertainty corresponds to 2σ.

2.8.2 GIA correction to GRACE-based ocean mass

GRACE observations of present-day gravity variations are sensitive to GIA, due to the sheer amount of rock material that is transported by GIA throughout the mantle and the resulting changes in surface topography, especially over the

formerly glaciated areas. The continuous change in the gravity field results in a nearly linear signal in GRACE observations. Since the gravity field is determined by global mass redistribution, GIA models used to correct GRACE data need to be global as well, especially when the region of interest is represented by all ocean areas. To date, the only global ice reconstruction publicly available is provided by the University of Toronto. Their latest product, named ICE-6G, has been published and distributed in 2015 (Peltier et al., 2015); note that the ice history has been simultaneously constrained with a specific Earth model, named VM5a. During the early period of the GRACE mission, the available Toronto model was ICE-5G (VM2) (Peltier, 2004). However, different groups have independently computed GIA model solutions based on the Toronto ice history reconstruction, by using different implementations of GIA codes and somehow different Earth models. The most widely used model is the one by Paulson et al. (2007), later updated by A et al. (2013). Both studies use a deglaciation history based on ICE-5G, but differ for the viscosity profile of the mantle: A et al. (2013) use a 3-D compressible Earth with VM2 viscosity profile and a PREM-based elastic structure used by Peltier (2004), whereas Paulson et al. (2007) use an incompressible Earth with self-gravitation, and a Maxwell 1-D multilayer mantle. Over most of the oceans, the GIA signature is much smaller than over the continents. However, once integrated over the global ocean, the signal w due to GIA is about $-1\,\mathrm{mm\,yr}^{-1}$ of equivalent sea-level change (Chambers et al., 2010), which is of the same order of magnitude as the total ocean mass

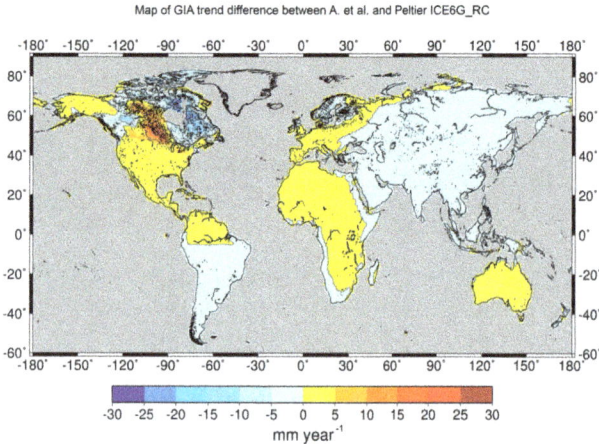

Figure 13. Difference map between two models of GIA correction to GRACE over land: A et al. (2013) versus Peltier et al. (2015). Units in millimeters per year of SLE.

change induced by increased ice melt (Leuliette and Willis, 2011). The main uncertainty in the GIA contribution to ocean mass change estimates, apart from the general uncertainty in ice history and Earth mechanical properties, originates from the importance of changes in the orientation of the Earth's rotation axis (Chambers et al., 2010; Tamisiea, 2011). Different choices in implementing the so-called "rotational feedback" lead to significant changes in the resulting GIA contribution to GRACE estimates. The issue of properly accounting for rotational effects has not been settled yet (Mitrovica et al., 2005; Peltier and Luthcke, 2009; Mitrovica and Wahr, 2011; Martinec and Hagedoorn, 2014). Table 9b summarizes the values of the mass-rate GIA contribution w according to the literature, where various models and averaging methods are employed. The weighted average of the values in Table 9b, for which an assessment of the standard deviation is available, is $w = (-1.44 \pm 0.36)$ mm yr^{-1} (the uncertainty is 2σ), which we assume represents the preferred estimate.

2.8.3 GIA correction to GRACE-based terrestrial water storage

As discussed in the previous section, the GIA correction to apply to GRACE over land is significant, especially in regions formerly covered by the ice sheets (Canada and Scandinavia). Over Canada, GIA models significantly differ. This is illustrated in Fig. 13, which shows difference between two models of GIA correction to GRACE over land, the A et al. (2013) and Peltier et al. (2009) models. We see that over the majority of the land areas, differences are small, except over northern Canada, in particular around the Hudson Bay, where differences larger than ± 20 mm yr^{-1} SLE are noticed. This may affect GRACE-based TWS estimates over Canadian river basins.

When averaged over the whole land surface as done in some studies to estimate the combined effect of land water storage and glacier melting from GRACE (e.g., Reager et al., 2016; see Sect. 2.7), the GIA correction ranges from ~ 0.5 to 0.7 mm yr^{-1} (in mm yr^{-1} SLE). Values for different GIA models are given in Table 9c.

2.8.4 GIA correction to GRACE-based ice sheet mass balance

The GRACE gravity field observations allow the determination of mass balances of ice sheets and large glacier systems with inaccuracy similar or superior to the input–output method or satellite laser and radar altimetry (Shepherd et al., 2012). However, GRACE ice-mass balances rely on successfully separating and removing the apparent mass change related to GIA. While the GIA correction is small compared to the mass balance for the Greenland ice sheet (ca. $< 10\%$), its magnitude and uncertainty in Antarctica is on the order of the ice-mass balance itself (e.g., Martín-Español et al., 2016). Particularly for today's glaciated areas, GIA remains poorly resolved due to the sparse data constraining the models, leading to large uncertainties in the climate history, the geometry and retreat chronology of the ice sheet, as well as the Earth structure. The consequences are ambiguous GIA predictions, despite fitting the same observational data. There are two principal approaches towards resolving GIA underneath the ice sheets. Empirical estimates can be derived that make use of the different sensitivities of satellite observations to ice-mass changes and GIA (e.g., Riva et al., 2009, 2010; Wu et al., 2010). Alternatively, GIA can be modeled numerically by forcing an Earth model with a fixed ice retreat scenario (e.g., Peltier, 2009; Whitehouse et al., 2012) or with output from a thermodynamic ice sheet model (Gomez et al., 2013; Konrad et al., 2015). Values of GIA-induced apparent mass change for Greenland and Antarctica as listed in the literature should be applied with caution (Table 9d) when applying them to GRACE mass balances. Each of these estimates may rely on a different GRACE postprocessing strategy and may differ in the approach used for solving the gravimetric inverse problem (mascon analysis, forward-modeling, averaging kernels). Of particular concern is the modeling and filtering of the pole tide correction caused by the rotational variations related to GIA, affecting coefficients of harmonic degree $l = 2$ and order $m = \pm 1$. As mentioned above, agreement on the modeling of the rotational feedback has not been reached within the GIA community. Furthermore, the pole tide correction applied during the determination gravity-field solutions differs between the GRACE processing centers and may not be consistent with the GIA correction listed. This inconsistency may introduce a significant bias in the ice-mass balance estimates (e.g., Sasgen et al., 2013, Supplement). Wahr et al. (2015) presented recommendations on how to treat the pole tides in GRACE analysis. However, a systematic intercomparison of the GIA predictions in terms of their low-degree coefficients and their consistency with the GRACE processing standards still need to be done.

The GRACE-based ocean mass, Antarctica mass and terrestrial water storage changes are very model dependent. As these GIA corrections cannot be assessed from independent information, they represent a large source of uncertainties to the sea-level budget components based on GRACE.

2.9 Ocean mass change from GRACE

Since 2002, GRACE satellite gravimetry has provided a revolutionary means for measuring global mass change and redistribution at monthly intervals with unprecedented accuracy, and offered the opportunity to directly estimate ocean mass change due to water exchange between the ocean and other components of the Earth (e.g., ice sheets, mountain glaciers, terrestrial water). GRACE time-variable gravity data have been successfully applied in a series of studies of ice mass balance of polar ice sheets (e.g., Velicogna and Wahr, 2006; Luthcke et al., 2006) and mountain glaciers (e.g., Tamisiea et al., 2005; J. Chen et al., 2007) and their contributions to global sea-level change. GRACE data can also be used to directly study long-term oceanic mass change or nonsteric sea-level change (e.g., Willis et al., 2008; Leuliette and Miller, 2009; Cazenave et al., 2009), and provide a unique opportunity to study interannual or long-term TWS change and its potential impacts on sea-level change (Richey et al., 2015; Reager et al., 2016).

GRACE time-variable gravity data can be used to quantify ocean mass change from three different main approaches. One is through measuring ice mass balance of polar ice sheets and mountain glaciers and variations of TWS, and their contributions to the GMSL (e.g., Velicogna and Wahr, 2006; Schrama et al., 2014). The second approach is to directly quantify ocean mass change using ocean basin mask (kernel) (e.g., A and Chambers, 2008; Llovel et al., 2010; Johnson and Chambers, 2013). In the ocean basin kernel approach, coastal ocean areas within certain distance (e.g., 300 or 500 km) from the coast are excluded, in order to minimize contaminations from mass change signal over the land (e.g., glacial mass loss and TWS change). The third approach solves mass changes on land and over ocean at the same time via forward modeling (e.g., Chen et al., 2013; Yi et al., 2015). The forward modeling is a global inversion to reconstruct the "true" mass change magnitudes over land and ocean with geographical constraint of locations of the mass change signals, and can help effectively reduce leakage between land and ocean (Chen et al., 2013).

Estimates of ocean mass changes from GRACE are subject to a number of major error sources. These include (1) leakage errors from the larger signals over ice sheets and land hydrology due to GRACE's low spatial resolution (of at least a few to several hundred kilometers) and the need for coastal masking, (2) spatial filtering of GRACE data to reduce spatial noise, (3) errors and biases in geophysical model correc-

tions (e.g., GIA, atmospheric mass) that need to be removed from GRACE observations to isolate oceanic mass change and/or polar ice sheets and mountain glaciers mass balance, and (4) residual measurement errors in GRACE gravity measurements, especially those associated with GRACE low-degree gravity changes. In addition, how to deal with the absent degree-1 terms, i.e., geocenter motion in GRACE gravity fields, is expected to affect estimates of GRACE-based oceanic mass rates and ice mass balances.

With a different treatment of the GRACE land–ocean signal leakage effect through global forward modeling, Chen et al. (2013) estimated ocean mass rates using GRACE RL05 time-variable gravity solutions over the period 2005–2011. They demonstrated that the ocean mass change contributes up to $1.80 \pm 0.47 \, \text{mm yr}^{-1}$ (over the same period), which is significantly larger than previous estimates over about the same period. Yi et al. (2015) further confirmed that correct calibration of GRACE data and appropriate treatment of GRACE leakage bias are critical to improve the accuracy of GRACE-estimated ocean mass rates. Table 10 summarizes different estimates of GRACE ocean mass rates. The uncertainty estimates of the listed studies (Table 10) are computed from different methods, with different considerations of error sources into the error budget, and represent different confidence levels.

As demonstrated in Chen et al. (2013), different treatments of just the degree-2 spherical harmonics of the GRACE gravity solution alone can lead to substantial differences in GRACE-estimated ocean mass rates (ranging from 1.71 to $2.17 \, \text{mm yr}^{-1}$). Similar estimates from GRACE gravity solutions from different data processing centers can also be different. In the meantime, long-term degree-1 spherical harmonics variation, representing long-term geocenter motion and neglected in some of the previous studies (due to the lack of accurate observations) are also expected to have a non-negligible effect on GRACE-derived ocean mass rates (Chen et al., 2013). Different methods for computing ocean mass change using GRACE data may also lead to different estimates (Chen et al., 2013; Johnson and Chambers, 2013; Jensen et al., 2013).

To help better understand the potential and uncertainty of GRACE satellite gravimetry in quantification of the ocean mass rate, Table 11 provides a comparison of GRACE-estimated ocean mass rates over the period January 2005 to December 2016 based on different GRACE data products and different data processing methods, including the CSR, GFZ and JPL GRACE RL05 spherical harmonic solutions (i.e., the so-called GSM solutions), as well as CSR, JPL and GSFC mascon solutions (the available GSFC mascons only cover the period up to July 2016). The three GRACE GSM results (CSR, GFZ and JPL) are updates from Johnson and Chambers (2013), with degree-2 zonal term replaced by satellite laser ranging results (Cheng and Ries, 2012), geocenter motion from Swenson et al. (2008), GIA model from A et al. (2013), an averaging kernel with a land mask

Table 9. Estimated contributions of GIA to the rate of absolute sea-level change observed by altimetry (a), to the rate of mass change observed by GRACE over the global oceans (b), to the rate of mass change observed by GRACE over land (c), and to Greenland and Antarctic ice sheets (c), during the altimetry era. The GIA corrections are expressed in millimeters per year SLE except over Greenland and Antarctica where values are given in gigatons per year (ice mass equivalent). Most of the GIA contributions are expressed as a value ± 1 standard deviation; a few others are given in terms of a plausible range, and for some the uncertainties are not specified.

(a) GIA correction to absolute sea level measured by altimetry

Reference	GIA ($mm\,yr^{-1}$ SLE)	Notes
Peltier (2009) (Table 3)	-0.30 ± 0.02 -0.29 ± 0.03 -0.28 ± 0.02	Average of three groups of four values obtained by variants of the analysis procedure, using ICE-5G(VM2), over a global ocean, in the range of latitudes 66° S to 66° N and 60° S to 60° N, respectively.
Tamisiea (2011) (Fig. 2)	-0.15 to -0.45 -0.20 to -0.50	Simple average over the oceans for a range of estimates obtained varying the Earth model parameters, over a global ocean and between latitudes 66° S and 66° N.
Huang (2013) (Table 3.6)	-0.26 ± 0.07 -0.27 ± 0.08	Average from an ensemble of 14 GIA models over a global ocean and between latitude from 66° S to 66° N.
Spada (2017) (Table 1)	-0.32 ± 0.08	Based on four runs of the sea-level equation solver SELEN (Spada and Stocchi, 2007) using model ICE-5G(VM2), with different assumptions in solving the SLE.

(b) GIA contribution to GRACE mass rate of change over the oceans

Reference	GIA ($mm\,yr^{-1}$ SLE)	Notes
Peltier (2009) (Table 3)	-1.60 ± 0.30	Average of values from 12 corrections for variants of the analysis procedure, using ICE-5G (VM2).
Chambers et al. (2010) (Table 1)	-1.45 ± 0.35	Average over the oceans for a range of estimates produced by varying the Earth models.
Tamisiea (2011) (Figs. 3 and 4)	-0.5 to -1.9 -0.9 to -1.5	Ocean average of a range of estimates varying the Earth model, and based on a restricted set, respectively.
Huang (2013) (Table 3.7)	-1.31 ± 0.40 -1.26 ± 0.43	Average from an ensemble of 14 GIA models over a global ocean and between latitude from 66° S to 66° N, respectively.

(c) GIA contribution to GRACE-based terrestrial water storage change

Reference	GIA correction ($mm\,yr^{-1}$ SLE) without Greenland, Antarctica, Iceland, Svalbard, Hudson Bay and the Black Sea
A et al. (2013)	0.63
Peltier ICE5G	0.68
Peltier ICE6G_rc	0.71
ANU_ICE6G	0.53

that extends out 300 km, and no destriping or smoothing, as described in Johnson and Chambers (2013). An update of GRACE ocean mass rate from Chen et al. (2013) is also included for comparisons, which is based on the CSR GSM solutions using forward modeling (a global inversion approach), with similar treatments of the degree-2 zonal term, geocenter motion and GIA effects.

The JPL mascon ocean mass rate is computed from all mascon grids over the ocean, and the GSFC mascon ocean mass rate is computed from all ocean mascons, with the Mediterranean, Black and Red seas excluded. A coastline resolution improvement (CRI) filter is already applied in the JPL mascons to reduce leakage (Wiese et al., 2016b), and in both the GSFC and JPL mascon solutions, the ocean and land are separately defined (Luthcke et al., 2013; Watkins et al., 2015). For the CSR mascon results, an averaging kernel with a land mask that extends out 200 km is applied to reduced leakage (Chen et al., 2017b). Similar treatments or corrections of degree-2 zonal term, geocenter motion and GIA effects are also applied in the three mascon solutions. When

Table 9. Continued.

(d) GIA contribution to GRACE mass rate of the ice sheets		
Reference	Greenland GIA ($Gt\,yr^{-1}$)	Notes
Simpson et al. (2009)[r]	-3 ± 12[m]	Thermodynamic sheet/solid Earth model, 1-D (uncoupled); constrained by geomorphology; inversion results in Sutterley et al. (2014).
Peltier (2009) (ICE-5G)[b]	-4[d]	Ice load reconstruction/solid Earth model, 1-D (ICE-5G/similar to VM2); Greenland component of ICE-5G ($13\,Gt\,yr^{-1}$) + Laurentide component of ICE-5G ($-17\,Gt\,yr^{-1}$); inversion results in Khan et al. (2016), Discussion.
Khan et al. (2016) (GGG-1D)[a]	15 ± 10[f]	Ice load reconstruction/solid Earth model, 1-D (uncoupled); constrained with geomorphology and GPS; Greenland component ($+32\,Gt\,yr^{-1}$) + Laurentide component of ICE-5G ($-17\,Gt\,yr^{-1}$); inversion results in Khan et al. (2016), Discussion.
Fleming and Lambeck (2004)[a] (Green1)	3[d]	Ice load reconstruction/solid Earth model, 1-D (uncoupled); constrained with geomorphology; Greenland component ($+20\,Gt\,yr^{-1}$) + Laurentide component of ICE-5G ($-17\,Gt\,yr^{-1}$); inversion in Sasgen et al. (2012, Supplement).
Wu et al. (2010)[b]	-69 ± 19[m]	Joint inversion estimate based on GPS, satellite laser ranging, and very long baseline interferometry, and bottom pressure from ocean model output; inversion results in Sutterley et al. (2014).
Reference	Antarctica GIA ($Gt\,yr^{-1}$)	Notes
Whitehouse et al. (2012) (W12a)[a]	60[f]	Thermodynamic sheet/solid Earth model, 1-D (uncoupled); constrained by geomorphology; inversion results in Shepherd et al. (2012), Supplement (Fig. S8).
Ivins et al. (2013) (IJ05_R2)[a]	40–65[f]	Ice load reconstruction/solid Earth model, 1-D; constrained by geomorphology and GPS uplift rates; Ivins et al. (2013); inversion results in Shepherd et al. (2012), Supplement (Fig. S8).
Peltier (2009) (ICE-5G)[b]	140–180[f]	Ice load reconstruction/solid Earth model ICE-5G(VM2); constrained by geomorphology; inversion results in Shepherd et al. (2012), Supplement (Fig. S8).
Argus et al. (2014) (ICE-6G)[b]	107[f]	Ice load reconstruction/solid Earth model ICE-6G(VM5a); constrained by geomorphology and GPS; theory recently corrected by Purcell et al. (2016); inversion results in Argus et al. (2014), conclusion 7.8.
Sasgen et al. (2017) (REGINA)[a]	55 ± 22[f]	Joint inversion estimate based on GRACE, altimetry, GPS and viscoelastic response functions; lateral heterogeneous Earth model parameters; inversion results in Sasgen et al. (2017), Table 1.
Gunter et al. (2014) (G14)[a]	ca. 64 ± 40[a] (multimodel uncert.)	Joint inversion estimate based on GRACE, altimetry, GPS and regional climate model output; conversion of uplift to mass using average rock density; inversion results in, Gunter et al. (2014) Table 1.
Martín-Español et al. (2016) (RATES)[a]	55 ± 8 45 ± 7*	Joint inversion estimate based on GRACE, altimetry, GPS and regional climate model output; inversion results in Sasgen et al. (2017), * is improved for GIA of smaller spatial scales; inversion results in Martin-Español et al. (2016), Fig. 6.

[a] Regional model. [b] Global model. [c] Mascon inversion. [d] Forward modeling inversion. [e] Averaging kernel inversion. [f] Inversion method not specified.

solving GRACE mascon solutions, the GRACE GAD fields (representing ocean bottom pressure changes, or combined atmospheric and oceanic mass changes) have been added back to the mascon solutions. To correctly quantify ocean mass change using GRACE mascon solutions, the means of the GAD fields over the oceans, which represents mean atmospheric mass changes over the ocean (as ocean mass is conserved in the GAD fields) need to be removed from GRACE mascon solutions. The removal of GAD average over the ocean in GRACE mascon solutions has very minor or negligible effect (of $\sim 0.02\,mm\,yr^{-1}$) on ocean mass rate

Table 10. Recently published (since 2013) estimates of GRACE-based ocean mass rates (GIA corrected). Most of the listed studies use either the A13 (A et al., 2013) or Paulson07 (Paulson et al., 2007) GIA model.

Data sources		Time period	Ocean mass trends (mm yr^{-1})
Chen et al. (2013)	(A13 GIA)	Jan 2005–Dec 2011	1.80 ± 0.47
Johnson and Chambers (2013)	(A13 GIA)	Jan 2003–Dec 2012	1.80 ± 0.15
Purkey et al. (2014)	(A13 GIA)	Jan 2003–Jan 2013	1.53 ± 0.36
Dieng et al. (2015a)	(Paulson07 GIA)	Jan 2005–Dec 2012	1.87 ± 0.11
Dieng et al. (2015b)	(Paulson07 GIA)	Jan 2005–Dec 2013	2.04 ± 0.08
Yi et al. (2015)	(A13 GIA)	Jan 2005–Jul 2014	2.03 ± 0.25
Rietbroek et al. (2016)		Apr 2002–Jun 2014	1.08 ± 0.30
Chambers et al. (2017)		2005–2015	2.11 ± 0.36

estimates, but is important for studying GMSL change at seasonal timescales.

Over the 12-year period (2005–2016), the three GRACE GSM solutions show pretty consistent estimates of ocean mass rate, in the range of 2.3 to 2.5 mm yr^{-1}. Greater differences are noticed for the mascon solutions. The GSFC mascons show the largest rate of 2.61 mm yr^{-1}. The CSR and JPL mascon solutions show relatively smaller ocean mass rates of 1.76 and 2.02 mm yr^{-1}, respectively, over the studied period. Based on the same CSR GSM solutions, the forward modeling and basin kernel estimates agree reasonably well (2.52 vs. 2.44 mm yr^{-1}). In addition to the degree-2 zonal term, geocenter motion, and GIA correction, the degree-2, order-1 spherical harmonics of the current GRACE RL05 solutions are affected by the definition of the reference mean pole in GRACE pole tide correction (Wahr et al., 2015). This mean pole correction, excluded in all estimates listed in Table 11 (for fair comparison), is estimated to contribute ~ -0.11 mm yr^{-1} to GMSL. How to reduce errors from the different sources plays a critical role in estimating ocean mass change from GRACE time-variable gravity data.

GRACE satellite gravimetry has brought a completely new era for studying global ocean mass change. Owing to the extended record of GRACE gravity measurements (now over 15 years), improved understanding of GRACE gravity data and methods for addressing GRACE limitations (e.g., leakage and low-degree spherical harmonics), and improved knowledge of background geophysical signals (e.g., GIA), GRACE-derived ocean mass rates from different studies in recent years show clearly increased consistency (Table 11). Most of the results agree well with independent observations from satellite altimeter and Argo floats, although the uncertainty ranges are still large. The GRACE Follow-On (FO) mission was launched in May 2018. The GRACE and GRACE-FO together are expected to provide at least over 2 (or even 3) decades of time-variable gravity measurements. Continuous improvements of GRACE data quality (in future releases) and background geophysical models are also expected, which will help improve the accuracy GRACE observed ocean mass change.

For the sea-level budget assessment over the GRACE period, we use the ensemble mean.

3 Sea-level budget results

In Sect. 2, we have presented the different terms of the sea-level budget equation, mostly based on published estimates (and in some cases, from their updates). We now use them to examine the closure of the sea-level budget. For all terms, we only consider ensemble mean values.

3.1 Entire altimetry era (1993–present)

3.1.1 Trend estimates over 1993–present

Because it is now clear that the GMSL and some components are accelerating (e.g., Nerem et al., 2018), we propose to characterize the long-term variations of the time series by both a trend and an acceleration. We start by looking at trends. Table 12 gathers the trends estimated in Sect. 2. The end year is not always the same for all components (see Sect. 2). Thus the word "present" means either 2015 or 2016 depending on the component. As no trend estimate is available for the entire altimetry era for the terrestrial water storage contribution, we do not consider this component. The residual trend (GMSL minus sum of components trend) may then provide some constraint on the TWS contribution.

Results presented in Table 12 are discussed in detail in Sect. 4.

3.1.2 Acceleration

The GMSL acceleration estimated in Sect. 2.2 using Ablain et al.'s (2017b) TOPEX-A drift correction amounts to 0.10 mm yr^{-2} for the 1993–2017 time span. This value is in good agreement with the Nerem et al. (2018) estimate (of 0.084 ± 0.025 mm yr^{-2}) over nearly the same period, after removal of the interannual variability of the GMSL. In Nerem et al. (2018), acceleration of individual components are also estimated as well as acceleration of the sum of components. The latter agrees well with the GMSL acceleration.

Table 11. Ocean mass trends (in mm yr^{-1}) estimated from GRACE for the period January 2005–December 2016 (the GSFC mascon solutions cover up to July 2016). The uncertainty is based on 2 times the sigma of least-squares fitting.

Data sources	Ocean mass trend (mm yr^{-1})
GSM CSR forward modeling (update from Chen et al., 2013)	2.52 ± 0.17
GSM CSR (update from Johnson and Chambers, 2013)	2.44 ± 0.15
GSM GFZ (update from Johnson and Chambers, 2013)	2.30 ± 0.15
GSM JPL (update from Johnson and Chambers, 2013)	2.48 ± 0.16
Mascon CSR (200 km)	1.76 ± 0.16
Mascon JPL	2.02 ± 0.16
Mascon GSFC (update from Luthcke et al., 2013)	2.61 ± 0.16
Ensemble mean	2.3 ± 0.19

Table 12. Trend estimates for individual components of the sea-level budget, sum of components and GMSL minus sum of components over 1993–present. Uncertainties of the sum of components and residuals represent rooted mean squares of components errors, assuming that errors are independent.

Component	Trends (mm yr^{-1}) 1993–present
1.GMSL (TOPEX-A drift corrected)	3.07 ± 0.37
2. Thermosteric sea level (full depth)	1.3 ± 0.4
3. Glaciers	0.65 ± 0.15
4. Greenland	0.48 ± 0.10
5. Antarctica	0.25 ± 0.10
6. TWS	/
7. Sum of components (without TWS → 2.+3.+4.+5.)	2.7 ± 0.23
8. GMSL minus sum of components (without TWS)	0.37 ± 0.3

Here we do not estimate the acceleration of the component ensemble means because time series are not always available. We leave this for a future assessment.

3.2 GRACE and Argo period (2005–present)

3.2.1 Sea-level budget using GRACE-based ocean mass

If we consider the ensemble mean trends for the GMSL, thermosteric and ocean mass components given in Sects. 2.2, 2.3 and 2.9 over 2005–present, we find agreement (within error bars) between the observed GMSL (3.5 ± 0.2 mm yr^{-1}) and the sum of Argo-based thermosteric plus GRACE-based ocean mass (3.6 ± 0.4 mm yr^{-1}) (see Table 13). The residual (GMSL minus sum of components) trend amounts to -0.1 mm yr^{-1}. Thus in terms of trends, the sea-level budget appears closed over this time span within quoted uncertainties.

3.2.2 Trend estimates over 2005–present from estimates of individual contributions

Table 13 gathers trends of individual components of the sea-level budget over 2005–present, as well as the trend of the sum of components and residuals (GMSL minus sum of components). As for the longer period, ensemble mean values are considered for each component.

As for Table 12, the results presented in Table 13 are discussed in detail in Sect. 4.

3.2.3 Year-to-year budget over 2005–present using GRACE-based ocean mass

We now examine the year-to-year sea-level and mass budgets. Table 14 provides annual mean values for the ensemble mean GMSL, GRACE-based ocean mass and Argo-based thermosteric component. The components are expressed as anomalies and their reference is arbitrary. So to compare with the GMSL, a constant offset for all years was applied to the thermosteric and ocean mass annual means. The reference year (where all values are set to zero) is 2003.

Figure 14 shows the sea-level budget over 2005–2015 in terms of an annual bar chart using values given in Table 14. It compares for years 2005 to 2016 the annual mean GMSL (blue bars) and annual mean sum of thermosteric and GRACE-based ocean mass (red bars). Annual residuals are also shown (green bars). These are either positive or negative depending on the years. The trend of these annual residuals is estimated to be 0.135 mm yr^{-1}.

In Fig. 15 is also shown the annual sea-level budget over 2005–2015 but now using the individual components for the mass terms. As we have no annual estimates for TWS, we ignore it, so that the total mass includes only glaciers, Greenland and Antarctica. The annual residuals thus include the TWS component in addition to the missing contributions (e.g., deep ocean warming). For years 2006 to 2011, the residuals are negative, an indication of a negative TWS to sea level as suggested by GRACE results (Reager et al., 2016; Scanlon et al., 2018). But as of 2012, the residuals become

Table 13. Trend estimates for individual components of the sea-level budget, sum of components and GMSL minus sum of components over 2005–present.

Component	Trend (mm yr^{-1}) 2005–present
1. GMSL	3.5 ± 0.2
2. Thermosteric sea level (full depth)	1.3 ± 0.4
3. Glaciers	0.74 ± 0.1
4. Greenland	0.76 ± 0.1
5. Antarctica	0.42 ± 0.1
6. TWS from GRACE (mean of Reager et al., 2016 and Scanlon et al., 2018)	-0.27 ± 0.15
7. Sum of components (2.+3.+4.+5.+6.)	2.95 ± 0.21
8. Sum of components (thermosteric full depth + GRACE-based ocean mass)	3.6 ± 0.4
9. GMSL minus sum of components (including GRACE-based TWS → 2.+3.+4.+5.+6.)	0.55 ± 0.3
10.GMSL minus sum of components (without GRACE-based TWS → 2.+3.+4.+5.)	0.28 ± 0.2
11. GMSL minus sum of components (thermosteric full depth + GRACE-based ocean mass)	-0.1 ± 0.3

Table 14. Annual mean values for the ensemble mean GMSL and sum of components (GRACE-based ocean mass and Argo-based thermosteric, full depth). Constant offset applied to the sum of components. The reference year (where all values are set to zero) is 2003.

Year	Ensemble mean GMSL (mm)	Sum of components (mm)	GMSL minus sum of components (mm)
2005	7.00	8.78	−0.78
2006	10.25	10.78	−0.53
2007	10.51	11.35	−0.85
2008	15.33	15.07	0.25
2009	18.78	18.88	−0.10
2010	20.64	20.53	0.11
2011	20.91	21.38	−0.48
2012	31.10	29.33	1.77
2013	33.40	33.87	−0.47
2014	36.65	36.22	0.43
2015	46.34	45.69	0.65

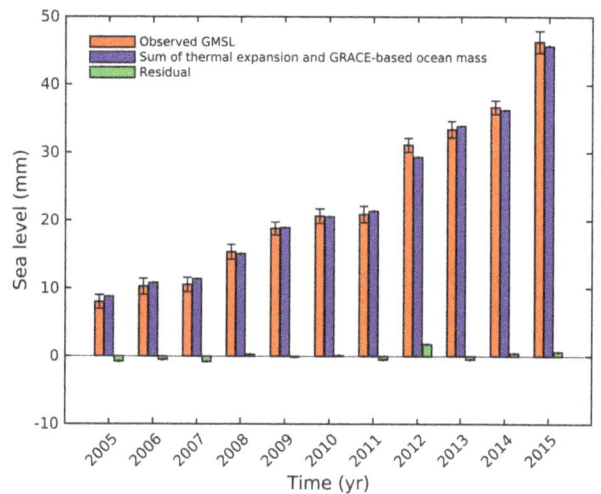

Figure 14. Annual sea level (blue bars) and sum of thermal expansion (full depth) and GRACE ocean mass component (red bars). Black vertical bars are associated uncertainties. Annual residuals (green bars) are also shown.

positive and on average over 2005–2015, the residual trend amounts to $+0.28 \, \text{mm yr}^{-1}$, a value larger than when using GRACE ocean mass.

Finally, Fig. 16 presents the mass budget. It compares annual GRACE-based ocean mass to the sum of the mass components, without TWS as in Fig. 15. The residual trend over the 2005–2015 time span is $0.14 \, \text{mm yr}^{-1}$. It may dominantly represent the TWS contribution. From one year to another residuals can be either positive or negative, suggesting important interannual variability in the TWS or even in the deep ocean.

4 Discussion

The results presented in Sect. 2 for the components of the sea-level budget are based on syntheses of the recently pub-

lished literature. When needed, the time series have been updated. In Sect. 3, we considered ensemble means for each component to average out random errors of individual estimates. We examined the closure or nonclosure of the sea-level budget using these ensemble mean values, for two periods: 1993–present and 2005–present (Argo and GRACE period). Because of the lack of observation-based TWS estimates for the 1993–present time span, we compared the observed GMSL trend to the sum of components excluding TWS. We found a positive residual trend of $0.37 \pm 0.3 \, \text{mm yr}^{-1}$, supposed to include the TWS contribution, plus other imperfectly known contributions (deep ocean warming) and data errors.

For the 2005–present time span, we considered both GRACE-based ocean mass and the sum of individual mass

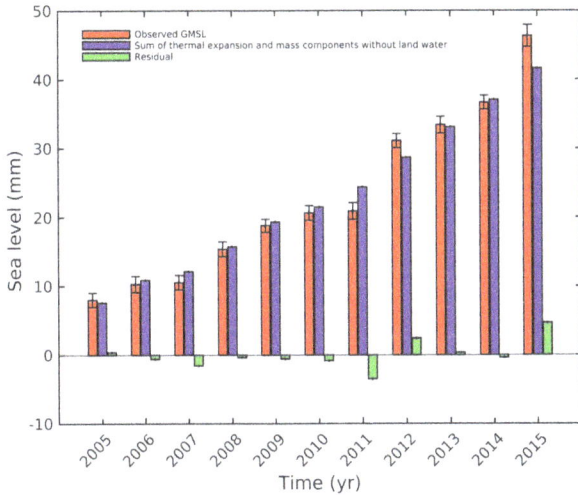

Figure 15. Annual global mean sea level (blue bars) and sum components without TWS (full depth thermal expansion + glaciers + Greenland + Antarctica) (red bars). Black vertical bars are associated uncertainties. Annual residuals (green bars) are also shown.

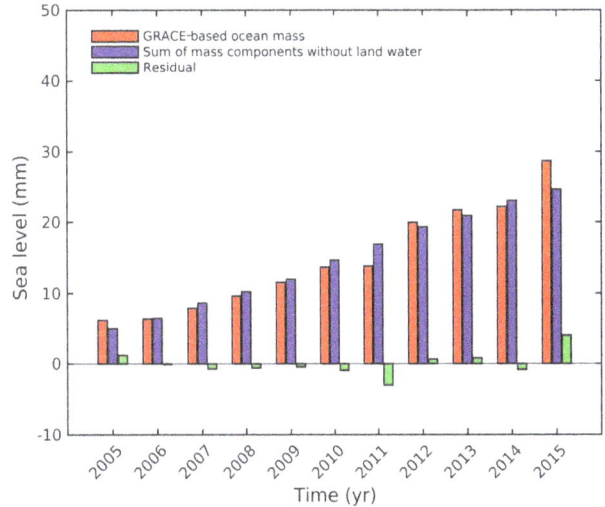

Figure 16. Annual GRACE-based ocean mass (red bars) and sum components without TWS (full depth thermal expansion + glaciers + Greenland + Antarctica) (blue bars). Annual residuals (green bars) are also shown.

components, allowing us to also look at the mass budget. For TWS, as discussed in Sect. 2.7, GRACE provides a negative trend contribution to sea level over the last decade (i.e., increase on water storage on land) attributed to internal natural variability (Reager et al., 2016), unlike hydrological models that lead to a small (possibly not significantly different from zero) positive contribution to sea level over the same period. Assuming that GRACE observations are perfect, such discrepancies could be attributed to the inability of models to correctly account for uncertainties in meteorological forcing and inadequate modeling of soil storage capacity (see discussion in Sect. 2.7). However, when looking at the sea-level budget over the GRACE time span and using the GRACE-based TWS, we find a rather large positive residual trend (> 0.5 mm yr^{-1}) that needs to be explained. Since GRACE-based ocean mass is supposed to represent all mass terms, one may want to attribute this residual trend to an additional contribution of the deep ocean to the abyssal contribution already taken into account here, but possibly underestimated because of incomplete monitoring by current observing systems. If such a large positive contribution from the deep ocean (meaning ocean warming) is real (which is unlikely, given the high implied heat storage), this has to be confirmed by independent approaches, e.g., using ocean reanalysis, and eventually model-based and top-of-the-atmosphere estimates of the Earth energy imbalance.

In addition to mean trends over the period, we also looked at the annual budget for all years, starting in 2005. For most components, annual mean values are provided during the Argo-GRACE era, except for the terrestrial water storage component. However, the sea-level budget based on GRACE ocean mass (plus ocean thermal expansion; Fig. 14) includes the TWS contribution. As shown in Fig. 14, yearly residuals are small, suggesting near closure of the sea-level budget. The residual trend amounts to 0.13 mm yr^{-1}. It could be interpreted as an additional deep ocean contribution not accounted by the SIO estimate (see Sect. 2.3). However, when looking at Fig. 14, we note that yearly residuals are either positive or negative, an indication of interannual variability that can hardly be explained by a deep ocean contribution. The residual trend derived from the difference (GMSL minus sum of components) (Table 13) amounts -0.1 ± 0.3 mm yr^{-1}, suggesting a sea-level budget closed within 0.3 mm yr^{-1} over 2005–present, with no substantial deep ocean contribution.

Figure 16 compares GRACE ocean mass to the sum of mass components (excluding TWS, for the reasons mentioned above). In principle, this mass budget may provide a constraint on the TWS contribution. The corresponding residual trend amounts to 0.14 mm yr^{-1} over the GRACE period, a value that disagrees with the above-quoted GRACE-based TWS estimates. However, it is worth noting that the GRACE-based TWS trend is very dependent on the considered time span because of the strong interannual variability; a recent study by Palanisamy et al. (2018), based on 347 land river basins, found GRACE-based TWS trend of zero over 2005–2015. Given the remaining data uncertainties, any robust conclusion can hardly be reached so far. That being said, more work is needed to clarify the sign discrepancy between GRACE-based and model-based TWS estimates.

5 Data availability

The data sets used in this study are freely available at https://doi.org/10.17882/54854. We provide annual mean time series (expressed in millimeters of equivalent sea level) between 2005 and 2015 for all components of the sea level budget: (1) global mean sea level (GMSL, GMSL.txt as data file) time series from multi-mission satellite altimetry; ensemble mean of six different sea level products (AVISO/CNES, CSIRO, University of Colorado, ESA SL_cci, NASA/GSFC, NOAA). (2) Global mean ocean thermal expansion (Steric.txt as data file) time series: ensemble mean from 10 processing groups (CORA, CSIRO, ACECRC/IMAS-UTAS, ICCES, ICDC, IPRC, JAMSTEC, MRI/JMA, NECI/NOAA, SIO). (3) Glacier contribution (Glaciers.txt) from 5 different products (update of Gardner et al., 2013, update of Marzeion et al., 2012, update of Cogley, 2009, update of Leclercq et al., 2011 and average of GRACE-based estimates of Marzeion et al., 2017). (4) Greenland ice sheet contribution (GreenlandIcesheet.txt as data file): ensemble mean from eight different products (Update from Barletta et al., 2013, Groh and Horwath, 2016, Update from Luthcke et al., 2013, Update from Sasgen et al., 2012, Update from Schrama et al., 2014, Update from van den Broeke et al., 2016, Wiese et al., 2016b, Update from Wouters et al., 2008). (5) Antarctica ice sheet contribution (AntarcticIcesheet.txt as data file): ensemble mean from 11 different products (Updated Martin-Espagnol et al., 2016, Updated Fosberg et al., 2017, Updated Groh and Horwath, 2016, Updated Luthcke et al., 2013, Updated Sasgen et al., 2013, Updated Velicogna et al., 2014, Updated Wiese et al., 2016, Updated from Wouters et al., 2013, Updated Rignot et al., 2011, Update Schrama et al., 2014 version 1, Update Schrama et al., 2014 version 2). We also provide the GRACE-based ocean mass time series that is an ensemble mean of seven different products (GSM CSR Forward Modeling (update from Chen et al., 2013), GSM CSR (update from Johnson and Chambers, 2013), GSM GFZ (update from Johnson and Chambers, 2013), GSM JPL (update from Johnson and Chambers, 2013), Mascon CSR (200 km), Mascon JPL, Mascon GSFC (update from Luthcke et al., 2013)).

6   Concluding remarks

As mentioned in the introduction, the global mean sea-level budget has been the subject of numerous previous studies, including successive IPCC assessments of the published literature. What is new in the effort presented here is that it involves the international community currently studying present-day sea level and its components. Moreover, it relies on a large variety of datasets derived from different space-based and in situ observing systems. The near closure of the sea-level budget, as reported here over the GRACE and Argo era, suggests that no large systematic errors affect these independent observing systems, including the satellite altimetry system. Study of the sea-level budget allows improved understanding of the different processes causing sea-level rise, such as ocean warming and land ice melt. When accuracy in-

creases, it will offer an integrative view of the response of the Earth system to natural and anthropogenic forcing and internal variability, and provide an independent constraint on the current Earth energy imbalance. Validation of climate models against observations is another important application of this kind of assessment (e.g., Slangen et al., 2017).

However, important uncertainties still remain, which affect several terms of the budget; for example the GIA correction applied to GRACE data over Antarctica or the net land water storage contribution to sea level. The latter results from a variety of factors but is dominated by groundwater pumping and natural climate variability. Both terms are still uncertain and accurately quantifying them remains a challenge.

Several ongoing international projects related to sea level should provide, in the near future, improved estimates of the components of the sea-level budget. This is the case, for example, of the ice sheet mass balance intercomparison exercise (IMBIE, second assessment), a community effort supported by NASA (National Aeronautics and Space Administration) and ESA, dedicated to reconciling satellite measurements of ice sheet mass balance (The IMBIE Team, 2018). This is also the case for the ongoing ESA Sea Level Budget Closure project (Horwath et al., 2018) that uses a number of space-based essential climate variables (ECVs) reprocessed during the last few years in the context of the ESA Climate Change Initiative project. The recently launched GRACE follow-on mission will lengthen the current mass component time series, with hopefully increased precision and resolution. Finally, the deep Argo project, still in an experimental phase, will provide important information on the deep ocean heat content in the coming years. Availability of this new dataset will provide new insights into the total thermosteric component of the sea-level budget, allowing other missing or poorly known contributions to be constrained from the evaluation of the budget.

The sea-level budget assessment discussed here essentially relies on trend estimates. But annual budget estimates have been proposed for the first time over the GRACE-Argo era. It is planned to provide updates of the global sea-level budget every year, as done for more than a decade for the global carbon budget (Le Queré et al., 2018). In the next assessments, updates of all components will be considered, accounting for improved evaluation of the raw data, improved processing and corrections, use of ocean reanalysis, etc. The need for additional information where gaps exist should also be considered. As a closing remark, study of the sea-level budget in terms of time series and not just trends, as done here, will be required.

Author contributions. This community assessment was initiated by AC and BM as a contribution to the Grand Challenge "Regional Sea Level and Coastal Impacts" of the World Climate Research Programme (WCRP). The results presented in this paper were prepared by nine different teams dedicated to the various terms of the sea-level budget (i.e., altimetry-based sea level, tide gauges, thermal expansion, glaciers, Greenland, Antarctica, terrestrial water storage, glacial isostatic adjustment, ocean mass from GRACE). Thanks to the team leaders (in alphabetic order) MA, JB, NC, JC, CD, SJ, JTR, KvS, GS, IV and RvdW, who interacted with their team members, collected all needed information, provided a synthesized assessment of the literature and when needed, updated the published results. The coordinators AC and BM collected those materials and prepared a first draft of the paper, but all authors contributed to its refinement and to the discussion of the results. Special thanks are addressed to JB, EB, GC, JC, GJ (PMEL Contribution Number 4776), BM, FP, RP and ES for improving the successive versions of the paper, and to HP for providing all figures presented in Sect. 3.

The views, opinions and findings contained in this paper are those of the authors and should not be construed as an official position, policy or decision of the NOAA, US Government or other institutions.

Competing interests. The authors declare that they have no conflict of interest.

Acknowledgements. We are grateful to the anonymous reviewer for his/her thorough comments that helped to improve the paper.

Edited by: Giuseppe M. R. Manzella

References

A, G. and Chambers, D. P.: Calculating trends from GRACE in the presence of large changes in continental ice storage and ocean mass, Geophys. J. Int., 272, https://doi.org/10.1111/j.1365-246X.2008.04012.x, 2008.

A, G., Wahr, J., and Zhong, S.: Computations of the viscoelastic response of a 3-D compressible Earth to surface loading: an application to Glacial Isostatic Adjustment in Antarctica and Canada, Geophys. J. Int., 192.2, 557–572, 2013.

Ablain, M., Cazenave, A., Valladeau, G., and Guinehut, S.: A New Assessment of the Error Budget of Global Mean Sea Level Rate Estimated by Satellite Altimetry over 1993–2008, Ocean Sci., European Geosciences Union, 2009, 5, 193–201, https://doi.org/10.5194/os-5-193-2009, 2009.

Ablain, M., Philipps, S., Urvoy, M., Tran, N., and Picot, N.: Detection of Long-Term Instabilities on Altimeter Backscatter Coefficient Thanks to Wind Speed Data Comparisons from Altimeters and Models, Mar. Geod., 35 (sup1), 258–75, https://doi.org/10.1080/01490419.2012.718675, 2012.

Ablain, M., Cazenave, A., Larnicol, G., Balmaseda, M., Cipollini, P., Faugère, Y., Fernandes, M. J., Henry, O., Johannessen, J. A., Knudsen, P., and Andersen, O.: Improved Sea Level Record over the Satellite Altimetry Era (1993–2010) from the Climate Change Initiative Project, Ocean Sci., 11, 67–82, https://doi.org/10.5194/os-11-67-2015, 2015.

Ablain, M., Legeais, J. F., Prandi, P., Marcos, M., Fenoglio-Marc, L., Dieng, H. B., Benveniste, J., and Cazenave, A.: Satellite Altimetry-Based Sea Level at Global and Regional Scales, Surv.

Geophys., 38, 7–31, https://doi.org/10.1007/s10712-016-9389-8, 2017a.

Ablain, M., Jugier, R., Zawadki, L., and Taburet, N.: The TOPEX-A Drift and Impacts on GMSL Time Series, AVISO Website, October 2017, available at: https://meetings.aviso.altimetry.fr/fileadmin/user_upload/tx_ausyclsseminar/files/Poster_OSTST17_GMSL_Drift_TOPEX-A.pdf, last access: October 2017b.

Abraham, J. P., Baringer, M., Bindoff, N. L., Boyer, T., Cheng, L. J., Church, J. A., Conroy, J. L., Domingues, C. M., Fasullo, J. T., Gilson, J., and Goni, G.: A review of global ocean temperature observations: Implications for ocean heat content estimates and climate change, Rev. Geophys., 51, 450–483, https://doi.org/10.1002/rog.20022, 2013.

Argus, D. F., Peltier, W. R., and Drummond, R.: The Antarctica component of postglacial rebound model ICE-6G_C (VM5a) based on GPS positioning, exposure age dating of ice thicknesses, and relative sea level histories, Geophys. J. Int., 198, 537–563, 2014.

Bahr, D. B. and Radić, V.: Significant contribution to total mass from very small glaciers, The Cryosphere, 6, 763–770, https://doi.org/10.5194/tc-6-763-2012, 2012.

Bahr, D., Pfeffer, W., Sassolas, C., and Meier, M.: Response time of glaciers as a function of size and mass balance: 1. Theory, J. Geophys. Res.-Solid Earth, 103, 9777–9782, 1998.

Bamber, J. L., Westaway, R. M., Marzeion, B., and Wouters, B.: The land ice contribution to sea level during the satellite era, Environ. Res. Lett., 13, 063008, https://doi.org/10.1088/1748-9326/aac2f0, 2018.

Barletta, V. R., Sørensen, L. S., and Forsberg, R.: Scatter of mass changes estimates at basin scale for Greenland and Antarctica, The Cryosphere, 7, 1411–1432, https://doi.org/10.5194/tc-7-1411-2013, 2013.

Bartnett, T. P.: The estimation of "global" sea level change: A problem of uniqueness, J. Geophys. Res., 89, 7980–7988, 1984.

Beck, H. E., van Dijk, A. I. J. M., de Roo, A., Dutra, E., Fink, G., Orth, R., and Schellekens, J.: Global evaluation of runoff from 10 state-of-the-art hydrological models, Hydrol. Earth Syst. Sci., 21, 2881–2903, https://doi.org/10.5194/hess-21-2881-2017, 2017.

Beckley, B. D., Callahan, P. S., Hancock, D. W., Mitchum, G. T., and Ray, R. D.: On the 'Cal-Mode' Correction to TOPEX Satellite Altimetry and Its Effect on the Global Mean Sea Level Time Series, J. Geophys. Res.-Oceans, 122, 8371–8384, https://doi.org/10.1002/2017jc013090, 2017.

Belward, A. S., Estes, J. E., and Kline, K. D.: The IGBP-DIS global 1-km land-cover data set DISCover: A project overview, Photogramm. Eng. Remote Sens., 65, 1013–1020, 1999.

Bolch, T., Sandberg Sørensen, L., Simonsen, S. B., Mölg, N., Machguth, H., Rastner, P., and Paul, F: Mass loss of Greenland's glaciers and ice caps 2003–2008 revealed from ICESat laser altimetry data, Geophys. Res. Lett., 40, 875–881, 2013.

Bosmans, J. H. C., van Beek, L. P. H., Sutanudjaja, E. H., and Bierkens, M. F. P.: Hydrological impacts of global land cover change and human water use, Hydrol. Earth Syst. Sci., 21, 5603–5626, https://doi.org/10.5194/hess-21-5603-2017, 2017.

Boyer, T., Domingues, C., Good, S., Johnson, G. C., Lyman, J.M., Ishii, M., Gouretski, V., Antonov, J., Bindoff, N., Church, J. A., Cowley, R., Willis, J., and Wijffels, S.: Sensitivity of global ocean heat content estimates to mapping methods, XBT bias corrections, and baseline climatology, J. Climate, 29, 4817–4842, https://doi.org/10.1175/JCLI-D-15-0801.1, 2016.

Box, J. E. and Colgan, W. T.: Sea level rise contribution from Arctic land ice: 1850–2100, Snow, Water, Ice and Permafrost in the Arctic (SWIPA) 2017, Oslo, Norway: Arctic Monitoring and Assessment Programme (AMAP), 2017.

Brun, F., Berthier, E., Wagnon, P., Kääb, A., and Treichler, D.: A spatially resolved estimate of High Mountain Asia mass balances from 2000 to 2016, Nat. Geosci., 10, 668–673, https://doi.org/10.1038/NGEO2999, 2017.

Butt, N., de Oliveira, P. A., and Costa, M. H.: Evidence that deforestation affects the onset of the rainy season in Rondonia, Brazil, J. Geophys. Res.-Atmos., 116, D11120, https://doi.org/10.1029/2010jd015174, 2011.

Calafat, F. M., Chambers, D. P., and Tsimplis, M. N.: On the ability of global sea level reconstructions to determine trends and variability, J. Geophys. Res., 119, 1572–1592, 2014.

Cazenave, A., Dominh, K., Guinehut, S., Berthier, E., Llovel, W., Ramillien, G., Ablain, M., and Larnicol, G.: Sea level budget over 2003–2008: A reevaluation from GRACE space gravimetry, satellite altimetry and Argo, Global Planet. Change, 65, 83–88, https://doi.org/10.1016/j.gloplacha.2008.10.004, 2009.

Cazenave, A., Dieng, H. B., Meyssignac, B., von Schuckmann, K., Decharme, B., and Berthier, E.: The Rate of Sea-Level Rise, Nat. Clim. Change, 4, 358–361, https://doi.org/10.1038/nclimate2159, 2014.

Cazenave, A., Champollion, N., Paul, F., and Benveniste, J.: Integrative Study of the Mean Sea Level and Its Components, Space Science Series of ISSI, Spinger, 416 pp., Vol. 58, 2017.

Chagnon, F. J. F. and Bras, R. L.: Contemporary climate change in the Amazon, Geophys. Res. Lett., 32, L13703, https://doi.org/10.1029/2005gl022722, 2005.

Chambers, D. P., Wahr, J., Tamisiea, M. E., and Nerem, R. S.: Ocean mass from GRACE and glacial isostatic adjustment, J. Geophys. Res.-Solid Earth, 115, L11415, https://doi.org/10.1029/2010JB007530, 2010.

Chambers, D. P., Cazenave, A., Champollion, N., Dieng, H. B., Llovel, W., Forsberg, R., von Schuckmann, K., and Wada, Y.: Evaluation of the Global Mean Sea Level Budget between 1993 and 2014, Surv. Geophys. 38, 309–327, https://doi.org/10.1007/s10712-016-9381-3, 2017.

Chao, B. F., Wu, Y. H., and Li, Y. S.: Impact of artificial reservoir water impoundment on global sea level, Science, 320, 212–214, https://doi.org/10.1126/science.1154580, 2008.

Chen, J. L., Wilson, C. R., Tapley, B. D., Blankenship, D. D., and Ivins, E. R.: Patagonia Icefield Melting Observed by GRACE, Geophys. Res. Lett., 34, L22501, https://doi.org/10.1029/2007GL031871, 2007.

Chen, J. L., Wilson, C. R., and Tapley, B. D.: Contribution of ice sheet and mountain glacier melt to recent sea level rise, Nat. Geosci., 9, 549–552, https://doi.org/10.1038/NGEO1829, 2013.

Chen, J. L., Wilson, C. R., Tapley, B. D., Save, H., and Cretaux, J.-F., Long-term and seasonal Caspian Sea level change from satellite gravity and altimeter measurements, J. Geophys. Res.-Solid Earth, 122, 2274–2290, https://doi.org/10.1002/2016JB013595, 2017a.

Chen, J., Famiglietti, J. S., Scanlon, B. R., and Rodell, M.: Groundwater Storage Changes: Present Status from GRACE Observations, Surv. Geophys., 37, 397–417, https://doi.org/10.1007/s10712-015-9332-4, 2017b.

Chen, X., Zhang, X., Church, J. A., Watson, C. S., King, M. A., Monselesan, D., Legresy, B., and Harig, C.: The Increasing Rate of Global Mean Sea-Level Rise during 1993–2014, Nat. Clim. Change, 7, 492–95, https://doi.org/10.1038/nclimate3325, 2017.

Cheng, L., Trenberth, K., Fasullo, J., Boyer, T., Abraham, J., and Zhu, J.: Improved estimates of ocean heat content from 1960–2015, Sci. Adv., 3, e1601545, https://doi.org/10.1126/sciadv.1601545, 2017.

Cheng, M. K. and Ries, J. R.: Monthly estimates of C20 from 5 SLR satellites based on GRACE RL05 models, GRACE Technical Note 07, The GRACE Project, Center for Space Research, University of Texas at Austin, 2012.

Choblet, G., Husson, L., and Bodin, T.: Probabilistic surface reconstruction of coastal sea level rise during the twentieth century, J. Geophys. Res., 119, 9206–9236, 2014.

Church, J. A. and White, N. J.: A 20th century acceleration in global sea-level rise, Geophys. Res. Lett., 33, L01602, https://doi.org/10.1029/2005GL024826, 2006.

Church, J. A. and White, N. J.: Sea-Level Rise from the Late 19th to the Early 21st Century, Surv. Geophys., 32, 585–602, https://doi.org/10.1007/s10712-011-9119-1., 2011.

Church, J. A., Gregory, J., White, N. J., Platten, S., and Mitrovica, J. X.: Understanding and Projecting Sea Level Change, Oceanography, 24, 130–143, https://doi.org/10.5670/oceanog.2011.33, 2011.

Church, J. A., Clark, P. U., Cazenave, A., Gregory, J. M., Jevrejeva, S., Levermann, A., Merrifield, M. A., Milne, G. A., Nerem, R. S., Nunn, P. D., Payne, A. J., Pfeffer, W. T., Stammer, D., and Unnikrishnan, A. S.: Sea level change, in: Climate Change 2013: The Physical Science Basis, edited by: Stocker, T. F., Qin, D., Plattner, G.-K., Tignor, M., Allen, S. K., Boschung, J., Nauels, A., Xia, Y., Bex, V., and Midgley, P. M., Contribution of Working Group I to the Fifth Assessment Report of the Intergovernmental Panel on Climate Change (Cambridge University Press, Cambridge, United Kingdom and New York, NY, USA), 2013.

Cogley, J.: Geodetic and direct mass-balance measurements: comparison and joint analysis, Ann. Glaciol., 50, 96–100, 2009.

Couhert, A., Cerri, L., Legeais, J. F., Ablain, M., Zelensky, N. P., Haines, B. J., Lemoine, F. G., Bertiger, W. I., Desai, S. D., and Otten, M.: Towards the 1mm/y Stability of the Radial Orbit Error at Regional Scales." Advances in Space Research: The Official J. Committ. Space Res., 55, 2–23, https://doi.org/10.1016/j.asr.2014.06.041, 2015.

Dangendorf, S., Marcos, M., Müller, A., Zorita, E., Riva, R., Berk, K., and Jensen, J.: Detecting anthropogenic footprints in sea level rise, Nat. Commun., 6, 7849, https://doi.org/10.1038/ncomms8849, 2015.

Dangendorf, S., Marcos, M., Wöppelmann, G., Conrad, C. P., Frederikse, T., and Riccardo Riva, R.: Reassessment of 20th Century Global Mean Sea Level Rise, Proc. Natl. Acad. Sci. USA, 114, 5946–5951, https://doi.org/10.1073/pnas.1616007114, 2017.

Darras, S.: IGBP-DIS wetlands data initiative, a first step towards identifying a global delineation of wetalnds, IGBP-DIS Office, Toulouse, France, 1999.

Davidson, N.C.: How much wetland has the world lost? Long-term and recent trends in global wetland area, Mar. Freshw. Res., 65, 934–941 https://doi.org/10.1071/Mf14173, 2014.

Decharme, B., Brun, E., Boone, A., Delire, C., Le Moigne, P., and Morin, S.: Impacts of snow and organic soils parameterization on northern Eurasian soil temperature profiles simulated by the ISBA land surface model, The Cryosphere, 10, 853–877, https://doi.org/10.5194/tc-10-853-2016, 2016.

Dieng, H. B., Champollion, N., Cazenave, A., Wada, Y., Schrama, E. J. O., and Meyssignac, B.: Total land water storage change over 2003–2013 estimated from a global mass budget approach, Environ. Res. Lett., 10, 124010, https://doi.org/10.1088/1748-9326/10/12/124010, 2015a.

Dieng, H. B., Cazenave, A., von Schuckmann, K., Ablain, M., and Meyssignac, B.: Sea level budget over 2005–2013: missing contributions and data errors, Ocean Sci., 11, 789–802, https://doi.org/10.5194/os-11-789-2015, 2015b.

Dieng, H. B., Palanisamy, H., Cazenave, A., Meyssignac, B., and von Schuckmann, K.: The Sea Level Budget Since 2003: Inference on the Deep Ocean Heat Content, Surv. Geophys., 36, 209–229, https://doi.org/10.1007/s10712-015-9314-6, 2015c.

Dieng, H. B., Cazenave, A., Meyssignac, B., and Ablain, M.: New estimate of the current rate of sea level rise from a sea level budget approach, Geophys. Res. Lett., 44, 3744–3751, https://doi.org/10.1002/2017GL073308, 2017.

Döll, P., Fritsche, M., Eicker, A., and Mueller, S. H.: Seasonal water storage variations as impacted by water abstractions: Comparing the output of a global hydrological model with GRACE and GPS observations, Surv. Geophys., 35, 1311–1331, https://doi.org/10.1093/gji/ggt485, 2014a.

Döll, P., Müller, S. H., Schuh, C., Portmann, F. T., and Eicker, A.: Global-scale assessment of groundwater depletion and related groundwater abstractions: Combining hydrological modeling with information from well observations and GRACE satellites, Water Resour. Res., 50, 5698–5720, https://doi.org/10.1002/2014WR015595, 2014b.

Döll, P., Douville, H., Güntner, A., Müller Schmied, H., and Wada, Y.: Modelling freshwater resources at the global scale: Challenges and prospects, Surv. Geophys., 37, 195–221, Special Issue: ISSI Workshop on Remote Sensing and Water Resources, 2017.

Domingues, C., Church, J., White, N., Gleckler, P. J., Wijffels, S. E., Barker, P. M., and Dunn, J. R.: Improved estimates of upper ocean warming and multidecadal sea level rise, Nature, 453, 1090–1093, https://doi.org/10.1038/nature07080, 2008.

Douglas, B.: Global sea level rise, J. Geophys. Res.-Oceans, 96, 6981–6992, 1991.

Douglas, B.: Global sea rise: a redetermination, Surv. Geophys., 18, 279–292, 1997.

Douglas, B. C.: Sea level change in the era of recording tide gauges, in: Sea Level Rise, History and Consequences, 37–64, edited by: Douglas, B. C., Kearney, M. S., and Leatherman, S. P., Academic Press, San Diego, CA, 2001.

Dutrieux, P., De Rydt, J., Jenkins, A., Holland, P. R., Ha, H. K., Lee, S. H., Steig, E. J., Ding, Q., Abrahamsen, E. P., and Schröder, M: Strong sensitivity of Pine Island ice shelf melting to climatic variability, Science, 343, 174–178, https://doi.org/10.1126/science.1244341, 2014.

Escudier, P., Ablain, M., Amarouche, L., Carrère, L., Couhert, A., Dibarboure, G., Dorandeu, J., Dubois, P., Mallet, A., Mercier, F., and Picard, B.: Satellite radar altimetry: principle, accuracy and precision, in: Satellite altimetry over oceans and land surfaces, edited by: Stammer, D. L. and Cazenave, A., 617 pp., CRC Press, Taylor and Francis Group, Boca Raton, New York, London, ISBN:13:978-1-4987-4345-7, 2018.

FAO: Global forest resources assessment 2015: how have the world's forests changed?, Rome, 2015.

Farinotti, D., Brinkerhoff, D. J., Clarke, G. K. C., Fürst, J. J., Frey, H., Gantayat, P., Gillet-Chaulet, F., Girard, C., Huss, M., Leclercq, P. W., Linsbauer, A., Machguth, H., Martin, C., Maussion, F., Morlighem, M., Mosbeux, C., Pandit, A., Portmann, A., Rabatel, A., Ramsankaran, R., Reerink, T. J., Sanchez, O., Stentoft, P. A., Singh Kumari, S., van Pelt, W. J. J., Anderson, B., Benham, T., Binder, D., Dowdeswell, J. A., Fischer, A., Helfricht, K., Kutuzov, S., Lavrentiev, I., McNabb, R., Gudmundsson, G. H., Li, H., and Andreassen, L. M.: How accurate are estimates of glacier ice thickness? Results from ITMIX, the Ice Thickness Models Intercomparison eXperiment, The Cryosphere, 11, 949–970, https://doi.org/10.5194/tc-11-949-2017, 2017.

Farrell, W. and Clark, J.: On postglacial sea level, Geophys. J. Int., 46.3, 647–667, 1976.

Fasullo, J. T., Boening, C., Landerer, F. W., and Nerem, R. S.: Australia's unique influence on global sea level in 2010–2011, Geophys. Res. Lett., 40, 4368–4373, https://doi.org/10.1002/grl.50834, 2013.

Felfelani, F., Wada, Y., Longuevergne, L., and Pokhrel, Y. N.: Natural and human-induced terrestrial water storage change: A global analysis using hydrological models and GRACE, J. Hydrol., 553, 105–118, 2017.

Fleming, K. and Lambeck, K.: Constraints on the Greenland Ice Sheet since the Last Glacial Maximum from sea-level observations and glacial-rebound models, Quatern. Sci. Rev., 23, 1053–1077, 2004.

Forsberg, R., Sørensen, L., and Simonsen, S.: Greenland and Antarctica Ice Sheet Mass Changes and Effects on Global Sea Level, Surv. Geophys., 38, 89–104, https://doi.org/10.1007/s10712-016-9398-7, 2017.

Foster, S. and Loucks, D. P. (Eds.): Non-Renewable Groundwater Resources: A guidebook on socially-sustainable management for water-policy makers, IHP-VI, Series on Groundwater No. 10, UNESCO, Paris, France, 2006.

Frederikse, T., Jevrejeva, S., Riva, R. E., and Dangendorf, S.: A consistent sea-level reconstruction and its budget on basin and global scales over 1958–2014, J. Climate, 31.3, 1267–1280, https://doi.org/10.1175/JCLI-D-17-0502.1, 2017.

Frey, H., Machguth, H., Huss, M., Huggel, C., Bajracharya, S., Bolch, T., Kulkarni, A., Linsbauer, A., Salzmann, N., and Stoffel, M.: Estimating the volume of glaciers in the Himalayan-Karakoram region using different methods, The Cryosphere, 8, 2313–2333, https://doi.org/10.5194/tc-8-2313-2014, 2014.

Gardner, A. S., Moholdt, G., Cogley, J. G., Wouters, B., Arendt, A. A., Wahr, J., Berthier, E., Hock, R., Pfeffer, W. T., Kaser, G., Ligtenberg, S. R. M., Bolch, T., Sharp, M. J., Hagen, J. O., van den Broeke, M. R., and Paul F.: A reconciled estimate of glacier contributions to sea level rise: 2003 to 2009, Science, 340, 852–857, https://doi.org/10.1126/science.1234532, 2013.

Gomez, N., Pollard, D., and Mitrovica, J. X.: A 3-D coupled ice sheet–sea level model applied to Antarctica through the last 40 ky, Earth Planet. Sci. Lett., 384, 88–99, 2013.

Gornitz, V.: Sea-level rise: A review of recent past and near-future trends, Earth Surf. Process. Landf., 20, 7–20, https://doi.org/10.1002/esp.3290200103, 1995.

Gornitz, V.: Sea Level Rise: History and Consequences, edited by: Douglas, B. C., Kearney, M. S., and Leatherman, S. P., 97–119, Academic Press, San Diego, CA, USA, 2001.

Gornitz, V., Lebedeff, S., and Hansen, J.: Global sea level trend in the past century, Science, 215, 1611–1614, 1982.

Gornitz, V., Rosenzweig, C., and Hillel, D.: Effects of anthropogenic intervention in the land hydrologic cycle on global sea level rise, Global Planet. Change, 14, 147–161, https://doi.org/10.1016/s0921-8181(96)00008-2, 1997.

Gouretski, V. and Koltermann, K.P.: How much is the ocean really warming?, Geophys. Res. Lett., 34, L01610, https://doi.org/10.1029/2006GL027834, 2007.

Gregory, J. M. and Lowe, J. A.: Predictions of global and regional sea-level rise using AOGCMs with and without flux adjustment, Geophys. Res. Lett., 27, 3069–3072, 2000.

Gregory, J. M., White, N. J., Church, J. A., Bierkens, M. F. P., Box, J. E., van den Broeke, M. R., Cogley, J. G., Fettweis, X., Hanna, E., Huybrechts, P., Konikow, L. F., Leclercq, P. W., Marzeion, B., Oerlemans, J., Tamisiea, M .E., Wada, Y., Wake, L. M., and van de Wal, R. S. W.: Twentieth-Century Global-Mean Sea Level Rise: Is the Whole Greater than the Sum of the Parts?, J. Climate, 26, 4476–4499, https://doi.org/10.1175/JCLI-D-12-00319.1, 2013.

Grinsted, A.: An estimate of global glacier volume, The Cryosphere, 7, 141–151, https://doi.org/10.5194/tc-7-141-2013, 2013.

Groh, A. and Horwath, M.: The method of tailored sensitivity kernels for GRACE mass change estimates, Geophys. Res. Abstract., 18, EGU2016–12065, 2016.

Gunter, B. C., Didova, O., Riva, R. E. M., Ligtenberg, S. R. M., Lenaerts, J. T. M., King, M. A., van den Broeke, M. R., and Urban, T.: Empirical estimation of present-day Antarctic glacial isostatic adjustment and ice mass change, The Cryosphere, 8, 743–760, https://doi.org/10.5194/tc-8-743-2014, 2014.

Haeberli, W. and Linsbauer, A.: Brief communication "Global glacier volumes and sea level – small but systematic effects of ice below the surface of the ocean and of new local lakes on land", The Cryosphere, 7, 817–821, https://doi.org/10.5194/tc-7-817-2013, 2013.

Hamlington, B. D., Leben, R. R., Nerem, R. S., Han, W., and Kim, K. Y.: Reconstructing sea level using cyclostationary empirical orthogonal functions, J. Geophys. Res., 116, C12015, https://doi.org/10.1029/2011JC007529, 2011.

Hamlington, B. D., Thompson, P., Hammond, W. C., Blewitt, G., and Ray, R. D.: Assessing the impact of vertical land motion on twentieth century global mean sea level estimates, J. Geophys. Res.-Oceans, 121, 4980–4993, https://doi.org/10.1002/2016JC011747, 2016.

Hay, C. C., Morrow, E., Kopp, R. E., and Mitrovica, J. X.: Probabilistic Reanalysis of Twentieth-Century Sea-Level Rise, Nature 517, 481–484, https://doi.org/10.1038/nature14093, 2015.

Henry, O., Ablain, M., Meyssignac, B., Cazenave, A., Masters, D., Nerem, S., and Garric, G.: Effect of the Processing Methodology on Satellite Altimetry-Based Global Mean Sea Level Rise over the Jason-1 Operating Period, J. Geod., 88, 351–361, https://doi.org/10.1007/s00190-013-0687-3, 2014.

Horwath, M., Novotny, K., Cazenave, A., Palanisamy, H., Marzeion, B., Paul, F., Döll, P., Cáceres, D., Hogg, A., Shepherd, A., Forsberg, R., Sørensen, L., Barletta, V. R., Andersen, O. B., Ranndal, H., Johannessen, J., Nilsen, J. E., Gutknecht, B. D., Merchant, Ch. J., MacIntosh, C. R., and von Schuckmann, K.: ESA Climate Change Initiative (CCI) Sea Level Budget Closure (SLBC_cci) Sea Level Budget Closure Assessment Report D3.1, Version 1.0, 2018.

Hosoda, S., Ohira, T., and Nakamura, T.: A monthly mean dataset of global oceanic temperature and salinity derived from Argo float observations, JAMSTEC Rep. Res. Dev., 8, 47–59, 2008.

Huang, Z.: The Role of glacial isostatic adjustment (GIA) process on the determination of present-day sea level rise, Report no 505, Geodetic Science, The Ohio State University, 2013.

Hurkmans, R. T. W. L., Bamber, J. L., Davis, C. H., Joughin, I. R., Khvorostovsky, K. S., Smith, B. S., and Schoen, N.: Time-evolving mass loss of the Greenland Ice Sheet from satellite altimetry, The Cryosphere, 8, 1725–1740, https://doi.org/10.5194/tc-8-1725-2014, 2014.

Huss, M. and Hock, R.: A new model for global glacier change and sea-level rise, Front Earth Sci., 3, 54, https://doi.org/10.3389/feart.2015.00054, 2015.

IMBIE Team (the): Mass balance of the Antarctic ice sheet from 1992 to 2017, Nature, 558, 219–222, https://doi.org/10.1038/s41586-018-0179-y, 2018.

IPCC: Climate Change 2013: The Physical Science Basis, Contribution of Working Group I to the Fifth Assessment Report of the Intergovernmental Panel on Climate Change, edited by: Stocker, T. F., Qin, D., Plattner, G.-K., Tignor, M., Allen, S. K., Boschung, J., Nauels, A., Xia, Y., Bex, V., and Midgley, P. M., Cambridge University Press, Cambridge, United Kingdom and New York, NY, USA, 1535 pp., 2013.

Ishii, M. and Kimoto, M.: Reevaluation of Historical Ocean Heat Content Variations with Time-varying XBT and MBT Depth Bias Corrections, J. Oceanogr., 65, 287–299, https://doi.org/10.1007/s10872-009-0027-7, 2009.

Ishii, M., Fukuda, Y., Hirahara, S., Yasui, S., Suzuki, T., and Sato, K.: Accuracy of Global Upper Ocean Heat Content Estimation Expected from Present Observational Data Sets, SOLA, 2017, 13, 163–167, https://doi.org/10.2151/sola.2017-030, 2017.

Ivins, E. R., James, T. S., Wahr, J., Schrama, E. J. O., Landerer, F. W., and Simon, K. M.: Antarctic contribution to sea level rise observed by GRACE with improved GIA correction, J. Geophys. Res.-Solid Earth, 118, 3126–3141, https://doi.org/10.1002/jgrb.50208, 2013.

Jacob, T., Wahr, J., Pfeffer, W. T., and Swenson, S.: Recent contributions of glaciers and ice caps to sea level rise, Nature, 482, 514–518, https://doi.org/10.1038/nature10847, 2012.

Jensen, L., Rietbroek, R., and Kusche, J.: Land water contribution to sea level from GRACE and Jason-1 measurements, J. Geophys. Res.-Oceans, 118, 212–226, https://doi.org/10.1002/jgrc.20058, 2013.

Jevrejeva, S., Grinsted, A., Moore, J. C., and Holgate, S.: Nonlinear trends and multi-year cycle in sea level records, J. Geophys. Res., 111, 2005JC003229, https://doi.org/10.1029/2005JC003229, 2006.

Jevrejeva, S., Moore, J. C., Grinsted, A., Matthews, A. P., and Spada, G.: Trends and Acceleration in Global and Regional

Sea Levels since 1807, Global Planet. Change J.C., 113, 11–22, https://doi.org/10.1016/j.gloplacha.2013.12.004, 2014.

Johannesson, T., Raymond, C., and Waddington, E.: Time-Scale for Adjustment of Glaciers to Changes in Mass Balance, J. Glaciol., 35, 355–369, 1989.

Johnson, G. C. and Chambers, D. P.: Ocean bottom pressure seasonal cycles and decadal trends from GRACERelease-05: Ocean circulation implications, J. Geophys. Res.-Oceans, 118, 4228–4240, https://doi.org/10.1002/jgrc.20307, 2013.

Johnson, G. C. and Birnbaum, A. N.: As El Niño builds, Pacific Warm Pool expands, ocean gains more heat, Geophys. Res. Lett., 44, 438–445, https://doi.org/10.1002/2016GL071767, 2017.

Kääb, A., Treichler, D., Nuth, C., and Berthier, E.: Brief Communication: Contending estimates of 2003–2008 glacier mass balance over the Pamir-Karakoram-Himalaya, The Cryosphere, 9, 557–564, https://doi.org/10.5194/tc-9-557-2015, 2015.

Kaser, G., Cogley, J., Dyurgerov, M., Meier, M., and Ohmura, A.: Mass balance of glaciers and ice caps: Consensus estimates for 1961–2004, Geophys. Res. Lett., 33, L19501, https://doi.org/10.1029/2006GL027511 2006.

Keenan, R. J., Reams, G. A., Achard, F., de Freitas, J. V., Grainger, A., and Lindquist, E.: Dynamics of global forest area: Results from the FAO Global Forest Resources Assessment 2015, Forest Ecol. Manag., 352, 9–20 https://doi.org/10.1016/j.foreco.2015.06.014, 2015.

Kemp, A. C., Horton, B., Donnelly, J. P., Mann, M. E., Vermeer, M., and Rahmstorf, S.: Climate related sea level variations over the past two millennia, Proc. Natl. Acad. Sci. USA, 108.27, 11017–11022, 2011.

Khan, S. A., Sasgen, I., Bevis, M., Van Dam, T., Bamber, J. L., Wahr, J., Willis, M., Kjær, K. H., Wouters, B., Helm, V., Csatho, B., Fleming, K., Bjørk, A. A., Aschwanden, A., Knudsen, P., and Munneke, P. K.: Geodetic measurements reveal similarities between post – Last Glacial Maximum and present-day mass loss from the Greenland ice sheet, Sci. Adv., 2, 465–507, https://doi.org/10.1007/s10712-010-9100-4, 2016.

King, M. A., Altamimi, Z., Boehm, J., Bos, M., Dach, R., Elosegui, P., Fund, F., Hernández-Pajares, M., Lavallee, D., Cerveira, P. J. M., and Penna, N.: Improved constraints on models of glacial isostatic adjustment: a review of the contribution of ground-based geodetic observations, Surv. Geophys., 31, 465–507, https://doi.org/10.1007/s10712-010-9100-4, 2010.

Konikow, L. F.: Contribution of global groundwater depletion since 1900 to sea-level rise, Geophys. Res. Lett., 38, L17401, https://doi.org/10.1029/2011GL048604, 2011.

Konrad, H., Sasgen, I., Pollard, D., and Klemann, V.: Potential of the solid-Earth response for limiting long-term West Antarctic Ice Sheet retreat in a warming climate, Earth Planet. Sci. Lett., 432, 254–264, 2015.

Lambeck, K.: Sea-level change from mid-Holocene to recent time: An Australian example with global implications, in: Ice Sheets, Sea Level and the Dynamic Earth, edited by: Mitrovica, J. X. and Vermeersen, L. L. A., Geodynam. Series, 29, 33–50, 2002.

Lambeck, K. and Chappell, J.: Sea Level Change Through the Last Glacial Cycle, Science, 292, 679–686, https://doi.org/10.1126/science.1059549, 2001.

Lambeck, K., Woodroff, C. D., Antonioli, F., Anzidei, M., Gehrels, W. R., Laborel, J., and Wright, A. J.: Paleoenvironmental records, geophysical modelling and reconstruction of sea level

trends and variability on centennial and longer time scales, in: Understanding sea level rise and variability, edited by: Church, J. A., Woodworth, P. L., Aarup, T., and Wilson, W. S., Wiley-Blackwell, 2010.

Leclercq, P., Oerlemans, J., and Cogley, J.: Estimating the glacier contribution to sea-level rise for the period 1800–2005, Surv. Geophys., 32, 519–535, 2011.

Legeais, J.-F., Ablain, M., Zawadzki, L., Zuo, H., Johannessen, J. A., Scharffenberg, M. G., Fenoglio-Marc, L., Fernandes, M. J., Andersen, O. B., Rudenko, S., Cipollini, P., Quartly, G. D., Passaro, M., Cazenave, A., and Benveniste, J.: An improved and homogeneous altimeter sea level record from the ESA Climate Change Initiative, Earth Syst. Sci. Data, 10, 281–301, https://doi.org/10.5194/essd-10-281-2018, 2018.

Le Quéré, C., Andrew, R. M., Friedlingstein, P., Sitch, S., Pongratz, J., Manning, A. C., Korsbakken, J. I., Peters, G. P., Canadell, J. G., Jackson, R. B., Boden, T. A., Tans, P. P., Andrews, O. D., Arora, V. K., Bakker, D. C. E., Barbero, L., Becker, M., Betts, R. A., Bopp, L., Chevallier, F., Chini, L. P., Ciais, P., Cosca, C. E., Cross, J., Currie, K., Gasser, T., Harris, I., Hauck, J., Haverd, V., Houghton, R. A., Hunt, C. W., Hurtt, G., Ilyina, T., Jain, A. K., Kato, E., Kautz, M., Keeling, R. F., Klein Goldewijk, K., Körtzinger, A., Landschützer, P., Lefèvre, N., Lenton, A., Lienert, S., Lima, I., Lombardozzi, D., Metzl, N., Millero, F., Monteiro, P. M. S., Munro, D. R., Nabel, J. E. M. S., Nakaoka, S.-I., Nojiri, Y., Padin, X. A., Peregon, A., Pfeil, B., Pierrot, D., Poulter, B., Rehder, G., Reimer, J., Rödenbeck, C., Schwinger, J., Séférian, R., Skjelvan, I., Stocker, B. D., Tian, H., Tilbrook, B., Tubiello, F. N., van der Laan-Luijkx, I. T., van der Werf, G. R., van Heuven, S., Viovy, N., Vuichard, N., Walker, A. P., Watson, A. J., Wiltshire, A. J., Zaehle, S., and Zhu, D.: Global Carbon Budget 2017, Earth Syst. Sci. Data, 10, 405–448, https://doi.org/10.5194/essd-10-405-2018, 2018.

Lettenmaier, D. P. and Milly, P. C. D.: Land waters and sea level, Nat. Geosci., 2, 452–454, https://doi.org/10.1038/ngeo567, 2009.

Leuliette, E. W. and Miller, L.: Closing the sea level rise budget with altimetry, Argo, and GRACE, Geophys. Res. Lett., 36, L04608, https://doi.org/10.1029/2008GL036010, 2009.

Leuliette, E. W. and Willis, J. K.: Balancing the sea level budget, Oceanography, 24, 122–129, https://doi.org/10.5670/oceanog.2011.32, 2011.

Leuschen, C.: IceBridge Geolocated Radar Echo Strength Profiles, Boulder, Colorado, NASA DAAC at the National Snow and Ice Data Center, https://doi.org/10.5067/FAZTWP500V70, last access: 15 June 2014.

Levitus, S., Antonov, J. I., Boyer, T. P., Baranova, O. K., Garcia, H. E., Locarnini, R. A., Mishonov, A. V., Reagan, J. R., Seidov, D., Yarosh E. S., and Zweng, M. M.: World ocean heat content and thermosteric sea level change (0–2000 m), 1955–2010, Geophys. Res. Lett., 39, L10603, https://doi.org/10.1029/2012GL051106, 2012.

Llovel, W., Becker, M., Cazenave, A., Crétaux, J. F., and Ramillien, G.: Global land water storage change from GRACE over 2002–2009; Inference on sea level, C. R. Geosci., 342, 179–188, https://doi.org/10.1016/j.crte.2009.12.004, 2010.

Llovel, W., Willis, J. K., Landerer, F. W., and Fukumori, I.: Deep-ocean contribution to sea level and energy budget not detectable over the past decade, Nat. Clim. Change 4, 1031–1035, https://doi.org/10.1038/nclimate2387, 2014.

Lo, M. H. and Famiglietti, J. S.: Irrigation in California's Central Valley strengthens the southwestern U.S. water cycle, Geophys. Res. Lett., 40, 301–306, https://doi.org/10.1002/grl.50108, 2013.

Loriaux, T. and Casassa, G.: Evolution of glacial lakes from the Northern Patagonia Icefield and terrestrial water storage in a sea-level rise context, Global Planet. Change, 102, 33–40, 2013.

Lovel, T. R. and Belward, A. S.: The IGBP-DIS global 1 km land cover data set, DISCover: first results, Int. J. Remote Sens., 18, 3291–3295, 1997.

Luthcke, S. B., Zwally, H. J., Abdalati, W., Rowlands, D. D., Ray, R. D., Nerem, R. S., Lemoine, F. G., McCarthy, J. J., and Chinn, D. S.: Recent Greenland Ice Mass Loss by Drainage System from Satellite Gravity Observations, Science, 314, 1286–1289, https://doi.org/10.1126/science.1130776, 2006.

Luthcke, S. B., Sabaka, T., Loomis, B., Arendt, A., Mccarthy, J., and Camp, J.: Antarctica, Greenland and Gulf of Alaska land-ice evolution from an iterated GRACE global mascon solution, J. Glaciol., 59, 613–631, 2013.

MacDicken, K. G.: Global Forest Resources Assessment, What, why and how?, Forest Ecol. Manage., 352, 3–8, https://doi.org/10.1016/j.foreco.2015.02.006, 2015.

Martinec, Z. and Hagedoorn, J.: The rotational feedback on linear-momentum balance in glacial isostatic adjustment, Geophys. J. Int., 199, 1823–1846, 2014.

Martín-Español, A., Zammit-Mangion, A., Clarke, P. J., Flament, T., Helm, V., King, M. A., and Wouters, B.: Spatial and temporal Antarctic Ice Sheet mass trends, glacio-isostatic adjustment, and surface processes from a joint inversion of satellite altimeter, gravity, and GPS data, J. Geophys. Res.-Earth Surf., 121, 182–200, 2016.

Marzeion, B., Jarosch, A. H., and Hofer, M.: Past and future sea-level change from the surface mass balance of glaciers, The Cryosphere, 6, 1295–1322, https://doi.org/10.5194/tc-6-1295-2012, 2012

Marzeion, B., Cogley, J., Richter, K., and Parkes, D.: Attribution of global glacier mass loss to anthropogenic and natural causes, Science, 345, 919–920, 2014.

Marzeion, B., Champollion, N., Haeberli, W., Langley, K., Leclercq, P., and Paul, F.: Observation-Based Estimates of Global Glacier Mass Change and Its Contribution to Sea-Level Change, Surv. Geophys., 28, 105–130, 2017.

Marzeion, B., Kaser, G., Maussion, F., and Champollion, N.: Limited influence of climate change mitigation on short-term glacier mass loss, Nat. Clim. Change, 8, 305–308, https://doi.org/10.1038/s41558-018-0093-1, 2018.

Masters, D., Nerem, R. S., Choe, C., Leuliette, E., Beckley, B., White, N., and Ablain, M.: Comparison of Global Mean Sea Level Time Series from TOPEX/Poseidon, Jason-1, and Jason-2, Mar. Geod., 35 (sup1), 20–41, https://doi.org/10.1080/01490419.2012.717862, 2012.

Matthews, E. and Fung, I.: Methane emission from natural wetlands: Global distribution, area, and environmental characteristics of sources, Global Biogeochem. Cy., 1, 61–86, 1987.

Maussion, F., Butenko, A., Eis, J., Fourteau, K., Jarosch, A. H., Landmann, J., Oesterle, F., Recinos, B., Rothenpieler, T., Vlug, A., Wild, C. T., and Marzeion, B.: The Open Global Glacier Model (OGGM) v1.0, Geosci. Model Dev. Dis-

cuss., https://doi.org/https://doi.org/10.5194/gmd-2018-9, in review, 2018.

McMillan, M., Shepherd, A., Sundal, A., Briggs, K., Muir, A., Ridout, A., Hogg, A., and Wingham, D.: Increased ice losses from Antarctica detected by Cryosat-2, Geophys. Res. Lett., 41, 3899–3905, 2014.

Merrifield, M. A., Merrifield, S. T., and Mitchum, G. T.: An anomalous recent acceleration of global sea level rise, J. Climate, 22, 5772–5781,https://doi.org/10.1175/2009JCLI2985.1, 2009.

Meyssignac, B., Becker, M., Llovel, W., and Cazenave, A.: An Assessment of Two-Dimensional Past Sea Level Reconstructions Over 1950–2009 Based on Tide-Gauge Data and Different Input Sea Level Grids, Surv. Geophys., 33, 945–972, https://doi.org/10.1007/s10712-011-9171-x, 2011.

Milly, P. C. D., Cazenave, A., and Gennero, M.C.: Contribution of climate-driven change in continental water storage to recent sea-level rise, Proc. Natl. Acad. Sci. USA, 100, 13158–13161, 2003.

Milly, P. C. D., Cazenave, A., Famiglietti, J. S., Gornitz, Vivien, Laval, Katia, Lettenmaier, D. P., Sahagian, D. L., Wahr, J. M., and Wilson, C. R.: Terrestrial water-storage contributions to sea-level rise and variability , in: Understanding Sea-Level Rise and Variability, 226–255, 2010.

Milne, G. A., Gehrels, W. R., Hughes, C. W., and Tamisiea, M. E.: Identifying the causes of sea-level change, Nat. Geosci., 2.7, 471–478, 2009.

Mitrovica, J. X. and Milne, G. A.: On post-glacial sea level: I. General theory, Geophys. J. Int., 154, 253–267, 2003.

Mitrovica, J. X. and Wahr, J.: Ice age Earth rotation, Annu. Rev. Earth Planet. Sci., 39, 577–616, 2011.

Mitrovica, J. X., Wahr, J., Matsuyama, I., and Paulson, A.: The rotational stability of an ice-age earth, Geophys. J. Int., 161.2, 491–506, 2005.

Mitrovica, J. X., Wahr, J., Matsuyama, I., Paulson, A., and Tamisiea, M. E.: Reanalysis of ancient eclipse, astronomic and geodetic data: A possible route to resolving the enigma of global sea-level rise, Earth Planet. Sci. Lett., 243, 390–399, https://doi.org/10.1016/j.epsl.2005.12.029, 2006.

Mouginot, J., Rignot, E., and Scheuchl, B.: Sustained increase in ice discharge from the Amundsen Sea Embayment, West Antarctica, from 1973 to 2013, Geophys. Res. Lett., 41, 1576–1584, 2014.

Munk, W.: Twentieth century sea level: An enigma, Proc. Natl. Acad. Sci. USA, 99, 6550–6555, https://doi.org/10.1073/pnas.092704599, 2002.

Natarov, S. I., Merrifield, M. A., Becker, J. M., and Thompson, P. R.: Regional influences on reconstructed global mean sea level, Geophys. Res. Lett., 44, 3274–3282, 2017.

Nerem, R. S., Chambers, D. P., Choe, C., and Mitchum, G. T.: "Estimating Mean Sea Level Change from the TOPEX and Jason Altimeter Missions.", Mar. Geod., 33 (sup1), 435–446, https://doi.org/10.1080/01490419.2010.491031, 2010.

Nerem, R. S., Beckley, B. D., Fasullo, J., Hamlington, B. D., Masters, D., and Mitchum, G. T.: Climate Change Driven Accelerated Sea Level Rise Detected In The Altimeter Era, Proc. Natl. Acad. Sci. USA, 115, 2022–2025, https://doi.org/10.1073/pnas.1717312115, 2018.

Nghiem, S., Hall, D., Mote, T., Tedesco, M., Albert, M., Keegan, K., Shuman, C., Digirolamo, N., and Neumann, G.: The extreme melt across the Greenland ice sheet in 2012, Geophys. Res. Lett., 39, L20502, https://doi.org/10.1029/2012GL053611, 2012.

Nobre, P., Malagutti, M., Urbano, D. F., de Almeida, R. A. F., and Giarolla, E.: Amazon Deforestation and Climate Change in a Coupled Model Simulation, J. Climate, 22, 5686–5697, https://doi.org/10.1175/2009jcli2757.1, 2009.

Oki, T. and Kanae, S.: Global hydrological cycles and world water resources, Science, 313, 1068–1072, https://doi.org/10.1126/science.1128845, 2006.

Ozyavas, A., Khan, S. D., and Casey, J. F.: A possible connection of Caspian Sea level fluctuations with meteorological factors and seismicity, Earth Planet Sci. Lett., 299, 150–158, https://doi.org/10.1016/j.epsl.2010.08.030, 2010.

Palanisamy, H., Cazenave, A., Blazquez, A., Döll, P., Caceres, D., and Decharme, B., Land water storage changes over world river basins from GRACE and global hydrological models, LEGOS internal report, August, 2018.

Paul, F., Huggel, C., and Kääb A.: Combining satellite multispectral image data and a digital elevation model for mapping of debris-covered glaciers, Remote Sens. Environ., 89, 510–518, 2004.

Paulson, A., Zhong, S., and Wahr, J.: Inference of mantle viscosity from GRACE and relative sea level data, Geophys. J. Int., 171, 497–508, https://doi.org/10.1111/j.1365-246X.2007.03556.x, 2007.

Peltier, W. R.: Global glacial isostatic adjustment and modern instrumental records of relative sea level history, in: Sea-Level Rise: History and Consequences, edited by: Douglas, B. C., Kearney, M. S., and Leatherman, S. P., Vol. 75, Academic Press, San Diego, 65–95, 2001.

Peltier, W. R.: Global glacial isostasy and the surface of the ice-age Earth: the ICE-5G (VM2) model and GRACE, Annu. Rev. Earth Planet. Sci., 32, 111–149, 2004.

Peltier, W. R.: Closure of the budget of global sea level rise over the GRACE era: the importance and magnitudes of the required corrections for global glacial isostatic adjustment, Quatern. Sci. Rev., 28, 1658–1674, 2009.

Peltier, W. R. and Luthcke, S. B.: On the origins of Earth rotation anomalies: New insights on the basis of both "paleo-geodetic" data and Gravity Recovery and Climate Experiment (GRACE) data, J. Geophys. Res.-Solid Earth, 114, B11405, https://doi.org/10.1029/2009JB006352, 2009.

Peltier, W. R., Argus, D. F., and Drummond, R.: Space geodesy constrains ice age terminal deglaciation: The global ICE-6G_C (VM5a) model, J. Geophys. Res.-Solid Earth, 120, 450–487, 2015.

Perera, J.: A Sea Turns to Dust, New Sci., 140, 24–27, 1993.

Pfeffer, W., Arendt, A., Bliss, A., Bolch, T., Cogley, J., Gardner, A., Hagen, J.-O., Hock, R., Kaser, G., Kienholz, C., Miles, E., Moholdt, G., Mölg, N., Paul, F., Radić, V., Rastner, P., Raup, B., Rich, J., and Sharp, M.: The Randolph Glacier Inventory: a globally complete inventory of glaciers, J. Glaciol., 60, 537–552, 2014.

Plag, H. P. and Juettner, H. U.: Inversion of global tide gauge data for present-day ice load changes, Memoir. Natl. Inst. Polar Res., 54, 301–317, 2001.

Pokhrel, Y. N., Hanasaki, N., Yeh, P. J. F., Yamada, T., Kanae, S., and Oki, T.: Model estimates of sea level change due to anthropogenic impacts on terrestrial water storage, Nat. Geosci., 5, 389–392, https://doi.org/10.1038/ngeo1476, 2012.

Purcell, A. P., Tregoning, P., and Dehecq, A.: An assessment of the ICE6G_C (VM5a) glacial isostatic adjustment model, J. Geophys. Res.-Solid Earth 121, 3939–3950, 2016.

Purkey, S. and Johnson, G. C.: Warming of global abyssal and deep southern ocean waters between the 1990s and 2000s: Contributions to global heat and sea level rise budget, J. Climate, 23, 6336–6351, https://doi.org/10.1175/2010JCLI3682.1, 2010.

Purkey, S. G., Johnson, G. C., and Chambers, D. P.: Relative contributions of ocean mass and deep steric changes to sea level rise between 1993 and 2013, J. Geophys. Res.-Oceans, 119, 7509–7522, https://doi.org/10.1002/2014JC010180, 2014.

Radic, V. and Hock, R.: Regional and global volumes of glaciers derived from statistical upscaling of glacier inventory data, J. Geophys. Res.-Earth Surf., 115, F01010, https://doi.org/10.1029/2009JF001373, 2010.

Ray, R. D. and Douglas, C.: Experiments in reconstructing twentieth-century sea levels, Prog. Oceanogr. 91, 495–515, 2011.

Reager, J. T., Thomas, B. F., and Famiglietti, J. S., River basin flood potential inferred using GRACE gravity observations at several months lead time, Nat. Geosci., 7, 588–592, https://doi.org/10.1038/ngeo2203, 2014.

Reager, J. T., Gardner, A. S., Famiglietti, J. S., Wiese, D. N., Eicker, A., and Lo, M. H.: A decade of sea level rise slowed by climate-driven hydrology, Science, 351, 699–703, https://doi.org/10.1126/science.aad8386, 2016.

Rhein, M. A., Rintoul, S. R., Aoki, S., Campos, E., Chambers, D., Feely, R. A., Gulev, S., Johnson, G. C., Josey, S. A., Kostianoy, A., and Mauritzen, C.: Observations: Ocean, in: Climate Change 2013: The Physical Science Basis. Contribution of Working Group I to the Fifth Assessment Report of the Intergovernmental Panel on Climate Change, edited by: Stocker, T. F., Qin, D., Plattner, G.-K., Tignor, M., Allen, S. K., Boschung, J., Nauels, A., Xia, Y., Bex, V., and Midgley, P. M., Cambridge University Press, Cambridge, United Kingdom and New York, NY, USA, 2013.

Richey, A. S., Thomas, B. F., Lo, M. H., Reager, J. T., Famiglietti, J. S., Voss, K., Swenson, S., and Rodell, M.: Quantifying renewable groundwater stress with GRACE, Water Resour. Res., 51, 5217–5238, https://doi.org/10.1002/2015WR017349, 2015.

Rietbroek, R., Brunnabend, S. E., Kusche, J., Schröter, J., and Dahle, C.: Revisiting the contemporary sea-level budget on global and regional scales, Proc. Natl. Acad. Sci. USA, 113, 1504–1509, https://doi.org/10.1073/pnas.1519132113, 2016.

Rignot, E. J., Velicogna, I., van den Broeke, M. R., Monaghan, A. J., and Lenaerts, J. T. M.: Acceleration of the contribution of the Greenland and Antarctic ice sheets to sea level rise, Geophys. Res. Lett., 38, L05503, https://doi.org/10.1029/2011GL046583, 2011a.

Rignot, E., Mouginot, J., and Scheuchl, B.: Ice flow of the Antarctic Ice Sheet, Science, 333, 1427–1430, https://doi.org/10.1126/science.1208336, 2011b.

Riser, S. C., Freeland, H. J., Roemmich, D., Wijffels, Troisi, S. A., Belbéoch, M., Gilbert, D., Xu, J., Pouliquen, S., Thresher, A., Le Traon, P. Y., Maze, G., Klein, B., M Ravichandran, M., Grant, F., Poulain, P. M., Suga, T., Lim, B., Sterl, A., Sutton, P., Mork, K. A., Vélez-Belchí, P. J., Ansorge, I., King, B., Turton, J., Baringer, M., and Jayne, S. R.: Fifteen years of ocean observations with the global Argo array, Nat. Clim. Change, 6, 145–153, https://doi.org/10.1038/NCLIMATE2872, 2016.

Riva, R. E., Gunter, B. C., Urban, T. J., Vermeersen, B. L., Lindenbergh, R. C., Helsen, M. M., Bamber, J. L., van de Wal, R. S., van den Broeke, M. R. and Schutz, B. E.: Glacial isostatic adjustment over Antarctica from combined ICESat and GRACE satellite data, Earth Planet. Sci. Lett., 288, 516–523, 2009.

Riva, R. E. M., Bamber, J. L., Lavallée, D. A., and Wouters, B.: Sea-level fingerprint of continental water and ice mass change from GRACE, Geophys. Res. Lett., 37, L19605, https://doi.org/10.1029/2010GL044770, 2010.

Rodell, M., Velicogna, I., and Famiglietti, J. S.: Satellite-based estimates of groundwater depletion in India, Nature, 460, 999–1002, https://doi.org/10.1038/nature08238, 2009.

Roemmich, D. and Gilson, J.: The 2004–2008 mean and annual cycle of temperature, salinity, and steric height in the global ocean from the Argo Program, Prog. Oceanogr., 82, 81–100, 2009.

Roemmich, D. and Gilson, J.: The global ocean imprint of ENSO, Geophys. Res. Lett., 38, L13606, https://doi.org/10.1029/2011GL047992, 2011.

Roemmich, D., Gould, W. J., and Gilson, J.: 135 years of global ocean warming between the Challenger expedition and the Argo Programme, Nat. Clim. Change, 2, 425–428, https://doi.org/10.1038/nclimate1461, 2012.

Roemmich, D., Church, J., Gilson, J., Monselesan, D., Sutton, P., and Wijffels, S.: Unabated planetary warming and its ocean structure since 2006, Nat. Clim. Change, 5, 240–245, https://doi.org/10.1038/NCLIMATE2513, 2015.

Roemmich, D., Gilson, J., Sutton, P., and Zilberman, N.: Multidecadal change of the South Pacific gyre circulation, J. Phys. Oceanography, 46, 1871–1883, https://doi.org/10.1175/jpo-d-15-0237.1, 2016.

Sahagian, D.: Global physical effects of anthropogenic hydrological alterations: sea level and water redistribution, Global Planet. Change, 25, 39–48, https://doi.org/10.1016/S0921-8181(00)00020-5, 2000.

Sahagian, D. L., Schwartz, F. W., and D. K. Jacobs, D. K.: Direct anthropogenic contributions to sea level rise in the twentieth century, Nature, 367, 54–57, https://doi.org/10.1038/367054a0, 1994.

Sasgen, I., Van Den Broeke, M., Bamber, J. L., Rignot, E., Sørensen, L. S., Wouters, B., Martinec, Z., Velicogna, I., and Simonsen, S. B.: Timing and origin of recent regional ice-mass loss in Greenland, Earth Planet. Sci. Lett., 333, 293–303, 2012.

Sasgen, I., Konrad, H., Ivins, E. R., Van den Broeke, M. R., Bamber, J. L., Martinec, Z., and Klemann, V.: Antarctic ice-mass balance 2003 to 2012: regional reanalysis of GRACE satellite gravimetry measurements with improved estimate of glacial-isostatic adjustment based on GPS uplift rates, The Cryosphere, 7, 1499–1512, https://doi.org/10.5194/tc-7-1499-2013, 2013.

Sasgen, I., Martín-Español, A., Horvath, A., Klemann, V., Petrie, E. J., Wouters, B., and Konrad, H.: Joint inversion estimate of regional glacial isostatic adjustment in Antarctica considering a lateral varying Earth structure (ESA STSE Project REGINA), Geophys. J. Int., 211, 1534–1553, 2017.

Scanlon, B. R., Jolly, I., Sophocleous, M., and Zhang, L.: Global impacts of conversions from natural to agricultural ecosystems on water resources: Quantity versus quality, Water Resour. Res., 43, W03437, https://doi.org/10.1029/2006WR005486 2007.

Scanlon, B. R., Zhang, Z., Save, H., Sun, A. Y., Schmied, H. M., van Beek, L. P., and Longuevergne, L.: Global models underestimate large decadal declining and rising water storage trends relative to GRACE satellite data, Proc. Natl. Acad. Sci. USA, https://doi.org/10.1073/pnas.1704665115, 2018.

Schellekens, J., Dutra, E., Martínez-de la Torre, A., Balsamo, G., van Dijk, A., Sperna Weiland, F., Minvielle, M., Calvet, J.-C., Decharme, B., Eisner, S., Fink, G., Flörke, M., Peßenteiner, S., van Beek, R., Polcher, J., Beck, H., Orth, R., Calton, B., Burke, S., Dorigo, W., and Weedon, G. P.: A global water resources ensemble of hydrological models: the eartH2Observe Tier-1 dataset, Earth Syst. Sci. Data, 9, 389–413, https://doi.org/10.5194/essd-9-389-2017, 2017.

Schrama, E. J., Wouters, B., and Rietbroek, R.: A mascon approach to assess ice sheet and glacier mass balances and their uncertainties from GRACE data, J. Geophys. Res.-Solid Earth, 119, 6048–6066, 2014.

Schwatke, C., Dettmering, D., Bosch, W., and Seitz, F.: DAHITI – an innovative approach for estimating water level time series over inland waters using multi-mission satellite altimetry, Hydrol. Earth Syst. Sci., 19, 4345–4364, https://doi.org/10.5194/hess-19-4345-2015, 2015.

Shamsudduha, M., Taylor, R. G., and Longuevergne, L.: Monitoring groundwater storage changes in the highly seasonal humid tropics: Validation of GRACE measurements in the Bengal Basin, Water Resour. Res., 48, W02508, https://doi.org/10.1029/2011WR010993, 2012.

Shepherd, A., Ivins, E. R., A, G., Barletta, V. R., Bentley, M. J., Bettadpur, S., Briggs, K. H., Bromwich, D. H., Forsberg, R., Galin, N., Horwath, M., Jacob, S., Joughin, I., King, M. A., Lenaerts, J. T. M., Li, J., Ligtenberg, S. R. M., Luckman, A., Luthcke, S. B., McMillan, M., Meister, R., Milne, G., Mouginot, J., Muir, A., Nicolas, J. P., Paden, J., Payne, A. J., Pritchard, H., Rignot, E., Rott, H., Sandberg Søorensen, L., Scambos, T. A., Scheuchl, B., Schrama, E. J. O., Smith, B., Sundal, A. V., van Angelen, J. H., van de Berg, W. J., van den Broeke, M. R., Vaughan, D. G., Velicogna, I., Wahr, J., Whitehouse, P. L., Wingham, D. J., Yi, D., Young, D., Zwally, H. J.: A reconciled estimate of ice-sheet mass balance, Science, 338, 1183–1189, https://doi.org/10.1126/science.1228102, 2012.

Shukla, J., Nobre, C., and Sellers, P.: Amazon Deforestation and Climate Change, Science, 247, 1322–1325, https://doi.org/10.1126/science.247.4948.1322, 1990.

Slangen, A. B. A., Meyssignac, B., Agosta, C., Champollion, N., Church, J. A., Fettweis, X., Ligtenberg, S. R. M., Marzeion, B., Melet, A., Palmer, M. D., Richter, K., Roberts, C. D., and Spada, G.: Evaluating model simulations of 20th century sea-level rise. Part 1: global mean sea-level change, J. Climate, 30, 8539–8563, https://doi.org/10.1175/jcli-d-17-0110.1, 2017.

Sloan, S. and Sayer, J. A.: Forest Resources Assessment of 2015 shows positive global trends but forest loss and degradation persist in poor tropical countries, Forest Ecol. Manage., 352, 134–145, https://doi.org/10.1016/j.foreco.2015.06.013, 2015.

Solomon, S., Qin, D., Manning, M., Averyt, K., and Marquis, M. (Eds.): Climate Change 2007: The Physical Science Basis. Contribution of Working Group I to the Fourth Assessment Report of the Intergovernmental Panel on Climate Change, Cambridge Univ. Press, Cambridge, UK, 2007.

Spada, G.: Glacial isostatic adjustment and contemporary sea level rise: An overview, Surv. Geophys., 38, 153–185, 2017.

Spada, G. and Galassi, G.: New estimates of secular sea level rise from tide gauge data and GIA modelling, Geophys. J. Int., 191, 1067–1094, 2012.

Spada, G. and Galassi, G.: Spectral analysis of sea level during the altimetry era, and evidence for GIA and glacial melting fingerprints, Global Planet. Change, 143, 34–49, 2016.

Spada, G. and Stocchi, P.: SELEN: A Fortran 90 program for solving the "sea-level equation", Comput. Geosci., 33.4, 538–562, 2007.

Spracklen, D. V., Arnold, S. R., and Taylor, C. M.: Observations of increased tropical rainfall preceded by air passage over forests, Nature, 489, 282–U127, https://doi.org/10.1038/nature11390, 2012.

Stammer, D. and Cazenave, A.: Satellite Altimetry Over Oceans and Land Surfaces, 617 pp., CRC Press, Taylor and Francis Group, Boca Raton, New York, London, ISBN:13:978-1-4987-4345-7, 2018.

Strassberg, G., Scanlon, B. R., and Rodell, M.: Comparison of seasonal terrestrial water storage variations from GRACE with groundwater-level measurements from the High Plains Aquifer (USA), Geophys. Res. Lett., 34, L14402, https://doi.org/10.1029/2007GL030139, 2007.

Sutterley, T. C., Velicogna, I., Csatho, B., van den Broeke, M., Rezvan-Behbahani, S., and Babonis, G.: Evaluating Greenland glacial isostatic adjustment corrections using GRACE, altimetry and surface mass balance data, Environ. Rese. Lett., 9, 014004, https://doi.org/10.1088/1748-9326/9/1/014004, 2014.

Swenson, S., Chambers, D., and Wahr, J.: Estimating geocenter variations from a combination of GRACE and ocean model output, J. Geophys. Res., 113, B08410, https://doi.org/10.1029/2007JB005338, 2008.

Tamisiea, M. E.: Ongoing glacial isostatic contributions to observations of sea level change, Geophys. J. Int., 186, 1036–1044, 2011.

Tamisiea, M. E., Leuliette, E. W., Davis, J. L., and Mitrovica, J. X.: Constraining hydrological and cryospheric mass flux in southeastern Alaska using space-based gravity measurements, Geophys. Res. Lett., 32, L20501, https://doi.org/10.1029/2005GL023961, 2005.

Tapley, B. D., Bettadpur, S., Ries, J. C., Thompson, P. F., and Watkins, M. M. L.: GRACE measurements of mass variability in the Earth system, Science, 305, 503–505, https://doi.org/10.1126/science.1099192, 2004a.

Tapley, B. D., Bettadpur, S., Ries, J. C., Thompson, P. F., and Watkins, M. M.: The Gravity Recovery and Climate Experiment; Mission Overview and Early Results, Geophy. Res. Lett., 31, L09607, https://doi.org/10.1029/2004GL019920, 2004b.

Taylor, R. G., Scanlon, B., Döll, P., Rodell, M., van Beek, R., Wada,Y., Longuevergne, L., LeBlanc, M., Famiglietti, J. S., Edmunds, M., Konikow, L., Green, T. R., Chen, J., Taniguchi, M., Bierkens, M. F. P., MacDonald, A., Fan, Y., Maxwell, R. M., Yechieli, Y., Gurdak, J. J., Allen, D. M., Shamsudduha, M., Hiscock, K., Yeh, P. J. F., Holman, I., and Treidel, H.:Groundwater and climate change, Nat. Clim. Change, 3, 322–329, https://doi.org/10.1038/nclimate1744, 2013.

Thompson, P. R. and Merrifield, M. A.: A unique asymmetry in the pattern of recent sea level change, Geophys. Res. Lett., 41, 7675–7683, 2014.

Tiwari, V. M., Wahr, J., and Swenson, S.: Dwindling groundwater resources in northern India, from satellite gravity observations, Geophys. Res. Lett., 36, L18401, https://doi.org/10.1029/2009GL039401, 2009.

Turcotte, D. L. and Schubert, G.: Geodynamics. Cambridge University Press, Cambridge, 2014, Mar. Geode., 35 (sup1), 42–60, https://doi.org/10.1080/01490419.2012.718226, 2012.

Valladeau, G., Legeais, J. F., Ablain, M., Guinehut, S., and Picot, N.: Comparing Altimetry with Tide Gauges and Argo Profiling Floats for Data Quality Assessment and Mean Sea Level Studies, Mar. Geodesy., 35, 42–60, https://doi.org/10.1080/01490419.2012.718226, 2012.

van den Broeke, M. R., Enderlin, E. M., Howat, I. M., Kuipers Munneke, P., Noël, B. P. Y., van de Berg, W. J., van Meijgaard, E., and Wouters, B.: On the recent contribution of the Greenland ice sheet to sea level change, The Cryosphere, 10, 1933–1946, https://doi.org/10.5194/tc-10-1933-2016, 2016.

van Wessem, J. M., van de Berg, W. J., Noël, B. P. Y., van Meijgaard, E., Amory, C., Birnbaum, G., Jakobs, C. L., Krüger, K., Lenaerts, J. T. M., Lhermitte, S., Ligtenberg, S. R. M., Medley, B., Reijmer, C. H., van Tricht, K., Trusel, L. D., van Ulft, L. H., Wouters, B., Wuite, J., and van den Broeke, M. R.: Modelling the climate and surface mass balance of polar ice sheets using RACMO2 – Part 2: Antarctica (1979–2016), The Cryosphere, 12, 1479–1498, https://doi.org/10.5194/tc-12-1479-2018, 2018.

Velicogna, I.: Increasing rates of ice mass loss from the Greenland and Antarctic ice sheets revealed by GRACE, Geophys. Res. Lett., 36, L19503, https://doi.org/10.1029/2009GL040222, 2009.

Velicogna, I. and Wahr, J.: Measurements of Time-Variable Gravity Show Mass Loss in Antarctica, Science, 311, 1754–1756, https://doi.org/10.1126/science.1123785, 2006.

Velicogna, I., Sutterley, T. C., and Van Den Broeke, M. R.: Regional acceleration in ice mass loss from Greenland and Antarctica using GRACE time-variable gravity data, Geophys. Res. Lett., 41, 8130–8137, 2014.

von Schuckmann, K., Palmer, M. D., Trenberth, K. E., Cazenave, A., Chambers, D., Champollion, N., Hansen, J., Josey, S. A., Loeb, N., Mathieu, P. P., Meyssignac, B., and Wild, M.: Earth's energy imbalance: an imperative for monitoring, Nat. Clim. Change, 26, 138–144, https://doi.org/10.1038/NCLIMATE2876, 2016.

Vörösmarty, C. J. and, Sahagian, D.: Anthropogenic disturbance of the terrestrial water cycle, Biosci., 50, 753–765, https://doi.org/10.1641/0006-3568(2000)050[0753:Adottw]2.0.Co;2, 2000.

Voss, K. A., Famiglietti, J. S., Lo, M., de Linage, C., Rodell, M., and Swenson, S. C.: Groundwater depletion in the Middle East from GRACE with implications for transboundary water management in the Tigris-Euphrates-Western Iran region, Water Resour. Res., 49, 904–914, https://doi.org/10.1002/wrcr.20078, 2013.

Wada, Y., Reager, J. T., Chao, B. F., Wang, J., Lo, M. H., Song, C., and Gardner, A. S.: Modelling groundwater depletion at regional and global scales: Present state and future prospects, Surv. Geophys., 37, 419-451, https://doi.org/10.1007/s10712-015-9347-x, Special Issue: ISSI Workshop on Remote Sensing and Water Resources, 2017.

Wada, Y., van Beek, L. P. H., and Bierkens, M. F. P.: Nonsustainable groundwater sustaining irrigation: A global assessment, Water Resour. Res., 48, W00L06, https://doi.org/10.1029/2011WR010562, Special Issue: Toward Sustainable Groundwater in Agriculture, 2012a.

Wada, Y., van Beek, L. P. H., Sperna Weiland, F. C., Chao, B. F., Wu, Y. H., and Bierkens, M. F. P.: Past and future contribution of global groundwater depletion to sea-level rise, Geophys. Res. Lett., 39, L09402, https://doi.org/10.1029/2012GL051230, 2012b.

Wada, Y., Lo, M. H., Yeh, P. J. F., Reager, J. T., Famiglietti, J. S., Wu, R. J., and Tseng, Y. H.: Fate of water pumped from underground causing sea level rise, Nat. Clim. Change, 6, 777–780, https://doi.org/10.1038/nclimate3001, 2016.

Wahr, J., Nerem, R. S., and Bettadpur, S. V.: The pole tide and its effect on GRACE time-variable gravity measurements: Implications for estimates of surface mass variations, J. Geophys. Res.-Solid Earth, 120, 4597–4615, 2015.

Wang, J., Sheng, Y., Hinkel, K. M., and Lyons, E. A.: Drained thaw lake basin recovery on the western Arctic Coastal Plain of Alaska using high-resolution digital elevation models and remote sensing imagery, Remote Sens. Environ., 119, 325–336, https://doi.org/10.1016/j.rse.2011.10.027, 2012.

Watkins, M. M., Wiese, D. N., Yuan, D.-N., Boening, C., and Landerer, F. W.: Improved methods for observing Earth's time variable mass distribution with GRACE using spherical cap mascons, J. Geophys. Res.-Solid Earth, 120, 2648–2671, https://doi.org/10.1002/2014JB011547, 2015.

Watson, C. S., White, N. J., Church, J. A., King, M. A., Burgette, R. J., and Legresy, B.: Unabated Global Mean Sea-Level Rise over the Satellite Altimeter Era, Nat. Clim. Change, 5, 565–568, https://doi.org/10.1038/nclimate2635, 2015.

Wenzel, M. and Schroter, J.: Reconstruction of regional mean sea level anomalies from tide gauges using neural networks, J. Geophys. Res., 115, C08013, https://doi.org/10.1029/2009JC005630, 2010.

Whitehouse, P. L., Bentley, M. J., Milne, G. A., King, M. A., and Thomas, I. D.: A new glacial isostatic adjustment model for Antarctica: calibrating the deglacial model using observations of relative sea-level and present-day uplift rates, Geophys. J. Int., 190, 1464–1482, 2012.

Wiese, D. N., Landerer, F. W., and Watkins, M. M.: Quantifying and reducing leakage errors in the JPL RL05M GRACE mascon solution, Water Resour. Res., 52, 7490–7502, https://doi.org/10.1002/2016WR019344, 2016a.

Wiese, D., Yuan, D., Boening, C., Landerer, F., and Watkins, M.: JPL GRACE Mascon Ocean, Ice, and Hydrology Equivalent Water Height RL05M. 1 CRI Filtered, Ver. 2, PO. DAAC, CA, USA. Dataset provided by Wiese in Nov/Dec 2017, 2016b.

Wijffels, S. E., Roemmich, D., Monselesan, D., Church, J., and Gilson, J.: Ocean temperatures chronicle the ongoing warming of Earth, Nat. Clim. Change, 6, 116–118, https://doi.org/10.1038/nclimate2924, 2016.

Willis, J. K., Chambers, D. T., and Nerem, R. S.: Assessing the globally averaged sea level budget on seasonal to interannual time scales, J. Geophys. Res., 113, C06015, https://doi.org/10.1029/2007JC004517, 2008.

Wöppelmann, G. and Marcos, M.: Vertical land motion as a key to understanding sea level change and variability, Rev. Geophys., 54, 64–92, https://doi.org/10.1002/2015RG000502, 2016.

Wouters, B., Chambers, D., and Schrama, E.: GRACE observes small-scale mass loss in Greenland, Geophys. Res. Lett., 35, L20501, https://doi.org/10.1029/2008GL034816, 2008.

Wouters, B., Bamber, J. Á., Van den Broeke, M. R., Lenaerts, J. T. M., and Sasgen, I.: Limits in detecting acceleration of ice sheet mass loss due to climate variability, Nat. Geosci., 6, 613–616, 2013.

Wu, X., Heflin, M. B., Schotman, H., Vermeersen, B. L., Dong, D., Gross, R. S., Ivins, E. I., Moore, A. W., and Owen, S.: Simultaneous estimation of global present-day water transport and glacial isostatic adjustment, Nat. Geosci., 39, 642–646, 2010.

Yi, S., Sun, W., Heki, K., and Qian, A., An increase in the rate of global mean sea level rise since 2010. Geophys. Res. Let., 42, 3998–4006, https://doi.org/10.1002/2015GL063902, 2015.

Zawadzki, L. and Ablain, M., Estimating a drift in TOPEX-A Global Mean Sea Level using Poseidon-1 measurements, paper presented at the OSTST meeting, October 2016, La Rochelle, 2016.

Zemp, M., Frey, H., Gärtner-Roer, I., Nussbaumer, S., Hoelzle, M., Paul, F., Haeberli, W., Denzinger, F., Ahlström, A., Anderson, B., Bajracharya, S., Baroni, C., Braun, L., Cáceres, B., Casassa, G., Cobos, G., Dávila, L., Delgado Granados, H., Demuth, M., Espizua, L., Fischer, A., Fujita, K., Gadek, B., Ghazanfar, A., Hagen, J., Holmlund, P., Karimi, N., Li, Z., Pelto, M., Pitte, P., Popovnin, V., Portocarrero, C., Prinz, R., Sangewar, C., Severskiy, I., Sigurdsson, O., Soruco, A., Usubaliev, R., and Vincent, C., Historically unprecedented global glacier decline in the early 21st century, J. Glaciol., 61, 745–762, 2015.

Zwally, J. H., Li, J., Robbins, J. W., Saba, J. L., Yi, D. H., and Brenner, A. C.: Mass gains of the Antarctic ice sheet exceed losses, J. Glaciol., 61, 1013–1036, https://doi.org/10.3189/2015JoG15J071, 2016.

Team list. Anny Cazenave (LEGOS, France, and ISSI, Switzerland), Benoit Meyssignac (LEGOS, France), Michael Ablain (CLS, France), Magdalena Balmaseda (ECMWF, UK), Jonathan Bamber (U. Bristol, UK), Valentina Barletta (DTU-SPACE, Denmark), Brian Beckley (SGT Inc./NASA GSFC, USA), Jérôme Benveniste (ESA/ESRIN, Italy), Etienne Berthier (LEGOS, France), Alejandro Blazquez (LEGOS, France), Tim Boyer (NOAA, USA), Denise Caceres (Goethe U., Germany), Don Chambers (U. South Florida, USA), Nicolas Champollion (U. Bremen, Germany), Ben Chao (IES-AS, Taiwan), Jianli Chen (U. Texas, USA), Lijing Cheng (IAP-CAS, China), John A. Church (U. New South Wales, Australia), Stephen Chuter (U. Bristol, UK), J. Graham Cogley (Trent U., Canada), Soenke Dangendorf (U. Siegen, Germany), Damien Desbruyères (IFREMER, France), Petra Döll (Goethe U., Germany), Catia Domingues (CSIRO, Australia), Ulrike Falk (U. Bremen, Germany), James Famiglietti (JPL/Caltech, USA), Luciana Fenoglio-Marc (U. Bonn, Germany), Rene Forsberg (DTU-SPACE, Denmark), Gaia Galassi (U. Urbino, Italy), Alex Gardner (JPL/Caltech, USA), Andreas Groh (TU-Dresden, Germany), Benjamin Hamlington (Old Dominion U., USA), Anna Hogg (U. Leeds, UK), Martin Horwath (TU-Dresden, Germany), Vincent Humphrey (ETHZ, Switzerland), Laurent Husson (U. Grenoble, France), Masayoshi Ishii (MRI-JMA, Japan), Adrian Jaeggi (U. Bern, Switzerland), Svetlana Jevrejeva (NOC, UK), Gregory Johnson (NOAA/PMEL, USA), Nicolas Kolodziejczyk (LOPS, France), Jürgen Kusche (U. Bonn, Germany), Kurt Lambeck (ANU, Australia, and ISSI, Switzerland), Felix Landerer (JPL/Caltech, USA), Paul Leclercq (UIO, Norway), Benoit Legresy (CSIRO, Australia), Eric Leuliette (NOAA, USA), William Llovel (LEGOS, France), Laurent Longuevergne (U. Rennes, France), Bryant D. Loomis (NASA GSFC, USA), Scott B. Luthcke (NASA GSFC, USA), Marta Marcos (UIB, Spain), Ben Marzeion (U. Bremen, Germany), Chris Merchant (U. Reading, UK), Mark Merrifield (UCSD, USA), Glenn Milne (U. Ottawa, Canada), Gary Mitchum (U. South Florida, USA), Yara Mohajerani (UCI, USA), Maeva Monier (Mercator-Ocean, France), Didier Monselesan (CSIRO, Australia), Steve Nerem (U. Colorado, USA), Hindumathi Palanisamy (LEGOS, France), Frank Paul (UZH, Switzerland), Begoña Perez (Puertos del Estados, Spain), Christopher G. Piecuch (WHOI, USA), Rui M. Ponte (AER inc., USA), Sarah G. Purkey (SIO/UCSD, USA), John T. Reager (JPL/Caltech, USA), Roelof Rietbroek (U. Bonn, Germany), Eric Rignot (UCI and JPL, USA), Riccardo Riva (TU Delft, The Netherlands), Dean H. Roemmich (SIO/UCSD USA), Louise Sandberg Sørensen (DTU-SPACE, Denmark), Ingo Sasgen (AWI, Germany), E.J.O. Schrama (TU Delft, The Netherlands), Sonia I. Seneviratne (ETHZ, Switzerland), C.K. Shum (Ohio State U., USA), Giorgio Spada (U. Urbino, Italy), Detlef Stammer (U. Hamburg, Germany), Roderic van de Wal (U. Utrecht, The Netherlands), Isabella Velicogna (UCI and JPL, USA), Karina von Schuckmann (Mercator-Océan, France), Yoshihide Wada (U. Utrecht, The Netherlands), Yiguo Wang (NERSC/BCCR, Norway), Christopher Watson (U. Tasmania, Australia), David Wiese (JPL/Caltech, USA), Susan Wijffels (CSIRO, Australia), Richard Westaway (U. Bristol, UK), Guy Woppelmann (U. La Rochelle, France), Bert Wouters (U. Utrecht, The Netherlands).

3

A 14-year dataset of in situ glacier surface velocities for a tidewater and a land-terminating glacier in Livingston Island, Antarctica

Francisco Machío[1], Ricardo Rodríguez-Cielos[2], Francisco Navarro[3], Javier Lapazaran[3], and Jaime Otero[3]

[1]Escuela Superior de Ingeniería y Tecnología, Universidad Internacional de La Rioja (UNIR), Calle Almansa, 101, 28040 Madrid, Spain
[2]Departamento de Señales, Sistemas y Radiocomunicaciones, ETSI de Telecomunicación, Universidad Politécnica de Madrid, Av. Complutense, 30, 20040 Madrid, Spain
[3]Departamento de Matemática Aplicada a las Tecnologías de la Información y las Comunicaciones, ETSI de Telecomunicación, Universidad Politécnica de Madrid, Av. Complutense, 30, 20040 Madrid, Spain

Correspondence to: Francisco Machío (francisco.machio@unir.net)

Abstract. We present a 14-year record of in situ glacier surface velocities determined by repeated global navigation satellite system (GNSS) measurements in a dense network of 52 stakes distributed across two glaciers, Johnsons (tidewater) and Hurd (land-terminating), located on Livingston Island, South Shetland Islands, Antarctica. The measurements cover the time period 2000–2013 and were collected at the beginning and end of each austral summer season. A second-degree polynomial approximation is fitted to each stake position, which allows estimating the approximate positions and associated velocities at intermediate times. This dataset is useful as input data for numerical models of glacier dynamics or for the calibration and validation of remotely sensed velocities for a region where very scarce in situ glacier surface velocity measurements have been available so far.

1 Introduction

In situ measured glacier surface velocities are an important source of information for the study of glacier dynamics. The strain field is defined in terms of velocity gradients, and the stresses are defined in terms of strains through the constitutive relationship (most often Nye's generalization of Glen's flaw law; e.g. Cuffey and Paterson, 2010, Sect. 3). The velocity field gradients are thus indicative of observed deformation patterns such as folding or foliation, and damage expressions such as fracturing, faulting, and crevassing (Hambrey and Lawson, 2000; Ximenis et al., 2000). Furthermore, observed surface velocities can give an insight into basal conditions. In particular, they have been used for a long time to infer basal drag (e.g. van der Veen and Whillans, 1989; Hooke et al., 1989).

Observed surface velocities are commonly used as input data for numerical models. In theory, they could be directly used as Dirichlet boundary conditions at the glacier surface for the velocity field. However, the usual practice is to impose a traction-free boundary (i.e. Neumann conditions) at the glacier surface, and the velocities are used instead for tuning the model's free parameters, such as the viscosity coefficient (ice hardness) in the constitutive relationship or the basal friction coefficient in the sliding law. Some models have treated these coefficients as constant in space (e.g. Hanson, 1995; Martín et al., 2004; Otero et al., 2010). Recently, it is becoming more and more common to establish the viscosity and/or the basal friction coefficients as functions of position. This is done by means of inversion procedures that heavily rely on observed velocities at the glacier surface. For instance, in the method introduced by Arthern and Gudmundsson (2010) and modified by Jay-Allemand et al. (2011), the surface velocities are used to solve the Dirichlet problem involved in the inverse Robin problem solving for the viscosity or basal friction coefficients. However, these inversion procedures require a large amount of measured velocities, which are seldom available from in situ measurements and thus require the use of remotely sensed velocities, such as differential interferometric synthetic-aperture radar (D-InSAR), SAR offset tracking, or SAR coherence tracking velocities (e.g. Strozzi et al., 2002; Rignot and Kanagaratnam, 2006; Joughin et al., 2010; Wuite et al., 2015). But even in these cases in situ measured glacier velocities are still of wide interest, since they provide a means for the calibration and validation of remotely sensed velocities (e.g. Strozzi et al., 2008; Schellenberger et al., 2015). This is of interest in view of the recent efforts to derive time series for regional or global ice-velocity fields such as those involved in the MEaSUREs program (https://nsidc.org/data/nsidc-0484/versions/2, accessed on 7 May 2017), the GoLIVE project (https://nsidc.org/data/golive, accessed on 7 May 2017), and the ENVEO CryoPortal (http://cryoportal.enveo.at/, accessed on 7 May 2017).

In this paper, we present a 14-year record of in situ glacier surface velocities determined by repeated global navigation satellite system (GNSS) measurements in a dense network of stakes on two glaciers, Johnsons and Hurd, located on Livingston Island, South Shetland Islands (Fig. 1). These islands, located off the north-western tip of the Antarctic Peninsula, previously had a scarce record of in situ velocity observations, which included measurements in the late 1980s on Nelson Island (Ren Jiaven et al., 1995), earlier measurements in the late 1990s on Johnsons Glacier (Ximenis et al., 1999), and measurements in the Arctowski Icefield, the Bellingshausen Dome, and the Central Dome of King George Island between 1999/2000 and 2008/09 (Blindow et al., 2010; Rückamp et al., 2010, 2011). Such in situ velocity measurements are critical for the validation of the estimates of remote-sensor-based studies of ice discharge in the region such as those by Osmanoğlu et al. (2013, 2014) for King George and Livingston islands (the present dataset has in fact been used in the latter paper with such purposes), as well as for tuning free parameters of glacier dynamics models, as done by Martín et al. (2004) and Otero et al. (2010) using an earlier (and shorter) version of the dataset presented. An added interest of the presented velocity record is that it corresponds to both a tidewater glacier and a land-terminating glacier, two glacier types that are typical in this region but very different in dynamical behaviour.

2 Geographical setting

Our study area is Hurd Peninsula ($62°39$–$42'$ S, $60°19$–$25'$ W), located in the south of Livingston Island, South Shetland Archipelago, Antarctica. This peninsula is the setting of Juan Carlos I Station (JCI), which provided the logistic sup-

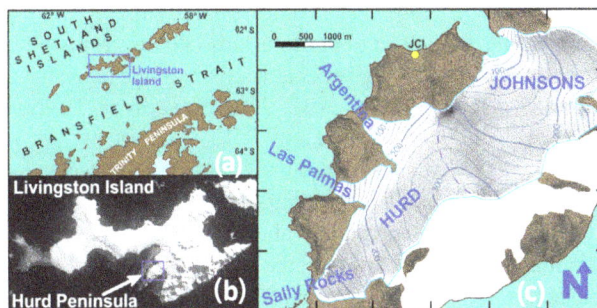

Figure 1. (a) Location of Livingston Island in the South Shetland Archipelago. (b) Location of Hurd Peninsula on Livingston Island (orthophoto generated from SPOT 1991 image by Universitat de Barcelona and Institut Cartogràfic de Catalunya, 1992). (c) Location and surface elevation map of Hurd and Johnsons glaciers, Hurd Peninsula, Livingston Island. The dashed blue line indicates the ice divide separating Hurd and Johnsons glaciers. Elevations and outline are based on a survey during summer 1998/99 and 2000/01. The yellow dot shows the position of Juan Carlos I Station (JCI).

port for our fieldwork (Fig. 1). Hurd Peninsula is covered by an ice cap that extends over an area of about $13.5 \, \text{km}^2$ and spans an altitude range from sea level to about 370 m a.s.l. It is partly surrounded by mountains ranging from 250 to 400 m in height.

This ice cap can be divided into two main glacier systems. The first main unit is Johnsons Glacier, a tidewater glacier, mostly flowing north-westwards, terminating at a 50 m height calving front of which just a few metres (typically < 10 m) are submerged. This calving front extends approximately 500–600 m along the coast. The second main unit is Hurd glacier, flowing mostly south-westwards and terminating on land, with three main lobes, named Sally Rocks (flowing south-westwards), Las Palmas (flowing westwards) and Argentina (flowing north-westwards). There are three additional smaller basins, all flowing eastwards to False Bay, which were excluded from this study because they contain heavily crevassed icefalls which prevent safe field measurements.

The local ice divide separating Johnsons and Hurd lies between 250 and 330 m a.s.l. (Fig. 1c). Hurd Glacier has an average surface slope of about $3°$, though the small westward flowing glacier tongues Argentina and Las Palmas are steeper, around $13°$. Typical surface slopes for Johnsons Glacier range between $10°$ in its northern areas and $6°$ in the southern ones.

The Hurd Peninsula ice cap is a polythermal ice mass, showing an upper layer of cold ice, several tens of metres thick, in the ablation zone. This layer is uniformly distributed in Hurd Glacier and shows a more irregular distribution for Johnsons Glacier (Navarro et al., 2009). In the snouts of Hurd Glacier (in Sally Rocks area) and its side lobes Argentina and Las Palmas, where the glacier thickness tapers to zero, the cold ice layer extends down to bedrock, so the glacier is frozen to the bed, implying a compressional stress regime. In contrast, the area close to the Johnsons calving front shows the extensional stress regime characteristic of the terminus of tidewater glaciers (Molina et al., 2007; Navarro et al., 2009; Otero et al., 2010).

The average ice thickness of the joint Hurd–Johnsons, determined from ground-penetrating radar data in 2000/01, was 93.6 ± 2.5 m, with maximum values about 200 m in the accumulation area of Hurd Glacier and only about 160 m in Johnsons Glacier (Navarro et al., 2009). The bed of Johnsons Glacier is smooth, with altitudes decreasing towards the ice front, where glacier bed elevation is slightly below sea level (typically < 10 m). The Hurd Glacier bed, however, is more irregular, with a clear overdeepening in the thickest ice area, close to the head of Argentina side lobe, and another smaller one near the head of Las Palmas side lobe.

The Hurd Peninsula ice cap is subjected to the maritime climate of the western Antarctic Peninsula (AP) region. The annual average temperature at JCI during the period 1994–2014 was $-1.2\,°C$, with average summer (December–January–February) and winter (June–July–August) temperatures of 1.9 and $-4.7\,°C$, respectively (Bañón and Vasallo, 2016). Summer, winter, and annual mass balances have been measured using the glaciological method on the same network of stakes used for the glacier velocity measurements and then integrated to the entire glacier basins. The mean surface mass balances over the period 2002–2011 have not been significantly different from zero for either glacier: 0.05 ± 0.30 m w.e. for Johnsons and -0.15 ± 0.44 m w.e. for Hurd. The ranges indicate the SDs, showing that the mass balances have a noticeable interannual variability. The estimated errors of the annual mass balances are lower (approximately ± 0.1 m w.e.). The slightly more negative balance for Hurd Glacier is due to its lower accumulation rates, attributed to snow redistribution by wind, together with higher ablation rates due to Hurd's hypsometry, which shows a much larger share of area at the lowermost altitudes (< 100 m) as compared with Johnsons (Navarro et al., 2013). The average accumulation area ratios over the same period were $44 \pm 24\,\%$ for Hurd Glacier and $61 \pm 21\,\%$ for Johnsons Glacier (in mean and SD). Their equilibrium line altitudes (ELA) for the same period were 228 ± 57 and 187 ± 37 m a.s.l., respectively (Navarro et al., 2013).

3 Methods

The glacier surface velocities were estimated based on repeated differential GNSS measurements in a network of stakes deployed by the authors on Johnsons and Hurd glaciers. The network consisted (at the end of 2013) of 22 stakes for Johnsons and 30 stakes for Hurd Glacier (Fig. 2). The location of the stakes was chosen to provide a wide spatial coverage of the glacier basins and their accumulation and ablation zones. Moreover, several sets of stakes

were installed along predefined glacier flow lines in order to facilitate possible glacier dynamics modelling studies. Ease of access for stake measurements was also taken into consideration. Over the 14-year time period, some of the stakes have been lost (e.g. by iceberg calving at Johnsons Glacier front, because they have fallen due to intensive melting, or because they have been buried by heavy snowfalls), and new ones have been added. Because of this, there are differences in the set of stakes shown in the various figures in this paper, as they correspond to different snapshots in time. Also, the set of stakes included in the PANGAEA database (see Sect. 4) is larger than that in any of the figures because it includes all of the stakes that have existed at any time within the complete measurement period.

The stakes were surveyed two to four times per summer campaign during the period 2000–2013. Measurements are restricted to the summer season because Juan Carlos I Station operates only during the austral summer. At least one measurement at the beginning and another at the end of each summer season have been performed. In this way, we are able to compute not only annual-averaged velocities but also summer velocities and "extended winter" (all year excluding the summer) velocities. In some cases additional measurements during the summer provide temporal velocity variations during summer. The GNSS measurements were carried out using a Trimble 5700 system, with data controller TSC2. The measurements were performed either in real-time kinematics (RTKs) or in fast-static (post-processed) mode; for the former, an occupation time of 10 s was set, and for the latter it was 3–5 min depending on the number of satellites available. In general, RTK mode was used, but in some cases a radio link to the base station was not available and fast-static mode was employed. The GNSS base station was located at Juan Carlos I Station (Fig. 1) at a distance of 2–4 km from each stake measurement point. The base station Juan Carlos I is a permanent GNSS station with coordinates determined with an accuracy better than 0.007 m in horizontal and 0.012 m in vertical directions (Ramírez-Rodríguez, 2007). The estimated horizontal accuracy for the stake positions lies between 0.07 and 0.60 m. The main contributor to this uncertainty is not the GNSS measurement error (which has average values of 0.07 and 0.10 m for horizontal and vertical positioning, respectively) but the estimated uncertainties in the correction for tilt of the stakes. The correction of the stake positions for tilt requires us to measure the tilt angle and the azimuth of the tilt. Throughout the 14-year measurement period, two different ways to correct for tilt were employed. One of them consisted in measuring the tilt using a clinometer and the azimuth using a compass and then making the corresponding geometric corrections. However, these measurements are especially difficult and can imply large uncertainties in the case of large tilts of stakes buried under snow. Moreover, the azimuth reading has to be corrected by magnetic declination and by a grid convergence factor, both of which involve uncertainties. A second method used was

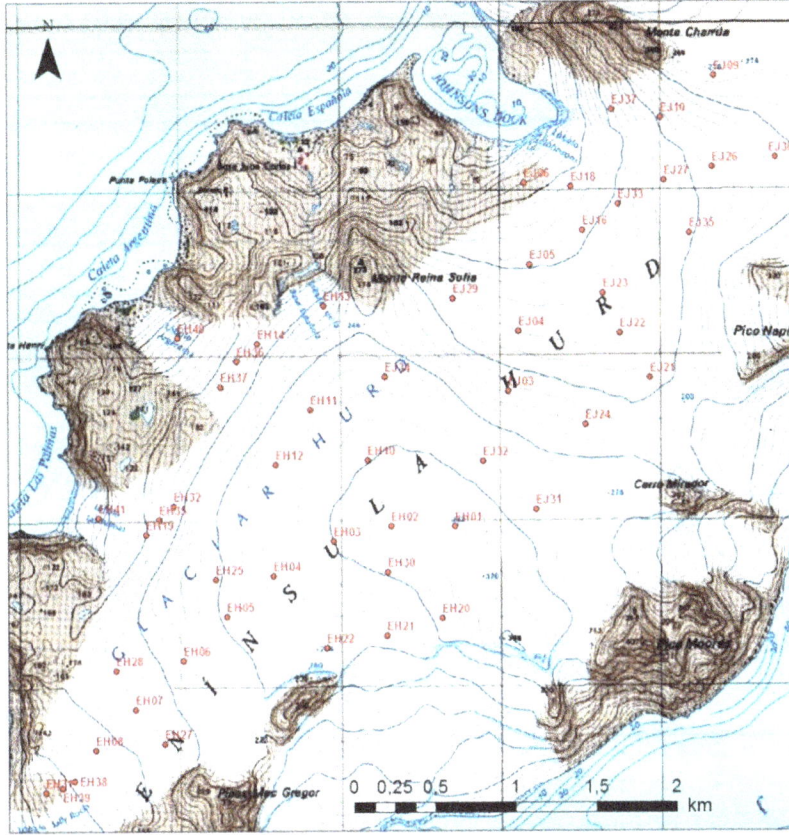

Figure 2. Network of stakes on Hurd and Johnsons glaciers at the end of the 2012–2013 Antarctic summer campaign. (Base map: SGE, 1991.)

to measure two different points on the stake and to compute tilt and azimuth from their coordinates. However, for stakes tilted and deeply buried under snow these two points are usually close to each other, which implies larger errors. The estimated coordinates of the stakes were projected into the UTM system for Zone 20S.

From the collected positions of the stakes at discrete times, the stake positions at any time can be estimated by applying the procedure described below. We will just focus on horizontal velocities, since the vertical component of the velocity is very small and prone to errors such as those of the tilt of the stake. From the known position (x_{t_i}, y_{t_i}) of a stake at a given time t_i (expressed in days since 1 January 1999 at 00:00), with the subscript i indicating the sequential number of the observation (from $i = 1$ to $i = n$), we define the planimetric position of a stake over time (i.e. its trajectory) by the discrete functions

$$X(t_i) = X(x_{t_1}, x_{t_2}, \ldots, x_{t_n}),$$ (1)
$$Y(t_i) = Y(y_{t_1}, y_{t_2}, \ldots, y_{t_n}).$$

We approximate the stake positions with a second-degree polynomial, which is equivalent to assuming that the stake moves with constant acceleration:

$$X_a(t_i) = a_x t_i^2 + b_x t_i + c_x,$$ (2)
$$Y_a(t_i) = a_y t_i^2 + b_y t_i + c_y.$$

The unknown coefficients are determined by the least-square fitting method, minimizing the residual vectors

$$
\boldsymbol{R}_x =
\begin{bmatrix}
t_1^2 & t_1 & 1 \\
t_2^2 & t_2 & 1 \\
\ldots & \ldots & \ldots \\
t_n^2 & t_n & 1
\end{bmatrix}
\begin{bmatrix}
a_x \\
b_x \\
c_x
\end{bmatrix}
-
\begin{bmatrix}
X(t_1) \\
X(t_2) \\
\ldots \\
X(t_n)
\end{bmatrix}
$$ (3)
$$= \boldsymbol{A}\boldsymbol{C}_x - \boldsymbol{X},$$

$$
\boldsymbol{R}_y =
\begin{bmatrix}
t_1^2 & t_1 & 1 \\
t_2^2 & t_2 & 1 \\
\ldots & \ldots & \ldots \\
t_n^2 & t_n & 1
\end{bmatrix}
\begin{bmatrix}
a_y \\
b_y \\
c_y
\end{bmatrix}
-
\begin{bmatrix}
Y(t_1) \\
Y(t_2) \\
\ldots \\
Y(t_n)
\end{bmatrix}
$$
$$= \boldsymbol{A}\boldsymbol{C}_y - \boldsymbol{Y}.$$

Minimization of $\boldsymbol{R}_x (\boldsymbol{R}_y)$ yields the coefficients for $X_a(Y_a)$ describing the stake positions over time. From the time derivatives of the positions, the horizontal velocity of a stake will be given by the expressions

$$v = v_x i + v_y j,$$
$$v_x = X_a{'}(t_i) = 2a_x t_i + b_x,$$
$$v_y = Y_a{'}(t_i) = 2a_y t_i + b_y,$$
$$\|v\| = \sqrt{v_x^2 + v_y^2}. \tag{4}$$

To obtain the error estimates e_x and e_y of the adjusted functions $X_a(t_i)$ and $Y_a(t_i)$ (from which we calculate the error in horizontal positioning as $e_{xy} = \sqrt{e_x^2 + e_y^2}$), we follow the parametric adjustment procedure (Ghilani, 2010), which has to be applied separately for X and Y (for brevity, we just describe it below for X). For a least square approximation, assuming observations of equal weight, these equations are

$$C_x = \left[A^{\mathrm{T}} A\right]^{-1} \left[A^{\mathrm{T}} X\right], \tag{5}$$
$$R_x = A C_x - X,$$
$$\widetilde{X} = X + R_x, s$$
$$e_x = \sqrt{\frac{R_x^{\mathrm{T}} R_x}{r}},$$

where X is the vector of observations, $\widetilde{X}$ is the vector of estimates, A is the matrix of coefficients, R_x is the vector of residuals, C_x is the vector of unknowns (the coefficients in the polynomial adjustment), e_x^2 is the reference variance, and r is the number of degrees of freedom ($r = n - 3$, with n the number of observations).

The above equations are solved for each stake trajectory. C_x is solved first to determine the coefficients of the second-degree polynomial adjustment. Then, the adjusted values $A C_x$ are calculated and the residuals R_x computed, and finally the error in position e_x is calculated. The process is repeated for the corresponding equations in the y direction.

We note that the above error estimates do not represent actual errors in the data points but the SDs of the data point positions with respect to their corresponding values (for the same time t) in the polynomial approximation defined by Eq. (2).

The velocity error for a given stake between two particular positions $x_i = (x_i, y_i)$, $x_{i+1} = (x_{i+1} y_{i+1})$, with positioning errors e_{x_i}, $e_{x_{i+1}}$, respectively, separated by a time interval $\Delta t = t_{i+1} - t_i$ (i.e. the error in $v_i = \frac{x_{i+1} - x_i}{\Delta t}$) is given by

$$e_{v_i} = \frac{1}{\Delta t} \sqrt{e_{x_i}^2 + e_{x_{i+1}}^2}. \tag{6}$$

The error estimate resulting from the polynomial approximation is given by the root-mean-square deviation of observed and approximated values:

$$e_{v_x} = \sqrt{\frac{1}{N} \sum_{i=1}^{N} \left(v_{x_i}^{\mathrm{obs}} - v_{x_i}^{\mathrm{pol}}\right)^2}, \tag{7}$$
$$e_{v_y} = \sqrt{\frac{1}{N} \sum_{i=1}^{N} \left(v_{y_i}^{\mathrm{obs}} - v_{y_i}^{\mathrm{pol}}\right)^2},$$
$$\|e_v\| = \sqrt{e_{v_x}^2 + e_{v_y}^2},$$

where $v_i^{\mathrm{obs}} = \frac{x_{i+1} - x_i}{\Delta t} = (v_{x_i}^{\mathrm{obs}} v_{y_i}^{\mathrm{obs}})$ is the average observed velocities calculated for each time interval and $v_i^{\mathrm{pol}} = (v_{x_i}^{\mathrm{pol}} v_{y_i}^{\mathrm{pol}})$ is the corresponding velocities calculated using Eq. (4) for time $\frac{t_i + t_{i+1}}{2}$, and N represents the number of velocity intervals ($N = n - 1$, with n the number of stake position observations). Note that the value given by Eq. (7) is a single value representing the average error for each polynomial approximation (i.e. a single error value for each stake), while the errors given by Eq. (6) are interval velocity errors between two consecutive positions of a given stake. The errors for the interval velocities are naturally much higher because they do not contain the smoothing from the polynomial approximation.

4   Results

The procedure described above was applied to every stake that has existed for any subperiod (perhaps the entire period) within the complete measurement period 2000–2013. The data are available at the PANGAEA database (http://doi.pangaea.de/10.1594/PANGAEA.846791) and are further described in Appendix A.

The results for the coefficients of the polynomial adjustments for the stake positions and the estimated horizontal positioning misfits for each stake are given in Table B1 of Appendix B. To illustrate the order of magnitude of the velocities and their associated errors, as well as their spatial variations, we have included in Table 1 the calculated values for 13 February 2013.

As an example, the detailed results for a particular stake, EJ14, are shown in Table 2 and Fig. 3. The latter shows the position changes of the stake over time.

In Figs. 4 and 5 we show the horizontal velocities (absolute values and directions) for all stakes of Hurd and Johnsons glaciers, respectively, for a given date (13 February 2013), calculated using the corresponding polynomial adjustments. We also show the corresponding contour lines of the absolute value of the velocities for the same date, calculated from the spatial interpolation of the velocity vector field. Maximum velocities on Hurd Glacier are only of a few metres per year, and approach $10\,\mathrm{m\,yr}^{-1}$ at the head of the unnamed glacier draining towards the south. Maximum velocities on Johnsons Glacier are much larger, up to several tens of metres per year, and reached $65\,\mathrm{m\,yr}^{-1}$ near the calving front. The location of

Table 1. Horizontal velocities for Hurd and Johnsons glacier stakes on 13 February 2013, calculated using the first-degree polynomial for velocity derived from the second-degree polynomial adjustment for stake positions. From left to right: stake name, X and Y components of horizontal velocity (v_x, v_y), absolute value of horizontal velocity ($\|v\|$), azimuth of horizontal velocity vector (θ), and error estimate for the horizontal velocity ($\|e_v\|$), calculated using the polynomial fit.

Stake	v_x (m yr^{-1})	v_y (m yr^{-1})	$\|v\|$ (m yr^{-1})	θ (°)	$\|e_v\|$ (m yr^{-1})
EH01	−0.4	−0.6	0.7	210	±0.4
EH02	−0.8	0.0	0.8	269	±0.5
EH03	−0.9	−0.3	1.0	249	±0.4
EH04	−1.7	−0.4	1.8	255	±0.3
EH05	−1.9	−0.8	2.1	247	±0.2
EH06	−2.9	−1.1	3.1	250	±0.3
EH07	−2.8	−1.8	3.4	238	±0.7
EH08	−2.3	−1.9	3.0	230	±0.3
EH10	−0.9	1.0	1.3	319	±0.2
EH11	−1.2	1.4	1.9	319	±0.2
EH12	−1.0	0.5	1.2	297	±1.6
EH13	−1.1	1.5	1.9	324	±0.2
EH14	−0.9	0.3	1.0	290	±0.8
EH19	−1.5	0.5	1.6	288	±0.3
EH20	−1.5	−0.6	1.6	249	±0.6
EH21	−2.3	−9.8	10.0	193	±0.6
EH22	−0.6	−0.8	1.0	216	±0.7
EH25	−2.1	−0.1	2.1	268	±0.2
EH27	−2.5	−1.3	2.8	242	±0.2
EH28	−1.3	−0.6	1.4	246	±0.4
EH30	−1.9	−2.8	3.4	213	±4.9
EH31	−1.1	−0.6	1.3	240	±0.9
EH32	−0.5	0.3	0.6	300	±1.0
EH35	−2.2	0.8	2.3	290	±0.5
EH36	−1.7	1.8	2.5	318	±1.1
EH37	−2.3	1.6	2.8	305	±1.4
EH38	−2.2	−1.9	2.9	229	±0.7
EH39	−2.2	−1.6	2.7	235	±0.4
EH40	−2.2	2.1	3.0	313	±0.3
EH41	−0.7	−0.7	1.0	223	±1.7
EJ03	2.4	6.6	7.0	20	±0.7
EJ04	0.9	7.3	7.3	7	±1.0
EJ05	0.6	10.9	11.0	3	±1.9
EJ06	−5.7	23.5	24.2	346	±0.5
EJ09	0.0	0.0	0.0	135	±0.3
EJ10	−4.1	−2.0	4.6	245	±1.1
EJ16	−7.4	14.0	15.8	332	±2.3
EJ18	−22.6	29.3	37.0	322	±0.2
EJ21	−0.6	1.0	1.2	328	±0.4
EJ22	−1.4	3.6	3.8	339	±0.6
EJ23	−1.8	6.5	6.8	344	±0.3
EJ24	1.0	5.2	5.3	11	±0.6
EJ26	−7.2	−2.7	7.7	250	±1.2
EJ27	−13.5	−2.4	13.7	260	±0.3
EJ29	3.5	3.2	4.8	47	±0.1
EJ30	−1.9	−1.6	2.4	230	±0.2
EJ31	1.4	2.7	3.1	26	±0.4
EJ32	2.0	2.2	3.0	41	±1.8
EJ33	−14.2	5.6	15.3	291	±0.3
EJ34	0.2	1.6	1.6	7	±7.1
EJ35	−6.3	−0.4	6.3	266	±0.4

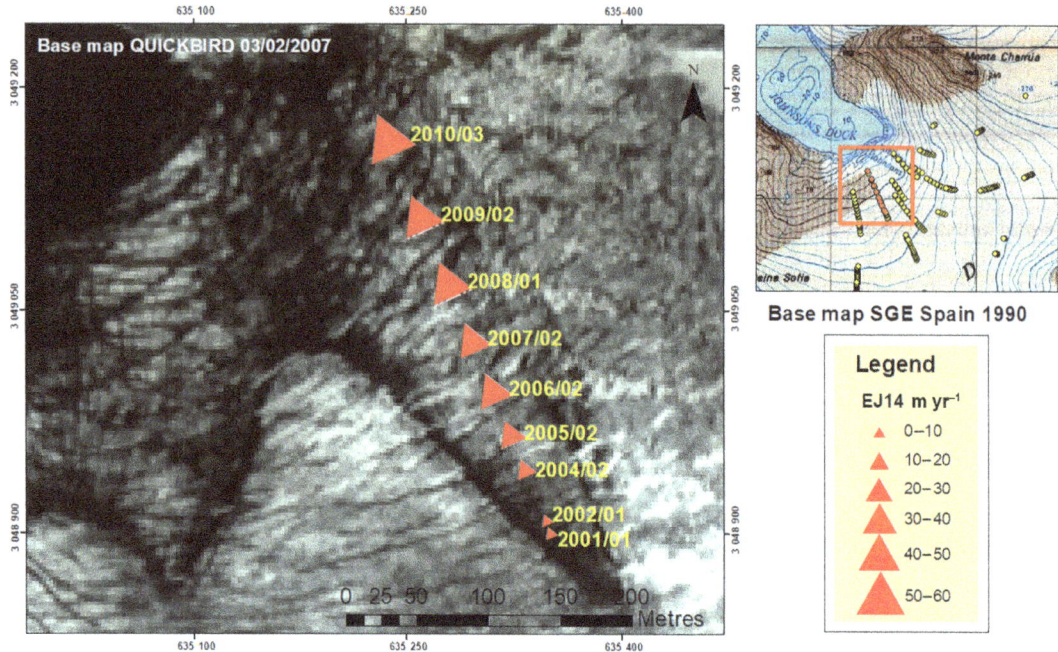

Figure 3. Map showing the time evolution of stake EJ14. Horizontal velocities and times for various positions are shown. The stake fell into a crevasse during 2010–2011, so it does not appear in Fig. 2. The inset shows the location of the main image (in the inset, EJ14 positions are shown in green). In this, and the following figures, UTM coordinates (sheet 20S) are indicated. The background satellite image is a photo from the QUICKBIRD system program (2007).

Table 2. Example of results for the adjustment by least squares of the position and the velocity of a stake (EJ14, near the calving front of Johnsons Glacier; see Fig. 3), together with the deviations from the polynomial approximation for the position, as well as the maximum horizontal velocity and its direction.

$X_a(t_i) = -8.3181 \times 10^{-6} t_i^2 + 5.7260572 \times 10^{-3} t_i + 635\,350.340$
$Y_a(t_i) = 1.90604 \times 10^{-5} t_i^2 - 1.12107159 t_i \times 10^{-2} + 3\,048\,898.260$
$v_x = -1.66362 \times 10^{-5} t_i + 5.7260572 \times 10^{-3}$
$v_y = 3.81208 \times 10^{-5} t_i - 1.12107159 \times 10^{-2}$
$e_x = \pm 1.7\,\text{m}$
$e_y = \pm 4.5\,\text{m}$
$\|e_x\| = \pm 4.8\,\text{m}$
$n = 25$
Maximum velocity: 57.3 m yr[1] on 1 March 2010
Maximum velocity azimuth: 336°

the main ice divides is apparent in the contour plots (zero velocity bands).

Discussion and summarizing conclusions

From the analysis of Figs. 4–5, we see that Johnsons and Hurd glaciers show two markedly different dynamical regimes. Since Johnsons is a tidewater glacier, it shows a pattern of velocities increasing from the ice divides (where horizontal velocities normal to the divide are zero by definition) towards its calving front, where yearly-averaged velocities up to 65 m yr^{-1} have been observed (Fig. 5). Hurd, on the other hand, is a land-terminating glacier, with much slower velocities (typically just a few metres per year), in which the largest velocities are reached in its middle to lower part (between stakes EH06 and EH08; see Fig. 4), where basal sliding likely occurs. The velocity field close to the land-terminating snouts shows a decreasing pattern (this is particularly noticeable in the snouts of Sally Rocks and Las Palmas lobes). Velocities are also high in the high-slope zones such as the Argentina lobe and the upper part of the Las Palmas lobe. Note that the high-velocity zone shown to the southeast of Hurd glacier, around stake EH21 (Fig. 4) does not really correspond to Hurd Glacier but to an unnamed glacier flowing southwards, towards False Bay, which has extremely steep slopes and is in fact a heavily crevassed icefall.

The decreasing velocities as we approach the land-terminating snouts have been attributed to the fact that the surficial cold ice layer reaches the bed in these zones, so the glacier is frozen to its bed and glacier movement is produced by internal deformation alone (no basal sliding). This is sup-

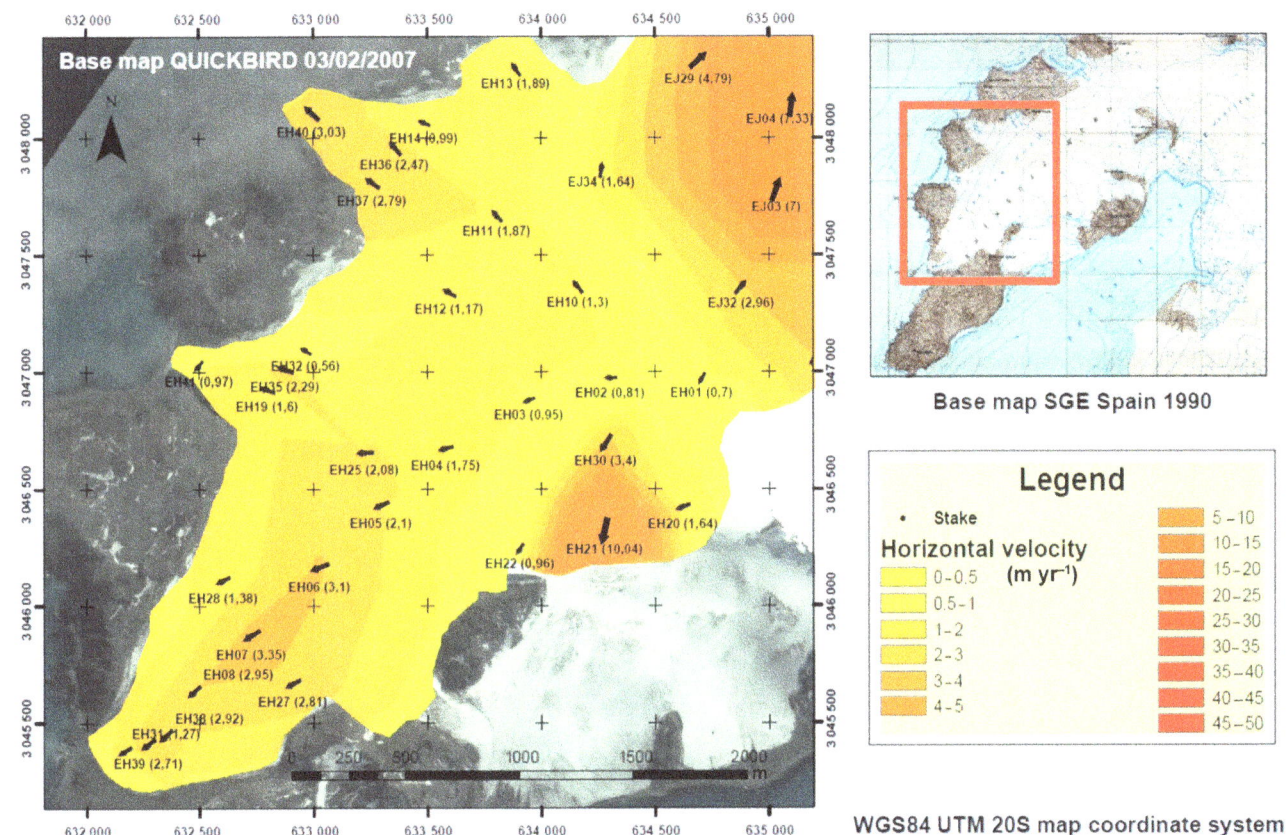

Figure 4. Map of contour lines of the absolute values of the horizontal velocity for Hurd Glacier, obtained from the spatial interpolation of the corresponding vector velocity field, calculated for 13 February 2013. The magnitude of velocity is denoted in brackets, arrows indicate direction. The yellow near-zero velocity band indicates the approximate location of the ice divides.

ported by both geomorphological observations, in particular the presence of compressional structures such as thrust faults close to the glacier termini (Molina et al., 2007; Molina, 2014) and to ground-penetrating radar studies that show that the cold ice layer reaches the bedrock in these zones (Navarro et al., 2009; Molina, 2014).

From the analysis of the polynomial interpolation of observed positions we see that a second-degree polynomial function (representing a uniformly accelerated motion) is sufficient to provide a fair adjustment to the observed position changes. The largest positioning error (misfit of the polynomial approximation), of 4.8 m, is found for stake EJ14, which had an estimated horizontal velocity of ca. $57\,\mathrm{m\,yr^{-1}}$ on 1 March 2010. Of course, one of the major drawbacks of the polynomial interpolation of the observed positions is that it does not allow us to represent seasonal variations in glacier velocities, which are known to occur for the glaciers in this region (e.g. Osmanoğlu et al., 2014). In fact, we tried to add a sinusoidal function to the polynomial fit, and the results did not improve the fit to the observations at all. This result was anticipated because the positioning measurements are mostly done only at the beginning and the end of each summer season and thus do not allow us to resolve yearly cycles. But the polynomial interpolation of all available positions for a given stake is just an example of what can be done with the available data. Calculations could be done for estimating, e.g., summer-averaged velocities or winter-averaged velocities (for the "extended winter", i.e. all of the year except for the summer season). However, this is still insufficient to study velocity variations on scales shorter than the seasonal one. For this reason, perhaps the highest interest of the dataset presented is its use for tuning the free parameters of numerical models of glacier dynamics (Martín et al., 2004; Otero et al., 2010), since these models represent averaged velocities on time-step scales, which are often of the order of weeks (especially for steady-state models such as those cited, in which a limited time evolution is applied to get the model to reach a steady-state configuration). But even for transient models, weekly time steps are usual (e.g. Otero et al., 2017). The available dataset is also useful for the validation of remotely sensed SAR velocities, with typical repeat cycles from a few days to several tens of days (up to 45 days for ALOS PALSAR).

Another shortcoming of the dataset presented is that it does not allow for an easy analysis of dynamical responses to climate changes (such as those regionally observed by Oliva

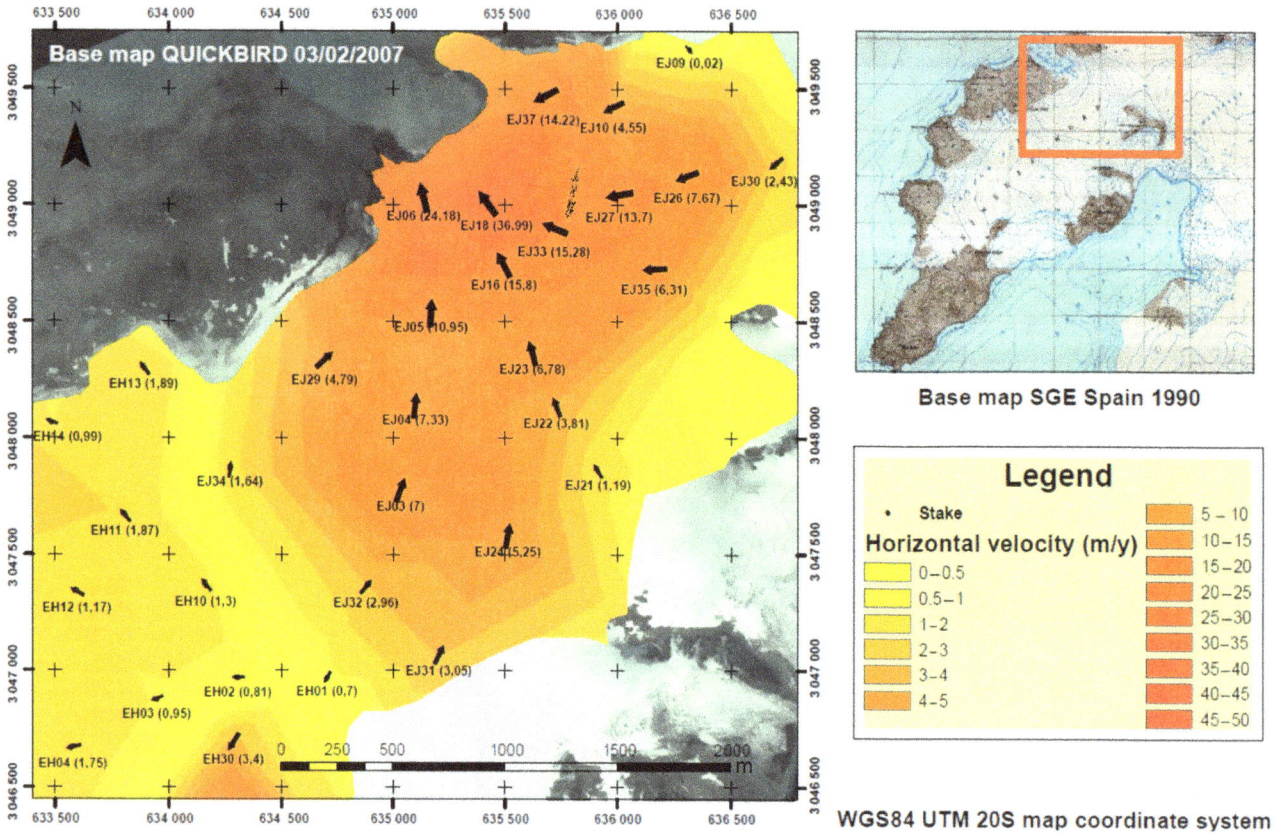

Figure 5. The magnitude of horizontal velocity for Johnsons Glacier, obtained from the spatial interpolation of the corresponding vector velocity field, calculated for 13 February 2013 using the first-degree polynomial derived from the second-degree polynomial adjustment of the stake positions. The numerical values for the absolute value of velocity at each stake (in brackets) and the vector velocity directions (arrows) are also represented. The yellow near-zero velocity band indicates the approximate location of the ice divides (except for the zone to the east, between UTM northing 3 048 000 and 3 048 500, which corresponds to a zone of thin ice frozen to the bed on the upper part of a nunatak).

et al., 2016) because what is available is a Lagrangian velocity field (velocities measured at stakes that change their position with time), while what is needed for studying glacier velocity variations in response to climate changes is an Eulerian velocity field (velocities measured at a fixed location in space).

From the above discussion, a desirable complement to the available in situ velocity dataset presented in this paper would be a continuous record of ice velocities at selected stakes.

Summarizing, the dataset presented is a useful source of input data for numerical models of glacier dynamics and for the calibration–validation of remotely sensed velocity data. It fills an observational data gap in the region peripheral to the Antarctic Peninsula, and it is thus expected that these data will contribute to the understanding of the dynamics of the ice masses in this region and their response to environmental changes.

Appendix A

The shape file CNDA-ESP_SIMRAD_VELOCITY.shp available in the PANGAEA database (http://doi.pangaea.de/10.1594/PANGAEA.846791), and its corresponding versions in Excel (.xlsx) and ASCII (.txt) formats, contain the position data for all stakes of Johnsons and Hurd glaciers for the period from 2000 to 2013. Below, we describe the contents of each individual field in the shape file, as described in file "fields_explanation.txt". We remind the reader that the set of stakes included in the data files is larger than that shown in the various figures in this paper, as it includes all stakes that have existed for any period within the entire measurement period, while the figures give snapshots in time. The PANGAEA data files also include a table (file "stake_dates.txt") indicating the dates of the start and the end of the measurement period for each stake.

- Field "t38_stake": the name of the stake under consideration (see stakes in Fig. 2).

- Field "t38_t0": the zero time of the time variable; we set it as 1 January 1999 at 00:00 GMT.

- Field "t38_fecha": the date and time of the measurement, in "YYYYMMDDHHMMSS" format.

- Field "t38_inct": the period of time in days from "t38_t0" to "t38_fecha" (t_n in the above equations).

- Field "t38_x": X coordinate in metres (UTM 20S) for the stake (considered in an ideal vertical position, after correction for tilt, if applicable) (x_{t_n} in Eq. 1).

- Field "t38_y": Y coordinate in metres (UTM 20S) for the stake (considered in an ideal vertical position, after correction for tilt, if applicable) (y_{t_n} in Eq. 1).

- Field "t38_x_ide": X coordinate in metres (UTM 20S) for the position of the stake for the given time, calculated using the second-degree polynomial adjustment ($X_a(t_n)$ in Eq. 2).

- Field "t38_y_ide": Y coordinate in metres (UTM 20S) for the position of the stake for the given time, calculated using the second-degree polynomial adjustment ($Y_a(t_n)$ in Eq. 2).

- Field "t38_vx": X component for horizontal velocity of the stake for the given time, expressed in metres per year, calculated from the second-degree polynomial adjustment (v_x in Eq. 4).

- Field "t38_vy": Y component for horizontal velocity of the stake for the given time, expressed in metres per year, calculated from the second-degree polynomial adjustment (v_y in Eq. 4).

- Field "t38_vxy": absolute value of horizontal velocity of the stake for the given time, expressed in metres per year, calculated from the X and Y components of the velocity obtained from the second-degree polynomial adjustment ($\|v\|$ in Eq. 4).

- Field "t38_v_aci": azimuth for horizontal velocity of the stake, expressed in sexagesimal degrees, at the date of the measurement.

- Field "t38_err_x": root-mean-squared deviation for the X position of the stake, expressed in metres (e_x).

- Field "t38_err_y": root-mean-squared deviation for the Y position of the stake, expressed in metres (e_y).

- Field "t38_max_x": maximum error obtained for the X position of the stake, expressed in metres.

- Field "t38_max_y": maximum error obtained for the Y position of the stake, expressed in metres.

- Field "t38_ax": the estimation of the "a_x" coefficient in the second-degree polynomial adjustment of the position X of the stake (a_x in Eq. 2).

- Field "t38_bx": the estimation of the "b_x" coefficient in the second-degree polynomial adjustment of the position X of the stake (b_x in Eq. 2).

- Field "t38_cx": the estimation of the "c_x" coefficient in the second-degree polynomial adjustment of the position X of the stake (c_x in Eq. 2).

- Field "t38_ay": the estimation of the "a_y" coefficient in the second-degree polynomial adjustment of the position Y of the stake (a_y in Eq. 2).

- Field "t38_by": the estimation of the "b_y" coefficient in the second-degree polynomial adjustment of the position Y of the stake (b_y in Eq. 2).

- Field "t38_cy": the estimation of the "c_y" coefficient in the second-degree polynomial adjustment of the position Y of the stake (c_y in Eq. 2).

- Field "dias": days after t38_t0 for a simulation (in this example, 5817 days).

- Field "prevista_x": example of X coordinate in metres (UTM 20S) for the stake (considered in an ideal vertical position, after correction for tilt) after 5817 days.

- Field "prevista_y": example of Y coordinate in metres (UTM 20S) for the stake (considered in an ideal vertical position, after correction for tilt) after 5817 days.

- Field "movxy": planimetric movement in metres (UTM 20S) for the stake (considered in an ideal vertical position, after correction for tilt) after 5817 days.

Appendix B

Table B1. Polynomial coefficients of the adjustment functions $X_a(t_i)$ and $Y_a(t_i)$, according to Eq. (2) for all the stakes of the glaciers under study. The units for the coefficients a, b, and c are $m\,yr^{-2}$, $m\,yr^{-1}$ and m, respectively. The table also shows the estimated horizontal positioning errors ($||e_x|| = \sqrt{e_x^2 + e_y^2}$, in metres) involved in the polynomial approximation of the position.

Stake	a_x	b_x	c_x	a_y	b_y	c_y	$||e_x||$ (m)
EH01	−0.0000000210	−0.0007499061	634 706.536	−0.0000001022	−0.0006149926	3 046 974.077	±0.4
EH02	−0.0000000260	−0.0019490138	634 314.937	0.0000000376	−0.0004418479	3 046 972.371	±0.3
EH03	0.0000000017	−0.0024338880	633 957.992	0.0000000387	−0.0013300512	3 046 883.749	±0.3
EH04	0.0000000208	−0.0048559670	633 602.733	0.0000000476	−0.0016952866	3 046 680.172	±0.2
EH05	0.0000000827	−0.0061748145	633 322.971	0.0000000176	−0.0023909505	3 046 440.393	±0.2
EH06	0.0000001329	−0.0093454794	633 066.469	0.0000001143	−0.0041399749	3 046 183.071	±0.3
EH07	0.0000002467	−0.0103102794	632 769.911	0.0000001116	−0.0060335306	3 045 901.732	±0.3
EH08	0.0000002202	−0.0084839082	632 511.087	0.0000002088	−0.0073298445	3 045 661.923	±0.3
EH10	−0.0000000115	−0.0021984650	634 172.352	0.0000000196	0.0024960084	3 047 352.461	±0.2
EH11	−0.0000000060	−0.0033251769	633 822.848	−0.0000000136	0.0040018668	3 047 646.355	±0.2
EH12	0.0000000058	−0.0029001077	633 610.294	0.0000000058	0.0014133530	3 047 331.978	±0.5
EH13	0.0000000989	−0.0040996188	633 908.319	0.0000000699	0.0034554549	3 048 276.555	±0.3
EH14	0.0000003949	−0.0066196722	633 507.694	−0.0000000176	0.0011105547	3 048 060.055	±0.7
EH16	0.0000005378	−0.0121620919	633 315.315	−0.0000003729	0.0088014677	3 047 778.378	±0.4
EH18	0.0000005266	−0.0079704077	632 901.615	0.0000004036	0.0003691018	3 047 006.754	±0.5
EH19	0.0000001954	−0.0062025268	632 821.167	0.0000000024	0.0013091532	3 046 916.708	±0.4
EH20	−0.0000000041	−0.0041607075	634 641.903	0.0000000207	−0.0018151514	3 046 428.116	±0.4
EH21	0.0000000702	−0.0069464784	634 311.036	−0.0000013781	−0.0125603214	3 046 415.468	±0.8
EH22	0.0000000418	−0.0019778722	633 913.716	0.0000000351	−0.0024914249	3 046 250.139	±0.3
EH23	0.0000000850	−0.0074476651	633 495.325	−0.0000000048	−0.0040643877	3 046 056.767	±0.3
EH25	0.0000001059	−0.0067918760	633 252.533	0.0000000337	−0.0005615131	3 046 656.096	±0.2
EH26	0.0000000665	−0.0061374875	633 283.461	0.0000000429	−0.0008986263	3 046 923.597	±0.3
EH27	0.0000001937	−0.0088012166	632 945.263	0.0000000846	−0.0044487636	3 045 685.828	±0.3
EH28	0.0000003036	−0.0065784553	632 626.203	0.0000002354	−0.0039697127	3 046 120.387	±0.4
EH30	−0.0000001863	−0.0031881664	634 304.297	−0.0000004025	−0.0036269810	3 046 722.181	±1.3
EH31	0.0000004603	−0.0077532371	632 191.166	0.0000002463	−0.0042862662	3 045 393.753	±0.5
EH32	0.0000004228	−0.0056874228	632 981.252	0.0000000427	0.0003263696	3 047 088.876	±0.8
EH34	0.0000028295	−0.0221847247	633 412.561	−0.0000004808	0.0070931604	3 047 929.344	±0.3
EH35	−0.0000001494	−0.0043610240	632 899.866	0.0000001839	0.0002225731	3 047 005.252	±0.3
EH36	−0.0000000071	−0.0044714224	633 378.639	−0.0000008964	0.0142598599	3 047 909.661	±1.4
EH37	0.0000001050	−0.0073407541	633 291.187	−0.0000008597	0.0132842232	3 047 763.133	±0.8
EH38	0.0000000544	−0.0065644463	632 376.878	−0.0000002724	−0.0024698302	3 045 462.670	±0.3
EH39	−0.0000000968	−0.0050881561	632 296.779	−0.0000001033	−0.0031726807	3 045 423.513	±0.2
EH40	0.0000001909	−0.0080001652	633 027.591	−0.0000000767	0.0064788163	3 048 074.317	±0.1
EH41	0.0000016791	−0.0191464499	632 550.165	−0.0000027307	0.0262539089	3 046 960.654	±0.6
EJ03r	−0.0000001531	0.0100735338	634 980.227	0.0000004252	0.0182063987	3 047 658.850	±0.8
EJ04	−0.0000003723	0.0061601161	635 075.122	−0.0000005812	0.0259414549	3 048 020.871	±1.2
EJ05	0.0000000492	0.0011547502	635 161.004	−0.0000003319	0.0333842531	3 048 375.872	±1.8

Table B1. Continued.

Stake	a_x	b_x	c_x	a_y	b_y	c_y	$||e_x||$ (m)
EJ05r	−0.0000004208	0.0031517495	635 155.477	−0.0000019291	0.0363184602	3 048 344.319	±1.6
EJ06	−0.0000011073	−0.0042784868	635 185.239	0.0000026943	0.0365926573	3 048 770.902	±3.1
EJ06r	0.0000016211	−0.0103198593	635 192.433	−0.0000178994	0.0932484671	3 048 692.082	±3.8
EJ09	0.0000000101	−0.0000640752	636 317.040	0.0000000306	−0.0003560897	3 049 669.857	±0.2
EJ10	−0.0000004307	−0.0068225266	636 026.014	0.0000000111	−0.0054558209	3 049 450.521	±0.5
EJ11	−0.0000056619	−0.0199142185	635 701.041	0.0000047076	−0.0040812151	3 049 278.770	±1.1
EJ14	−0.0000083181	0.0057260572	635 350.340	0.0000190604	−0.0112107159	3 048 898.259	±4.8
EJ14r	0.0000026146	−0.0265096848	635 395.318	−0.0000095539	0.0757322689	3 048 785.930	±1.5
EJ15	−0.0000115713	−0.0150161625	635 587.960	0.0000177466	−0.0141862506	3 049 134.762	±4.5
EJ16	−0.0000012393	−0.0075342532	635 564.261	0.0000014961	0.0227882933	3 048 586.798	±1.7
EJ16r	0.0000001769	−0.0118134174	635 579.144	0.0000003110	0.0244004790	3 048 564.554	±1.5
EJ17	−0.0000073534	−0.0169745300	635 820.867	0.0000070748	−0.0119983441	3 049 058.204	±1.8
EJ17r	−0.0000017878	−0.0295492517	635 853.941	0.0000041007	−0.0069866948	3 049 052.604	±1.1
EJ18	−0.0000047503	−0.0128379987	635 611.869	0.0000072099	0.0059753470	3 048 787.784	±3.6
EJ18r	−0.0000020714	−0.0198100707	635 635.496	0.0000032380	0.0163894918	3 048 764.951	±2.1
EJ19r	−0.0000027156	−0.0347723467	635 509.766	0.0000039813	0.0474550026	3 048 954.395	±2.9
EJ21	−0.0000000445	−0.0012778820	635 920.791	0.0000000648	0.0020946021	3 047 848.965	±0.1
EJ22	0.0000000237	−0.0039724530	635 745.947	0.0000000882	0.0088414448	3 048 083.628	±0.5
EJ23	−0.0000000349	−0.0046353861	635 644.992	0.0000001090	0.0167557974	3 048 276.292	±0.9
EJ24	−0.0000000521	0.0032607734	635 493.978	−0.0000001090	0.0152334223	3 047 502.873	±0.5
EJ26	−0.0000009008	−0.0103905425	636 381.563	−0.0000001309	−0.0059871063	3 049 160.852	±0.7
EJ27	−0.0000014544	−0.0219763759	636 156.950	0.0000004109	−0.0107406457	3 049 090.100	±1.5
EJ28	−0.0000031887	0.0011132529	636 126.270	0.0000006713	0.0002700461	3 048 619.696	±1.3
EJ29	−0.0000001806	0.0115365135	634 635.956	−0.0000000833	0.0097340264	3 048 288.181	±0.6
EJ30	0.0000002245	−0.0074398001	636 727.973	0.0000000834	−0.0051238313	3 049 205.806	±0.1
EJ31	−0.0000002608	0.0064071706	635 177.240	−0.0000004367	0.0119751394	3 047 018.091	±0.1
EJ32	−0.0000002940	0.0083991531	634 841.075	−0.0000007905	0.0142317135	3 047 310.149	±0.3
EJ33	0.0000040527	−0.0808182396	636 025.782	0.0000004695	0.0104154497	3 048 835.009	±0.7
EJ34	−0.0000001959	0.0026058368	634 258.842	−0.0000003753	0.0083361641	3 047 831.927	±0.1
EJ35	−0.0000030794	0.0145204365	636 169.067	−0.0000019491	0.018 988 6987	3 048 680.544	±1.5
EJ36	0.0000098129	−0.1574078351	636 082.069	−0.0000007515	0.0481336713	3 048 977.865	±1.0
EJ37	−0.0000076218	0.0437382981	635 654.292	0.0000010401	−0.0281496644	3 049 584.987	±2.8

Competing interests. The authors declare that they have no conflict of interest.

Acknowledgements. This work was supported by grant CTM2014-56473-R from the Spanish National Plan of R&D. The suggestions by Christof Völksen and an anonymous reviewer, as well as the topical editor, Reinhard Drews, greatly contributed to improve the manuscript.

Edited by: Reinhard Drews

References

Arthern, R. J. and Gudmundsson, G. H.: Initialization of ice-sheet forecasts viewed as an inverse Robin problem, J. Glaciol., 56, 527–533, 2011.

Bañón, M. and Vasallo, F.: AEMET en la Antártida, Climatología y meteorología sinóptica en las estaciones meteorológicas españolas en la Antártida, AEMET, Madrid., 152 pp., 2016.

Blindow, N., Suckro, S., Rückamp, M., Braun, M., Schindler, M., Breuer, B., Saurer, H., Simões, J. C., and Lange, M.: Geometry and status of the King George Island ice cap (South Shetland Islands, Antarctica), Ann. Glaciol., 51, 103–109, 2010.

Cuffey, K. M. and Paterson, W. S. B.: The Physics of Glaciers, 4th edn., Elsevier, Amsterdam, 2010.

Ghilani, C. D.: Adjustment Computations, Spatial Data Analysis, John Wiley and Sons, New Jersey, 2010.

Hambrey, M. and Lawson, W.: Structural styles and deformation fields in glaciers: a review, in: Deformation of Glacier Materials, edited by: Maltman, A. J., Hubbard, B., and Hambrey, M., Glaciological Society, London, Special Publications, 176, 147–157, 2000.

Hanson, B.: A fully three-dimensional finite-element model applied to velocities on Storglaciären, Sweden, J. Glaciol., 41, 91–102, 1995.

Hooke, R. LeB., Calla, P., Holmlund, P., Nilsson, M., and Stroeven, A.: A three-year record of seasonal variations in surface velocity, Storglaciären, Sweden, J. Glaciol., 35, 235–247, 1989.

Jay-Allemand, M., Gillet-Chaulet, F., Gagliardini, O., and Nodet, M.: Investigating changes in basal conditions of Variegated Glacier prior to and during its 1982–1983 surge, The Cryosphere, 5, 659–672, https://doi.org/10.5194/tc-5-659-2011, 2011.

Joughin, I, Smith, B., and Abdalati, W.: Glaciological advances made with interferometric synthetic aperture radar, J. Glaciol., 56, 1026–1042, 2010.

Martín, C., Navarro, F. J., Otero, J., Cuadrado, M. L., and Corcuera, M. I.: Three-dimensional modelling of the dynamics of Johnsons glacier (Livingston Island, Antarctica), Ann. Glaciol., 39, 1–8, 2004.

Molina, C.: Caracterización dinámica del glaciar Hurd combinando observaciones de campo y simulaciones numérica, PhD thesis, Universidad Politécnica de Madrid, 2014.

Molina, C., Navarro, F., Calvet, J., García-Selles, D., and Lapazaran, J.: Hurd Peninsula glaciers, Livingston Island, Antarctica, as indicators of regional warming: ice-volume changes during period 1956–2000, Ann. Glaciol., 46, 43–49, 2007.

Navarro, F. J., Otero, J., Macheret, Y. Y., Vasilenko, E. V., Lapazaran, J. J., Ahlstrøm, A. P., and Machío, F.: Radioglaciological studies on Hurd Peninsula glaciers, Livingston Island, Antarctica, Ann. Glaciol., 50, 17–24, 2009.

Navarro, F., Jonsell, U., Corcuera, M., and Martín-Español, A.: Decelerated mass loss of Hurd and Johnsons glaciers, Livingston Island, Antarctic Peninsula, J. Glaciol., 59, 115–128, 2013.

Oliva, M., Navarro, F., Hrbáček, F., Hernández, A., Nývlt, D., Pereira, P., Ruiz-Fernández, J., and Trigo, R.: Recent regional cooling of the Antarctic Peninsula and its impacts on the cryosphere, Sci. Total Environ., 580, 210–223, https://doi.org/10.1016/j.scitotenv.2016.12.030, 2016.

Osmanoğlu, B., Braun, M., Hock, R., and Navarro, F. J.: Surface velocity and ice discharge of the ice cap on King George Island, Antarctica, Ann. Glaciol., 54, 111–119, https://doi.org/10.3189/2013AoG63A517, 2013.

Osmanoglu, B., Navarro, F. J., Hock, R., Braun, M., and Corcuera, M. I.: Surface velocity and mass balance of Livingston Island ice cap, Antarctica, The Cryosphere, 8, 1807–1823, https://doi.org/10.5194/tc-8-1807-2014, 2014.

Otero, J., Navarro, F. J., Lapazaran, J. J., Welty, E., Puczko, D., and Finkelnburg, R.: Modeling the Controls on the Front Position of a Tidewater Glacier in Svalbard, Front. Earth Sci., 5, 29, https://doi.org/10.3389/feart.2017.00029, 2017.

Otero, J., Navarro, F. J., Martín, C., Cuadrado, M. L., and Corcuera, M. I.: A three-dimensional calving model: numerical experiments on Johnsons Glacier, Livingston Island, Antarctica, J. Glaciol., 56, 200–214, 2010.

Ramírez-Rodríguez, M. E.: Modelización de la deformación superficial en áreas volcánicas mediante la teoría de wavelets, Aplicación al Volcán Decepción. PhD Thesis, Facultad de Ciencias, Universidad de Cádiz, 2007.

Ren, J., Qin, D., Petit, J. R., Jouzel, J., Wang, W., Liu, C., Wang, S., Qian, S., and Wang, X.: Glaciological studies in Nelson Island, Antarctica, J. Glaciol., 41, 408–412, 1995.

Rignot, E. and Kanagaratnam, P.: Changes in the velocity structure of the Greenland Ice Sheet, Science, 311, 986–990, 2006.

Rodríguez Cielos, R. and Navarro Valero, F.: Continuous velocity model for Johnsons and Hurd glaciers from 1999 to 2013, with link to model results in shapefile and MS Excel format, Universidad Politécnica de Madrid, PANGAEA, https://doi.org/10.1594/PANGAEA.846791, 2015.

Rückamp, M., Blindow, N., Suckro, S., Braun, M., and Humbert, A.: Dynamics of the ice cap on King George Island, Antarctica: field measurements and numerical simulations, Ann. Glaciol., 51, 80–90, 2010.

Rückamp, M., Braun, M., Suckro, S., and Blindow, N.: Observed glacial changes on the King George Island ice cap, Antarctica, in the last decade, Global Planet. Change, 79, 99–109, https://doi.org/10.1016/j.gloplacha.2011.06.009, 2011.

Schellenberger, T., Dunse, T., Kääb, A., Kohler, J., and Reijmer, C. H.: Surface speed and frontal ablation of Kronebreen and Kongsbreen, NW Svalbard, from SAR offset tracking, The Cryosphere, 9, 2339–2355, https://doi.org/10.5194/tc-9-2339-2015, 2015.

Servicio Geográfico del Ejército (SGE): Livingston Island. Hurd Peninsula, (Scale 1 : 25 000), Madrid, Servicio Geográfico del Ejército, 1991.

Strozzi, T., Luckman, A., Murray, T., Wegmüller, U., and Werner, C.: Glacier motion estimation using SAR offset-tracking procedures, IEEE T. Geosci. Remote, 40, 2384–2391, 2002.

Strozzi, T., Kouraev, A., Wiesmann, A., Wegmüller, U., Sharov, A., and Werner, C.: Estimation of Arctic glacier motion with satellite L-band SAR data, Remote Sens. Environ., 112, 636–645, https://doi.org/10.1016/j.rse.2007.06.007, 2008.

Van der Veen, C. and Whillans, I.: Force Budget: I. Theory and Numerical Methods, J. Glaciol., 35, 53–60, 1989.

Wuite, J., Rott, H., Hetzenecker, M., Floricioiu, D., De Rydt, J., Gudmundsson, G. H., Nagler, T., and Kern, M.: Evolution of surface velocities and ice discharge of Larsen B outlet glaciers from 1995 to 2013, The Cryosphere, 9, 957–969, https://doi.org/10.5194/tc-9-957-2015, 2015.

Ximenis, L., Calvet, J., Enrique, J., Corbera, J., Fernández de Gamboa, C., and Furdàda, G.: The measurement of ice velocity, mass balance and thinning-rate on Johnsons Glacier, Livingston Island, South Shetland Islands, Antarctica, Acta Geológica Hispánica, 34, 403–409, 1999.

Ximenis, L., Calvet, J., García, D., Casas, J. M., and Sàbat, F.: Folding in the Johnsons Glacier, Livingston Island, Antarctica, in: Deformation of Glacier Materials, edited by: Maltman, A. J., Hubbard, B., and Hambrey, M., Glaciological Society, London, Special Publications, 176, 147–157, 2000.

4

The Rofental: a high Alpine research basin (1890–3770 m a.s.l.) in the Ötztal Alps (Austria) with over 150 years of hydrometeorological and glaciological observations

Ulrich Strasser[1], Thomas Marke[1], Ludwig Braun[3], Heidi Escher-Vetter[3], Irmgard Juen[2],
Michael Kuhn[2], Fabien Maussion[2], Christoph Mayer[3], Lindsey Nicholson[2], Klaus Niedertscheider[4],
Rudolf Sailer[1], Johann Stötter[1], Markus Weber[5], and Georg Kaser[2]

[1]Department of Geography, University of Innsbruck, Innsbruck, 6020, Austria
[2]Department of Atmospheric and Cryospheric Sciences, University of Innsbruck, Innsbruck, 6020, Austria
[3]Geodesy and Glaciology, Bavarian Academy of Sciences and Humanities, Munich, 80539, Germany
[4]Hydrographic Service of Tyrol, Innsbruck, 6020, Austria
[5]Photogrammetry and Remote Sensing, Technical University of Munich, Munich, 80333, Germany

Correspondence: Ulrich Strasser (ulrich.strasser@uibk.ac.at)

Abstract. A comprehensive hydrometeorological and glaciological data set is presented, originating from a multitude of glaciological, meteorological, hydrological and laser scanning recordings at research sites in the Rofental (1891–3772 m a.s.l., Ötztal Alps, Austria). The data sets span a period of 150 years and hence represent a unique time series of rich high-altitude mountain observations. Their collection was originally initiated to support scientific investigation of the glaciers Hintereisferner, Kesselwandferner and Vernagtferner. Annual mass balance, glacier front variation, flow velocities and photographic records of the status of these glaciers were recorded. Later, additional measurements of meteorological and hydrological variables were undertaken, and over time a number of autonomous weather stations and runoff gauges were brought into operation; the available data now comprise records of temperature, relative humidity, short- and longwave radiation, wind speed and direction, air pressure, precipitation, and river water levels. Since 2001, a series of distributed (airborne and terrestrial) laser scans is available, along with associated digital surface models. In 2016 a permanent terrestrial laser scanner was installed on "Im hintern Eis" (3244 m a.s.l.) to continuously observe almost the entire area of Hintereisferner. The data and research undertaken at the sites of investigation in the Rofental area enable combined research of cryospheric, atmospheric and hydrological processes in complex terrain, and support the development of several state-of-the-art glacier mass balance and hydroclimatological models. The institutions taking part in the Rofental research framework promote their site in several international research initiatives. In INARCH (International Network for Alpine Research Catchment Hydrology, http://words.usask.ca/inarch), all original research data sets are now provided to the scientific community according to the Creative Commons Attribution License by means of the PANGAEA repository.

1 Introduction

Glaciers in the Rofental, Ötztal Alps, have been under observation since the early 17th century; in particular the Vernagtferner (VF) was known for its dangerous surge-type advances (Richter, 1892) with flow velocities up to $11.5 \, \mathrm{m \, day}^{-1}$ (Nicolussi, 2013). In 1599, the tongue of VF reached the valley floor of the Rofental, blocking the valley and forming an ice-dammed lake which burst, causing a catastrophic lake outburst flood (GLOF) on 20 July in 1600. The painting of VF with its lake that was formed again in summer 1601 is the oldest known image of a glacier worldwide (Fig. 1). The lake was $\sim 1700 \, \mathrm{m}$ in length, and the lake level was at $\sim 2260 \, \mathrm{m \, a.s.l.}$ (Nicolussi, 1990). Further advances of VF are documented for the periods around 1680, 1770 (the maximum extent of VF in the little ice age, LIA), 1820 and 1848. Prior to the 19th century, each advance of VF formed the ice dam and lake in the Rofental, and the outflow of this lake was observed due to its dangerous nature (Nicolussi, 2013). The glacier changes were monitored with irregular glacier-front variation recordings.

A milestone in glacier cartography was the map "Der Vernagtferner im Jahre 1889 1 : 10 000", constructed by means of terrestrial photogrammetry (Finsterwalder, 1897). This map showed, for the first time, an entire glacier in large scale with unprecedented details, including the area that became ice-free since the last glacier advance period (Fig. 2a). The first map of Hochjochferner (HJF) (1883) by Blümcke and Hess (1895) and Hintereisferner (HEF) (1894) by Blümcke and Hess (1899) followed shortly afterwards (Fig. 2b). After that, geodetic glacier maps were generated frequently for HEF, KWF and VF (Brunner, 2013; Charalampidis, 2018)[1]. These maps were accompanied by frequent terrestrial photographs of VF between 1897 and 1928, documenting, for example, the last surge of the glacier (1897–1903). From these photographs Weber (2013) and Lindmayer (2015) reconstructed the extent and dynamics of VF for the respective periods. The monitoring of the geometry of the glaciers in the Rofental was accompanied by ice drilling experiments to reconstruct the glacier bed and dynamics (Blümcke and Hess, 1899), and first glacier flow theories were developed (Finsterwalder, 1897; Hess, 1904). With the foundation of the "International Commission for Snow and Ice" (ICSI) (as "International Glacier Commission" in Zürich 1894, renamed in 1948) the observation of glaciers became systemized and internationally coordinated, and the continuous observations of meteorological and hydrological variables began: the first totalizing rain gauge in the Rofental was in-

---

[1]HEF: 1850, 1894, 1920, 1939, 1953, 1962, 1964, 1967, 1969, 1979, 1991, 1997, 2001–2008
KWF: 1939, 1967, 1969, 1979, 1991, 1997, 2001–2008
VF: 1846, **1889**, 1897, 1899, 1901, 1904, 1912, 1938, 1954, 1966, 1969, 1972, **1979**, **1982**, **1990**, 1994, **1999**, 2002, 2003, **2006**, 2009 (bold years available at http://geo.badw.de/vernagtferner-digital/karten.html)

stalled in the village of Vent (1900 m a.s.l.) in 1905, followed by continuous measurements further up in the valley since 1952. The "Combined Water, Ice and Heat Balance Project in the Rofental" became an official initiative of the UNESCO International Hydrological Decade (IHD, 1964–1974) (Hoinkes et al., 1974). Later, it was continued as part of the UNESCO International Hydrological Program (IHP) (http://en.unesco.org/themes/water-security/hydrology). Ablation stakes and pits for mass balance monitoring have been continuously maintained at HEF and Kesselwandferner (KWF) (since 1952), at VF (since 1965), and for short discontinuous periods at HJF. As of 2017, the glacier mass balance time series of HEF, VF and KWF are among the longest uninterrupted series worldwide (Fischer et al., 2015; Mayer et al., 2013a), and several automatic weather stations (AWSs) as well as runoff gauges at VF and in the village of Vent are in continuous operation. These are complemented by a network of historical rain gauges (totalizators) and modern precipitation gauges.

The comprehensive pool of long-term observations available for the Rofental provides the basis for (i) manifold process studies on energy balance, ice dynamics, glacier hydrology and hydraulics (e.g., Kuhn et al., 1985a, b; Kuhn, 1987), (ii) new ground-based and remote sensing monitoring methods (Escher-Vetter and Siebers, 2013; Juen et al., 2013; Helfricht et al., 2014a), (iii) model development and application in glaciological and regional hydrological research (Kaser et al., 2010; Escher-Vetter and Oerter, 2013; Schöber et al., 2014, 2016; Hanzer et al., 2016; Schmieder et al., 2016, 2018), (iv) the evaluation of potential future glacier evolution and changes of the hydrological regime in a changing climate (Weber et al., 2009; Marke et al., 2013; Marzeion and Kaser, 2014; Weber and Prasch, 2015a, b; Hanzer et al., 2017), (v) attributing observed glacier changes to different drivers (Painter et al., 2013; Marzeion et al., 2014a, b) and, finally, (vi) as calibration and validation site for estimating the contribution of glaciers to global sea level rise (Marzeion et al., 2012a, b; Marzeion and Levermann, 2014). For VF, a comprehensive collection of 50 years of significant scientific work of the Commission of Glaciology of the Bavarian Academy of Sciences and Humanities has been edited by Braun and Escher-Vetter (2013). Historical elevation and area changes of HEF, KWF and VF are documented in the Austrian glacier inventories, available for 1969, 1997 and 2006 (Abermann et al., 2009): Between 1969 and 1997, the glacier area in the Rofental decreased from 42.9 to 37.7 km², corresponding to 12 % (Kuhn et al., 2006); recently, the retreat of the glaciers has accelerated. The key glaciological results for HEF, KWF and VF are reported annually to the World Glacier Monitoring Service (WGMS, http://wgms.ch). The state of HEF and KWF in 2014 is shown in Fig. 3.

In addition to monitoring glacier mass balance with glaciological, geodetical and hydrological methods as well as the long-term recordings of respective variables, the Rofental has been an open laboratory for the development of new method-

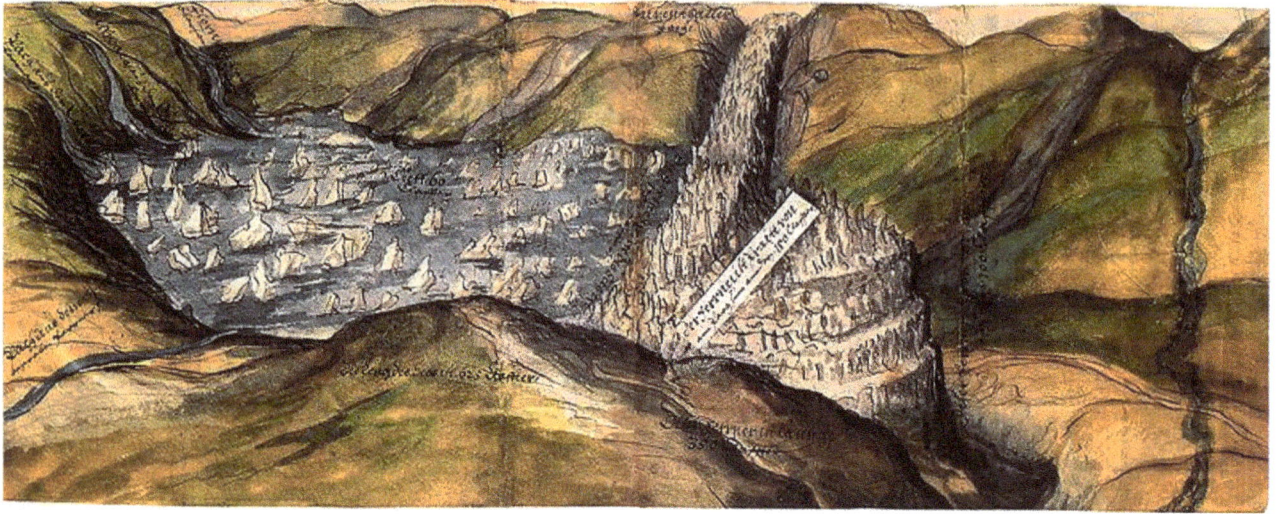

Figure 1. The "Rofentaler Eissee" in 1601, the first known image of a glacier worldwide. Painted in water colors by Abraham Jäger. The original is in the Tiroler Landesmuseum Ferdinandeum in Innsbruck. From Nicolussi (1990).

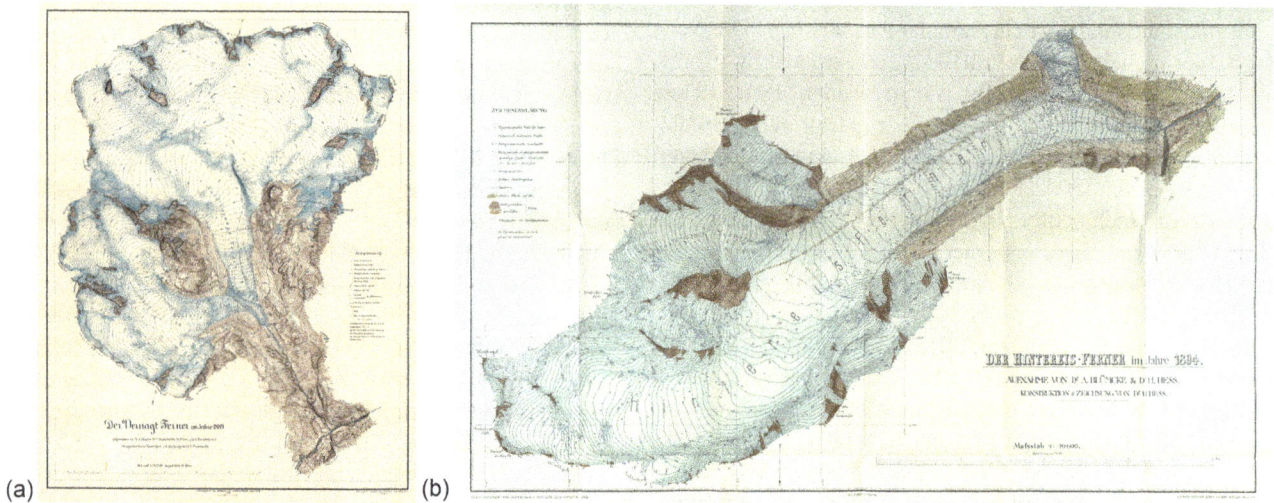

(a)   (b)

Figure 2. (a) The 1889 map of Vernagtferner showing the entire glacier in large scale: "Der Vernagtferner im Jahre 1889 1 : 10 000" (Finsterwalder, 1897). **(b)** "Der Hintereisferner im Jahre 1894" (Blümcke and Hess, 1899).

ologies. For example, a series of airborne lidar-derived high-resolution digital terrain models (DTMs) of HEF and its surroundings has been processed spanning 2001–2011 (Geist and Stötter, 2002; Helfricht et al., 2014b; Klug et al., 2017). They are subject to ongoing evaluations and method comparison studies as well as the monitoring and study of periglacial morphodynamics (Sailer et al., 2012, 2014). Since 2016, a permanent terrestrial laser scanning station is operating at Im hintern Eis (3244 m a.s.l.), allowing for high-resolution, on-demand monitoring of almost the entire surface area of HEF.

All available data for the Rofenal area are placed on a PANGAEA repository (https://doi.org/10.1594/PANGAEA.876120). Recordings of devices which still (as of fall 2017) undergo a test

and development phase will be continuously made available in PANGAEA. These data will also be documented as continuation of this publication by means of the ESSD "living data process". The data collection that is already accessible comprises (i) image files, e.g., glacier topography maps, photographs or animations, and diagrams; (ii) raster grids, e.g., spatial data derived from laser scanning, as tab-delimited text files; (iii) time series of point observations, e.g., meteorological and hydrological recordings, as well as calculated results for points or areas such as mass balances, as tab-delimited text files; and (iv) the glacier inventories in ArcGIS shape file format. All data are comprised with the period of coverage, and the exact coordinates of the locations (in latitudes and longitudes). Areas are defined by

Figure 3. State of Hintereisferner (HEF), Kesselwandferner (KWF) and Guslarferner (GF) on 28 September 2014. Aerial photo by Christoph Mayer, view to the south.

latitudes and longitudes of their bounds. The data described in the present publication have been collected by several institutions operationally measuring cryospheric, atmospheric and hydrological variables along with their changes in the framework of their monitoring programs in the Rofental in the Ötztal Alps, Austria. The institutions involved are the University of Innsbruck with the Department of Atmospheric and Cryospheric Sciences (formerly Institute of Meteorology and Geophysics, http://acinn.uibk.ac.at), the Department of Geography (http://uibk.ac.at/geographie), the Bavarian Academy of Sciences and Humanities in Munich (geo.badw.de) and the Hydrographic Service of Tyrol in Innsbruck (http://tirol.gv.at/umwelt/wasser/wasserkreislauf) which is a section of the Federal Ministry of Agriculture, Forestry, Environment and Water Management (BMLFUW, http://bmlfuw.gv.at).

In this paper, we document the available data from the Rofental area. It is structured in (i) glaciological data, i.e., recordings of glacier volume and geometry changes for HEF, KWF, VF and HJF; (ii) meteorological data as recorded by temporally installed or permanent AWSs; (iii) hydrological data characterizing the water balance of the respective glaciated (sub) catchment; and (iv) airborne and terrestrial laser scanning data. The link to the respective PANGAEA repository parent is https://doi.org/10.1594/PANGAEA.876120. This parent comprises all DOI links to download the data described. Nevertheless, the respective direct DOI links are explicitly referred to here as well.

The selection of data, documented here and available for download from PANGAEA, is only a portion of all the observations that have been collected. Countless documents, photographs, tables and analogue measuring tapes await digitization, and many older digital data still have to be processed and correctly documented. According to the purpose of the

INARCH special issue to which this paper belongs we have concentrated our efforts on providing (i) a mostly complete picture of the water balance components of the Rofental – the mass balances of the observed glaciers being an important highlight of these – and (ii) the meteorological data to force a typical hydrological catchment model. Particular attention is paid to the glaciological, meteorological and hydrological processes in the complex Alpine topography and their spatiotemporal variations in the valley.

2 The Rofental – site description

The Rofental (98.1 km^2, Fig. 4) is a glaciated headwater catchment in the central eastern Alps, namely in the upper Ötztal Alps (Tyrol, Austria): as of 2008 approximately one-third of its area is ice-covered (Müller et al., 2009). The valley floor is a narrow discontinuous riparian zone typically less than 100 m in width. The Rofental stretches from 1891 m a.s.l. at the gauge at Vent, the lowest point of the catchment, to 3772 m a.s.l. at the summit of Wildspitze, the highest summit of Tyrol. The average slope is 25° and the average elevation is 2930 m a.s.l.. The Rofenache is a tributary to the Venter Ache, Ötztaler Ache and the Inn and as such contributes to the Danube system (i.e., the Black Sea). The river gauge at Vent (1891 m a.s.l., 46.85694° N, 10.82361° E) has been operated continuously from the Hydrographic Service of Tyrol since 1967. The characteristic water discharges (m^3 s^{-1}, 1971–2013) are NQ = 0.09 (lowest discharge), MQ = 4.6 (mean discharge) and HQ = 109 (highest discharge); further characteristic data are published as annual review in the "Hydrographisches Jahrbuch von Österreich" (e.g., BMLFUW, 2011; available at: http://bmlfuw.gv.at,data-downloadatehyd.gv.at). The runoff regime of the Rofenache has not been modified by any measures of hydropower generation and is dominated by the melt of snow

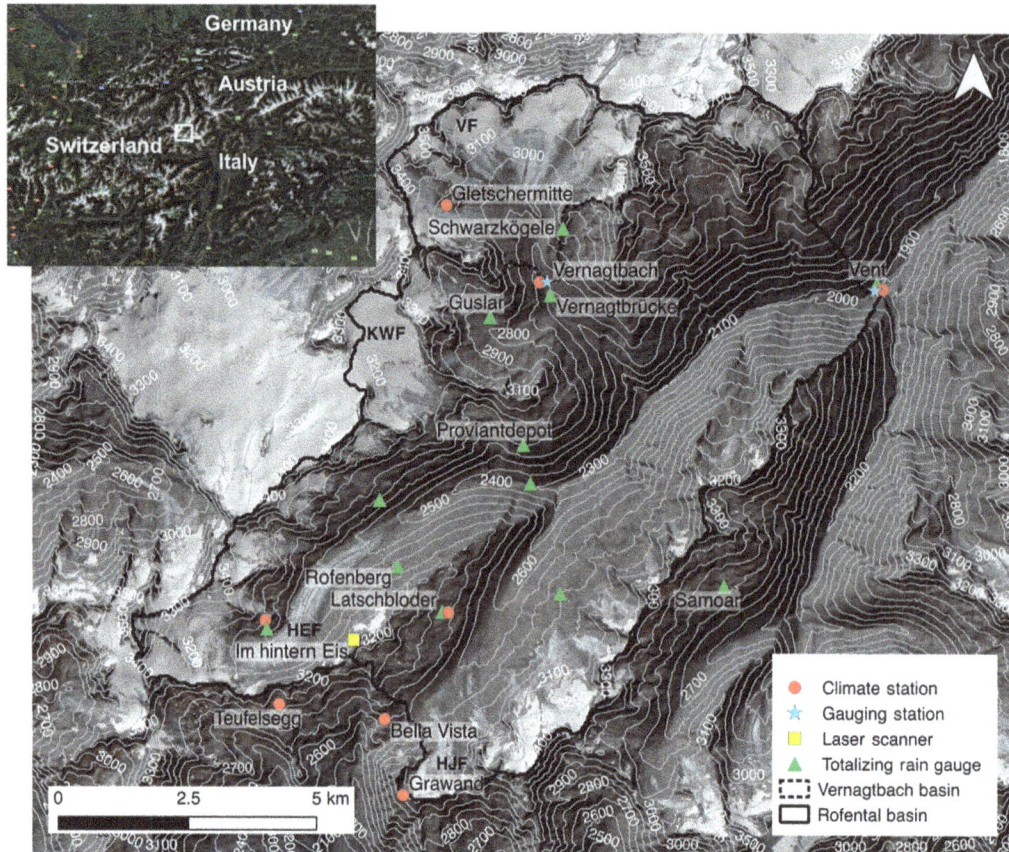

Figure 4. The Rofenache (98.1 km^2) and Vernagtbach (11.44 km^2) catchments with permanent meteorological stations and the runoff gauges.

and ice during spring and summer, respectively. The early melt-season onset is typically in April. The gauge at Vernagtbach (2635 m a.s.l.) has been operationally maintained by the Bavarian Academy of Sciences and Humanities since 1973 and is the highest streamflow recording site in Austria with measurements also documented at http://ehyd.gv.at (see above) since 2003. The Vernagtbach catchment stretches from 2635 m a.s.l. at the gauge to 3635 m a.s.l. at the summit of Hinterer Brochkogel. According to the glacier inventory of 2006 (https://doi.org/10.1594/PANGAEA.844985), the ice coverage of the Vernagtbach at that time was 71 %. The current rapid decrease of the glaciated area in the Rofental is documented in the WGMS database (http://wgms.ch).

The climate of the Rofental is characterized as an inner Alpine dry type (Fig. 5). The mean annual temperature at the station in Vent (1900 m a.s.l., 46.85833° N, 10.91250° E) is 2.5 °C, and total annual precipitation varies between 797 mm in Vent (1982–2003, Kuhn et al., 2006) and > 1500 mm in the higher altitudes around 3000 m a.s.l., confirmed by the recordings at the various totalisators (see Sect. 3.2.3, Table 4). In these higher regions, seasonal snow cover lasts from October until the end of June. Figure 5 shows temper-

atures and precipitation of the station at Vent at 1900 m a.s.l. for the period 1969–2006.

The geological bedrock in the Rofental area mainly consists of biotite-plagioclase, biotite and muscovite gneisses, variable mica schists, and gneissic schists of the Austroalpine Ötztal nappe (Kreuss, 2012; Moser, 2012). Subordinate lithologies are quartzites and graphite schists. Granitic gneisses, amphibolites and diabase occur as layers ranging from a few meters to a few hundred meters thick within the metasedimentary sequence. Land cover in the Rofental is dominated by mountain pastures and coniferous forests in the lower areas, but these only cover little of the area (source: "Land Tirol", http://data.tirol.gv.at). Permafrost is likely to occur at north-facing slopes at higher altitudes (Klug et al., 2016).

Infrastructure in the valley is very good for monitoring instrumentation and fieldwork. A research station at HEF (built in 1966 in 3026 m a.s.l.) and one at Vernagtbach (built in 1973 in 2637 m a.s.l.) serve as logistic bases for fieldwork on the two glaciers. Several mountain huts are located in the Rofental, namely the "Vernagthütte/Würzburger Haus" (2755 m a.s.l.), the "Hochjoch-Hospiz" (2413 m a.s.l.), the "Brandenburger Haus" (3277 m a.s.l.) and close by the

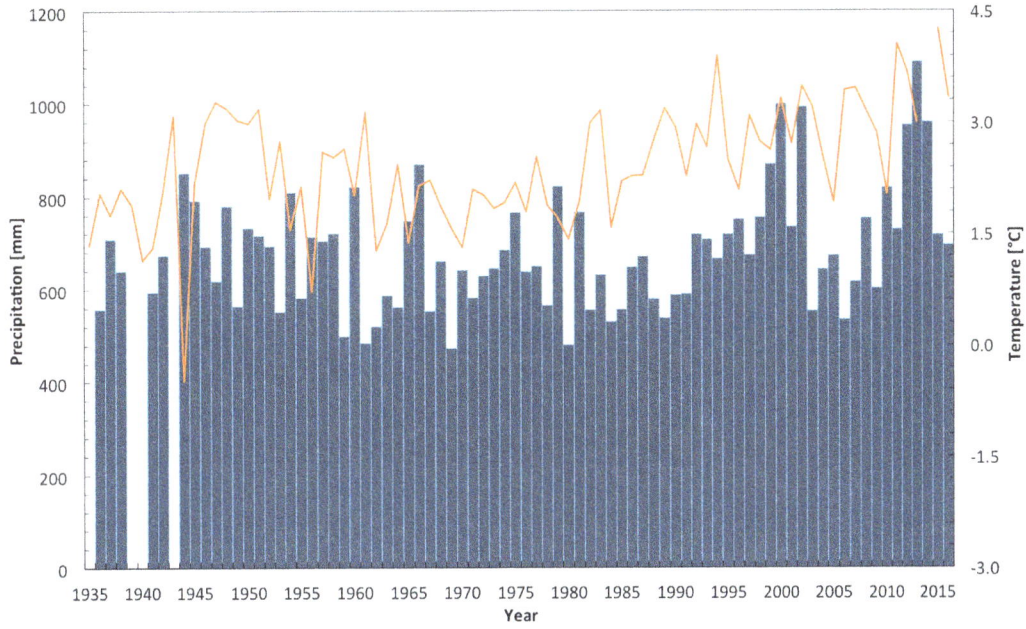

Figure 5. Annual precipitation sums (blue bars) and mean annual temperatures (orange line) 1935–2016 for the valley station "Vent" (1900 m a.s.l.). Data from PANGAEA (1935–2011: https://doi.org/10.1594/PANGAEA.806582 and 2012–2016: https://doi.org/10.1594/PANGAEA.876595).

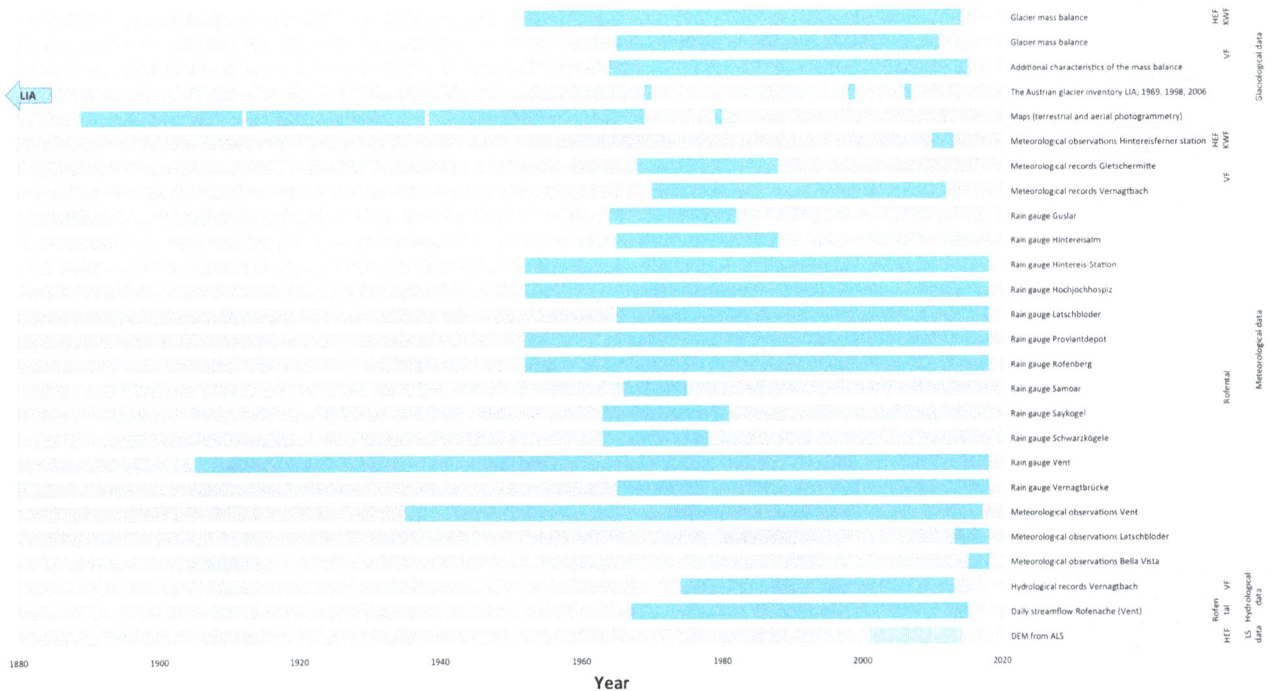

Figure 6. Available data time series for the Rofental in PANGAEA, structured in glaciological data, meteorological data, hydrological data and laser scanning data. LIA = little ice age.

Austrian–Italian borderline at the Hochjoch the "Schöne Aussicht" (also known as "Bella Vista", 2845 m a.s.l.), within the Schnalstal glacier ski resort (http://schnalstal.com/en/ glacier). The "Rofenhöfe" (2014 m a.s.l.), the highest permanently settled mountain farm in Austria, is well situated as base camp in the lower valley floor.

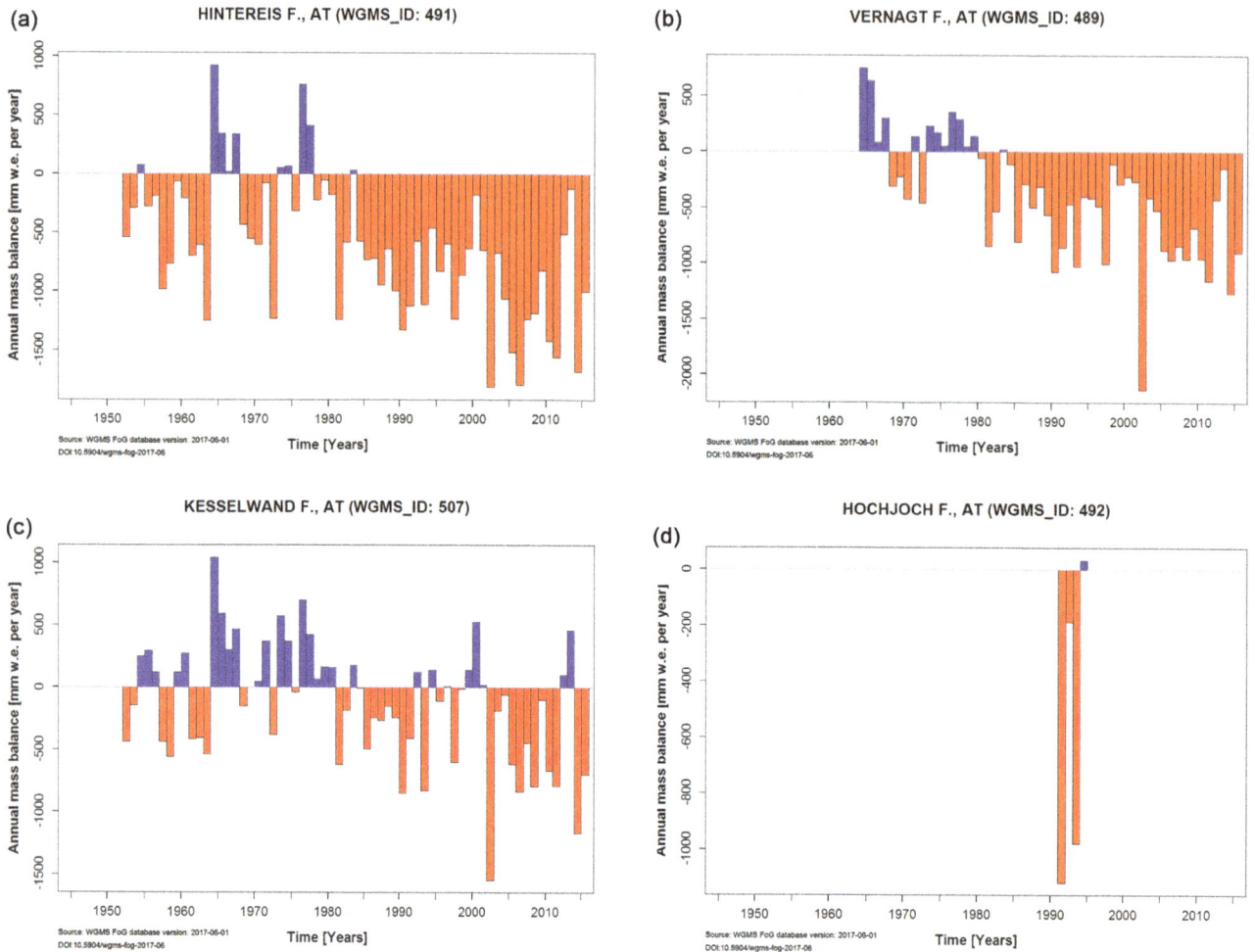

Figure 7. Annual mass balances for Hintereisferner (**a**), Vernagtferner (**b**), Kesselwandferner (**c**) and Hochjochferner (**d**) after the glaciological method using stake readings and snow pit data.

3 The data

In the following chapter the Rofental data are presented in the chronological order in which it has been recorded (Fig. 6). First, we describe the long time series of glaciological data (Sect. 3.1). HEF and KWF are monitored by the University of Innsbruck (Austria), whereas VF is monitored by the Bavarian Academy of Sciences and Humanities (Munich, Germany). Next, we describe the meteorological data, again structured by location, i.e., in the order HEF and KWF, VF, and then the valley area (Sect. 3.2). Following that, we describe the hydrological data, firstly in the HEF and KWF catchments, then the VF catchment, and finally, the Rofental as a whole (Sect. 3.3). In the last section (Sect. 3.4), the airborne and terrestrial laser scanning data, which serves glaciological, geomorphological and wider modeling purposes, is described.

3.1 Glaciological data

All glaciers in the Rofental are included in the three Austrian glacier inventories carried out in 1969, 1998 and 2006 (http://glaziologie.at/gletscherinventar.html) and also in the inventory of reconstructed glaciers at the time of the LIA (Kuhn et al., 1999, 2009, 2012; Lambrecht and Kuhn, 2007; Fischer et al., 2015), allowing detailed studies of the deglaciation in the catchments. The parent directory on PANGAEA for the Austrian glacier inventory is https://doi.org/10.5194/tc-9-753-2015, including the LIA maximum (https://doi.org/10.1594/PANGAEA.844987), and the years 1969 (https://doi.org/10.1594/PANGAEA.844983), 1998 (https://doi.org/10.1594/PANGAEA.844984) and 2006 (https://doi.org/10.1594/PANGAEA.844985). Annual reports of the variations of HEF (ID 491, since 1952), KWF (ID 507, since 1966), VF (ID 489, since 1965) and HJF (ID 492, 1991–1995) are provided to the World Glacier Monitoring Service (WGMS; http://wgms.ch) (Fig. 7).

3.1.1 Hintereisferner and Kesselwandferner

Changes in the areas of HEF and KWF have been documented on the basis of maps, aerial photos, and more recently satellite and airborne derived digital elevation models (DEMs) since the early 19th century (Lambrecht and Kuhn, 2007; see also the introduction). The traditional glaciological method of determining glacier-wide mass balance involves spatial extrapolation of local measurements of ablation and accumulation to provide values of the climatic mass balance, encompassing changes at the glacier surface and in the near subsurface (Cogley et al., 2011). Uncertainties in the methods are discussed in Zemp et al. (2013). Since 1952, a network of measurement stakes and pits was continuously maintained at HEF to directly measure the mass balance of the glacier (Hoinkes, 1970). Summer and winter mass balances have been measured separately. Interpretation of the surface mass changes of HEF is supported by the availability of daily images from an automatic camera which views the upper part of the glacier to below the ELA, providing useful information on the distribution of snow cover and the pattern of snow melt over the glacier surface. From 1952–2013 the mass balance of KWF has also been recorded, but with a much smaller number of observations and the assumption that the spatial distribution of the mass balance is analogous to the one of the adjacent HEF. Since summer 2013 a full network of observations is also undertaken and maintained at KWF. The mass balance values of HEF and KWF are available at WGMS (ID 491 and 507, since 1952 and 1966; see Fig. 7) and are also archived in the PANGAEA database (https://doi.org/10.1594/PANGAEA.803830, https://doi.org/10.1594/PANGAEA.803829, https://doi.org/10.1594/PANGAEA.818898 and https://doi.org/0.1594/PANGAEA.818757).

Geodetic determinations of glacier mass balance involve determining the volume change of the whole glacier body, encompassing englacial and basal volume changes. The volume change must then be converted into a mass change which is complicated in the case of a rapidly changing glacier whose surface type can change significantly over the monitoring interval. Apart from the geodetic mass balances on the basis of the glacier inventories, airborne laser scanning images are available for HEF since 2001 (see Sect. 3.4). From these, a time series of geodetic mass balances was derived for 2001–2002 to 2010–2011 (Fig. 8). This method allows for pixel-by-pixel correction of method-inherent discrepancies in the classical glaciological method (Klug et al., 2017).

3.1.2 Vernagtferner

Glaciological mass balance measurements for VF have been available since 1965 (Mayer et al., 2013a). Since the beginning, annual and winter balance was measured separately in order to discriminate ice melt and snow accumulation. The

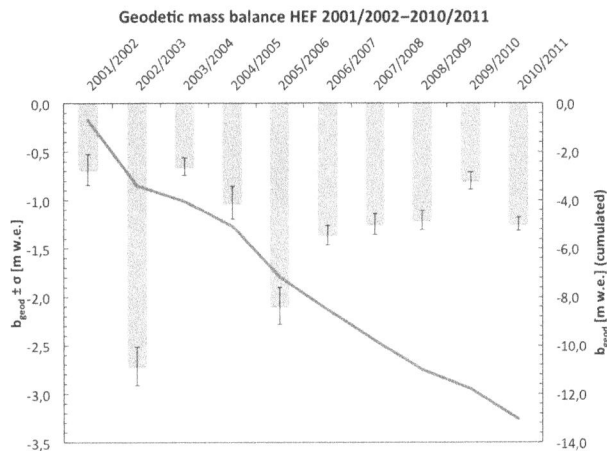

Figure 8. Available geodetic mass balances $b_{\mathrm{geod}} \pm \sigma$ [m w.e.] for Hintereisferner from 2001–2011 and the cumulated balance 2001–2011 (b_{geod} cumulated) as derived from airborne laser scanning measurements (Klug et al., 2017; see also Sect. 3.4).

mass balance values are available at WGMS (ID 489, since 1965; see Fig. 7) and are also archived in the PANGAEA database (https://doi.org/10.1594/PANGAEA.853832). The measurements are based on stake readings and snow pits for the annual balance, and snow depth probing and snow pits for the winter balance. The point data are then interpolated on the temporally closest map of the glacier. A summary of the mass balance series for VF is given in Fig. 9. Additional characteristics of the Vernagtferner mass balance for the period 1964–2014 are available at https://doi.org/10.1594/PANGAEA.854639.

In 1976, an automatic analogue camera has been installed on "Schwarzkögele" (3075 m a.s.l., 46.86575° N, 10.83245° E), capturing one picture of VF and its surrounding per day during the ablation period. Since 2010, three digital pictures per day are produced throughout the year (Weber, 2013). A time series of maps is available for VF since 1889, derived from terrestrial and aerial photogrammetry, aerial laser scanning and optical line scanner images; these maps are used to determine area and volume changes of the glacier for longer periods (Table 1 and Mayer et al., 2013b).

3.2 Meteorological data

Due to its complex topography, the Rofental is characterized by steep environmental gradients and large spatiotemporal variations of meteorological conditions. An ongoing effort has been undertaken to supplement data available from the lower regions with additional AWS installations in the higher elevations. As of 2017, the Rofental offers a comprehensive pool of valuable observations and model forcing data for mountain catchment hydrology research (Fig. 4; see also the introduction in Sect. 1).

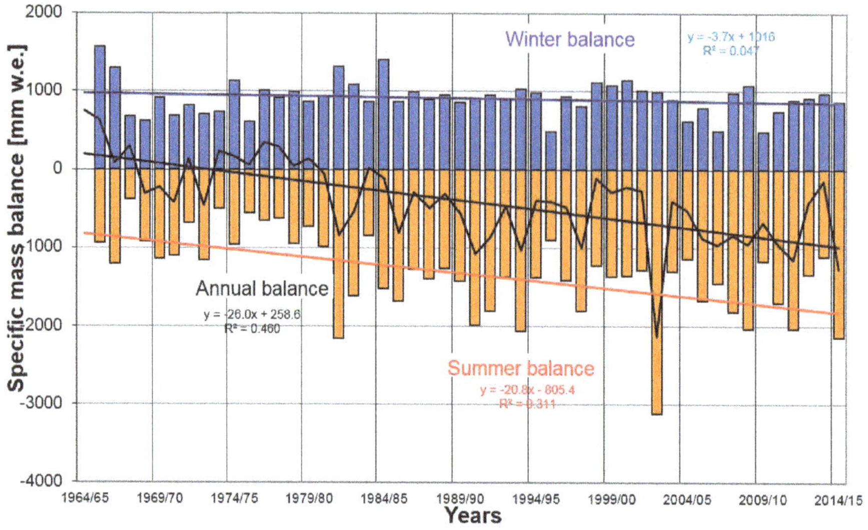

Figure 9. Graphical summary of the seasonal and annual mass balances of Vernagtferner from 1965 until 2014. The vertical bars represent seasonal values and the black line annual values.

Table 1. Available maps of Vernagtferner.

Year	Type	Area (km^2)	Volume change (km^3)	DOI
1889	Terrestrial photogrammetry	11 549		https://doi.org/10.1594/PANGAEA.834873
1912	Terrestrial photogrammetry	11 509	−2.137	https://doi.org/10.1594/PANGAEA.834873
1938	Terrestrial photogrammetry	10 410	−4.382	https://doi.org/10.1594/PANGAEA.834873
1954	–	9474	−4.543	*
1969	Aerial photogrammetry	9466	0.634	https://doi.org/10.1594/PANGAEA.834873
1979	Aerial photogrammetry	9397	1.840	https://doi.org/10.1594/PANGAEA.771301
1990	Aerial photogrammetry	8982	−4.931	*
1999	Aerial photogrammetry	8680	−7.888	*
2003	Aerial photogrammetry	8430	−3.355	*
2006	Optical line scanner	8173	−8.970	*
2009	Optical line scanner	7748	−8.403	*

* Data set will be uploaded to PANGAEA as soon as it is processed, quality checked and documented.

For all stations, the height of the sensors above ground is at least 1.5 m; in winter, the distance between the snow surface and the sensors can become much smaller, and in extreme snow-rich periods the instruments even can become completely snow-covered. Such periods can be recognized in the data by typical recordings of zero wind speed and increasing dampening of the other meteorological variables.

3.2.1 Hintereisferner and Kesselwandferner

The first meteorological observations at HEF and its surrounding began in 1968 and are documented in Kuhn et al. (1979). Short term projects, dedicated to measuring the surface mass balance of snow and ice on the glacier involved temporary installations of AWSs on the glacier (e.g., Siogas, 1977a; Harding et al., 1989; Obleitner, 1994). Since 2010, automatic meteorological measurements have been carried out at station "Hintereisferner" (3026 m a.s.l., 46.79867° N, 10.76042° E) (Fig. 10). The installation as of 2017 is detailed in Table 2.

Since 2014 an AWS has been seasonally operated on the glacier terminus, providing the meteorological data for surface energy balance assessments of the ice surface, complemented by surface height change observations with a sonic ranger.

From summer 2017, the permanent meteorological measurements at Hintereisferner will be complemented by recordings of the sensors on a 6 m high tower that has been equipped for detailed turbulent flux measurements (see Sect. 3.2.3). This new station is situated close to the permanent terrestrial laser scanner on Im hintern Eis (3244 m a.s.l., 46.79586° N, 10.78277° E; see Sect. 3.4.2).

Figure 10. The Hintereisferner AWS and research station in June 2013 (3026 m a.s.l., 46.79867° N, 10.76042° E), view to the northeast. Photo by Christian Wild.

Figure 11. The Vernagtbach AWS in July 2017 (2640 m a.s.l., 46.85663° N, 10.82857° E), view to the southeast. Photo by Ludwig Braun.

For KWF, only the meteorological observations on the glacier from 1958 by Ambach and Hoinkes (1963) are documented.

3.2.2 Vernagtferner

After the start of the glacier monitoring program by the Bavarian Academy of Sciences and Humanities, meteorological observations were initiated at the glacier forefield with the installation of a precipitation gauge in 1970. With the completion of the gauging station "Vernagtbach" (2635 m a.s.l., 46.85675° N, 10.82886° E; see Sect. 3.3.2), additional meteorological parameters have been observed since 1974 at the Vernagtbach climate station close by (2640 m a.s.l., 46.85663° N, 10.82857° E) (https://doi.org/10.1594/PANGAEA.775113). In 1975 hourly measurements of air temperature, relative humidity, air pressure, wind speed and direction were started. The same instruments were installed at the station "Gletschermitte" (3078 m a.s.l.), situated on a rock outcrop in the western part of the glacier at 46.86894° N, 10.80299° E, where hourly meteorological observations were collected during the summer months from 1968 until 1987 (with a varying beginning and end from year to year). The observed parameters were air temperature, relative humidity, wind speed and direction, and precipitation (https://doi.org/10.1594/PANGAEA.832562). Radiation sensors were installed at Vernagtbach in 1976. However, especially during winter, data gaps frequently occurred. The situation was considerably improved by installation of a first digital data logger in 1984. Since then, all-year data records are available from the Vernagtbach station (Escher-Vetter and Siebers, 2013). The meteorological observations were revised in 2002 with the installation of a modern AWS (Fig. 11), and since then all data are automatically transferred to the Bavarian Academy of Sciences and Humanities via GSM and a satellite network. In August 2010, the

Hydrographic Service of Tyrol extended the installation with a separate temperature sensor and an unheated Pluvio 2 pluviometer (2630 m a.s.l., 46.85667° N, 10.82861° E). Details about the most recent sensor configurations are given in Table 3.

On "Schwarzkögele" (3075 m a.s.l., 46.86575° N, 10.83245° E), a summit in the vicinity of VF, an autonomous climate station has been in operation since 1976 (Braun et al., 2013). Data recorded there comprise air temperature, relative humidity, global radiation, wind speed and direction as well as precipitation. After digitization these measurements will be made available in PANGAEA. Experiments and special investigations in the catchment of VF are listed in Escher-Vetter and Siebers (2013); since 2003, the meteorological observations have been extended to the ice surface of the glacier itself. Whereas in the first years these data have gaps (mainly in winter), they are mostly continuous since 2011.

3.2.3 Rofental apart from the glaciers

In the Rofental several totalizing rain gauges have are used to collect precipitation data, the first of which was installed in 1905 (see Fig. 6). The totalizing rain gauges are operated by the Department of Atmospheric and Cryospheric Sciences of the University of Innsbruck with financial support by the Hydrographic Service of Tyrol. These totalizing rain gauges provide a valuable picture of the historical evolution and temporal variability of precipitation, and they support the development of precipitation fields derived from interpolation of the recordings (Hoinkes and Steinacker, 1975), which is particularly important for distributed modeling exercises. Evaporation and freezing of the devices is inhibited by annual additions of oil and salt to the gauge reservoir. Readings of totals are undertaken every 2 months in summer with a 4-month break in winter. These totals are then redistributed to monthly values using the recordings of the weighing rain gauge in Vent. Altitude, geographical location and the period

Table 2. Weather and snow variables recorded by the sensors installed at the station Hintereisferner (3026 m a.s.l., 46.79867° N, 10.76042° E) in 2010, 2011 and 2012. Accuracy according to technical data sheets of the manufacturers. Original temporal resolution of the data records is 10 min.

Variable	Sensor	Period of operation	Accuracy	Unit
Air temperature	Vaisala HMP45AC	Since October 2010–present	±0.13 °C	°C
Relative humidity	Vaisala HMP45AC	Since October 2010–present	±2 % RH for 0–90 % RH, ±3 % RH for 90–100 % RH	%
Wind speed and direction	Young Wind Monitor	Since October 2010–present	0.5–1 m s^{-1}; ±3°	m s^{-1} and °
Shortwave and longwave radiative fluxes	Kipp & Zonen CNR 4	Since October 2010–present	±10 % (outgoing) < 10 % (incoming)	W m^{-2}
Atmospheric pressure	Setra CS 100	Since October 2010–present	±0.1 hPa	hPa
Soil and snow temperature	BetaTherm 100K6A	Since October 2010–present	±0.3 °C	°C
Snow depth	Campbell SR50A	Since October 2010–present	1 cm (or 0.4 % of distance)	cm

Data 2010: https://doi.org/10.1594/PANGAEA.809091; 2011: https://doi.org/10.1594/PANGAEA.809094; 2012: https://doi.org/10.1594/PANGAEA.809095.

Table 3. Sensors and sampling intervals of the AWS Vernagtbach (2640 m a.s.l., 46.85663° N, 10.82857° E)*.

Variable	Sensor	Period of operation	Interval	Unit
Air temperature (ventilated)	Thies PT-100	Since 2002	5 s 10 min^{-1}	°C
Air temperature (unventilated)	PT-100	Since 2002	5 s 10 min^{-1}	°C
Relative humidity	Thies hair hygrometer	Since 2002	20 s 10 min^{-1}	%
Wind speed	Thies cup anemometer	Since 2002	5 s 10 min^{-1}	m s^{-1}
Wind direction	Thies wind vane	Since 2002	5 s 10 min^{-1}	°
Shortwave downward radiation	Kipp & Zonen CM7B unventilated	Since 2002	5 s 10 min^{-1}	W m^{-2}
Shortwave upward radiation	Kipp & Zonen CM7B unventilated	Since 2002	5 s 10 min^{-1}	W m^{-2}
Longwave downward radiation	Schenk Pyradiometer 8111 unventilated	Since 2002 (summer only)	5 s 10 min^{-1}	W m^{-2}
Longwave upward radiation	Schenk Pyradiometer 8111 unventilated	Since 2002 (summer only)	5 s 10 min^{-1}	W m^{-2}
Precipitation sum	Belfort weighing gauge	Since 2002	5 s 10 min^{-1}	mm
Precipitation difference	Gertsch tipping bucket, unheated	Since 2002	Sum in 10 min	mm
Air pressure	Druck RPT 410	Since 2002	20 s 10 min^{-1}	hPa
Snow depth	Campbell SR50	Since 2002	120 s 10 min^{-1}	mm

* Further technical details can be found in Escher-Vetter and Siebers (2012). The additional temperature and rainfall recordings of the Hydrographic Service of Tyrol have been separately available since August 2010, visualized online at http://apps.tirol.gv.at/hydro/#/Niederschlag/?station=197075; data upon request.

of operation of these totalizing rain gauges are given in Table 4.

Meteorological observations have been made in close proximity to the village of Vent since 1934 (Lauffer, 1966; Siogas, 1977b). This long-term station provides a valuable reference for shorter series of meteorological observations and also the lower boundary conditions on the likely variation of meteorological variables with elevation across the catchment. Data are available for 1935–2011 at https://doi.org/10.1594/PANGAEA.806582, and for 2012–2016 at https://doi.org/10.1594/PANGAEA.876595. In September 2015 the weather station installation was updated and the position changed by a horizontal displacement of 102 m (new position: 1907 m a.s.l., 46.85745° N, 10.91288° E). The recordings of this station comprise the meteorological variables air temperature, relative humidity, wind speed and direction (since February 2016), and atmospheric pressure (since September 2016; same instruments

and specifications as for station Hintereisferner, Table 2). Precipitation is recorded with a heated Ott Pluvio 2 in millimeters per hour with an accuracy of 0.1 mm h^{-1}.

In the uppermost parts of the Hochjoch valley, two AWSs have been more recently brought into operation: "Latschbloder" (2919 m a.s.l., 46.80106° N, 10.80659° E), installed in September 2013, and Bella Vista (2805 m a.s.l., 46.78284° N, 10.79138° E), installed in June 2015. The data of these two fully automatic stations complement the spatial picture of the meteorological variables in the upper zones of the Rofental. Both stations collect 10 min values of temperature, precipitation (unheated, but also recording by the type of precipitation), wind (mean and maximum speed and direction), relative humidity, radiative fluxes (incoming and outgoing short- and longwave) and air pressure (Table 5). The Latschbloder station is about 25 m from the totalizing rain gauge at the same site (see Table 4). Three continuous years of consistent records are available for 2014,

Table 4. Totalizing rain gauges in the Rofental.

Station	Altitude (m a.s.l.)	Lat. (° N)	Long. (° E)	Period of operation	DOI
Vent	1900	46.85766	10.91127	1905–present	https://doi.org/10.1594/PANGAEA.876532
Hochjochhospiz	2360	46.82310	10.82616	1952–present	https://doi.org/10.1594/PANGAEA.876525
Vernagtbrücke	2600	46.85461	10.82979	1965–present	https://doi.org/10.1594/PANGAEA.876533
Proviantdepot	2737	46.82951	10.82407	1952–present	https://doi.org/10.1594/PANGAEA.876527
Rofenberg	2827	46.80847	10.79344	1952–present	https://doi.org/10.1594/PANGAEA.876528
Latschbloder	2910	46.80118	10.80561	1965–present	https://doi.org/10.1594/PANGAEA.876526
Hintereis station	2964	46.79727	10.76096	1952–present	https://doi.org/10.1594/PANGAEA.876523
Saykogel	2990	46.80491	10.83459	1963–1980	https://doi.org/10.1594/PANGAEA.876529
Schwarzkögele	3075	46.86575	10.83245	1963–March 1977	https://doi.org/10.1594/PANGAEA.876531
Guslar	2920	46.85060	10.81489	October 1964–September 1981	https://doi.org/10.1594/PANGAEA.876522
Samoar	2650	46.80708	10.87539	1966–1974	https://doi.org/10.1594/PANGAEA.876530
Hintereisalm	2900	46.81941	10.78842	1965–September 1987	https://doi.org/10.1594/PANGAEA.876524

2015 and 2016 (mean annual temperature −1.3, −1.5 and −2.2 °C (WXT520), and annual precipitation: 1590, 1311 and 1118 mm (Pluvio 2 unheated)).

The weather station at Bella Vista includes a heated Ott Pluvio 2, supplied with main power from the "Schöne Aussicht-Hütte" approximately 90 m away. This site also includes fuller snow instrumentation (Table 6) to measure: snow water equivalent (by means of a snow pillow), snow depth (by means of an ultrasonic ranger) and snow temperature profile (by means of a series of temperature sensors at different height levels). These snow sensors are still undergoing technical examination and development, and as yet no check for consistency has been undertaken (e.g., by using the data as input in a snow model as in Morin et al., 2012). In 2016, mean annual temperature at the station was −0.4 °C, and annual precipitation was 1605 mm. The Bella Vista weather and snow monitoring station is one of the highest of its kind in the Alps. During summer 2017 an automatic camera that has the station in its field of view has been installed (Fig. 12).

In summer 2017, a 6 m high tower close to the permanent terrestrial laser scanner on Im hintern Eis (3244 m a.s.l., 46.79586° N, 10.78277° E; see Sect. 3.4.2) was equipped for detailed turbulent flux measurements with the following sensors: three Lufft Ventus 2-D Sonic wind sensors at 1.5, 3 and 6 m altitude; two ventilated Rotronic HC2-S3 temperature–humidity sensors at 3 and 6 m altitude; a Campbell SR50 AH heated ultrasonic snow depth sensor in 1.5 m altitude; a Campbell Krypton hygrometer at 3 m altitude (only for specific campaigns); a Kipp & Zonen CNR4 net radiometer at 1.5 m altitude; a Metek USA-1 3-D sonic turbulence sensor at 3 m altitude; and a Setra 278 barometric pressure sensor in the logger box. The data of these sensors will be made available in PANGAEA as soon as first tests have proven the installation to be reliable, and a time series of at least a year of data is available.

Figure 12. Webcam picture of the Bella Vista weather station (2805 m a.s.l., 46.78284° N, 10.79138° E), view to the East. Left: Ott Pluvio 2 with wind shelter. Center: temperature profiler. Right: snow pillow (foreground) and mast with sensors (background) for wind speed and direction, snow height, temperature/humidity and radiative fluxes. The most recent picture is available at http://alpinehydroclimatology.net.

Three additional AWSs are located south of the Rofental, in the Italian Schnalstal: "Teufelsegg" (3035 m a.s.l., 46.7847° N, 10.7647° E) close to the accumulation area of HEF, "Grawand" (3220 m a.s.l., 46.7703° N, 10.7966° E) in the Schnalstal glacier ski resort and "Vernagt" (1950 m a.s.l., 46.7357° N, 10.8493° E) close to the village and the lake with the same name. These stations are maintained by the Hydrographic Office of the Civil Protection Agency of the Autonomous Province of Bolzano – South Tyrol. Their data are available at http://daten.buergernetz.bz.it/de/dataset/misure-meteo-e-idrografiche.

Table 5. Climate and snow variables recorded by the sensors installed at the station Latschbloder (2919 m a.s.l., 46.80106° N, 10.80659° E). Accuracy according to technical data sheets of the manufacturers. Original temporal resolution of the data records is 10 min.

Variable	Sensor	Period of operation	Resolution and accuracy	Unit
Air temperature	Vaisala WXT520	Since September 2013	$0.1\,°C \pm 0.3\,°C$	°C
Relative humidity	Vaisala WXT520	Since September 2013	$0.1\,\% \pm 3\,\%$ RH for 0–90 % RH, $0.1\,\% \pm 5\,\%$ RH for 90–100 % RH	%
Wind speed and direction	Vaisala WXT520	Since September 2013	$0.1\,\mathrm{m\,s^{-1}} \pm 3\,\%$ (speed) $1° \pm 3\,\%$ for $10\,\mathrm{m\,s^{-1}}$ (direction)	$\mathrm{m\,s^{-1}}$ and °
Radiative fluxes (short- and longwave)	Kipp & Zonen CNR 4	Since September 2013	10–$20\,\mathrm{W\,m^{-2}}$ (incoming) 5–$15\,\mathrm{W\,m^{-2}}$ (outgoing)	$\mathrm{W\,m^{-2}}$
Precipitation	Vaisala WXT520 Friedmann tipping bucket Ott Pluvio 2 v. 200 with wind shelter	Since September 2013 September 2013 to June 2014 Since July 2014	$0.01\,\mathrm{mm\,h^{-1}} \pm 5\,\%$* (not yet known) $0.01\,\mathrm{mm\,h^{-1}} \pm 1\,\%$	mm
Atmospheric pressure	Vaisala WXT520	Since September 2013	$0.1\,\mathrm{hPa} \pm 0.5\,\mathrm{hPa}$ for 0–30 °C $0.1\,\mathrm{hPa} \pm 1.0\,\mathrm{hPa}$ for −52–60 °C	hPa

* for hailstorm: $0.1\,\mathrm{hit\,cm^{-2}}$. The Vaisala WXT520 records rain and hail as well as their durations and intensities, but cannot recognize snowfall.
Data 2013: https://doi.org/10.1594/PANGAEA.879215; 2014: https://doi.org/10.1594/PANGAEA.879216; 2015: https://doi.org/10.1594/PANGAEA.879217; 2016: https://doi.org/10.1594/PANGAEA.879218; 2017: https://doi.org/10.1594/PANGAEA.879219.

Table 6. Weather and snow variables recorded by the sensors installed at the station Bella Vista (2805 m a.s.l., 46.78284° N, 10.79138° E). Accuracy according to technical data sheets of the manufacturers. Original temporal resolution of the data records is 10 min.

Variable	Sensor	Period of operation	Resolution and accuracy	Unit
Air temperature	E+E EE08 Vaisala WXT520	Since July, 2015	$< 0.5\,°C$[1] $0.1\,°C \pm 0.3\,°C$	°C
Relative humidity	E+E EE08 Vaisala WXT520	Since July, 2015	$\pm 2\,\%$ RH for 0–90 % RH, $\pm 3\,\%$ RH for 90–100 % RH $0.1\,\% \pm 3\,\%$ RH for 0–90 % RH, $0.1\,\% \pm 5\,\%$ RH for 90–100 % RH	%
Wind speed and direction	Vaisala WXT520 Kroneis 262	Since July 2015	$0.1\,\mathrm{m\,s^{-1}} \pm 3\,\%$ (speed) $1° \pm 3\,\%$ for $10\,\mathrm{m\,s^{-1}}$ (direction)	$\mathrm{m\,s^{-1}}$ and °
Radiative fluxes	Kipp & Zonen CNR 4	Since July 2015	10–$20\,\mathrm{W\,m^{-2}}$ (incoming) 5–$15\,\mathrm{W\,m^{-2}}$ (outgoing)	$\mathrm{W\,m^{-2}}$
Precipitation	Vaisala WXT520 Ott Pluvio 2 v. 200 with wind shelter	Since July, 2015 Since July, 2015	$0.01\,\mathrm{mm\,h^{-1}} \pm 5\,\%$[2] $0.01\,\mathrm{mm\,h^{-1}} \pm 1\,\%$	mm
Atmospheric pressure	Vaisala WXT520	Since July, 2015	$0.1\,\mathrm{hPa} \pm 0.5\,\mathrm{hPa}$ for 0–30 °C $0.1\,\mathrm{hPa} \pm 1.0\,\mathrm{hPa}$ for −52–60 °C	hPa
Snow water equivalent	Sommer snow pillow 3×3	(still experimental)[3]	(still experimental)	mm
Snow depth	Sommer USH-8	(still experimental)[3]	$1\,\mathrm{mm} \pm 0.1\,\%$	mm
Snow temperature	Pilz temperature profiler	(still experimental)[3]	(still experimental)	°C

[1] depending on air temperature; see technical data sheet of the manufacturer. [2] for hailstorm: $0.1\,\mathrm{hit\,cm^{-2}}$. The Vaisala WXT520 records rain and hail as well as their durations and intensities, but cannot recognize snowfall. [3] data not yet downloadable from PANGAEA. Data 2015: https://doi.org/10.1594/PANGAEA.879210; 2016: https://doi.org/10.1594/PANGAEA.879211; 2017: https://doi.org/10.1594/PANGAEA.879212.

3.3 Hydrological data

3.3.1 Hintereisferner and Kesselwandferner catchment

During the International Geophysical years 1957 to 1959 a gauging station was in operation at "Steg Hospiz" (2287 m a.s.l.), registering the combined streamflow from HEF and KWF (Lang, 1966). The measurements of this campaign are described here as an example for the many short- and longer-term monitoring activities carried out in the Rofental. The Steg Hospiz catchment is 26.6 km^2 in size, with a fraction of 58 % being covered by the glaciers at that time. Mean annual recorded streamflow for the catchment area amounted to 1848 mm (1957–1958) and 1770 mm (1958–1959), respectively (millimeters are equivalent to liters per square meter per year). Winter runoff (October through March) only was 5 and 10 % of the annual amount, whereas the three summer months (July through September) provided 76 and 72 %. Highest mean monthly streamflow amounts were registered in August 1958 (575 mm) and July 1959 (559 mm). The frequency distribution of daily stream-flow for 1957–1958 shows a period of daily low flows of less than 0.5 m^3 s^{-1} (18.8 L s^{-1} km^{-2}) for 217 days (October through May). Higher daily streamflow > 6.0 m^3 s^{-1} (225 L s^{-1} km^{-2}) only occurred during July and August. The observed maximum daily streamflow was 16.9 m^3 s^{-1}. The glacier contribution, determined as the fraction of (observed) negative mass balance to recorded streamflow, was relatively high in this period: 24 and 20 % of annual streamflow, respectively. The exponential decrease of the hydrograph in fall to the minimum in spring suggests that the winter streamflow mainly originates as delayed meltwater of the previous season from the glaciers. The increase in the water flows after the beginning of the snow melt period occurs with a certain time delay – due to the refreezing of meltwater, and its retention in the snow cover (Lang, 1966).

As of 2017, neither HEF nor KWF are equipped with a permanently registering discharge gauge; this is projected for future initiatives.

3.3.2 Vernagtferner catchment

The VF catchment is one of the very few glacierized catchments where simultaneous measurements of glacier mass balance and discharge exist for several decades since 1974 (Escher-Vetter and Reinwarth, 2013). Discharge is measured at the gauge Vernagtbach (46.85675° N, 10.82886° E), which at 2635 m a.s.l. is the highest streamflow gauge in Austria (Fig. 13). The water level is continuously monitored in the gauge since 1974, and water-level-to-discharge calibrations are regularly conducted by the salt injection method. The water level is simultaneously determined by three sonic rangers distributed across the runoff channel in order to detect the 2-D surface geometry of the water flow, and surface velocity is monitored by a Doppler system. The total catchment area covers 11.44 km^2, 7.3 km^2 of

Figure 13. The Vernagtbach gauging station in 2006 (2635 m a.s.l., 46.85675° N, 10.82886° E), view to the northwest. Photo by Ludwig Braun.

which was glacierized in 2015 (63.8 %). The vertical extent ranges from 2635 m a.s.l. at the gauge to 3635 m a.s.l. at the summit of "Hinterer Brochkogel" (see also Fig. 4). The discharge values are available as 5 min values for 2002 to 2012 (https://doi.org/10.1594/PANGAEA.829530). Hourly hydrological records for the Vernagtbach catchment are available for the period 1974 to 2001 at https://doi.org/10.1594/PANGAEA.775113. Monthly averages of discharge are available at https://doi.org/10.1594/PANGAEA.832432, and yearly values are available for the period 1974 to 2012 at https://doi.org/10.1594/PANGAEA.832429.

3.3.3 Rofental catchment

Streamflow in the Rofental catchment is recorded at the gauge "Vent–Rofenache" (1891 m a.s.l., 46.85722° N, 10.91083° E, 98.1 km^2) at the outlet of the catchment (https://doi.org/10.1594/PANGAEA.876119). Vent–Rofenache is one of the highest operational observation sites of the Hydrographic Service of Tyrol, providing a continuous time series of streamflow and sediment transport recordings since 1967 and 1999, respectively. The regime is dominated by snow melt and ice melt with a significant maximum in July and August (Fig. 14); the glaciated area has decreased from 44 % (in 1969) to 38 % (in 2009; Müller et al., 2009) and is expected to drop to almost zero during the course of the 21st century, due to the rapidly changing climatic conditions (Hanzer et al., 2017). The coexistence of the Vernagtbach gauge in the VF head watershed allows for combined hydrological investigations. Episodical recordings at smaller tributaries of the Rofenache were obtained during the spring snow melt season in 2014 (Schmieder et al., 2016), and during the glacier melt season in 2016 (Schmieder et al., 2018).

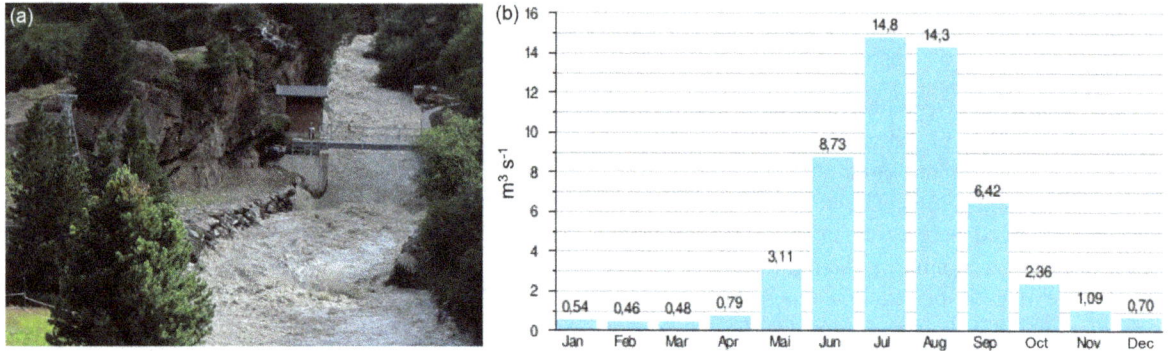

Figure 14. The rainfall totalisator and the gauge at Vent (1891 m a.s.l., 46.85722° N, 10.91083° E) (**a**; Photo: Hydrographic Service of the Tyrol), and mean monthly streamflow of the Rofenache 1971–2009 (**b**). Data from the Hydrographisches Jahrbuch (BMLFUW 2011).

3.4 Laser scanning data

Laser scanning is an active remote sensing technique which uses a laser beam to acquire 3-D point data, representing the surface and objects on that surface in high spatial resolution. In addition to high accuracy and resolution, additional information on the spectral properties intensity and reflectance of the scanned surface in the wavelength of the specific scanner is recorded. The DTMs derived from laser scanning measurements are becoming increasingly important for glaciological and geomorphological studies. DTM differencing is an important method for detection and quantification of surface changes and mass budget calculations (Sailer et al., 2014; Klug et al., 2017). Airborne laser scanning (ALS) and terrestrial laser scanning (TLS) are the most common methods for lidar data collection (Telling et al., 2017). The Rofental has been an intensive experimental site for both types of laser scanning data processing and analysis.

3.4.1 Airborne laser scans

ALS is well-suited for remote mountain areas, because no external light source is required and, due to overlapping flight strips, the entire surface is captured, even in very steep terrain (Höfle and Rutzinger, 2011). The vertical accuracy is in the order of 0.05 to 0.20 m, mainly dependent on the slope angle (Beraldin et al., 2010; Bollmann et al., 2011). Low vertical errors of 0.05 m were observed in the HEF catchment for areas with slope angle < 40° by comparison of an ALS-derived DTM with dGNSS (differential global navigation satellite system) measurements. For the areas with slope angle > 40°, an exponential increase of the vertical error was observed (1.0 m for 80°, Bollmann et al., 2011; Geist et al., 2005; Sailer et al., 2012).

The HEF data set is built from 21 separate ALS acquisitions from 2001 to 2013, covering an area of 32 km², including HEF (approx. 6.8 km² in 2011), KWF (approx. 3.8 km² in 2011), "Langtaufererjochferner" (approx. 1.1 km² in 2011) and "Stationsferner" (approx. 0.3 km² in 2011).

DEMs of 1 × 1 m were generated by calculating the z value (altitude) from the mean z value inside the respective grid cell by excluding 5 % of the smallest and largest observations of the ALS points. For the provision of the GeoTIFF (UTM32N) raster files the high-resolution DEMs were resampled to DEMs with a cell size of 10 × 10 m, applying bilinear interpolation. The resulting DEMs are available at https://doi.org/10.1594/PANGAEA.875889.

The available ALS measurements for HEF were used to study the volume changes of the glacier complementary to direct glaciological surface mass balance measurements. At least one scan was performed at the end of every glacier mass balance year (end of September). From these, a time series of geodetic mass balances was derived for 2001–2002 to 2010–2011 (see Fig. 8). In 2001–2002, 2002–2003, 2007–2008, 2008–2009 and 2010–2011 additional project-based intermediate campaigns were carried out. An overview of the data (UTM32N, WGS84) including the technical details is given in Table 7. Klug et al. (2017) corrected the surface models as well as glacier mass balance data by considering method-inherent uncertainties originating from snow cover, survey dates and density assumptions to calculate corrected annual geodetic mass balances. The most negative balance year is 2002–2003, with a mean specific mass balance of −2.713 ± 0.20 m water equivalent (w.e.). In the subsequent mass balance year 2003–2004, the smallest mass loss is observed (mean specific mass balance −0.654 ± 0.09 m w.e.). For the entire observation period (2001 to 2011) a mean mass balance of −1.3 m w.e. was calculated with the HEF ALS data set (Klug et al., 2017).

With the same ALS data sets as described in Table 7, Helfricht et al. (2014b) investigated the spatial snow distribution and its interannual persistence for a partly glacierized mountain area including HEF and KWF (∼ 36 km²).

Table 7. Overview of the available HEF ALS data with flight date, used Optech sensor, mean flight height above ground (m), maximum scanning angle (°), pulse repetition rate (Hz), across-track overlap (%), mean point density (points m^{-2}) and vertical accuracy (m). Data are available at https://doi.org/10.1594/PANGAEA.875889.

Flight date	Optech Sensor	Mean height above ground (m)	Maximum scanning angle (°)	Pulse repetition rate (Hz)	Across-track overlap (%)	Mean point density (points m^{-2})	Vertical accuracy (m)
11 Oct 2001	ALTM 1225	900	20	25 000	24	1.1	0.11
9 Jan 2002	ALTM 1225	900	20	25 000	24	1.2	0.14
7 May 2002	ALTM 1225	900	20	25 000	24	1.2	0.14
15 Jun 2002	ALTM 1225	900	20	25 000	24	1.3	0.14
8 Jul 2002	ALTM 1225	900	20	25 000	24	1.4	0.15
19 Aug 2002	ALTM 1225	900	20	25 000	24	1.4	0.10
18 Sep 2002	ALTM 3033	900	20	33 000	24	1.0	0.10
4 May 2003	ALTM 2050	1150	20	50 000	40	0.8	0.10
12 Aug 2003	ALTM 2050	1150	20	50 000	40	0.8	0.10
26 Sep 2003	ALTM 1225	900	20	25 000	24	1.0	0.06
5 Oct 2004	ALTM 2050	1000	20	50 000	24	2.0	0.07
12 Oct 2005	ALTM 3100	1000	22	70 000	50–70	3.4	0.07
8 Oct 2006	ALTM 3100	1000	20	70 000	37–75	2.0	0.08
11 Oct 2007	ALTM 3100	1000	20	70 000	37–75	3.4	0.06
9 Sep 2008	ALTM 3100	1000	20	70 000	40-45	2.2	0.06
7 May 2009	ALTM 3100	–	–	–	–	–	–
30 Sep 2009	ALTM 3100	1100	20	70 000	31–66	2.7	0.05
8 Oct 2010	ALTM Gemini	1000	25	70 000	62	3.6	0.03
4 Oct 2011	ALTM 3100	1100	20	70 000	25–75	2.9	0.04
11 May 2012	ALTM 3100	1200	20	70 000	31–66	2.8	0.06
3 Sep 2013	ALTM Gemini	1200	20	70 000	59–80	4.2	0.04

3.4.2 Terrestrial laser scans – the permanent laser scanner on Im hintern Eis

The utility of TLS in performing high-resolution glacier observations was tested for HJF from 2013–2015 when both glaciological and geodetic mass balances were measured. Likewise, a Riegl VZ-6000 TLS was used to produce surface models within 0.1 m of coincident high-accuracy ALS data for approx. 80 % of the surface of HJF. The TLS surface models are of higher spatial resolution and surface details than the ALS. The TLS data were used – along with coincident optical imagery from the onboard camera – to produce high-resolution surface classifications to map snow-cover extent, and to explore the spatial patterns of the surface mass balance of HJF (Prantl et al., 2017).

In 2016, a permanent TLS was installed in a climate controlled container at 3244 m a.s.l. close to the summit Im hintern Eis (46.79586° N, 10.78277° E). The 3-D laser scanner VZ-6000 (manufactured by RIEGL Laser Measurement Systems GmbH) offers a long measurement range of more than 6000 m and operates with beam divergence of 0.12 mrad at a wavelength of 1050 nm, particularly suited for snow- and ice-related applications. The device operates with an angular step width of 0.01° for a field of view of 60° vertically and 120° horizontally and a laser pulse repetition rate of 30 kHz, equivalent to an effective measurement rate of 23 000 measurements s^{-1}, leading to a point density of approx. 10 pts m^{-2} for a distance of 1000 m, and to approx. 1 to 2 pts m^{-2} for the remote areas at a distance of > 3500 m. Hence, the instrument is ideal for static topographic applications such as monitoring of glaciological and geomorphological processes in high mountain terrain. Due to the large field of view, nearly the entire HEF catchment area can be covered by the instrument (Fig. 15); only a very small part of the terminus and small flat areas in the upper glacier zones cannot be sampled. The TLS surface point cloud is accompanied by high-quality optical imagery (5 megapixel). This installation not only allows high temporal and spatial resolution glacier volume changes for HEF to be determined; in addition, the high spatial resolution supports the monitoring of surface features such as crevasses, evolving surface roughness, the supraglacial drainage network and geomorphodynamically induced surface changes (e.g., debris flows) or snow avalanches. Seven scans have been carried out during the test phase (September 2016 to April 2017).

Figure 15. The terrestrial laser scanner in its container housing at Im hintern Eis (3244 m a.s.l., 46.79586° N, 10.78277° E). View to the west over the upper areas of Hintereisferner to the summit of Weisskugel (3739 m a.s.l.) and Langtauferer Spitze (3529 m a.s.l.). Photo by Rudolf Sailer.

tivities. These data sets, however, represent only a small fraction of what has been observed and collected during the past few decades. Many measurements have been conducted during special field campaigns and are not yet documented for online data publication. For example, hydrological investigations have been conducted in the valley of the Hochjochbach, including streamflow observations with pressure sensors and tracer experiments since 2014 (Schmieder et al., 2016, 2018). Other data await digitization and processing. Originally published in analogue printwork, these data include 3 years of hourly ice temperatures from HEF (Markl and Wagner, 1978), observations of sublimation of ice and snow at HEF (Kaser, 1982, 1983), firn investigations at KWF (Ambach et al., 1978), remote sensing experiments and many energy balance investigations at HEF (Jaffé, 1958, 1960; Hoinkes and Untersteiner, 1952; Hoinkes, 1953a, b; Ambach and Hoinkes, 1963; Wendler, 1967; Kuhn et al., 1979, Wagner, 1979, 1980), and many others. For VF, the situation is similar; a comprehensive overview of the research work has been published by Braun and Escher-Vetter (2013), including various descriptions and documentations of the used data, methods and models. Older long-term field experiments at VF are described in Moser et al. (1986). Runoff data of the gauge "Vent" have been widely used in hydrological studies, the most recent ones including Schmieder et al. (2016) and Hanzer et al. (2016, 2017). On a regional scale, the Rofental data contributed to modeling exercises for future scenario climatic conditions and their effect on simulated streamflow discharge (Marke et al., 2011, 2013). The Hydrographic Service of Tyrol visualizes its data online at apps.tirol.gv.at/hydro, and provides their download at http://ehyd.gv.at. As of 2017, data of the sites

Latschbloder and Schöne Aussicht are used in several ongoing research projects (listed at http://alpinehydroclimatology.net), and also provide valuable experimental material for application in various student courses.

Several initiatives are also ongoing at the international research network level. Apart from its long history within UNESCO IHP (http://en.unesco.org/themes/water-security/hydrology), the Rofental recently became a research basin in the framework of the GEWEX INARCH project (http://words.usask.ca/inarch). It is a research catchment of the ERB Euro-Mediterranean Network of Experimental and Representative Basins (http://erb-network.simdif.com), and a regular complex site in the LTSER platform Tyrolean Alps (http://lter-austria.at/ta-tyrolean-alps), which belongs to the national and international long term ecological research network (LTER Austria, LTER Europe and ILTER). Hintereisferner station is part of the EU Horizon 2020 INTERACT framework of Arctic (and a few Alpine) research stations (https://eu-interact.org/field-sites/station-hintereis/).

The efforts to provide the Rofental data to the scientific community will continue in all of the currently involved institutions.

Conclusions and outlook

The Rofental in the Ötztal Alps (Austria) is a unique, high Alpine research basin (98.1 km², 1890–3770 m a.s.l.) with available time series of 150 years of glaciological and hydrometeorological observations. The glaciers in the Rofental attracted early attention and were – for centuries – observed due to the dangerous nature of frequently occurring glacier lake outburst floods. Over the last 100 years, the glacier monitoring has been accompanied by systematic recordings of meteorological and hydrological variables. Today, a glaciological and hydrometeorological data set is available for the Rofental that is without comparison worldwide, with regard to both its amount and temporal coverage. The Rofental data sets support manifold investigations in the context of coupled climate and glacier evolution, snow and glacier hydrology, water resources availability in mountainous regions, or method development like laser scanning or modeling. The scales of such research range from local, like particular micrometeorological assessments at a single observation site, to global, e.g., the estimation of the glacier contribution to sea level rise. This paper gives an overview of what has been measured and what is available already. All the data described are comprehensively documented and made freely available according to the Creative Commons Attribution License by means of the PANGAEA repository (https://doi.org/10.1594/PANGAEA.876120).

This paper covers availability of data as of fall 2017. There are still many manual and analogue historical recordings which are being digitized, processed and documented. After completion they will be made available via PANGAEA and will add to the extension of the records back in time (left side of the bars in Fig. 6). New and future data will continue to further build on the available Rofental database (right side of the bars in Fig. 6). This process of continuous enlargement of the data described here is ensured by the living data process as conceived by the journal.

Competing interests. The authors declare that they have no conflict of interest.

Special issue statement. This article is part of the special issue "Hydrometeorological data from mountain and alpine research catchments". It is not associated with a conference.

Acknowledgements. Many of the instruments and the monitoring activities presented here have been supported by the institutions to which the authors are affiliated to, and by countless research programs. These have been funded by the Bavarian Academy of Sciences and Humanities (BAdW), the Deutsche Forschungsgemeinschaft (DFG), the Austrian Academy of Sciences (ÖAW; project HydroGeM^3), the European Region Tyrol – South Tyrol – Trentino (project CRYOMON-SciPro – IPN 10-N33), the Autonomous Province of Bolzano – South Tyrol (project hiSnow – 23/40.3), the Austrian Federal Ministry of Agriculture, Forestry, Environment and Water Management (BMLFUW, section IV/4-water cycle), and others. The laser data acquisition and processing has been funded by the European Union (project OMEGA – EVK2-CT-2000-00069), the Austrian Research Promotion Agency FFG ASAP (Austrian Space Applications Programme; projects ALS-X – 815527 and SE.MAP – 840109), the Austrian Research Promotion Agency FFG COMET (Competence Centers for Excellent Technologies) in cooperation with the alpS GmbH (project MUSICALS – 826388), the Austrian Climate Research Programme (ACRP; project C4AUSTRIA – A963633), the Tyrolean Science Foundation (TWF) and the Department of Geography, University of Innsbruck.

All these funding institutions provide the support to continue our efforts in the monitoring of the water balance and climate elements of the Rofental in the long term. The authors gratefully acknowledge all the contributions and support from their countless colleagues in maintaining the instruments in the field, being so many that they cannot be listed here by name. Without their engagement, the long-term monitoring in the Rofental would never have been possible.

Edited by: Danny Marks

References

Abermann, J., Lambrecht, A., Fischer, A., and Kuhn, M.: Quantifying changes and trends in glacier area and volume in the Austrian Ötztal Alps (1969–1997–2006), The Cryosphere, 3, 205–215, https://doi.org/10.5194/tc-3-205-2009, 2009.

Ambach, W. and Hoinkes, H. C.: The heat balance of an Alpine snow field (Kesselwandferner, 3240 m, Oetztal Alps, 1958), IAHS Publ., 61, 24–36, 1963.

Ambach, W., Blumenthaler, M., Eisner, H., Kirchlechner, P., Schneider, H., Behrens, H., Moser, H., Oerter, H., Rauert, W., and Bergmann, H.: Untersuchungen der Wassertafel am Kesselwandferner (Ötztaler Alpen) an einem 30 Meter tiefen Firnschacht, Z. Gletscherkd. Glazialgeol., 14, 61–71, 1978.

Beraldin, J.-A., Blais, F., and Lohr U.: Laser scanning technology, in: Airborne and Terrestrial Laser Scanning, edited by: Vosselman, G. and Maas, H.-G., Whittles Publishing, Dunbeath, UK, 1–42, 2010.

Blümcke, A. and Hess, H.: Der Hochjochferner im Jahre 1883, Zeitschrift des Deutschen und Österreichischen Alpenvereins, Verlag des D. u. Ö. AV, München, 26 pp., 1895.

Blümcke, A. and Hess, H.: Untersuchungen am Hintereisferner, Wissenschaftliche Ergänzungshefte zur Zeitschrift des Deutschen und Österreichischen Alpenvereins, Verlag des D. u. Ö. AV, München, 87 pp., 1899.

BMLFUW: Hydrographisches Jahrbuch von Österreich 2009, Bundesministerium für Land- und Forstwirtschaft, Umwelt und Wasserwirtschaft, Wien, 2011.

Bollmann, E., Sailer, R., Briese, C., Stötter, J., and Fritzmann, P.: Potential of airborne laser scanning for geomorphologic feature and process detection and quantifications in high alpine mountains, Z. Geomorph., 55, 83–104, 2011.

Braun, L. and Escher-Vetter, H. (Eds.): Gletscherforschung am Vernagtferner, Themenband zum fünfzigjährigen Gründungsjubiläum der Kommission für Glaziologie der Bayerischen Akademie der Wissenschaften München, Z. Gletscherkd. Glazialgeol., 45/46, 389 pp., 2013.

Braun, L., Reinwarth, O., and Weber, M.: Der Vernagtferner als Objekt der Gletscherforschung, Z. Gletscherkd. Glazialgeol., 45/46, 85–104, 2013.

Brunner, K.: Karten und Ansichten des Vernagtferners seit 1600, Z. Gletscherkd. Glazialgeol., 45/46, 235–257, 2013.

Charalampidis, C., Braun, L., Fischer, A., Kuhn, M., Lambrecht, A., Mayer, C., Thomaidis, K. and Weber, M.: Effect of glacier geometry on mass loss in a warming climate, J. Geophys. Res.-Earth, in review, 2018.

Cogley, J. G., Hock, R., Rasmussen, L. A., Arendt, A. A., Bauder, A., Braithwaite, R. J., Jansson, P., Kaser, G., Müller, M., Nicholson, L., and Zemp, M.: Glossary of Glacier Mass Balance and Related Terms, IHP-VII Technical Documents in Hydrology No. 86, IACS Contribution No. 2, UNESCO-IHP, Paris, 2011.

Escher-Vetter, H. and Oerter, H.: Das Energiebilanz- und Abflussmodel PEV – frühe Modellansätze, Erweiterungen und ausgewählte Ergebnisse, Z. Gletscherkd. Glazialgeol., 45/46, 129–142, 2013.

Escher-Vetter, H. and Reinwarth, O.: Meteorologische und hydrologische Registrierungen an der Pegelstation Vernagtbach – Charateristika und Trends ausgewählter Parameter, Z. Gletscherkd. Glazialgeol., 45/46, 117–128, 2013.

Escher-Vetter, H. and Siebers, M.: Technical comments on the data records from the Vernagtbach station for the period 2002 to 2012, Commission for Geodesy and Glaciology, Section Glaciology Bavarian Academy of Sciences and Humanities, Munich, 2012.

Escher-Vetter, H. and Siebers, M.: Vom Registrierstreifen zur Satellitenübertragung – zur Entwicklung des Mess– und Aufzeichnungsgeräte an der Pegelstation Vernagtbach, Z. Gletscherkd. Glazialgeol., 45/46, 105–116, 2013.

Finsterwalder, S.: Der Vernagtferner, seine Geschichte und seine Vermessung in den Jahren 1888 und 1898, Wissenschaftliche Ergänzungshefte zur Zeitschrift des Deutschen und Österreichischen Alpenvereins, Verlag des D. u. Ö. AV, Graz, 112 pp., 1897.

Fischer, A., Seiser, B., Stocker Waldhuber, M., Mitterer, C., and Abermann, J.: Tracing glacier changes in Austria from the Little Ice Age to the present using a lidar-based high-resolution glacier inventory in Austria, The Cryosphere, 9, 753–766, https://doi.org/10.5194/tc-9-753-2015, 2015.

Geist, T. and Stötter, J.: First results on airborne laser scanning technology as a tool for the quantification of glacier mass balance, Proceedings of EARSeL-LISSIG-Workshop Observing our Cryosphere from Space, Bern, Switzerland, 11–13 March 2002.

Geist, T., Elvehøy, H., and Jackson, M.: Investigations on intra-annual elevation changes using multi-temporal airborne laser scanning data: case study Engabreen, Norway, Ann. Glaciol. 42, 195–201, https://doi.org/10.3189/172756405781812592, 2005.

Hanzer, F., Helfricht, K., Marke, T., and Strasser, U.: Multilevel spatiotemporal validation of snow/ice mass balance and runoff modeling in glacierized catchments, The Cryosphere, 10, 1859–1881, https://doi.org/10.5194/tc-10-1859-2016, 2016.

Hanzer, F., Förster, K., Nemec, J., and Strasser, U.: Projected cryospheric and hydrological impacts of 21st century climate change in the Ötztal Alps (Austria) simulated using a physically based approach, Hydrol. Earth Syst. Sci. Discuss., https://doi.org/10.5194/hess-2017-309, in review, 2017.

Harding, R. J., Entrasser, N., Escher-Vetter, H., Jenkins, A., Kaser, M., Kuhn, M., Morris, E. M., and Tanzer, G.: Energy and mass balance studies in the firn area of the Hintereisferner, in: Glacier fluctuations and climatic change, Glaciology and Quaternary Geology, Kluwer, Dordrecht, 325–341, 1989.

Helfricht, K., Kuhn, M., Keuschnig, M., and Heilig, A.: Lidar snow cover studies on glaciers in the Ötztal Alps (Austria): comparison with snow depths calculated from GPR measurements, The Cryosphere, 8, 41–57, https://doi.org/10.5194/tc-8-41-2014, 2014a.

Helfricht, K., Schöber, J., Schneider, K., Sailer, R., and Kuhn, M.: Interannual persistence of the seasonal snow cover in a glacierized catchment, J. Glaciol., 60, 889–904, https://doi.org/10.3189/2014JoG13J197, 2014b.

Hess, H.: Die Gletscher, Nabu Press, Braunschweig, 426 pp., 1904.

Höfle, B. and Rutzinger, M.. Topographic airborne LiDAR in geomorphology: A technological perspective, Z. Geomorph., 55, 1–29, https://doi.org/10.1127/0372-8854/2011/0055S2-0043, 2011.

Hoinkes, H.: Zur Mikrometeorologie der eisnahen Luftschicht, Arch. Met. Geoph. Biokl., 4, 451–458, 1953a.

Hoinkes, H.: Wärmeumsatz und Ablation auf Alpengletschern: II – Hornkees (Zillertaler Alpen), Geogr. Ann., 35, 116–140, https://doi.org/10.1080/20014422.1953.11880853, 1953b.

Hoinkes, H.: Methoden und Möglichkeiten von Massenhaushaltsstudien auf Gletschern, Ergebnisse der Messreihe Hintereisferner (Ötztaler Alpen) 1953–1968, Z. Gletscherkd. Glazialgeol., 6, 37–90, 1970.

Hoinkes H. und Steinacker, R.: Zur Parametrisierung der Beziehung Klima – Gletscher, Rivista Italiana di Geofisica, I, 97–104, 1975.

Hoinkes, H. and Untersteiner, N.: Wärmeumsatz und Ablation auf Alpengletschern I, Vernagtferner (Ötztaler Alpen), August 1950, Geogr. Ann., 34, 99–158, 1952.

Hoinkes, H., Dreiseitl, E., and Wagner, H. P.: Mass Balance of Hintereisferner and Kesselwandferner 1963/64 to 1972/73 in Relation to the Climatic Environment, Preliminary results of the combined water, ice and heat balances project in the Rofental, IHD-Activities in Austria 1965–1974, Report of the Int. Conference on the Results of the IHD, Paris, 2–14 September 1974, 42–53, 1974.

Jaffé, A.: Neuere Albedo- und Extinktionsmessungen an Gletschereisplatten, Ber. Dt. Wetterdienstes, 54, 273–274, 1958.

Jaffé, A.: Über Strahlungseigenschaften des Gletschereises, Archiv Met. Geoph. Biokl., Edn. 10, 376–395, https://doi.org/10.1007/BF02243201, 1960.

Juen, M., Mayer, C., Lambrecht, A., Eder, K., Wirbel, A., and Stilla, U.: Einsatz einer Thermalkamera und von Strahlungssensoren zur Oberflächenklassifizierung am Vernagtferner, Z. Gletscherkd. Glazialgeol., 45/46, 185–201, 2013.

Kaser, G.: Measurements of Evaporation from Snow, Archiv Met. Geoph. Biokl., Ser. B, 30, 333–340, 1982.

Kaser, G.: Über die Verdunstung auf dem Hintereisferner, Z. Gletscherk. Glazialgeol., 19, 149–162, 1983.

Kaser, G., Grosshauser, M., and Marzeion, B.: Contribution potential of glaciers to water availability in different climate regimes, P. Natl. Acad. Sci. USA, 107, 20223–20227, 2010.

Klug, C., Rieg, L., Ott, P., Mössinger, M., Sailer, R., and Stötter, J.: A Multi-Methodological Approach to Determine Permafrost Occurrence and Ground Surface Subsidence in Mountain Terrain, Tyrol, Austria, Permafrost Periglac., 28, 249–265, https://doi.org/10.1002/ppp.1896, 2016.

Klug, C., Bollmann, E., Galos, S. P., Nicholson, L., Prinz, R., Rieg, L., Sailer, R., Stötter, J., and Kaser, G.: A reanalysis of one decade of the mass balance series on Hintereisferner, Ötztal Alps, Austria: a detailed view into annual geodetic and glaciological observations, The Cryosphere Discuss., https://doi.org/10.5194/tc-2017-132, in review, 2017.

Kreuss, O.: Compiled geological map of ÖK50 Sheet 173 – Sölden, preliminary GEOFAST 1 : 50 000 edition 2016/03, Geologische Bundesanstalt Wien Publishing House, 2012.

Kuhn, M.: Micro-meteorological conditions for snow melt, J. Glaciol., 33, 24–26, 1987.

Kuhn, M., Kaser, G., Markl, G., Wagner, H. P., and Schneider, H. (Eds.): 25 Jahre Massenhaushaltsuntersuchungen am Hintereisferner, Auszug aus den glazialmeteorologischen Arbeiten im Gebiet des Hintereisferners in den Ötztaler Alpen, Institut für Meteorologie und Geophysik der Universität Innsbruck, 80 pp., 1979.

Kuhn, M., Kaser, G., Markl, G., Nickus, U., and Obleitner, F.: Fluctuations of Mass Balance and Climate: The 30 Years Record of Hinterreis- and Kesselwandferner, Z. Gletscherkd. Glazialgeol., 22, 490–416, 1985a.

Kuhn, M., Markl, G., Kaser, G., Nickus, U., Obleitner, F., and Schneider, H.: Fluctuations of climate and mass balances: Different responses of two adjacent glaciers, Z. Gletscherkd. Glazialgeol., 21, 409–416, 1985b.

Kuhn, M., Dreiseitl, E., Hofinger, S., Markl, G., Span, N., and Kaser, G.: Measurements and models of the mass balance of Hintereisferner, Geogr. Ann. A, 81, 659–670, https://doi.org/10.1111/1468-0459.00094, 1999.

Kuhn, M., Abermann, J., Olefs, M., Fischer, A., and Lambrecht, A.: Gletscher im Klimawandel: Aktuelle Monitoring- programme und Forschungen zur Auswirkung auf den Gebietsabfluss im Ötztal, Mitt. hydr. Dienst Österr., 86, 31–47, 2006.

Kuhn, M., Lambrecht, A., Abermann, J., Patzelt, G., and Groß, G.: Die österreichischen Gletscher 1998 und 1969, Flächen- und Volumenänderungen, Verlag der österr. Akad. Wiss., Wien, available at: hw.oeaw.ac.at/6616-0, 128 pp., 2009.

Kuhn, M., Lambrecht, A., Abermann, J., Patzelt, G., and Groß, G.: The Austrian Glaciers 1998 and 1969, Area and Volume Changes, Z. Gletscherkd. Glazialgeol., 43/44, 3–107, 2012.

Lambrecht, A. and Kuhn, M.: Glacier changes in the Austrian Alps during the last three decades, derived from the new Austrian glacier inventory, Ann. Glaciol., 46, 177–184, https://doi.org/10.3189/172756407782871341, 2007.

Lang, H.: Hydrometeorologische Ergebnisse aus Abflußmessungen im Bereich des Hintereisferners (Ötztaler Alpen) in den Jahren 1957–1959, Archiv Met. Geoph. Biokl. B, 14, 280–302, 1966.

Lauffer, I.: Das Klima von Vent, Dissertation, University of Innsbruck, available at: http://acinn.uibk.ac.at/sites/default/files/ Diss_Lauffer_Ingrid_1966.pdf, 1966.

Lindmayer, A.: Untersuchung zum Einfluss der Veränderung der Topographie der Eisoberfläche auf die spezifische Massenbilanz eines Gletschers am Beispiel des Vernagtferners im Zeitraum 1889–1938, Master Thesis, Catholic University of Eichstätt-Ingolstadt, Faculty of Mathematics and Geography, 102 pp., available at: http://kegglaziologie.de/download/MSc_ A_Lindmayer2015.pdf (last access: 26 November 2017), 2015.

Marke, T., Mauser, W., Pfeiffer, A., and Zängl, G.: A pragmatic approach for the downscaling and bias correction of regional climate simulations: evaluation in hydrological modeling, Geosci. Model Dev., 4, 759–770, https://doi.org/10.5194/gmd-4-759-2011, 2011.

Marke, T., Mauser, W., Pfeiffer, A., Zängl, G., Jacob, D., and Strasser, U.: Application of a hydrometeorological model chain to investigate the effect of global boundaries and downscaling on simulated river discharge, Environ. Earth Sci., 71/11, 4849–4846, https://doi.org/10.1007/s12665-013-2876-z, 2013.

Markl, G. and Wagner, H. P.: Messungen von Eis- und Firntemperaturen am Hintereisferner (Ötztaler Alpen), Symposium über die Dynamik temperierter Gletscher, Z. Gletscherkd. Glazialgeol., 13, 261–265, 1978.

Marzeion, B. and Levermann, A.: Loss of cultural world heritage and currently inhabited places to sea-level rise, Environ. Res. Lett., 9, 051001, https://doi.org/10.1088/1748-9326/9/3/034001, 2014.

Marzeion, B. and Kaser, G.: Peak Water: An Unsustainable Increase in Water Availability From Melting Glaciers, Mountains and climate change: A global concern, Centre for Development and Environment (CDE), Swiss Agency for Development and Cooperation (SDC) and Geographica Bernensia, Bern, 47–50, 2014.

Marzeion, B., Jarosch, A. H., and Hofer, M.: Past and future sea-level change from the surface mass balance of glaciers, The Cryosphere, 6, 1295–1322, https://doi.org/10.5194/tc-6-1295-2012, 2012a.

Marzeion, B., Hofer, M., Jarosch, A. H., Kaser, G., and Mölg, T.: A minimal model for reconstructing interannual mass balance

variability of glaciers in the European Alps, The Cryosphere, 6, 71–84, https://doi.org/10.5194/tc-6-71-2012, 2012b.

Marzeion, B., Cogley, J. G., Richter, K., and Parkes, D.: Attribution of global glacier mass loss to anthropogenic and natural causes, Science, 345, 919–921, https://doi.org/10.1126/science.1254702, 2014a.

Marzeion, B., Jarosch, A. H., and Gregory, J. M.: Feedbacks and mechanisms affecting the global sensitivity of glaciers to climate change, The Cryosphere, 8, 59–71, https://doi.org/10.5194/tc-8-59-2014, 2014b.

Mayer, C., Escher-Vetter, H., and Weber, M.: 46 Jahre glaziologische Massenbilanz des Vernagtferners, Z. Gletscherkd. Glazialgeol., 45/46, 219–234, 2013a.

Mayer, C., Lambrecht, A., Blumthaler, U., and Eisen, O.: Vermessung und Eisdynamik des Vernagtferners, Ötztaler Alpen, Z. Gletscherkd. Glazialgeol., 45/46, 259–280, 2013b.

Morin, S., Lejeune, Y., Lesaffre, B., Panel, J.-M., Poncet, D., David, P., and Sudul, M.: An 18-yr long (1993–2011) snow and meteorological dataset from a mid-altitude mountain site (Col de Porte, France, 1325 m alt.) for driving and evaluating snowpack models, Earth Syst. Sci. Data, 4, 13–21, https://doi.org/10.5194/essd-4-13-2012, 2012.

Moser, H., Escher-Vetter, H., Oerter, H., Reinwarth, O., and Zunke, D.: Abfluß in und von Gletschern, Technical University of Munich (TUM), Special Research Report 81, Final Report Project A1, GSF-Bericht, 41/86, 1–397, 1986.

Moser, M.: Compiled geological map of ÖK50 Sheet 172 – Weißkugel, preliminary GEOFAST 1 : 50 000 edition 2016/03, Geologische Bundesanstalt Wien, Publishing House, 2012.

Müller, G., Godina, R., and Gattermayer, W.: Der Pegel Vent/Rofenache – Herausforderungen für eine hydrographische Messstelle in einem vergletscherten Einzugsgebiet, Mitt. hydr. Dienst Österr., 86, 131–135, 2009.

Nicolussi, K.: Bilddokumente zur Geschichte des Vernagtferners im 17. Jahrhundert, Z. Gletscherkd. Glazialgeol., 26, 97–119, 1990.

Nicolussi, K.: Die historischen Vorstöße und Hochstände des Vernagtferners, Z. Gletscherkd. Glazialgeol., 45/46, 9–23, 2013.

Obleitner, F.: Climatological features of glacier and valley winds at the Hintereisferner (Ötztal Alps, Austria), Theor. Appl. Climatol., 49, 225–239, 1994.

Painter, T. H., Flanner, M. G., Kaser, G., Marzeion, B., VanCuren, R. A., and Abdalati, W.: End of the Little Ice Age in the Alps forced by industrial black carbon, P. Natl. Acad. Sci. USA, 110, 15216–15221, https://doi.org/10.1073/pnas.1302570110, 2013.

Prantl, H., Nicholson, L., Sailer, R., Hanzer, F., Juen, I., and Rastner, P.: Glacier snowline determination from terrestrial laser scanning intensity data, Geosciences, 7, 1–21, https://doi.org/10.3390/geosciences7030060, 2017.

Prasch, M., Marke, T., Strasser, U., and Mauser, W.: Large scale integrated hydrological modelling of the impact of climate change on the water balance with DANUBIA, Adv. Sci. Res., 7, 61–70, https://doi.org/10.5194/asr-7-61-2011, 2011.

Richter, E.: Urkunden über die Ausbrüche des Vernagt- und Gurglergletschers im 17. und 18. Jahrhundert, Forschungen zur deutschen Landes- und Volkskunde, 6, edited by: Engelhorn, J., Stuttgart, 345–440, 1892.

Sailer, R., Bollmann, E., Hoinkes, S., Rieg, L., Sproß, M., and Stötter, J.: Quantification of Geomorphodynamics in glaciated and recently deglaciated terrain based on airborne laserscanning data, Geogr. Ann. A, 94, 17–32, https://doi.org/10.1111/j.1468-0459.2012.00456.x, 2012.

Sailer, R., Rutzinger, M., Rieg, L., and Wichmann, V.: Digital elevation models derived from airborne laser scanning point clouds: appropriate spatial resolutions for multitemporal characterization and quantification of geomorphological processes, Earth Surf. Proc. Landf. 39, 272–284, https://doi.org/10.1002/esp.3490, 2014.

Schmieder, J., Hanzer, F., Marke, T., Garvelmann, J., Warscher, M., Kunstmann, H., and Strasser, U.: The importance of snowmelt spatiotemporal variability for isotope-based hydrograph separation in a high-elevation catchment, Hydrol. Earth Syst. Sci., 20, 5015–5033, https://doi.org/10.5194/hess-20-5015-2016, 2016.

Schmieder, J., Garvelmann, J., Marke, T., and Strasser, U.: Spatiotemporal tracer variability in the glacier melt end-member – How does it affect hydrograph separation results?, Hydrol. Process., in review, 2018.

Schöber, J., Schneider, K., Helfricht, K., Schattan, P., Achleitner, S., Schöberl, F., and Kirnbauer, R.: Snow cover characteristics in a glacierized catchment in the Tyrolean Alps – Improved spatially distributed modelling by usage of Lidar data, J. Hydrol., 519, 3492–3510, 2014.

Schöber, J., Achleitner, S., Bellinger, J., Kirnbauer, R., and Schöberl, F.: Analysis and modelling of snow bulk density in the Tyrolean Alps, Hydrol. Res., 47, 419–441, https://doi.org/10.2166/nh.2015.132, 2016.

Siogas, L.: Die Windverhältnisse an der Station Hintereis (3026 m) in den Ötztaler Alpen, Archiv Met. Geoph. Biokl. B, 25, 79–89, 1977a.

Siogas, L.: Die Luftdruckreihe Vent 1935–1970, Eine Analyse des jahres- u. tagesperiodischen sowie des aperiodischen Schwankungsverhaltens an einer inneralpinen Talstation im Vergleich zu anderen Stationen des Alpenraumes, Dissertation, University of Innsbruck, 150 pp., 1977b.

Telling, J., Lyda, A., Hartzell, P., and Glennie, C.: Review of Earth science research using terrestrial laser scanning, Earth-Sci. Rev. 169, 35–68, https://doi.org/10.1016/j.earscirev.2017.04.007, 2017.

Wagner, H.-P.: Strahlungshaushaltsuntersuchungen an einem Ostalpengletscher während der Hauptablationsperiode, Teil 1, Kurzwellige Strahlung, Archiv Met. Geoph. Biokl. B, 27, 297–324, 1979.

Wagner, H.-P.: Strahlungshaushaltsuntersuchungen an einem Ostalpengletscher während der Hauptablationsperiode, Teil 2, Langwellige Strahlung und Strahlungsbilanz, Archiv Met. Geoph. Biokl. B, 28, 41–62, 1980.

Weber, M.: Dokumentation der Veränderungen des Vernagtferners und des Guslarferners anhand von Fotografien, Z. Gletscherkd. Glazialgeol., 45/46, 49–84, 2013.

Weber, M., Braun, L., Mauser, W., and Prasch, M.: Die Bedeutung der Gletscherschmelze für den Abfluss der Donau gegenwärtig und in der Zukunft, Mitt. hydr. Dienst Österr., 86, 1–29, 2009.

Weber, M. and Prasch, M.: Influence of the glaciers on runoff regime and its change, chap. 56, in: Regional assessment of global change impacts, edited by: Mauser, W. and Prasch, M., Springer, Berlin, 493–509, 2015a.

Weber, M. and Prasch, M.: The influence of snow cover on runoff regime and its change, chap. 60, in: Regional assessment of global change impacts, edited by: Mauser, W. and Prasch, M., Springer, Berlin, 533–539, 2015b.

Wendler, G.: Die Vergletscherung in Abhängigkeit von Exposition und Höhe und der Gebietsniederschlag im Einzugsgebiet des Pegels Vent in Tirol, Archiv Met. Geoph. Biokl. B, 15, 260–273, 1967.

Zemp, M., Thibert, E., Huss, M., Stumm, D., Rolstad Denby, C., Nuth, C., Nussbaumer, S. U., Moholdt, G., Mercer, A., Mayer, C., Joerg, P. C., Jansson, P., Hynek, B., Fischer, A., Escher-Vetter, H., Elvehøy, H., and Andreassen, L. M.: Reanalysing glacier mass balance measurement series, The Cryosphere, 7, 1227–1245, https://doi.org/10.5194/tc-7-1227-2013, 2013.

5

Historical glacier outlines from digitized topographic maps of the Swiss Alps

Daphné Freudiger[1,2]**, David Mennekes**[1]**, Jan Seibert**[2]**, and Markus Weiler**[1]

[1]Chair of Hydrology, University of Freiburg, 79098, Freiburg, Germany
[2]Hydrology and Climate Unit, Department of Geography, University of Zurich, 8057, Zurich, Switzerland

Correspondence: Daphné Freudiger (daphne.freudiger@geo.uzh.ch)

Abstract. Since the end of the Little Ice Age around 1850, the total glacier area of the central European Alps has considerably decreased. In order to understand the changes in glacier coverage at various scales and to model past and future streamflow accurately, long-term and large-scale datasets of glacier outlines are needed. To fill the gap between the morphologically reconstructed glacier outlines from the moraine extent corresponding to the time period around 1850 and the first complete dataset of glacier areas in the Swiss Alps from aerial photographs in 1973, glacier areas from 80 sheets of a historical topographic map (the Siegfried map) were manually digitized for the publication years 1878–1918 (further called first period, with most sheets being published around 1900) and 1917–1944 (further called second period, with most sheets being published around 1935). The accuracy of the digitized glacier areas was then assessed through a two-step validation process: the data were (1) visually and (2) quantitatively compared to glacier area datasets of the years 1850, 1973, 2003, and 2010, which were derived from different sources, at the large scale, basin scale, and locally. The validation showed that at least 70 % of the digitized glaciers were comparable to the outlines from the other datasets and were therefore plausible. Furthermore, the inaccuracy of the manual digitization was found to be less than 5 %. The presented datasets of glacier outlines for the first and second periods are a valuable source of information for long-term glacier mass balance or hydrological modelling in glacierized basins. The uncertainty of the historical topographic maps should be considered during the interpretation of the results.

1 Introduction

The total glacier area of the central European Alps has considerably decreased during the last decades with differences of change in certain sub-periods (e.g. Fischer et al., 2014). Long-term glacier datasets are of great importance for understanding and assessing glacier changes (Fischer et al., 2015; Huss and Fischer, 2016) as well as for hydrological modelling of past and future streamflow (Huss, 2011; Stahl et al., 2016; Viviroli et al., 2011). Some glaciers of the central European Alps have been regularly monitored since nearly the end of the Little Ice Age (ca. 1850), but the majority were only recently or sporadically monitored and long time series of glacier data are rarely available (GLAMOS, 2015; WGMS, 2015). Remote sensing offers unique opportunities

to derive glacier outlines, areas, and glacier mass balance at the large scale. Several manual and (semi-)automated algorithms have been developed in recent decades to identify from remotely sensed data the entire glacier area of the central European Alps, leading to several glacier inventories starting from 1973 (e.g. Maisch et al., 2000; Paul et al., 2011; Fischer et al., 2014; Kääb et al., 2002). Assuming that the end of the Little Ice Age represents the largest glacier extent (Collins, 2008; Ivy-Ochs et al., 2009; Vincent et al., 2005), the outlines of the moraines correspond to the glacier cover from this recent maximum glacier extension around 1850. Mapping the moraines based on historical topographic maps, field observations, and aerial photographs from the years 1973, 1988, and 1989 therefore made it possible to also create a glacier inventory for the whole of Switzerland around 1850 (Maisch, 1992; Maisch et al., 2000, 2004; Müller et al., 1976). Between 1850 and 1973 no information on glacier area can be obtained from satellite images analysis and aerial photographs are only available locally. Nevertheless, other sources exist in Switzerland, such as historical topographic maps, where glacier areas have been surveyed and drawn manually.

The first topographic surveys started in 1809 in Switzerland, leading to the publication of the first topographic map for the whole of Switzerland (the Dufour map) based on geometric measurements at a scale of 1 : 100 000. It was subsequently published between 1845 and 1864. During the second half of the 19th century cartographic techniques were improved. For example, triangulation with angles was introduced (in ca. 1870), the absolute elevation of the "Pierre du Niton" was measured (in 1879), and the depth of the major Swiss lakes was assessed for the first time (in ca. 1870). These improvements made it possible to map glaciers in remote regions more accurately (Imhof, 1927). As a result, the Siegfried map was produced between 1868 and 1949 using the Dufour map as a baseline. The aim was to create homogenous maps for the whole of Switzerland for the Topographic Atlas of Switzerland at a scale of 1 : 50 000 for the Alps and 1 : 25 000 for the rest of Switzerland. The project started under the direction of the Chief of Staff, Hermann Siegfried, but most of the mapping was done by cartographers and topographers from the private sector. To ensure homogeneity, precise mapping instructions were set from the beginning (Imhof, 1927; Swisstopo, 2017). At that time, the Siegfried map was considered the most advanced topographic map ever produced; especially impressive was the drawing in the mountainous regions and the representation of rocks e.g. in glacierized areas (Imhof, 1927). Such historical topographic maps provide unique information on large-scale glacier areas for the time period 1868–1949 and are therefore valuable to fill the data gap between 1850 and 1973. They are linked, however, to uncertainties due to the mapping methods available at the time and possible errors in geo-referencing. Such uncertainties may sometimes lead to inaccuracies when glacier areas from historical maps are compared to other products, for example glacier areas from remotely sensed data (Imhof, 1927; Hall et al., 2003; Racoviteanu et al., 2009).

The aim of our study was (1) to digitize the historical Siegfried map at two time slices between 1892 and 1944; (2) to validate the digitized glacier areas through their comparison with glacier areas of different time periods and from different data sources in order to assess their accuracy at the large scale and locally; and (3) finally to create a dataset useful for long-term studies of glacier changes or hydrological modelling.

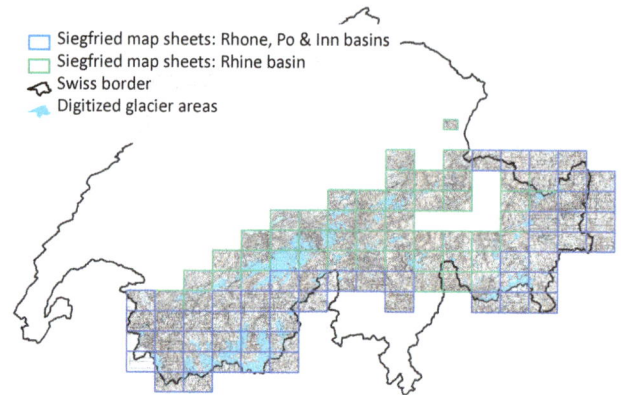

Figure 1. The 80 sheets of the Siegfried map covering the glacierized area of the Swiss Alps.

2 Data

2.1 Description of the Siegfried map of the Swiss Alps

The Siegfried map consists of a total of ca. 550 sheets that were revised at different publication years. Each sheet covers an area of 210 km² at a scale of 1 : 50 000 (Alpine regions) or 52.5 km² at a scale of 1 : 25 000. Elevation contours are represented every 30 and 10 m respectively (Fig. 1). The glacierized part of the Swiss Alps is covered by 80 sheets that we digitized for the publication years 1878–1918, with the highest frequency around 1900 (further called the first period) and 1917–1944, with the highest frequency around 1935 (further called the second period) as shown in Fig. 2. Only the original publication year is available for the sheets of the Siegfried map, which might differ from the survey year (see Sect. 3.5). The Siegfried map was digitized, geo-referenced, and made available by the Swiss Federal Office of Topography (Swisstopo).

The arithmetic precision requested from the topographers of the Siegfried map was 0.7 mm in the projection on the map (corresponding to 35 m in nature) for survey stations in the Alpine region (1 : 50 000). The contour lines are biased because the reference point "Pierre du Niton" is found to be 3.26 m higher than what was assumed at the time (Imhof, 1927). Errors can be up to 18 m because of this reference bias (Imhof, 1927). Furthermore, the measurement directives changed during the creation of the maps. At the beginning (around 1880) 300–500 survey points were needed for the creation of one sheet, while at the end of the 19th century, up to 6000 measurement points were prescribed (Imhof, 1927). Unfortunately, no information on the exact number of surveying stations was provided for the individual sheets (Swisstopo, Brigitte Schmied, personal communication, 24 January 2018). While the vertical accuracy of the Siegfried map has been estimated (Imhof, 1927; Rastner et al., 2016), large regional differences exist in the horizontal accuracy of the different sheets. These may relate to the number of surveying points (Caminada, 2003) and are therefore difficult to exactly estimate (Hall et al., 2003; Rastner et al., 2016).

2.2   Glacier areas and outlines for validation

We use four datasets of glacier areas and outlines covering the Swiss Alps for the years 1850, 1973, 2003, and 2010 (Fischer et al., 2014; Maisch et al., 2000; Müller et al., 1976; Paul et al., 2011) for the validation of the digitized glacier areas of the Siegfried map at the large scale for the first and second periods (around 1900 and around 1935). These four glacier inventories were produced with different technologies and methodologies summarized in Table 1. Furthermore, the outlines of seven glaciers (Silvretta, Oberaar, Unteraar, Limmern, Untergrindelwald, Damma, and Clariden) digitized by Andreas Bauder (ETH Zurich) from different historical maps from several years between 1864 and 1959 were available for local validation. The glacier outlines from years earlier than 1930 were digitized from the first publications of the Dufour and Siegfried maps and later than 1930 from the first publication of the National Map (e.g. Bauder et al., 2007, 2017; Huss et al., 2010). The glacier outlines used for local validation are visible in Fig. 4.

3   Digitization and validation of the Siegfried map

3.1   Digitization

All glacier areas of the 80 sheets from the Siegfried map were manually digitized using ArcMap 10.2.2 to create two shape files with the digitized glacier outlines of the first (around 1900) and second (around 1935) periods. Outcrops within the glaciers were removed. For the digitization, the study area was divided into two regions, the Rhine basin and the Rhone, Po, and Inn basins that were digitized by two different persons (Fig. 1) at a scale of 1 : 10 000. A third person finally controlled all digitized areas. Altogether, more than 500 000 nodes corresponding to an average of 28 nodes per kilometre of glacier outline and 250 working hours were needed to create the polygons and resulting shape files.

3.2   Data validation

3.2.1   Large-scale validation

To assess the quality and accuracy of the glacierized area from the Siegfried map at the large scale, the digitized glacier outlines of the first and second periods were compared with the glacier outlines of four available glacier inventories (Table 1, for the years ca. 1850, ca. 1973, 2003, and ca. 2010) in a two-step validation process. The accuracy of the Siegfried map is difficult to assess at the large scale (Hall et al., 2003; Rastner et al., 2016), as no contemporary data are available for comparison. The two-step validation process presented below, however, allowed us to assess whether the digitized

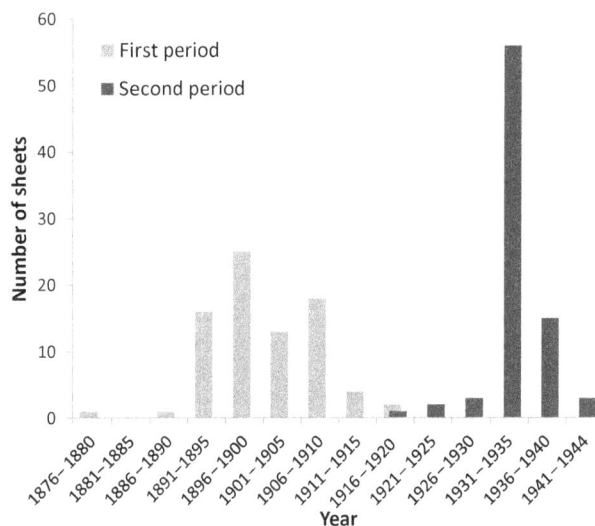

Figure 2. Frequency distribution of the publication years of the 80 sheets from the Siegfried map for the first and second periods.

glacier areas were consistent with the other available products, meaning that the digitized glacier area of the first and second periods followed a logical evolution compared with the other products.

In a first step, the shapes of the digitized glacier areas from the first and second periods were visually compared to the glacier shapes of the four inventories in order to ensure that they were consistent. During this comparison, the digitized glacier outlines from the first and second digitized periods that appeared in none of the other products were removed as the existence of a glacier in this location could not be verified. This was the case for ca. 0.03 % of the digitized area of the first and second periods (61.6 and 49.8 km^2 respectively).

To allow comparison between the glacier areas of the different data sources and available years, the digitized glacier outlines from the Siegfried maps and from the four inventories were divided into 957 glacier basins with a unique identity number, based on the river basin delineations given by the Federal Office for the Environment (FOEN). This method follows the recommendation of the GLIMS Analysis Tutorial (Racoviteanu et al., 2009). The total glacierized area was calculated for each of the 957 glacier basins and is further referred to as $A_{S,first}$ and $A_{S,second}$ for the two periods of the digitized Siegfried maps and as A_{1850}, A_{1973}, A_{2003}, and A_{2010} for the four glacier inventories. The basins were chosen to cover as many individual glaciers as possible. However, this delimitation was only used for the aim of comparison and does not represent the real delineation of a glacier area.

Assuming that all glaciers reached their maximum extent at the end of the Little Ice Age around 1850 in the central European Alps (Collins, 2008; Ivy-Ochs et al., 2009; Vincent et al., 2005), the glacier areas from 1850 should be the largest.

Table 1. Glacier inventories used for large-scale validation.

Year	Description	Format	References
ca. 2010	Aerial ortho-imageries acquired between 2008 and 2011	shp	Fischer et al. (2014)
2003	Landsat TM scenes acquired in autumn 2003	shp	Paul et al. (2011)
ca. 1973	Aerial photographs from September 1973	gif/tif	Müller et al. (1976), Maisch et al. (2000)
ca. 1850	Glacier outlines morphologically reconstructed from moraine extents of retreated glaciers from aerial photographs of 1973, 1988, and 1989, historical topographic maps, and field observation.	gif/tif	Müller et al. (1976), Maisch (1992), Maisch et al. (2000, 2004)

In the second validation step, $A_{S,first}$ and $A_{S,second}$ were therefore compared to A_{1850}, a product derived from the extent of the moraines identified from aerial photographs (Maisch et al., 2000; Müller et al., 1976). We then set the following conditions to assess the accuracy of $A_{S,first}$ and $A_{S,second}$.

- Highly consistent: $A_S < A_{1850}$

- Consistent: $(A_S - A_{1850}) / A_{1850} < 0.1$

- Poorly consistent: $0.1 > (A_S - A_{1850}) / A_{1850} < 0.5$

- Not consistent: $(A_S - A_{1850}) / A_{1850} > 0.5$

As land cover classification in remotely sensed data is not unequivocal (e.g. Racoviteanu et al., 2009) and definition and recognition of moraine partly rely on interpretation (Clark et al., 2004), the 1850 glacier inventory also shows uncertainties and we therefore considered A_S to be consistent with A_{1850} if A_S was up to 10 % larger than A_{1850}. In case of "poor consistency" or "no consistency" between the datasets, A_S and A_{1850} were further compared to the glacier areas of the further available products (A_{1973}, A_{2003}, and A_{2010}) to decide which one of the two products was more plausible. For this comparison, the shape of each glacierized area from $A_{S,first}$ and $A_{S,second}$ with poor or no consistency (in total 314 glacier basins) was visually compared to the shape of the corresponding glacierized area from A_{1973}, A_{2003}, and A_{2010}. It was assumed that glacier area decreased between 1850 and 2010 and that A_S or A_{1850} was more likely to be exact if its shape was most corresponding and overlapping the shape of the more recent years (A_{1973}, A_{2003}, and A_{2010}). This evaluation process was done by one person and allowed us to further assess the accuracy of the digitized maps. Two new categories were introduced for this comparison:

- A_S more consistent than A_{1850}: when the shape and area of the digitized maps were more in agreement with A_{1973}, A_{2003}, and A_{2010} than A_{1850}; and

- not consistent but plausible: when it could not be decided from the glacier shape of A_S or A_{1850} which one was more plausible. In this case, both datasets provided plausible glacier shapes, but their areas were not comparable.

The results of the validation are presented in Table 2 and Fig. 3. As the results were very similar for $A_{S,first}$ and $A_{S,second}$, only the results for $A_{S,first}$ are presented in Table 2 and Fig. 3. Overall 71 % of the digitized glaciers of $A_{S,first}$, and even 88 % in terms of glacier area, were consistent compared with the datasets. For 13 % of the glacier basins it was not possible to assess whether $A_{S,first}$ or A_{1850} was more plausible (Table 2). The results for $A_{S,second}$ were similar, with 70 % of the glaciers and 89 % of the total glacier area being consistent with the other products. The difference between the percentage of digitized glacier basins (> 70 %) and glacier area (> 88 %) indicates that small glaciers have a higher probability of inaccuracies than larger glaciers. This can also be observed in the spatial representation of the validation (Fig. 3, as an example for $A_{S,first}$).

3.2.2 Basin-based validation

Mercanton (1958) calculated the total glacier area for the main Swiss river basins and for two time periods based on an early edition of the Siegfried map (published between 1869 and 1895 – ca. 1876) and the first National Map (surveyed between 1917 and 1945 – ca. 1934). We assessed the total glacier area for the same river basins with the digitized glacier areas of the Siegfried maps for the first and second periods (published ca. 1900 and ca. 1935) and with the four glacier inventories (Table 3). Next, we calculated the mean relative change in glacier area (ΔA) per year according to Eq. (1) between each successive dataset.

$$\Delta A_i = \frac{1 - A_i / A_{i-1}}{y_i - y_{i-1}} \cdot 100, \tag{1}$$

with ΔA the relative yearly change in glacier area (% yr^{-1}), A_i the glacierized area (km^2) of the dataset i, and y the given year of the dataset. The results are presented in Table 3 and compared to the values estimated by Mercanton (1958).

Overall, the total glacier area of the different river basins decreased between 1850 and 2010 for the eight compared datasets and the glacierized area of the Swiss Alps decreased by a total of ca. 53 %. The yearly changes in glacier area are overall higher for the period after 1973 than for the period before 1973, reflecting the observed increases in glacier

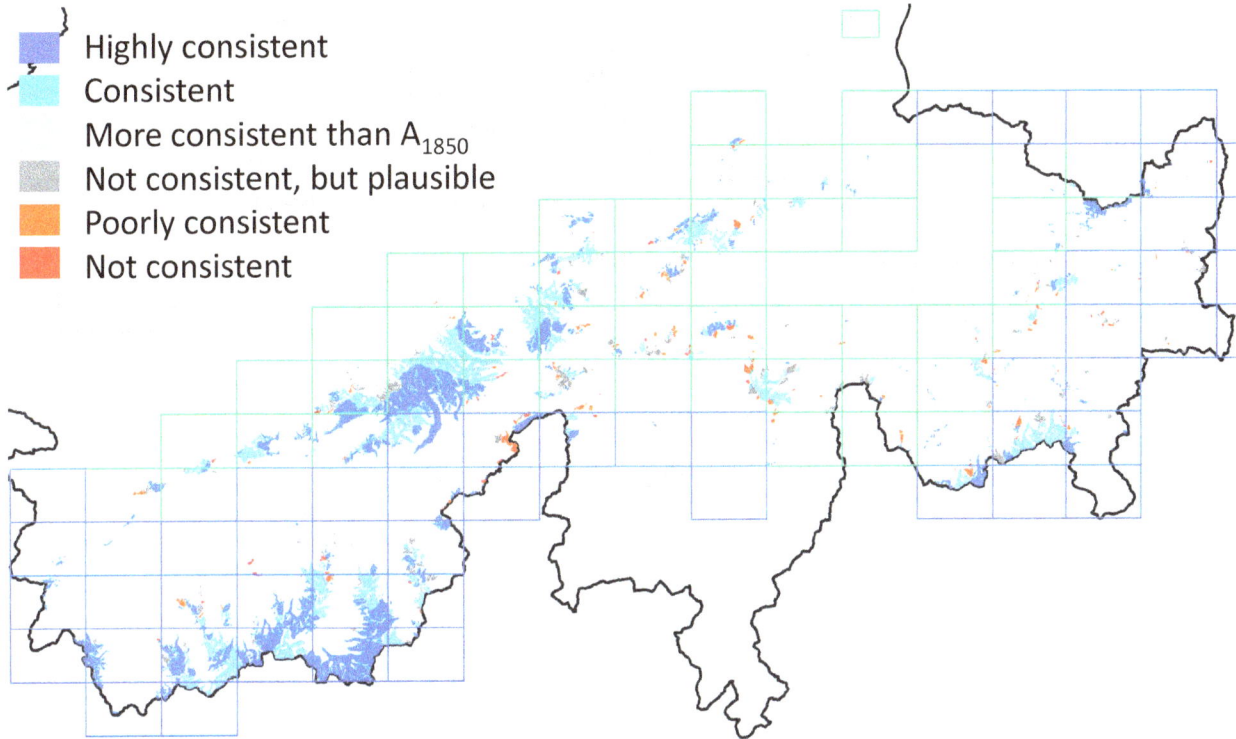

Figure 3. Large-scale validation of the digitized glacier outlines $A_{S,first}$ (around 1900).

Table 2. Large-scale validation of the digitized glacier outlines shown, as an example, for $A_{S,first}$.

	Number of glacier basins	%	Glacierized area (km^2)	%
Highly consistent	478	49.95	887.07	50.09
Consistent	165	17.24	660.54	37.30
More consistent than A_{1850}	38	3.97	16.65	0.94
Total "consistent"	681	71.16	1564.26	88.34
Not consistent with A_{1850}, but plausible	123	12.85	121.83	6.88
Poorly consistent	113	11.81	68.16	3.85
Not consistent	40	4.18	16.55	0.93
Total "not consistent"	153	15.99	84.71	4.78
Total	957		1770.80	

area loss in the last decades (e.g. Huss et al., 2008). However, some anomalies can be observed between the datasets. The total glacierized area of the Swiss Alps from the reconstructed glacierized area of ca. 1850 was 1781 km^2, compared to 1856 km^2 using the oldest Siegfried map (ca. 1876). The largest differences can be found in the Rhine (up to the Aare) and Inn River basins. These differences can be explained by an underestimation of the glacier areas in 1850 due to the difficult interpretation of the moraine geometry from remotely sensed data (Clark et al., 2004). On the other

hand, an overestimation of the glacier outlined in the first edition of the Siegfried map is possible due to the surveying methods available at the time, the low number of surveying stations used for the first Siegfried maps, and also the approach used for the calculation of the total glacier area by Mercanton (1958). Mercanton (1958) suggested two different methods to calculate the total glacier area and came to differences of 2.3 %. The total glacierized area for the first period (around 1900) was 1771 and 1711 km^2 for the second period (around 1935) for the digitized Siegfried map from

Table 3. Comparison of the total glacier area (km^2) for several Swiss river basins calculated from different datasets. The superscripts correspond to the mean yearly relative change in glacier area ΔA compared to the prior period (‰ yr^{-1}). As 1934 corresponds to the surveyed year of the National map and 1935 corresponds to the publication year of the Siegfried map and was certainly surveyed before 1935, the National map is ordered in the table after the Siegfried map.

Catchment	ca. 1850[1]	Siegfried ca. 1876[3]		Siegfried ca. 1900[2]		Siegfried ca. 1935[2]		Nat. map ca. 1934[3]		ca. 1973[1]		2003[1]		ca. 2010[1]	
Aare (up to Rhine, without Reuss and Linth/Limmat)	288.0	+1‰ yr^{-1}	296.5	0‰ yr^{-1}	294.2	−1‰ yr^{-1}	286.3	−30‰ yr^{-1}	277.5	−5‰ yr^{-1}	230.1	−5‰ yr^{-1}	197.4	−18‰ yr^{-1}	172.4
Reuss (up to Aare)	137.8	−1‰ yr^{-1}	134.4	0‰ yr^{-1}	134.9	−2‰ yr^{-1}	124.0	−100‰ yr^{-1}	112.1	−6‰ yr^{-1}	88.8	−7‰ yr^{-1}	69.9	−15‰ yr^{-1}	62.5
Linth/Limmat (up to Aare)	35.9	+3‰ yr^{-1}	38.6	+2‰ yr^{-1}	36.5	−1‰ yr^{-1}	35.5	−70‰ yr^{-1}	33.0	−6‰ yr^{-1}	26.1	−9‰ yr^{-1}	19.1	−14‰ yr^{-1}	17.3
Aare (up to Rhine)	461.8	+1‰ yr^{-1}	469.5	0‰ yr^{-1}	465.6	−1‰ yr^{-1}	445.8	−50‰ yr^{-1}	422.5	−5‰ yr^{-1}	345.0	−6‰ yr^{-1}	286.4	−17‰ yr^{-1}	252.1
Rhine (up to Aare)	150.2	+11‰ yr^{-1}	193.9	−9‰ yr^{-1}	154.0	−3‰ yr^{-1}	137.4	−90‰ yr^{-1}	125.4	−10‰ yr^{-1}	76.3	−17‰ yr^{-1}	37.8	−16‰ yr^{-1}	33.7
Rhine (up to Basel)	612.0	+3‰ yr^{-1}	663.4	−3‰ yr^{-1}	619.6	−2‰ yr^{-1}	583.2	−60‰ yr^{-1}	547.9	−6‰ yr^{-1}	421.3	−8‰ yr^{-1}	324.2	−17‰ yr^{-1}	285.9
Rhône (up to lake Geneva)	936.8	0‰ yr^{-1}	934.0	−1‰ yr^{-1}	909.3	0‰ yr^{-1}	898.4	−60‰ yr^{-1}	843.3	−3‰ yr^{-1}	740.6	−5‰ yr^{-1}	622.4	−11‰ yr^{-1}	573.6
Ticino (only Switzerland)	55.3	+3‰ yr^{-1}	60.1	−2‰ yr^{-1}	56.9	−3‰ yr^{-1}	50.9	−140‰ yr^{-1}	43.7	−9‰ yr^{-1}	29.4	−14‰ yr^{-1}	17.0	−23‰ yr^{-1}	14.3
Adda (only Switzerland)	42.5	+2‰ yr^{-1}	44.9	−3‰ yr^{-1}	42.0	0‰ yr^{-1}	41.6	−80‰ yr^{-1}	38.3	−5‰ yr^{-1}	31.4	−10‰ yr^{-1}	22.0	−16‰ yr^{-1}	19.5
Inn (up to Swiss border)	134.2	+6‰ yr^{-1}	153.7	−3‰ yr^{-1}	142.9	−1‰ yr^{-1}	137.2	−220‰ yr^{-1}	107.6	−7‰ yr^{-1}	80.5	−11‰ yr^{-1}	53.3	−17‰ yr^{-1}	46.8
Total	1780.8	+1‰ yr^{-1}	1856.1	−2‰ yr^{-1}	1770.7	−1‰ yr^{-1}	1711.4	−80‰ yr^{-1}	1580.7	−5‰ yr^{-1}	1303.2	−7‰ yr^{-1}	1038.9	−14‰ yr^{-1}	940.1

[1] Glacier inventory (Table 1). [2] Digitized glacier areas from the Siegfried maps. [3] Mercanton (1958).

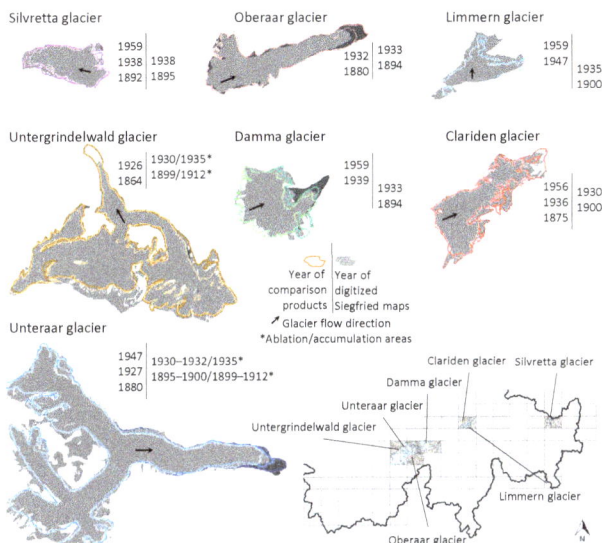

Figure 4. Comparison of the digitized outlines of the Siegfried map for seven glaciers with contemporary products from other sources.

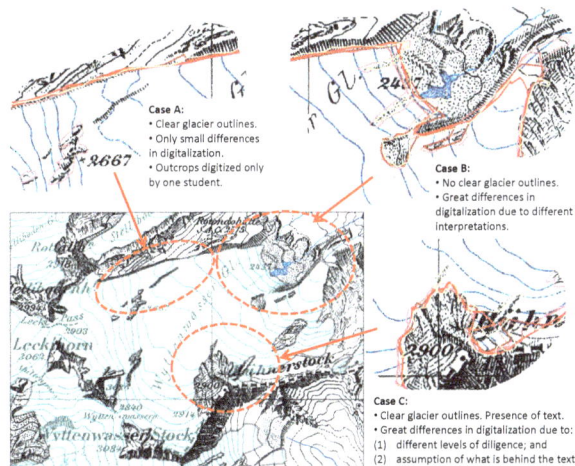

Figure 5. Examples of conflicts encountered during digitization of historical maps. The map shows the glacier area digitized for the end product (blue area with black outlines) and in the background the sheet of the Siegfried map for the publication year 1934 for the Wyttenwasser glacier. In cases A to C the outlines of the same glacier area digitized by five students are shown (coloured lines).

later editions, corresponding better to the glacier area from the 1850 product.

The comparison between the glacier area of the National map (surveyed ca. 1934) and the Siegfried map (published ca. 1935) shows the largest differences with ΔA up to $22\,\%\,\mathrm{yr}^{-1}$, although both datasets are only separated by 1 year. This discrepancy can be explained by the difference between the surveying year given with the National map and the publication year given with the Siegfried map, as the sheets could have been published several years after surveying. Furthermore, the two digitized Siegfried maps (published ca. 1900 and ca. 1935) are in many river basins similar, with small ΔA. Even if the relative changes in glacier area were expected to be small between 1900 and 1935 due to the positive mass balances observed in the Alps in the 1920s (e.g. Huss et al., 2008), these minimal changes between the first and second periods of the digitized Siegfried map are more likely due to the fact that some sheets were only re-edited without updating the glacier areas. The comparison of the datasets shows that this must have been the case for several glacier areas and the digitized glacier areas of the second period (1935) are often not representative for the year 1935. This is especially the case in the river basins Linth/Limmat, Rhône, and Adda. We found that the glacier outlines of 28 out of 80 sheets of the second period were identical to the corresponding sheet of the first period.

3.2.3 Local validation

In Fig. 4, the glacier outlines of the glacier areas from seven sheets of the digitized Siegfried map for the first and second periods are compared to glacier outlines digitized from the Dufour and Siegfried maps (earlier than 1930) and

from the National Map (later than 1930) for local validation. For the Limmern, Clariden, Untergrindelwald, and Silvretta glaciers, only little differences were observed between the first and second digitized periods, meaning that the Siegfried map of the second period was probably only re-edited (see Sect. 3.2.1). The comparison between the datasets shows for all glaciers (with the exception of the Limmern glacier) good consistency and logical evolution in shape, especially in the ablation area. For the accumulation zone, differences can be observed between the datasets, especially for the Untergrindelwald and Unteraar glaciers. These differences are due to different delineations of the glacier area between the different products. The additional glacierized area from the accumulation zone in the digitized Siegfried map is also present in the four glacier inventories and therefore consistent. The shape of the Limmern glacier is in the ablation area different for the digitized Siegfried map and for the comparison products from the National map. The Siegfried map seems inaccurate for this glacier.

3.3 Accuracy of the digitization

To assess the accuracy of the digitization, five hydrology masters students (age 20–25 years) digitized all glaciers over a $23\,\mathrm{km^2}$ area from a sheet of the Siegfried map for the publication years 1894 (within the first period) and 1934 (within the second period). All of them had the same rules for digitizing the glacier outlines. In other studies (e.g. Paul et al., 2013), the errors introduced by different interpreters increased with decreasing glacier area when glacier outlines are derived from remotely sensed data. We, therefore, chose the small Wyttenwasser glacier for the cross-comparison.

In Fig. 5 the digitized area of the Wyttenwasser glacier is shown for the publication year 1934 and several sources of conflict for the digitization are pointed out (cases A–C). In case A, the glacier outline in the map is clearly drawn and the differences in the digitized outlines are small and depend only on the diligence of the students. In case B, larger differences are observed, as the map drawing had to be interpreted, e.g. where does the glacier area stop – at the blue topographic line or at the limit between the white area and the black dots? Are the black dots on the glacier area covering ice or is this rock? Where does the glacier tongue end between blue and black topographic lines? In case C, one part of the glacier area is overprinted with text, which leads to different interpretations of the glacier outline behind the text. Cases B and C illustrate well the different assumptions that need to be made during digitization of historical topographic maps, leading to uncertainties. The comparison of all digitized glacier areas for the years 1894 and 1934 resulted in differences of up to 5 % between the five students. These results are comparable to the differences in standard deviation of 2 to 18 % for small glaciers ($< 1\,\mathrm{km}^2$) and smaller than 5 % for larger glaciers ($> 1\,\mathrm{km}^2$) observed by Fischer et al. (2014) and Paul et al. (2013) while deriving glacier outlines from remotely sensed data.

3.4 Accuracy of the digitized glacier areas

We estimate the precision of digitization to be ca. 5 %. However, it is more difficult to estimate the accuracy of the Siegfried map itself. As the uncertainty is different for each sheet (see Sects. 2.1 and 3.2.1), large regional differences can be found in the accuracy of the glacier outlines. On some sheets, the inaccuracy of the Siegfried map might be much higher than the interpretation bias of the digitization. However, the large-scale and basin-scale validations allowed us to assess which ones of the digitized glacier areas followed a logical evolution in shape and area and were therefore plausible compared with the other available products of glacier outlines for different years; 71 % of the glaciers and 88 % of the glacier area were considered consistent through the analysis. The local validation furthermore showed that the shape of the seven analysed glaciers was well represented in the Siegfried map. This analysis however is only valid for the studied glaciers and not for the entire area. While the presented product of glacier outlines contains all digitized glacier areas from the Siegfried map for Switzerland (Sect. 4), we recommend only using the glacier areas that were stated as "consistent", "highly consistent", or "more consistent than A_{1850}" by the validation process. If the other glacier areas are used, their large uncertainty should be considered in the interpretation of the results.

During the creation of the Siegfried map, the time span from measurements to publication extended up to several years, due to the material available at that time and to the complex topography. The Siegfried sheets are unfortunately only given with the publication year and no further information can be found on the surveying year. Therefore, one should keep in mind that the year given with the digitized glacier outlines from the Siegfried map is only representative for a period of time and cannot be taken as an exact date.

4 Data use and application

The digitized glacier areas of the Rhine River basins were used to develop the glacier routine of the HBV-light model in order to implement transient changes in glacier area and volume from 1900 to date (Seibert et al., 2018). This model was then used within the ASG-Rhine project with the aim of calculating the snowmelt, glacier melt, and rainfall contribution to the Rhine discharge for the time period 1900–2006 (Stahl et al., 2017). The glacier areas and glacier mass balances of several glaciers calculated within the ASG-Rhine project for the beginning of the 20th century showed comparable results to contemporary analyses or observations from other studies (Stahl et al., 2017). These different applications show that the digitized Siegfried map brings important information on glacier area for large-scale and long-term analysis and can be successfully used to better understand and model glacier area changes.

5 Data availability

The datasets of glacier area for the first and second digitized periods (around 1900 and around 1935) presented in this paper are freely available from the FreiDok plus data repository (https://freidok.uni-freiburg.de/data/15008) and have the DOI https://doi.org/10.6094/UNIFR/15008. For both digitized periods, two shape files are available. The first shape file contains the glacierized areas delineated from the digitized sheets themselves with the name of the sheet as identification and the year of publication. The sheets that are identical for both periods are identified in a comment field in the shape file (in total 28 of 80 sheets). The second shape file contains the digitized glacierized areas delineated by glacier basins as described above for the first digitized period (in total 957) and the second digitized period (in total 948) with unique identification numbers. As some glacier extents overlap several sheets and might therefore contain several publication years, the information of both shape files cannot be resumed in a single file. For each digitized glacier, the results of the validation are given in the shape file to enable the use of the different categories (see Table 2) depending on the need of the study (see Sect. 3.2). The basin outlines used for glacier delineation are also available in a separate shape file. All shape files are in the CH1903_LV03 (EPSG:21781) projection.

6   Conclusions

We digitized glacier outlines from the Siegfried map for the Swiss Alps for two periods around 1900 and 1935. We dealt with the challenges of digitization of historic maps (e.g. uncertainties in georeferencing, time of measurement vs. time of publication) with two validation schemes at the large scale, basin scale, and locally. Comparison to four existing glacier inventories covering different time periods revealed that at least 70 % of the digitized glaciers and 88 % of the total glacier area were comparable for both digitized periods to the glacier areas and shape of the glacier inventories and therefore plausible. Further comparison at the river basin and glacier scale showed reliable glacier representation for most of the areas. The uncertainty of the digitization itself was assessed separately and was less than 5 %, which is comparable to the accuracy of deriving glacier outlines and areas from remotely sensed data. The presented datasets for a first period around 1900 and a second period around 1935 are valuable information for the glacier extent in the Swiss Alps at the beginning of the 20th century where no other data source is available covering the entire Swiss Alps. The dataset closes the gap between the reconstruction of the glacier areas at around 1850 from the moraine extent and the first complete dataset of glacier areas in the Swiss Alps from aerial photographs in 1973. Under consideration of the data uncertainty, the use of the digitized datasets in combination with other existing glacier inventories can provide important information about changes in glacier areas for the last 120 years, which is essential for long-term and accurate glacier mass balance or hydrological modelling in glacierized basins.

Author contributions. DF homogenized and validated the presented datasets and prepared the manuscript with contributions from all co-authors.

Competing interests. The authors declare that they have no conflict of interest.

Acknowledgements. The authors thank Swisstopo for providing the historical topographic Siegfried map and Matthias Huss for providing several glacier outlines for validation. We are grateful to Damaris De for the digitization of part of the glacier areas. We furthermore thank Mirko Mälicke and his students Ruben Beck, Daniela Boru, Helena Böddecker, Verena Lang, Lukas Maier, and Miranda Perrone for assessing the accuracy of the digitization. We also want to thank two anonymous referees and Rheinhard Drews for their valuable comments and suggestions that helped to improve our manuscript. The glacier areas of the Rhine basin were digitized within the ASG-Rhein project (snow and glacier melt components of the streamflow of the River Rhine and its tributaries considering the influence of climate change) funded by the International Commission for the Hydrology of the Rhine Basin (CHR). The remaining glacier areas were digitized within the Hydro-CH2018 project funded by the Federal Office for the Environment (FOEV). The first author was funded by the German Federal Environmental Foundation (DBU).

Edited by: Reinhard Drews

References

Bauder, A., Funk, M., and Huss, M.: Ice-volume changes of selected glaciers in the Swiss Alps since the end of the 19th century, Ann. Glaciol., 46, 145–149, 2007.

Bauder, A., Fischer, M., Funk, M., Gabbi, J., Hoelzle, M., Huss, M., Kappenberger, G., and Steinegger, U.: The Swiss Glaciers 2013/14 and 2014/15, Glaciological Report No. 135/136, Zurich, ISSN 1424-2222, 2017.

Caminada, P.: Pioniere der Alpentopographie: Die Geschichte der Schweizer Kartenkunst, VS-Verlag, Zurich, 2003.

Clark, C. D., Evans, D. J. A., Khatwa, A., Bradwell, T., Jordan, C. J., Marsh, S. H., Mitchell, W. A., and Bateman, M. D.: Map and GIS database of glacial landforms and features related to the last British Ice Sheet, Boreas, 33, 359–375, https://doi.org/10.1111/j.1502-3885.2004.tb01246.x, 2004.

Collins, D. N.: Climatic warming, glacier recession and runoff from Alpine basins after the Little Ice Age maximum, Ann. Glaciol., 48, 119–124, https://doi.org/10.3189/172756408784700761, 2008.

Fischer, M., Huss, M., Barboux, C., and Hoelzle, M.: The new Swiss Glacier Inventory SGI2010: Relevance of using high-resolution source data in areas dominated by very small glaciers, Arctic, Antarct. Alp. Res., 46, 933–945, 2014.

Fischer, M., Huss, M., and Hoelzle, M.: Surface elevation and mass changes of all Swiss glaciers 1980–2010, The Cryosphere, 9, 525–540, https://doi.org/10.5194/tc-9-525-2015, 2015.

Freudiger, D., Mennekes, D., Seibert, S., and Weiler, M.: Historical glacier outlines from digitized topographic maps of the Swiss Alps for the years ca. 1900 and ca. 1935, FreiDok, available at: https://freidok.uni-freiburg.de/data/15008, https://doi.org/10.6094/UNIFR/15008, 2017.

GLAMOS: "The Swiss Glaciers", Reports of the Glaciological Commission of the Swiss Academy of Science (SAS), published by the Laboratory of Hydraulics, Hydrology and Glaciology (VAW) of ETH Zurich, No. 1-132, 1881–2015, 2015.

Hall, D. K., Bayr, K. J., Schöner, W., Bindschadler, R. A., and Chien, J. Y. L.: Consideration of the errors inherent in mapping historical glacier positions in Austria from the ground and space (1893–2001), Remote Sens. Environ., 86, 566–577, https://doi.org/10.1016/S0034-4257(03)00134-2, 2003.

Huss, M.: Present and future contribution of glacier storage change to runoff from macroscale drainage basins in Europe, Water Resour. Res., 47, 1–14, https://doi.org/10.1029/2010WR010299, 2011.

Huss, M. and Fischer, M.: Sensitivity of Very Small Glaciers in the Swiss Alps to Future Climate Change, Front. Earth Sci., 4, 1–17, https://doi.org/10.3389/feart.2016.00034, 2016.

Huss, M., Bauder, A., Funk, M., and Hock, R.: Determination of the seasonal mass balance of four Alpine glaciers since 1865, J. Geophys. Res., 113, F01015, https://doi.org/10.1029/2007JF000803, 2008.

Huss, M., Usselmann, S., Farinotti, D., and Bauder, A.: Glacier mass balance in the south-eastern Swiss Alps since 1900 and perspectives for the future, Erdkunde, 2010, 119–140, https://doi.org/10.3112/erdkunde.2010.02.02, 2010.

Imhof, E.: Unsere Landeskarten und ihre weitere Entwicklung, Schweizerische Zeitschrift für Vermessungswesen und Kulturtechnik, 4, 81–178, 1927

Ivy-Ochs, S., Kerschner, H., Maisch, M., Christl, M., Kubik, P. W., and Schlüchter, C.: Latest Pleistocene and Holocene glacier variations in the European Alps, Q. Sci. Rev., 28, 2137–2149, https://doi.org/10.1016/j.quascirev.2009.03.009, 2009.

Kääb, A., Paul, F., Maish, M., Hoelze, M., and Haeberli, W.: The new remote sensing derived Swiss glacier inventory: II. First results, Ann. Glaciol., 34, 362–366, https://doi.org/10.3189/172756402781817941, 2002.

Maisch, M.: Die Gletscher Graubündens. Rekonstruktion und Auswertung der Gletscher und deren Veränderungen seit dem Hochstand von 1850 in Gebiet der Östlichen Schweizer Alpen, Teile A und B, Phys. Geogr. Zurich, 33, 1992.

Maisch, M., Wipf, A., Denneler, B., Battaglia, J., and Benz, C.: Die Gletscher der Schweizer Alpen: Gletscherstand 1850. Aktuelle Vergletscherung, Gletscherschwundszenarien, End report NFP 31, Second Edition, Zurich, vdf Hochschulverlag, ETH Zurich, p. 373, 2000.

Maisch, M., Paul, F., and Kääb, A.: Kerngrössen, Flächen- und Volumenänderungen der Gletscher 1850–2000, Hydrologischer Atlas der Schweiz (HADES), Bern, 2004.

Mercanton, P. L.: Aires englacée et cotes frontales des glaciers suisses – Leurs changements de 1876 à 1934 d'après l'Atlas Siegfried et la Carte Nationale et quelques indications sur les cariations de 1934 à 1957, Wasser und Energiewirtschaft, 12, 347–351, 1958.

Müller, F., Caflish, T., and Müller, G.: Firn und Eis der Schweizer Alpen, Gletscherinventar, Zurich vdf Hochschulverlag ETH Zurich, p. 373, 1976.

Paul, F., Frey, H., and Le Bris, R.: A new glacier inventory for the European Alps from Landsat TM scenes of 2003: Challenges and results, Ann. Glaciol., 59, 144–152, 2011.

Paul, F., Barrand, N. E., Baumann, S., Berthier, E., Bolch, T., Casey, K., Frey, H., Joshi, S. P., Konovalov, V., Le Bris, R., Mölg, N., Nosenko, G., Nuth, C., Pope, A., Racoviteanu, A., Rastner, P., Raup, B., Scharrer, K., Steffen, S., and Winsvold, S.: On the accuracy of glacier outlines derived from remote-sensing data, Ann. Glaciol., 54, 171–182, https://doi.org/10.3189/2013AoG63A296, 2013.

Racoviteanu, A. E., Paul, F., Raup, B., Khalsa, S. J. S., and Armstrong, R.: Challenges and recommendations in mapping of glacier parameters from space: Results of the 2008 global land ice measurements from space (GLIMS) workshop, Boulder, Colorado, USA, Ann. Glaciol., 50, 53–69, https://doi.org/10.3189/172756410790595804, 2009.

Rastner, P., Joerg, P. C., Huss, M., and Zemp, M.: Historical analysis and visualization of the retreat of Findelengletscher, Switzerland, 1859–2010, Global Planet. Change, 145, 67–77, https://doi.org/10.1016/j.gloplacha.2016.07.005, 2016.

Seibert, J., Vis, M. J. P., Kohn, I., Weiler, M., and Stahl, K.: Technical note: Representing glacier geometry changes in a semi-distributed hydrological model, Hydrol. Earth Syst. Sci., 22, 2211–2224, 2018 https://doi.org/10.5194/hess-22-2211-2018, 2018.

Stahl, K., Weiler, M., Kohn, I., Freudiger, D., Seibert, J., Vis, M., and Gerlinger, K.: The snow and glacier melt components of streamflow of the river Rhine and its tributaries considering the influence of climate change – Synthesis report, Lelystad, The Netherlands, available at: www.chr-khr.org/en/publications (last access: 4 July 2017), 2016.

Stahl, K., Weiler, M., Freudiger, D., Kohn, I., Seibert, J., Vis, M., Gerlinger, K., and Böhm, M.: The snow and glacier melt components of streamflow of the river Rhine and its tributaries considering the influence of climate change, Final report to the International Commission for the Hydrology of the Rhine (CHR), available at: www.chr-khr.org/en/publications, last access: 4 July, 2017.

Swisstopo: Federal Office of Topography – Siegfried maps 1 : 50 000, available at: https://shop.swisstopo.admin.ch/en/products/maps/historical/DIGIT_SIEGFRIED50, last access: 15 May, 2017.

Swisstopo: Federal Office of Topography – Background information on the Dufour Map, available at: https://www.swisstopo.admin.ch/en/knowledge-facts/maps-and-more/historical-maps/dufour-map.html, last access: 20 February, 2018.

Vincent, C., Le Meur, E., Six, D., and Funk, M.: Solving the paradox of the end of the Little Ice Age in the Alps, Geophys. Res. Lett., 32, 1–4, https://doi.org/10.1029/2005GL022552, 2005.

Viviroli, D., Archer, D. R., Buytaert, W., Fowler, H. J., Greenwood, G. B., Hamlet, A. F., Huang, Y., Koboltschnig, G., Litaor, M. I., López-Moreno, J. I., Lorentz, S., Schädler, B., Schreier, H., Schwaiger, K., Vuille, M., and Woods, R.: Climate change and mountain water resources: overview and recommendations for research, management and policy, Hydrol. Earth Syst. Sci., 15, 471–504, https://doi.org/10.5194/hess-15-471-2011, 2011.

WGMS: Global Glacier Change Bulletin No. 1 (2012–2013), edited by: Zemp, M. I., Nussbaumer, S. U., Hüsler, F., Machguth, H., Mölg, N., Paul, F., and Hoelzle, M., World Glacier Monitoring Service ICSU(WDS) IUGG(IACS) UNEP UNESCO WMO, Zurich, 2015.

6

A global, high-resolution data set of ice sheet topography, cavity geometry, and ocean bathymetry

Janin Schaffer[1], Ralph Timmermann[1], Jan Erik Arndt[1], Steen Savstrup Kristensen[2], Christoph Mayer[3], Mathieu Morlighem[4], and Daniel Steinhage[1]

[1]Alfred Wegener Institute, Helmholtz Centre for Polar and Marine Research, Bremerhaven, Germany
[2]DTU Technical University of Denmark, 2800 Lyngby, Denmark
[3]Bavarian Academy of Sciences and Humanities, Commission for Geodesy and Glaciology, Munich, Germany
[4]University of California, Irvine, Department of Earth System Science, Croul Hall, Irvine, California 92697-3100, USA

Correspondence to: J. Schaffer (janin.schaffer@awi.de) and R. Timmermann (ralph.timmermann@awi.de)

Abstract. The ocean plays an important role in modulating the mass balance of the polar ice sheets by interacting with the ice shelves in Antarctica and with the marine-terminating outlet glaciers in Greenland. Given that the flux of warm water onto the continental shelf and into the sub-ice cavities is steered by complex bathymetry, a detailed topography data set is an essential ingredient for models that address ice–ocean interaction. We followed the spirit of the global RTopo-1 data set and compiled consistent maps of global ocean bathymetry, upper and lower ice surface topographies, and global surface height on a spherical grid with now 30 arcsec grid spacing. For this new data set, called RTopo-2, we used the General Bathymetric Chart of the Oceans (GEBCO_2014) as the backbone and added the International Bathymetric Chart of the Arctic Ocean version 3 (IBCAOv3) and the International Bathymetric Chart of the Southern Ocean (IBCSO) version 1. While RTopo-1 primarily aimed at a good and consistent representation of the Antarctic ice sheet, ice shelves, and sub-ice cavities, RTopo-2 now also contains ice topographies of the Greenland ice sheet and outlet glaciers. In particular, we aimed at a good representation of the fjord and shelf bathymetry surrounding the Greenland continent. We modified data from earlier gridded products in the areas of Petermann Glacier, Hagen Bræ, and Sermilik Fjord, assuming that sub-ice and fjord bathymetries roughly follow plausible Last Glacial Maximum ice flow patterns. For the continental shelf off Northeast Greenland and the floating ice tongue of Nioghalvfjerdsfjorden Glacier at about 79 N, we incorporated a high-resolution digital bathymetry model considering original multibeam survey data for the region. Radar data for surface topographies of the floating ice tongues of Nioghalvfjerdsfjorden Glacier and Zachariæ Isstrøm have been obtained from the data centres of Technical University of Denmark (DTU), Operation Icebridge (NASA/NSF), and Alfred Wegener Institute (AWI). For the Antarctic ice sheet/ice shelves, RTopo-2 largely relies on the Bedmap-2 product but applies corrections for the geometry of Getz, Abbot, and Fimbul ice shelf cavities.

1 Introduction

Mass loss from the Greenland ice sheet presently accounts for about 10 % of the observed global mean sea-level rise (Church et al., 2013). The ocean plays an important role in modulating the flow of ice by delivering heat to the marine-terminating outlet glaciers around Greenland (e.g. Seale et al., 2011; Straneo et al., 2012). The warming and accumulation of Atlantic Water in the subpolar North Atlantic has been suggested to be the driver of the glaciers' retreat around the coast of Greenland (e.g. Straneo and Heimbach, 2013). The complex bathymetry in this region is thought to steer the flux of warm water of Atlantic origin from the open ocean onto the continental shelf towards the calving fronts of outlet glaciers and into the cavities below floating ice tongues. One of the key regions here is the Northeast Greenland continental shelf, where a system of troughs supports the flux of warm water towards the floating ice tongues of Nioghalvfjerdsfjorden Glacier (also referred to as 79 North Glacier) and Zachariæ Isstrøm (Arndt et al., 2015; Wilson and Straneo, 2015). Recently, these glaciers were observed to retreat and melt rapidly (Mouginot et al., 2015). In such regions detailed bathymetry data and consistent data sets of ice topographies are essential ingredients for studying the interaction between the ocean and the cryosphere.

Around Antarctica, research into ocean–cryosphere interaction has been an established field of science for several decades. Many aspects of water mass modification in the Southern Ocean's marginal seas can only be understood if the fluxes of heat and freshwater at the base of the ice shelves surrounding the Antarctic continent are considered (e.g. Foldvik et al., 1985). Scientific interest has increased further with growing evidence that mass loss from the Antarctic ice sheet is accelerating (e.g. McMillan et al., 2014) and driven by enhanced ice shelf basal melting (Pritchard et al., 2012). There again, a well-constrained rendition of ocean bathymetry and cavity geometry is key to a successful analysis of field data and to a realistic representation of the relevant processes in numerical models.

The Refined Topography data set (RTopo-1; Timmermann et al., 2010) provides consistent maps of the global ocean bathymetry and the upper and lower ice surface topographies of the Antarctic ice sheet and shelves. Horizontal grid spacing of these maps is 1 arcmin. Based on RTopo-1, ocean general circulation models have successfully been used e.g. to simulate Southern Ocean warming and increased ice shelf basal melting around Antarctica (Timmermann and Hellmer, 2013; Kusahara and Hasumi, 2013), the flow of Circumpolar Deep Water onto the Amundsen Sea continental shelf (Assmann et al., 2013; Nakayama et al., 2014a), and pathways of basal meltwater from Antarctic ice shelves (Kusahara and Hasumi, 2014; Nakayama et al., 2014b). Parts of RTopo-1 were used to compile improved maps of bedrock and ice topographies for Antarctica in Bedmap2 (Fretwell et al., 2013).

The Greenland ice sheet, however, has remained a blank area in RTopo-1.

The aim of this paper is to present the newly compiled global topography data set RTopo-2, which provides a detailed bathymetry for the continental shelf around Greenland and contains ice and bedrock surface topographies for Greenland and Antarctica as part of a global, self-consistent data set with a horizontal grid spacing of 30 arcsec. In the following sections, we introduce the data used, the processing applied to each data set and the strategies followed for merging the data sets in a self-consistent way. We demonstrate the improvements achieved in RTopo-2 compared to previous products and discuss the most relevant caveats.

2 Data sets and processing

2.1 Overview of RTopo-2 maps

We followed the spirit of RTopo-1 and compiled global fields for

1. bedrock topography (ocean bathymetry, surface topography of continents, bedrock topography under grounded or floating ice);

2. surface elevation (upper ice surface height for Antarctic and Greenland ice sheets/ice shelves, bedrock elevation for ice-free continent, zero for ocean);

3. ice base topography for the Antarctic and Greenland ice sheets/ice shelves (ice draft for ice shelves and floating glaciers, zero in absence of ice);

4. a surface type mask that indicates open ocean, grounded ice (ice sheets), floating ice (ice shelves/floating glaciers), and bare land surface;

5. positions of coastlines and ice shelf/floating glacier front lines.

The bedrock topography is identical to the surface elevation for ice-free land surface and identical to the ice base topography for grounded ice (Fig. 1). Ice not connected to the Greenland or Antarctic ice sheet is not covered in our data set. Glaciers on subantarctic and Greenland islands are thus labelled as bare land surface with the surface elevation preserved. In contrast to RTopo-1, we now provide all maps with a horizontal grid spacing of 30 arcsec.

2.2 Data sources and merging procedure

RTopo-2 has been compiled by combining various gridded data sets (Table 1) with different grid spacings, projections, and coverages (Fig. 2) into global maps. For data handling, interpolation, and blending we developed a command script written in Interactive Data Language (IDL). Using a global bathymetry data set (see below for details) as a backbone, regional grids have been created from various source data sets tinuous fields (an obvious example here is the discontinuity at ice shelf fronts) is another source for the creation of local inconsistencies that need to be cured. For RTopo-2, the term *consistency* implies that

- ice thickness > 0 in the ice-covered region (with ice thickness = surface height minus lower ice topography);

- water column thickness > 0 in the open ocean and sub-ice cavities (with water column thickness = lower ice topography minus bedrock topography);

- water column thickness is zero, i.e. lower ice and bedrock topographies are identical, for grounded ice;

- lower ice topography is negative (below sea level) and surface height is positive (above sea level) for ice shelves;

- ice draft and thickness are zero outside ice-covered regions;

- bedrock topography is below sea level in the ocean;

- there are no enclosed gaps ("holes") in the ice sheet/ice shelves other than those associated with rock outcrops;

- there are no water areas south of the coastline of Antarctica.

These points may all seem trivial, but they are in fact not universally ensured in the gridded data sets available to date. Note that the surface type mask plays a key role in our algorithm; instead of being a merely diagnostic property, the surface type determines the conditions to which consistency is enforced. Choices that needed to be made include deciding which of the topographies should be trusted more – e.g. whether bedrock from one source or lower ice topography from another source is more reliable. These decisions were not always straightforward and are somewhat subjective; we give some of the reasoning in the sections discussing specific regional data sets below. In general, consistency and continuity have been valued higher than an exact rendition of the source data sets in RTopo-2.

2.3 Bedrock and bathymetry data sets

2.3.1 World Ocean bathymetry

As the nucleus of RTopo-2 we used an updated version of the General Bathymetric Chart of the Oceans (GEBCO) 30 arcsec data set (Becker et al., 2009), namely the GEBCO_2014 (20150318) grid that was released in March 2015 (Weatherall et al., 2015). The global grid of seafloor elevations is based on quality-controlled ship depth soundings. In between soundings the interpolation was guided by satellite-derived gravity data (Sandwell and Smith, 2009). In some areas, GEBCO_2014 furthermore uses information from regional undersea mapping projects (see next section).

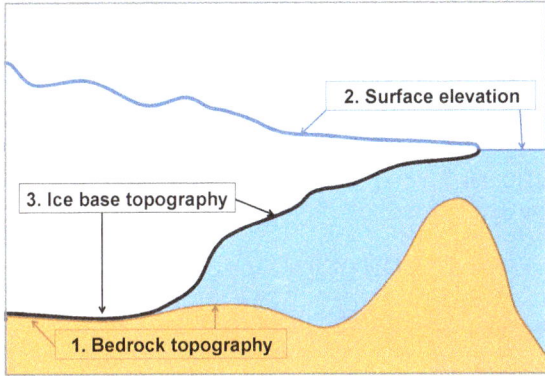

Figure 1. Sketch of a 2-D vertical section along a floating glacier tongue/ice shelf with grounded and floating ice (white), sub-ice bedrock/ocean seafloor (brown), and water in a subglacial cavity and the open ocean (blue). Lines indicate the bedrock topography (brown), the surface elevation (dark blue), and the ice base topography (black).

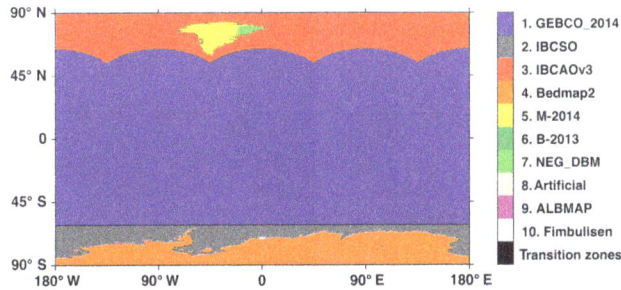

Figure 2. Data sources for the ocean bathymetry in RTopo-2. Black areas denote transition zones between the data sets (source flag 20). Further explanations are given in Table 1.

and subsequently merged into the existing fields. Interpolation from different projections to our geographic grid was based on Delaunay triangulation and subsequent linear interpolation. The regional "patches" have been blended into the existing fields using weight functions that – depending on the distance from the boundaries of the regional grids – vary between 0 and 1 and ensure a smooth transition between the two data sets without smoothing the topographies. As weight functions we used hyperbolic tangent functions with empirically derived length scales that have been cut off below values of 0.05 and above 0.95 to avoid overly long tails. This approach yields good results only when the two grids to be merged do not differ too strongly in the area of overlap, but with the data sets used here it was always possible to choose the location and width of the transition zone in a way that ensured a smooth blending.

For each of the newly incorporated regional grids, we had to ensure or enforce consistency with the existing topographies. The necessity for this step is quite obvious when independent data sets are combined; interpolation of discon-

Table 1. Data sources for individual regions merged in RTopo-2. The index numbers correspond to the source flags in Fig. 2.

	Region	Data obtained from
1.	World Ocean bathymetry	GEBCO_2014 (Weatherall et al., 2015)
2.	Southern Ocean bathymetry	IBCSO (Arndt et al., 2013)
3.	Arctic Ocean bathymetry	IBCAOv3 (Jakobsson et al., 2012)
4.	Antarctic ice sheet/shelf surface height and thickness and bedrock topography	Bedmap2 (Fretwell et al., 2013)
5.	Greenland ice sheet/glacier surface height and thickness and bedrock topography	Morlighem et al. (2014) (M-2014)
6.	Fjord and shelf bathymetry close to the Greenland coast	Bamber et al. (2013) (B-2013)
7.	Bathymetry on Northeast Greenland continental shelf	Arndt et al. (2015) (NEG_DBM)
8.	Bathymetry in several narrow Greenland fjords and on parts of the Greenland continental shelf	artificial, see Merging strategy and Data corrections in Sect. 2.2.3 for details
9.	Bathymetry for Getz and western Abbot Ice Shelf cavities	ALBMAP (Le Brocq et al., 2010)
10.	Bathymetry for Fimbulisen cavity	Nøst (2004), Smedsrud et al. (2006)
11.	Ice thickness for Nioghalvfjerdsfjorden Glacier and Zachariæ Isstrøm	DTU (Seroussi et al., 2011) Operation Icebridge (Allen et al., 2010, updated 2015) Alfred Wegener Institute (AWI) Mayer et al. (2000)
12.	Contour of iceberg A-23A in Weddell Sea	Paul et al. (2015)

2.3.2 Southern and Arctic Ocean bathymetries

GEBCO_2014 includes the International Bathymetric Chart of the Southern Ocean (IBCSO) Version 1.0 (Arndt et al., 2013) south of 60° S and the latest version of the International Bathymetric Chart of the Arctic Ocean (IBCAOv3) (Jakobsson et al., 2012) north of 64° N. Ice sheets, floating ice shelves, and glaciers in GEBCO_2014 are represented by their surface elevation, which in the case of the Antarctic ice sheet has been adopted from the IBCSO "ice surface" grid. Given that we aim at a continuous representation of the sub-ice cavities as part of the global ocean, we replaced the GEBCO_2014 data by the IBCSO "bedrock topography" grid south of 61.5° (Fig. 2). Towards the Arctic Ocean, we re-combined GEBCO_2014 with IBCAOv3 (Fig. 2) in order to keep the high-resolution information from multibeam surveys off the southern tip of Greenland (south of 64° N) that are included in IBCAOv3 (Jakobsson et al., 2012) but have not been adopted in GEBCO_2014. Both digital bathymetry products, IBCAOv3 and IBCSO, have a horizontal grid spacing of 500 m × 500 m and were constructed from a combination of all multibeam, dense single beam, and land surface height data available for these regions. We achieved a smooth blending by hyperbolic tangent functions with 50 km/20 km length scale along the transition lines between GEBCO_2014 and IBCSO/IBCAOv3, respectively.

2.3.3 Ocean bathymetry in Greenland fjords and continental shelf regions

The bathymetry of the continental shelf along the Greenland coast is crucial to ice–ocean studies in this region and thus there is a rising interest in a good representation of these areas. Nevertheless, away from the commonly used ship routes and especially in ice-covered areas, the depth of the sea floor is only weakly constrained. Data coverage maps of IBCAOv3 show that many shelf and fjord areas around the coast of Greenland are not covered by soundings (Jakobsson et al., 2012). To achieve a more detailed representation of Greenland continental shelf bathymetry, we included additional data sources (Table 1).

B-2013: Bedrock topography from Bamber et al. (2013)

Based on surface elevation maps from the Greenland Iceland Mapping Project (GIMP, Howat et al., 2014) and ice thickness data from multiple airborne surveys between 1970 to 2012, Bamber et al. (2013) compiled a data set of ice thickness and bedrock elevation on and around Greenland (B-2013). Using the ocean bathymetry from IBCAOv3 and with plausible assumptions for historic glacier ice flow pathways, the bottom topography was modified in several places to achieve a better representation of the fjord structures and of the troughs that connect the fjords to the continental shelf break. While this was clearly an important step towards a better representation of the relevant processes in models, the B-2013 data set has three major weaknesses.

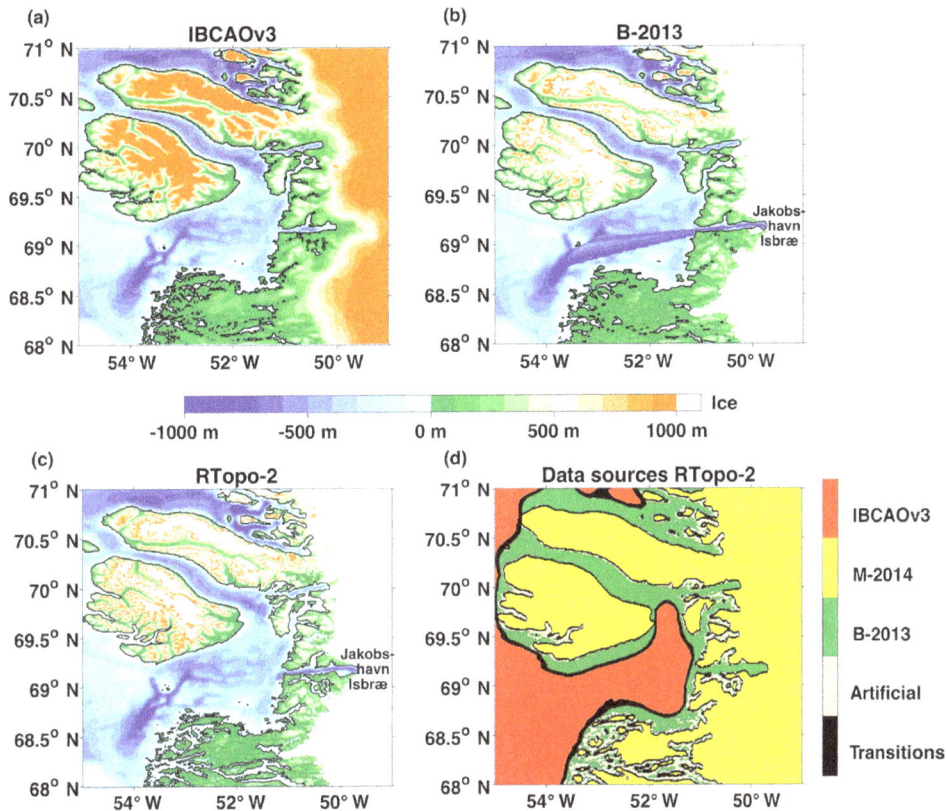

Figure 3. Coastal and shelf region in the west of Greenland, including Jakobshavn Isbræ. Maps show the ocean bathymetry/surface elevation in IBCAOv3 **(a)**, ocean bathymetry/bedrock elevation in B-2013 **(b)**, ocean bathymetry/bedrock elevation in RTopo-2 **(c)**, and the data sources for RTopo-2 **(d)**. The colour scale is identical for all bathymetry maps. White shading indicates grounded ice and floating ice tongues; black lines mark the coastline. The colour flags of the different data sources are identical to Fig. 2.

First, lines of very steep gradients still indicate an unrealistic bedrock topography where the fjord's bathymetry was interpolated/extrapolated across the continental shelf (e.g. Fig. 3b). Second, many of the smaller fjords are not resolved due to the 1 km grid spacing of the data set. Lastly, the different maps in this data set are not fully consistent with each other: combining surface elevation and ice thickness maps yields an ice bottom topography that is not identical to the bedrock topography grid provided for grounded ice areas.

M-2014: Bedrock topography from Morlighem et al. (2014)

In addition to the airborne ice thickness survey data and the surface elevation obtained from GIMP (Howat et al., 2014), the Morlighem et al. (2014) (M-2014) data set also considers satellite-derived ice motion data and applies a mass conservation scheme to derive an ice thickness distribution of the Greenland ice sheet that is consistent with the observed flow lines. Like in B-2013, the bedrock elevation was calculated by subtracting the ice thickness from the surface elevation. While the resulting topography data set in M-2014 does not contain any information for ocean areas, it still provides very useful guidance for fjord structure and topography. The distribution of grounded/floating ice and bare land in M-2014 follows the GIMP coastline and thus represents even the smaller fjords with a lot of detail. Morlighem et al. (2014, Supplement) showed that many ice-covered and open-ocean fjords are not resolved in the Bamber et al. (2013) data set. We used the land/sea/ice mask from M-2014 as the most important criterion for merging the different bathymetry data sets and applying modifications to the data around Greenland.

Merging strategy

To benefit from the best parts of each data set, we used

- the bedrock elevation from M-2014 for all locations with grounded ice (see below),

- the bedrock elevation from B-2013 within the fjords and in a narrow band of about 25 km width along the Greenland coast, and

– the bathymetry from IBCAOv3 further away from the coast, with transition zones of 10 km width.

Consequently, the large areas of continental shelf around Greenland are mainly determined by IBCAOv3 data while the fjord topographies are given by the B-2013 bedrock (e.g. Fig. 3d).

Before merging the B-2013 bathymetry with the M-2014 bedrock elevation, we smoothed the B-2013 data set by using an unweighted moving average with a 1 km footprint. Smoothing was necessary to avoid artefacts arising from differing grid spacings and/or from steep unrealistic gradients in the B-2013 bedrock (see above). In regions where the GIMP coastline demands ocean but B-2013 gives land values, we prescribed small patches with negative topography values (source flag 8, e.g. Fig. 3d). The depth of these artificial points was chosen to be 10 m for grid points right next to land and 100 m for grid points along the centre of the fjords. These small patches of artificial values were smoothed with their surroundings (using a moving average with 1 km smoothing radius) to obtain plausible shapes of bedrock topography (e.g. Fig 3c).

Data modifications

Three sectors turned out to be particularly difficult to handle:

1. the region around Petermann and Ryder glaciers (North Greenland),

2. the North Greenland fjords system off Hagen Bræ and Marie Sophie and Academy glaciers, and

3. the Sermilik Fjord in front of Helheim, Fenris and Midgaard glaciers, and Køge Bugt (Southeast Greenland).

Observations at the front of Petermann Glacier's floating ice tongue imply that the subglacial fjord is about 900 m deep (Johnson et al., 2011) as opposed to about 400 m in B-2013. The observations cover only a small area at the glacier front; no information for the subglacial bathymetry towards the grounding line of the floating ice tongue is available. We deepened the centre of the trough by prescribing a depth of 500 m, which is more likely to under- than to overestimate the true depth. Subsequently we smoothed over the artificial depth values by applying a moving average with a smoothing radius of 4 km, taking into account the surrounding data points and the surface type mask.

Further to the east, the B-2013 representation of the continental shelf area in front of Ryder Glacier features a very steep gradient and a deep trough close to the coast, which appears unrealistic. We replaced some of the interpolated deep and adjacent shallow parts with a smooth deep fjord/shelf bathymetry (Fig. 4a and 4b). In practice, we prescribed small patches with depths values of 750, 800, and 900 m. Afterwards we smoothed the bedrock elevation within these areas by applying a moving average with a smoothing radius

of 4 km. We think that this gives a more plausible representation compared to B-2013, although it needs to be kept in mind that there is no observational evidence for either case.

For the fjord system in front of the Hagen Bræ, Marie Sophie, and Academy glaciers, we defined small patches of artificial topography (with depth values ranging between 50 and 250 m) to achieve a smooth transition between the subglacial bedrock and the fjord bathymetry at the glacier fronts. We connected the under-ice bathymetry with the ocean bathymetry following plausible Last Glacial Maximum (LGM) ice flow patterns. The LGM ice sheet margin was approximately located at the continental shelf break in this region (Funder et al., 2011). In addition, we smoothed the ocean bathymetry by using a moving average with a smoothing radius of 2.5 km for the fjord system to remove steep artificial gradients arising from B-2013.

For the Sermilik Fjord off Helheim Glacier and for Køge Bugt, the bathymetry data from B-2013 show very deep troughs and steep gradients on the continental shelf (Fig. 4c). We inserted artificial (mainly shallower) depth values in several locations in the fjords and smoothed over the relevant part of the grid applying a moving average with a smoothing radius of 1.5 km. The result (Fig. 4d) is to a large extent consistent with the observations of Andresen et al. (2012) in Sermilik Fjord.

All regions with inserted artificial values were marked with the data source flag 8.

2.3.4 Bathymetry of the Northeast Greenland continental shelf

Bottom topography on the continental shelf northeast of Greenland is poorly resolved and contains a number of artefacts in IBCAOv3. Reprocessing and combining multi- and single-beam echo sounding data from more than 2 decades resulted in a significantly improved digital bathymetry model (NEG_DBM) (Arndt et al., 2015). In addition to the echo sounding data, maximum depths from CTD profiles were included in areas with no other available information.

We included the NEG_DBM bedrock elevation in the continental shelf area between the Greenland coast in the west and the continental shelf break (600 m depth contour) in the east, from 75 to about 80.5° N (Fig. 2). The coastline of the mainland remains based on M-2014/GIMP, while the topography and coastline of the islands in this area were adopted from the NEG_DBM.

2.4 Ice sheet topography and cavity geometry

2.4.1 Greenland ice and bedrock topographies

As discussed in Sect. 2.3.3, Morlighem et al. (2014) combined (1) the surface elevation obtained within GIMP, (2) data from airborne ice thickness surveys, and (3) satellite-derived ice motion data to provide high-resolution maps of ice thickness and bedrock topography for the Greenland ice

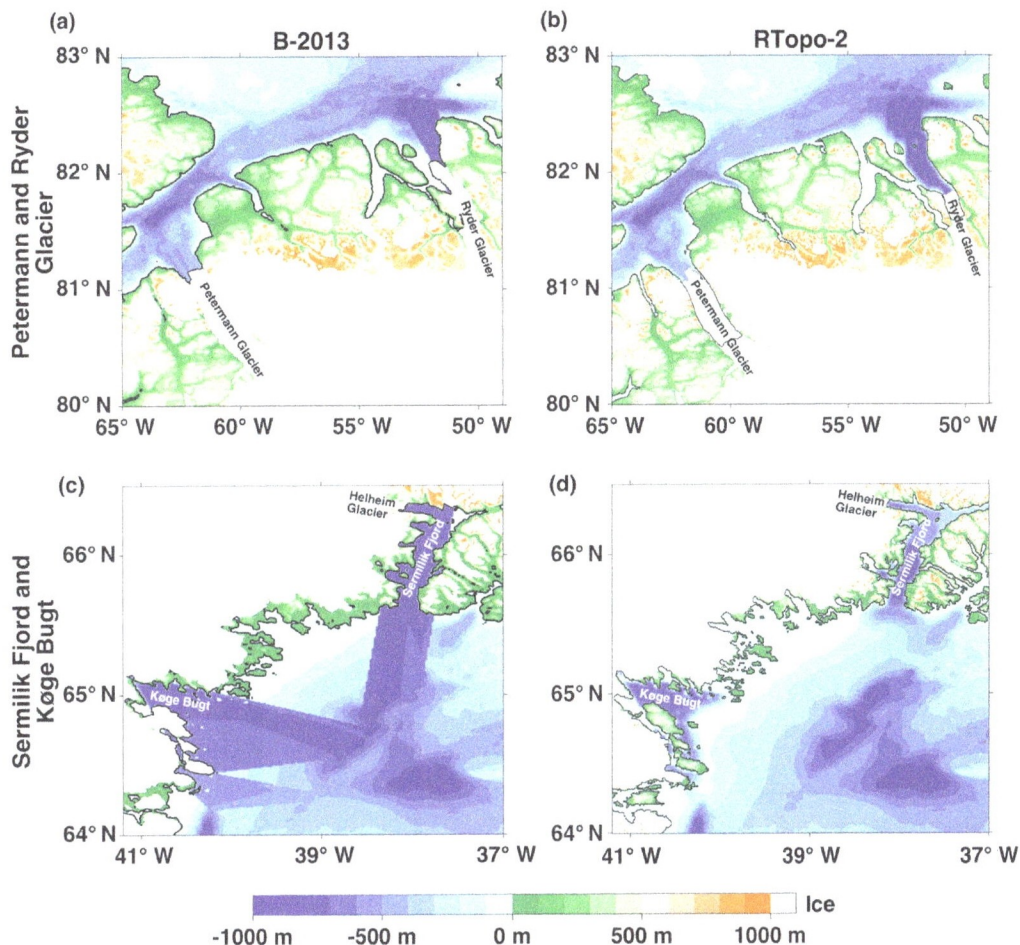

Figure 4. Maps of the ocean bathymetry/bedrock elevation in B-2013 **(a, c)** and RTopo-2 **(b, d)**. The upper panel **(a, b)** shows the coastal and shelf region in the northern sector of Greenland including the fjord system in front of Petermann Glacier and Ryder Glacier. The lower panel **(c, d)** gives the coastal and shelf region in the southeastern sector of Greenland including the Sermilik Fjord and Køge Bugt. White shading indicates grounded ice and floating ice tongues; black lines mark the coastline.

sheet. Using mass conservation as a constraint in the optimization, an ice thickness distribution that is consistent with the observed flow lines was proposed. Given that ice bottom topography was calculated by subtracting the ice thickness from the surface elevation, this also affects the representation of bedrock for grounded ice areas. The M-2014 maps extend to the ice front in case of grounded ice and to the coastline for bare land. For floating ice, the bedrock elevation extends only to the grounding lines while ice thickness and surface elevation data cover the full area to the ice front. With a horizontal grid spacing of 150 m, the data set resolves many more fine-scale structures than the 1 km B-2013 product (Morlighem et al., 2014, Supplement). All maps provided by the M-2014 data set are fully consistent with each other.

We use M-2014 as the backbone representation of the Greenland ice sheet geometry and as the basis for the ice/land/sea mask within the perimeter of the Greenland continent. These are based on ocean and ice masks from GIMP,

while the ice shelves were added by using InSAR mapping (differential satellite radar interferometry) following Rignot et al. (2011).

2.4.2 Northeast Greenland glacier topographies

Given that the floating ice tongue of Nioghalvfjerdsfjorden Glacier is one of the very few places in the Arctic where ice and ocean interact at an ice shelf base that covers more than just a very small area, this is a region of particular scientific interest. We therefore decided to enhance the ice thickness data in this area by using recently obtained airborne radar data as well as seismic soundings (Mayer et al., 2000) and airborne data from 1998 (Fig. 5). We used spherical triangulation to interpolate the ice thickness data in the area of the floating ice tongues of Nioghalvfjerdsfjorden Glacier and Zachariæ Isstrøm to our regular grid. The resulting ice thickness map was smoothed along the flow lines to avoid inter-

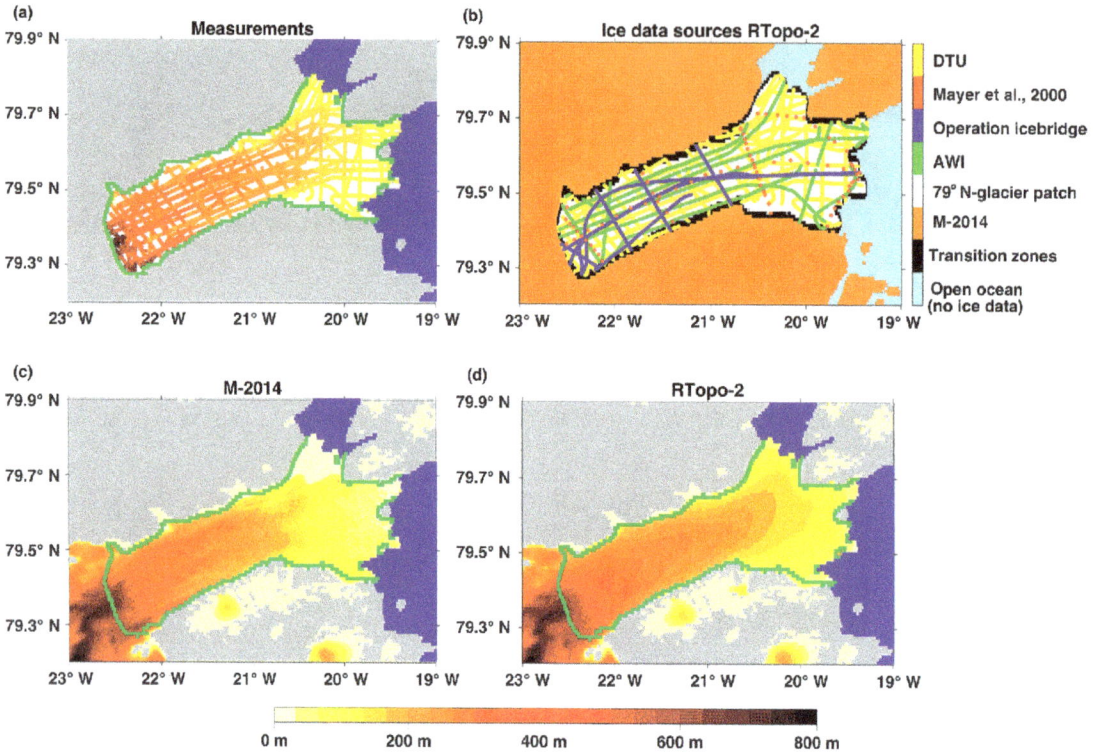

Figure 5. Ice thickness maps of the floating ice tongue of Nioghalvfjerdsfjorden Glacier. The maps show the coverage of ice thickness measurements from radar and seismic soundings **(a)**, data sources used in RTopo-2 **(b)**, and ice thicknesses in M-2014 **(c)** and RTopo-2 **(d)**. Shaded in grey is land; blue shaded is the open ocean. The dark green line in panels **(a)**, **(c)**, and **(d)** indicates the grounding line.

polation artefacts. Compared to M-2014, the main benefit of our grid is that it also covers the thickness of floating ice in Dijmphna Sund (Fig. 5c and d).

Assuming hydrostatic equilibrium, we calculated the surface height (ζ) of floating ice from the gridded ice thickness (H) using

$$\zeta = H \frac{\rho_{\text{water}} - \rho_{\text{ice}}}{\rho_{\text{water}}}, \tag{1}$$

with densities (ρ) of $\rho_{\text{water}} = 1023 \, \text{kg m}^{-3}$ and $\rho_{\text{ice}} = 917 \, \text{kg m}^{-3}$. The ice draft results from $\zeta - H$.

We combined the newly gridded glacier topographies with the surrounding surface height and ice draft maps with a transition zone of 2 km width. Corrections needed to be applied in areas where the newly gridded ice thickness exceeded the water depth. In regions where the surface type mask derived from M-2014 proposes the existence of floating ice, bedrock topography was corrected by applying a minimum water column thickness of 1 m. This procedure is justified by the fact that ice thickness observations for the floating ice tongues are much more densely spaced than the very sparse sub-ice bathymetry measurements obtained from seismics.

In comparison, the RTopo-2 ice thickness map derived from measurements deviates from M-2014 mostly towards the glacier front (Fig. 5c and d). East of 20.5° W, the ice is

up to 80 m thicker in RTopo-2. For the part of the glacier front which extends northward into Dijmphna Sund the ice thickness in M-2014 is only 1 m with a classification as grounded ice. In contrast, based on the observations from e.g. Mayer et al. (2000), RTopo-2 shows floating ice with thicknesses up to 150 m in this area.

For the sub-ice cavity of Nioghalvfjerdsfjorden Glacier, the NEG_DBM provides a bathymetry grid that has been interpolated from seismic observations of Mayer et al. (2000). We expect the sub-ice bathymetry to roughly follow plausible LGM ice flow stream lines (Evans et al., 2009) which we inferred from the seismic data points. We adjusted the interpolated bathymetry accordingly to achieve a more realistic representation of the sub-ice cavity geometry.

2.4.3   Antarctic ice and bedrock topographies

As discussed in Sect. 2.3.2, we used the bedrock topography from IBCSO Version 1.0 (Arndt et al., 2013) (polar stereographic grid with true scale at 65° S) for the bathymetry of the Southern Ocean, including the sub-ice shelf cavities. In the north, a 50 km wide transition zone along 61.5° S connects the IBCSO data to the GEBCO_2014 grid. On the Antarctic continent, bedrock topography is derived from the

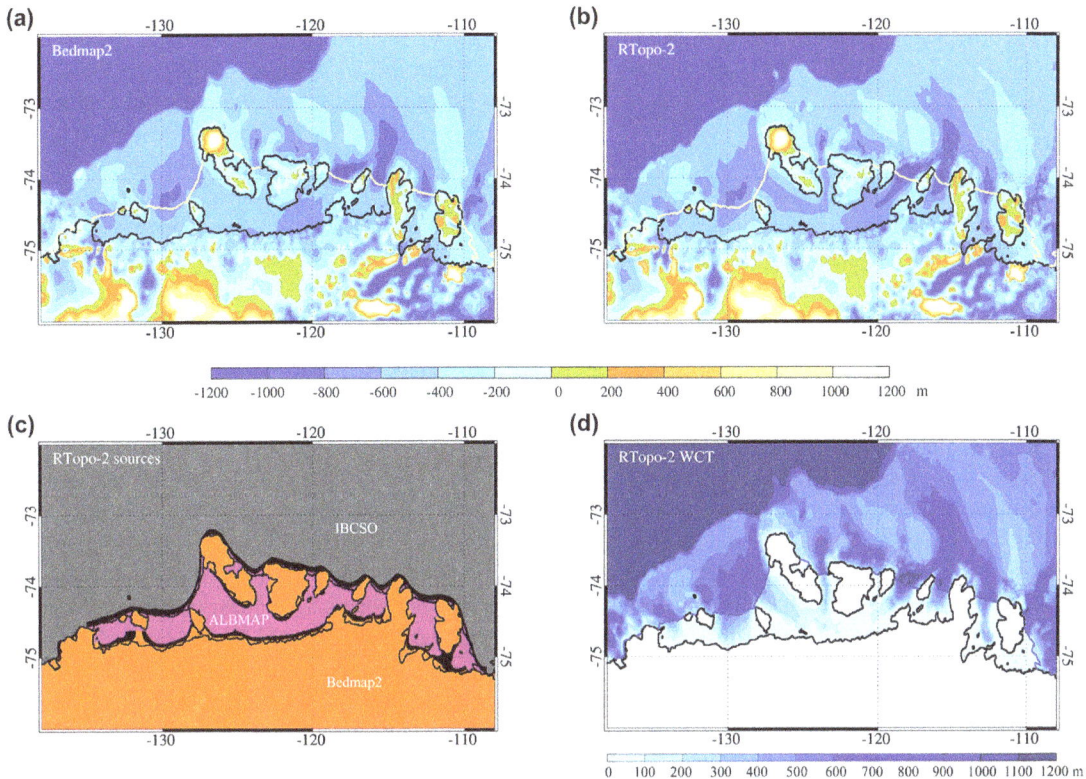

Figure 6. Bathymetry for Getz Ice Shelf cavity and its surrounding in Bedmap2 **(a)** and the merged bathymetry in RTopo-2 **(b)**. **(c)** Indication of data sources for the RTopo-2 bathymetry grid, with colours corresponding to the global source map in Fig. 2. **(d)** Water column thickness obtained from RTopo-2 ice shelf draft and bathymetry.

Bedmap2 (Fretwell et al., 2013) data set, as are the surface and ice bottom topographies.

Where the coastline of the Antarctic continent is formed by a transition from grounded ice or bare land to open ocean, we join the Bedmap2 and IBCSO topographies in a narrow band directly at the coast. Along the grounding lines of sub-ice cavities, the transition between the IBCSO and Bedmap2 topographies is in a roughly 8 km wide band 10 km off the grounding line (i.e. within the sub-ice cavity). In any case, a smooth transition between the IBCSO and Bedmap2 grids is easy to ensure due to the fact that Bedmap2 bedrock topography data have been incorporated in the generation of IBCSO. Small inconsistencies that still arise from the interpolation (mainly due to the discontinuity of ice draft along the ice front) were cured by enforcing grounded ice bottom topography to be identical to bedrock topography.

Given that the IBCSO data set incorporates not only bedrock relief but also ice surface topography from Bedmap2, it may seem better to use the IBCSO products throughout Antarctica and thus avoid the stitching between the two grids along the Antarctic continent. We decided not to follow this approach because IBCSO does not provide information about the thickness of floating ice shelves. Given that the compilation of RTopo-2 has been targeted towards

studies of ice dynamics and ice–ocean interaction at the interfaces between ice sheets and ocean, we decided that discontinuities of ice thickness across the grounding lines are to be avoided as far as possible. Therefore, ice surface and bottom topographies for grounded and floating ice are to be adopted from one self-consistent data set, which is possible only with Bedmap2. Similar consistency arguments apply to the bedrock relief under grounded ice; again we decided to use the original Bedmap2 product here to avoid introducing inconsistencies.

2.4.4 Local corrections for Antarctic sub-ice shelf bathymetry

For Filchner-Ronne Ice Shelf and the ice streams in its catchment basin, as well as for the ice topographies in many other regions, the benefit of a largely improved data coverage and grid spacing in Bedmap2 is very obvious and quite substantial. However, with regard to the representation of sub-ice cavity bathymetry, the transition from RTopo-1 to Bedmap2 does not universally yield an improvement. Although Fig. 6 in Fretwell et al. (2013) indicates that sub-ice shelf bathymetry for most of the ice shelves in Bedmap2 goes back to RTopo-1, many details of cavity bathymetry that appear plausible and are in some cases well covered by original data have vanished in the transition. This section reports on the local data corrections or reconstruction procedures we applied.

Getz and Abbot ice shelves

According to Fretwell et al. (2013), sub-ice bathymetry for Getz Ice Shelf cavity in Bedmap2 (Fig. 6a) has been derived from the topography grid of Nitsche et al. (2007). While this data set provides an excellent bathymetry map for the open Amundsen Sea, it suffers from missing data for the sub-ice shelf cavities. As a result, Bedmap-2 suggests a very shallow water column in large parts of the cavity. For RTopo-2 (Fig. 6b), we decided to go back to the submarine trough structure that RTopo-1 inherited from ALBMAP (Le Brocq et al., 2010). Upper and lower ice surface topographies and the surface type mask (locations of coast and grounding line) continue to be derived from Bedmap2. A smooth transition of sub-ice shelf bathymetry to the bedrock topography under grounded ice (from Bedmap2) and to open-ocean bathymetry (from IBCSO) is achieved using tanh functions in blending zones of $\approx 15\,$km width (Fig. 6c). As a result, water column thickness in the sub-ice cavity (Fig. 6d) features continuous troughs extending from the open-ocean continental shelf across the ice front towards the grounding line. The high basal melt rates suggested for Getz Ice Shelf (e.g. Depoorter et al., 2013) make the existence of such transport pathways for warm water seem plausible. A strict evaluation, however, is made very difficult by the lack of sub-ice bathymetry data, and there is no proof that the structures we suggest are correct.

A similar case can be made for Abbot Ice Shelf. For the eastern part of Abbot Ice Shelf, sub-ice bathymetry in Bedmap2 is derived from the Graham et al. (2011) data set, which incorporates ALBMAP bedrock topography in the sub-ice cavity. For the western part of the Abbot Ice Shelf, however, Bedmap2 utilizes the bathymetry map of Nitsche et al. (2007), which again leads to a very small water column thickness with virtually no connection to the open ocean in this sector of the ice shelf. We decided to restore the structure of a sub-ice trough connected to the eastern Amundsen Sea from ALBMAP for the Abbot Ice Shelf cavity west of 98° W. Also here, it should be kept in mind that bathymetry under this ice shelf is only weakly constrained.

Larsen C Ice Shelf

For Larsen C Ice Shelf cavity, Fig. 6 in Fretwell et al. (2013) indicates that bedrock topography was derived from RTopo-1. On a closer look, however, substantial differences between Bedmap2 and RTopo-1 can be seen in this area. Specifically, many of the deep troughs that had been inferred from ice draft at the grounding line in RTopo-1 have been removed (or are far less pronounced) in Bedmap2. Revisiting all the data sets in question here, we found that the RTopo-1 ice and

bedrock topographies along the southern part of Larsen C grounding line imply an ice thickness maximum to be found not within the ice streams feeding the ice shelf but shortly downstream from where the ice comes afloat. This pattern is most obvious for the ice stream between the Joerg and Kenyon peninsulas in the southwestern corner of Larsen C Ice Shelf and not likely to be a good approximation to reality. Given that it can be safely assumed that the ice topographies in Bedmap2 are more reliable than the combination of data sets used in RTopo-1, we conclude that there is no sufficient evidence for the existence of the deep throughs suggested near the Larsen C grounding line in RTopo-1. RTopo-2 thus simply adopts the Larsen C cavity geometry from Bedmap2.

Fimbulisen

Bedmap2 bathymetry under the floating Fimbulisen is claimed to be derived from RTopo-1 but in fact deviates from the latter substantially. Specifically, the deep troughs between the islands (ice rises) in the eastern part of the cavity (i.e. between 1 and 5° E) that Nøst (2004) inferred from original seismic data are no longer there. We decided to go back to the Nøst (2004)/Smedsrud et al. (2006) data set that was already incorporated in RTopo-1 for the Fimbulisen cavity between 1° and 5° E.

2.5 Tabular iceberg in Weddell Sea

In August 1986, three giant icebergs (A-22, A-23, and A-24), each one between 3000 and 4000 km^2 in area, calved from the Filchner Ice Shelf front (Ferrigno and Gould, 1987) and grounded at the eastern Berkner Bank. Iceberg A-24 came ungrounded in March 1990 and drifted northward through the Weddell and Scotia seas; icebergs A-22 and A-23 broke in two in 1994 and 1991, respectively. One of their remnants, A-23A, is still grounded on the eastern slope of Berkner Bank and continues to form a barrier to the sea ice drifting in this region, frequently creating a polynya in its lee (Markus, 1996). When modelling sea ice in this area, a comparison between modelled and observed (mostly remote sensing) data is strongly complicated if this effect is omitted in the model (see e.g. Haid and Timmermann, 2013). A similar case can be made for the iceberg's effect on the ocean currents on the continental shelf. Despite the fact that Grosfeld et al. (2001) showed that iceberg calving and grounding does change the circulation and hydrography in the Filchner Ice Shelf–ocean system, it is not common for today's ocean models to take this into account.

To enable high-resolution modelers to do the model-to-data comparison in a more consistent way – especially given that data coverage is about to improve considerably in the framework of ongoing and planned field activities in the area – and to achieve a more realistic representation of the local ocean currents in hindcast simulations, we decided to include the signature of A-23A in RTopo-2. The area cov-

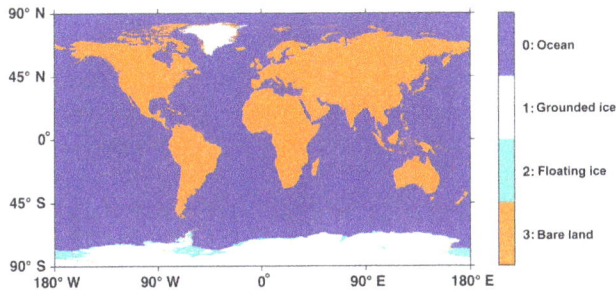

Figure 7. Global surface type mask (compare with Fig. 7 in Timmermann et al. (2010), but Greenland ice sheet included).

ered by the iceberg was picked from a composite of 2013 MODIS images (Paul et al., 2015) and defined to be covered with grounded ice, i.e. ice with a lower surface topography identical to the ocean bathymetry (which clearly is only a schematic representation of the real situation). Freeboard of the iceberg is represented as a constant value in RTopo's surface height field; it is computed from Archimedes' principle assuming densities of ocean and ice to be 1028 and 910 kg m^{-3}, respectively. Thickness of the iceberg in this equation is assumed to be such that its draft is equal to the minimum water depth in the area that is covered by the iceberg. Eventually, this procedure yields a freeboard (surface height) value of about 42 m over the iceberg area, which is consistent with the freeboard derived from SAR interferometry applied to TanDEM-X image pairs of June 2013 (M. Rankl, personal communication, 2016).

Note that the original bathymetry grid under the iceberg is fully preserved so that the whole feature can be removed without any loss of information in case this seems desirable in any particular application of the data set.

2.6 Surface type mask

In addition to the maps of the bedrock and ice topographies we provide a global mask which distinguishes between open ocean, bare land, grounded ice, and floating ice (Fig. 7).

On the polar continents, the mask largely follows M-2014 for Greenland and Bedmap2 for Antarctica. Ice caps not connected to the Antarctic ice sheet or Greenland mainland have been removed from the mask and classified as bare land. Ice surface height in these cases has been adopted as bedrock surface height. In contrast to RTopo-1, the surface type mask in RTopo-2 contains information about rock outcrops (surface type "bare land") in Antarctica.

Lakes and enclosed seas outside Antarctica and Greenland are marked as bare land in the mask but are still present in the bathymetry data set adopted from GEBCO_2014. This was done to avoid the tedious procedure of manually removing features with a topography below sea level and no connection to the world ocean when setting up an ocean general circulation model.

For the Northeast Greenland continental shelf, the NEG_DBM bathymetry map provides bedrock elevation with a very high data coverage. We used this data set to adjust the surface type mask: grid points with an elevation > 0 were classified as bare land surface, and negative elevation obviously enforced a classification as ocean. Within the perimeter of the Greenland continent, the surface type mask remained being defined by M-2014.

3 Error estimates

In the transition corridors between different data sets and in regions where values have been inferred from consistency arguments, errors are hard to quantify. Here we can only give an overview on error estimates provided by the authors of some of our source data sets.

3.1 Bathymetry

In GEBCO_2014 approximately 18 % of the non-land grid cells are constrained by bathymetric control data, which consist of echo sounding data as well as pre-prepared bathymetric grids that may contain interpolated areas (Weatherall et al., 2015). In the current IBCAO version, the Arctic Ocean is mapped by multibeam surveys covering 11 % of the area and an additional vast amount of single beam data (Jakobsson et al., 2012). In IBCSO around 17 % of grid cells in the Southern Ocean are directly constrained with data; 15.4 % of data points are from multibeam bathymetry (Arndt et al., 2013). In areas with no direct measurements, the ocean bathymetry was interpolated between measurements and/or plausible bathymetry.

In general, the accuracy of echo sounding systems can be expected to be about one percent of the water depth. However, in the areas between the sounding tracks uncertainties can be much higher.

3.2 Ice and bedrock topographies

For ice surface heights of Greenland the overall root-mean-square deviation between the GIMP digital elevation model and ICESat elevation is ± 9.1 m (Howat et al., 2014). The deviation varies strongly with region (Fig. 6 in Howat et al., 2014).

The technical error of ice thicknesses derived from radio echo sounding depends mainly on the sampling interval and transmitted signal length, both of which vary from system to system. The vertical resolution in ice thickness of the various employed RES systems varies between 5.05 and 8.45 m; the sampling precision is higher, usually in the order of 1 m. Thus an uncertainty of about 15 to 35 m for the ice thickness is realistic. However, complex geometries and steep topography confining the investigated glaciers and ice tongues can cause side and multiple reflections which mask the subglacial reflections, especially in airborne measurements.

Ice thickness and ice draft mapped by Morlighem et al. (2014) are subject to errors of about 35 m for areas with a dense radar sounding coverage. In areas which are less well constrained, errors can exceed 50 m (Morlighem et al., 2014).

Ice thickness maps derived from the available observations for Nioghalvfjerdsfjorden Glacier and Zachariæ Isstrøm reveal distinct differences between data sets from different years. All data across Zachariæ Isstrøm are based on radar data from 2010 to 2014 (obtained from Operation Icebridge and AWI flights). Based on Landsat optical imagery, Mouginot et al. (2015) observed an accelerated retreat in the ice front position in 2013/14 and estimated a mass loss of $5 \, \mathrm{Gt} \, \mathrm{yr}^{-1}$. The extent of Zachariæ Isstrøm in RTopo-2 thus represents the state prior to its decay and not the present state.

Ice thickness data covering Nioghalvfjerdsfjorden Glacier include additionally a large number of radar tracks and seismic data obtained 10 years earlier in 1997/98 (DTU/Seroussi et al., 2011; Mayer et al., 2000) (see Fig. 5b). Data from the two time slices differ by about 50 m in several places, especially within 5 km from the grounding line where the floating ice tongue is subject to strong basal melt. The differences are in the same range as the along-track noise in some of the radar tracks (see above). Such discrepancies were smoothed out by our interpolation procedure.

Next to the uncertainties related to data interpretation and processing, the representation of the firn layer ("firn correction") is an issue that requires serious attention. While in B-2013 a firn layer thickness of 10 m in all ablation regions around Greenland is assumed, there is no firn correction applied in M-2014. Snow depth varies strongly over the Greenland ice sheet and within the seasonal cycle (Nghiem et al., 2005), with most of the snow melting during the summer season close to the coast. A constant-value firn correction is therefore bound to be a rather crude approximation. We keep the M-2014 assumption of zero firn layer thickness and note that for determining a regionally varying firn correction the local depth–density relationship, respectively, the variation of the dielectric properties with depth needs to be known.

4   Summary and outlook

We compiled a global 30 arcsec data set for World Ocean bathymetry and Greenland/Antarctic ice sheet/shelf topography. High-resolution data from Greenland floating glaciers and of bathymetry on the Northeast Greenland continental shelf were compiled into a synthesis of gridded bathymetry products including the Morlighem et al. (2014) Greenland ice and bedrock topographies and the Bedmap2 Antarctic topography data sets. Similar to RTopo-1 (Timmermann et al., 2010), the RTopo-2 data set contains maps of global bedrock topography and the upper and lower surface heights of the Antarctic and Greenland ice sheet/ice shelf system. Consistent with the topography maps, a surface type mask for open ocean, grounded ice, floating ice, and bare land surface is provided.

This new data set provides enough local detail for a wide range of global or regional studies. Our main target group are ocean modelers who aim at a realistic representation of ice–ocean interaction in an ocean general circulation or climate model. In the current version, particular attention has been paid to the floating glaciers and the continental shelf in the Northeast Greenland sector. Other Greenland fjord regions are of similar interest but suffer from a lack of data. We encourage users who are specifically interested in one of those fjords to carefully review the data using information unused by us as a benchmark. Additional contributions of (gridded or ungridded) fjord/shelf bathymetry and/or glacier/ice shelf/cavity geometry are welcome and will be used to update the data set as soon as possible.

5   Data availability

The RTopo-2 data set has been published by Schaffer and Timmermann (2016) on the PANGAEA database in four variants.

1. The complete global 30 arcsec data set has been split into four files:

   - RTopo-2.0.1_30sec_bedrock_topography.nc (3.7 GB),
   - RTopo-2.0.1_30sec_ice_base_topography.nc (3.7 GB),
   - RTopo-2.0.1_30sec_surface_elevation.nc (3.7 GB), and
   - RTopo-2.0.1_30sec_aux.nc (2.8 GB), which contains auxiliary maps for data sources and the surface type mask.

2. A regional 30 arcsec subset that covers all variables around Greenland in the interval 80° E–0°, 55–85° N is available in RTopo-2.0.1_30sec_Greenland.nc (0.5 GB).

3. A regional 30 arcsec subset for the Antarctic region south of 50° S has been split into two files:

   - RTopo-2.0.1_30sec_Antarctica_data.nc (2.5 GB) contains bedrock topography, ice base topography, and surface elevation.
   - RTopo-2.0.1_30sec_Antarctica_aux.nc (0.6 GB) contains auxiliary maps for data sources and the surface type mask.

4. A complete global 1 arcmin data set that has been split into two files:

   - RTopo-2.0.1_1min_data.nc (2.8 GB) contains maps of bedrock topography, ice bottom topography, and surface elevation.

- RTopo-2.0.1_1min_aux.nc (0.7 GB) contains auxiliary maps for data sources and the surface type mask.

Data sets for the location of coastlines (RTopo-2.0.1_coast.asc, 50 MB) and the ice shelf/floating glacier front lines (RTopo-2.0.1_isf.asc, 1.4 MB) have been prepared in ASCII format. Grounding lines are represented as parts of the coastline. To enable communication in case of errors or updates, we would appreciate a notification from users of our data set.

Author contributions. R. Timmermann (Southern Hemisphere) and J. Schaffer (Northern Hemisphere) designed the merging strategies and processed the RTopo-2 data sets. J. E. Arndt and M. Morlighem provided the latest versions of bathymetry and ice thickness maps for Greenland and the continental shelf in its vicinity. S. S. Kristensen, C. Mayer, and D. Steinhage provided pre-processed ice thickness data for Nioghalvfjerdsfjorden Glacier. J. Schaffer prepared the manuscript with contributions from all co-authors.

Acknowledgements. The authors would like to thank S. Paul and R. Zentek for extracting the iceberg position from the MODIS data, X. Asay-Davis, S. Coers, B. K. Galton-Fenzi, H. Gudmundsson, H. H. Hellmer, D. Jansen, L. Padman, and D. Martin for helpful discussions, and W. Cohrs, H. Liegmahl-Pieper, and C. Wübber for providing excellent computing facilities at AWI. GEBCO_2014 Grid (version 20150318) data were obtained from http://www.gebco.net. Some data used in this paper were acquired by NASA's Operation IceBridge project. Funding by the Helmholtz Climate Initiative REKLIM (Regional Climate Change), a joint research project of the Helmholtz Association of German research centres (HGF) is gratefully acknowledged. We thank M. Jakobsson and one anonymous reviewer for their help in improving the manuscript.

Edited by: D. Carlson

References

Allen, C., Leuschen, C., Gogineni, P., Rodriguez-Morales, F., and Paden, J.: IceBridge MCoRDS L2 Ice Thickness, Version 1. [2010–2012], Boulder, Colorado USA, NASA National Snow and Ice Data Center Distributed Active Archive Center, doi:10.5067/GDQ0CUCVTE2Q, 2010, updated 2015.

Andresen, C. S., Straneo, F., Ribergaard, M. H., Bjørk, A. A., Andersen, T. J., Kuijpers, A., Nørgaard-Pedersen, N., Kjær, K. H., Schjøth, F., Weckström, K., and Ahlstrøm, A. P.: Rapid response of Helheim Glacier in Greenland to climate variability over the past century, Nat. Geosci., 5, 37–41, doi:10.1038/NGEO1349, 2012.

Arndt, J. E., Schenke, H. W., Jakobsson, M., Nitsche, F. O., Buys, G., Goleby, B., Rebesco, M., Bohoyo, F., Hong, J., Black, J.,

Greku, R., Udintsev, G., Barrios, F., Reynoso-Peralta, W., Taisei, M., and Wigley, R.: The International Bathymetric Chart of the Southern Ocean (IBCSO) Version 1.0 – A new bathymetric compilation covering circum-Antarctic waters, Geophys. Res. Lett., 40, 3111–3117, doi:10.1002/grl.50413, 2013.

Arndt, J. E., Jokat, W., Dorschel, B., Myklebust, R., Dowdeswell, J. A., and Evans, J.: A new bathymetry of the Northeast Greenland continental shelf: Constraints on glacial and other processes, Geochem. Geophy. Geosy., 16, 3733–3753, doi:10.1002/2015GC005931, 2015.

Assmann, K. M., Jenkins, A., Shoosmith, D. R., Walker, D. P., Jacobs, S. S., and Nicholls, K. W.: Variability of Circumpolar Deep Water transport onto the Amundsen Sea continental shelf through a shelf break trough, J. Geophys. Res.-Oceans, 118, 6603–6620, doi:10.1002/2013JC008871, 2013.

Bamber, J. L., Griggs, J. A., Hurkmans, R. T. W. L., Dowdeswell, J. A., Gogineni, S. P., Howat, I., Mouginot, J., Paden, J., Palmer, S., Rignot, E., and Steinhage, D.: A new bed elevation dataset for Greenland, The Cryosphere, 7, 499–510, doi:10.5194/tc-7-499-2013, 2013.

Becker, J. J., Sandwell, D. T., Smith, W. H. F., Braud, J., Binder, B., Depner, J., Fabre, D., Factor, J., Ingalls, S., Kim, S.-H., Ladner, R., Marks, K., Nelson, S., Pharaoh, A., Trimmer, R., Rosenberg, J. V., Wallace, G., and Weatherall, P.: Global Bathymetry and Elevation Data at 30 Arc Seconds Resolution: SRTM30_PLUS, Mar. Geod., 32, 355–372, doi:10.1080/01490410903297766, 2009.

Church, J. A., Clark, P. U., Cazenave, A., Gregory, J. M., Jevrejeva, S., Levermann, A., Merrifield, M. A., Milne, G. A., Nerem, R. S., Nunn, P. D., Payne, A. J., Pfeffer, W. T., Stammer, D., and Unnikrishnan, A. S.: Climate Change 2013: The Physical Science Basis. Contribution of Working Group I to the Fifth Assessment Report of the Intergovernmental Panel on Climate Change, chap. Sea Level Change, Cambridge University Press, Cambridge, United Kingdom and New York, NY, USA, 2013.

Depoorter, M. A., Bamber, J. L., Griggs, J. A., Lenaerts, J. T. M., Ligtenberg, S. R. M., van den Broeke, M. R., and Moholdt, G.: Calving fluxes and basal melt rates of Antarctic ice shelves, Nature, 502, 89–92, doi:10.1038/nature12567, 2013.

Evans, J., Cofaigh, C. Ø., Dowdeswell, J. A., and Wadhams, P.: Marine geophysical evidence for former expansion and flow of the Greenland Ice Sheet across the north-east Greenland continental shelf, J. Quaternary Sci., 24, 279–293, doi:10.1002/jqs.1231, 2009.

Ferrigno, J. G. and Gould, W. G.: Substantial changes in the coastline of Antarctica revealed by satellite imagery, Polar Rec., 23, 577–583, doi:10.1017/S003224740000807X, 1987.

Foldvik, A., Gammelsrød, T., and Tørresen, T.: Circulation and water masses on the southern Weddell Sea shelf, Antarct. Res. Ser., 43, 5–20, 1985.

Fretwell, P., Pritchard, H. D., Vaughan, D. G., Bamber, J. L., Barrand, N. E., Bell, R., Bianchi, C., Bingham, R. G., Blankenship, D. D., Casassa, G., Catania, G., Callens, D., Conway, H., Cook, A. J., Corr, H. F. J., Damaske, D., Damm, V., Ferraccioli, F., Forsberg, R., Fujita, S., Gim, Y., Gogineni, P., Griggs, J. A., Hindmarsh, R. C. A., Holmlund, P., Holt, J. W., Jacobel, R. W., Jenkins, A., Jokat, W., Jordan, T., King, E. C., Kohler, J., Krabill, W., Riger-Kusk, M., Langley, K. A., Leitchenkov, G., Leuschen, C., Luyendyk, B. P., Matsuoka, K., Mouginot, J., Nitsche, F. O.,

Nogi, Y., Nost, O. A., Popov, S. V., Rignot, E., Rippin, D. M., Rivera, A., Roberts, J., Ross, N., Siegert, M. J., Smith, A. M., Steinhage, D., Studinger, M., Sun, B., Tinto, B. K., Welch, B. C., Wilson, D., Young, D. A., Xiangbin, C., and Zirizzotti, A.: Bedmap2: improved ice bed, surface and thickness datasets for Antarctica, The Cryosphere, 7, 375–393, doi:10.5194/tc-7-375-2013, 2013.

Funder, S., Kjeldsen, K. K., Kjær, K. H., and Cofaigh, C.: The Greenland Ice Sheet during the past 300,000 years: A review, Developments in Quaternary Science, 15, 699–713, doi:10.1016/B978-0-444-53447-7.00050-7, 2011.

Graham, A. G. C., Nitsche, F. O., and Larter, R. D.: An improved bathymetry compilation for the Bellingshausen Sea, Antarctica, to inform ice-sheet and ocean models, The Cryosphere, 5, 95–106, doi:10.5194/tc-5-95-2011, 2011.

Grosfeld, K., Schröder, M., Fahrbach, E., Gerdes, R., and Mackensen, A.: How iceberg calving and grounding change the circulation and hydrography in the Filchner Ice Shelf-Ocean System, J. Geophys. Res., 106, 9039–9055, doi:10.1029/2000JC000601, 2001.

Haid, V. and Timmermann, R.: Simulated heat flux and sea ice production at coastal polynyas in the southwestern Weddell Sea, J. Geophys. Res., 118, 2640–2652, doi:10.1002/jgrc.20133, 2013.

Howat, I. M., Negrete, A., and Smith, B. E.: The Greenland Ice Mapping Project (GIMP) land classification and surface elevation data sets, The Cryosphere, 8, 1509–1518, doi:10.5194/tc-8-1509-2014, 2014.

Jakobsson, M., Mazer, L., Coakley, B., Dowdeswell, J. A., Forbes, S., Fridman, B., Hodnesdal, H., Noormets, R., Pedersen, R., Rebesco, M., Schenke, H. W., Yarayskaya, Z., Accettella, D., Armstrong, A., Anderson, R. M., Bienhoff, P., Camerlenghi, A., Church, I., Edwards, M., Gardner, J. V., Hall, J. K., Hell, B., Hestvik, O., Kristoffersen, Y., Marcussen, C., Mohammed, R., Mosher, D., Nghiem, S. V., Pedrosa, M. T., Travaglini, P. G., and Weatherall, P.: The International Bathymetric Chart of the Arctic Ocean (IBCAO) Version 3.0, Geophys. Res. Lett., 39, L12609, doi:10.1029/2012GL052219, 2012.

Johnson, H. L., Münchow, A., Falkner, K. K., and Melling, H.: Ocean Circulation and properties in Petermann Fjord, Greenland, J. Geophys. Res., 116, C01003, doi:10.1029/2010JC006519, 2011.

Kusahara, K. and Hasumi, H.: Modeling Antarctic ice shelf responses to future climate changes and impacts on the ocean, J. Geophys. Res.-Oceans, 118, 2454–2475, doi:10.1002/jgrc.20166, 2013.

Kusahara, K. and Hasumi, H.: Pathways of basal meltwater from Antarctic ice shelves: A model study, J. Geophys. Res.-Oceans, 119, 5690–5704, doi:10.1002/2014JC009915, 2014.

Le Brocq, A. M., Payne, A. J., and Vieli, A.: An improved Antarctic dataset for high resolution numerical ice sheet models (ALBMAP v1), Earth Syst. Sci. Data, 2, 247–260, doi:10.5194/essd-2-247-2010, 2010.

Markus, T.: IGARSS '96: Remote Sensing for a Sustainable Future, chap. The effect of the grounded tabular icebergs in front of the Berkner Island on the Weddell Sea ice drift as seen from satellite passive microwave sensors, 1791–1793, IEEE Press, Piscataway, New Jersey, USA, doi:10.1109/IGARSS.1996.516802, 1996.

Mayer, C., Reeh, N., Jung-Rothenhäusler, F., Huybrechts, P., and Oerter, H.: The subglacial cavity and implied dynamics under Nioghalvfjerdsfjorden Glacier, NE-Greenland, J. Geophys. Res., 27, 2289–2292, doi:10.1029/2000GL011514, 2000.

McMillan, M., Shepherd, A., Sundal, A., Briggs, K., Muir, A., Ridout, A., Hogg, A., and Wingham, D.: Increased ice losses from Antarctica detected by CryoSat-2, Geophys. Res. Lett., 41, 3899–3905, doi:10.1002/2014GL060111, 2014.

Morlighem, M., Rignot, E., Mouginot, J., Seroussi, H., and Larour, E.: Deeply incised submarine glacial valley beneath the Greenland ice sheet, Nat. Geosci., 7, 418–422, doi:10.1038/NGEO2167, 2014.

Mouginot, J., Rignot, E., Scheuchl, B., Fenty, I., Khazendar, A., Morlighem, M., Buzzi, A., and Paden, J.: Fast retreat of Zachariæ Isstrøm, northeast Greenland, Science, 350, 1357–1361, doi:10.1126/science.aac7111, 2015.

Nakayama, Y., Timmermann, R., Schröder, M., and Hellmer, H. H.: On the difficulty of modeling Circumpolar Deep Water intrusions onto the Amundsen Sea continental shelf, Ocean Model., 84, 26–34, doi:10.1016/j.ocemod.2014.09.007, 2014a.

Nakayama, Y., Timmermann, R., Rodehacke, C. B., Schröder, M., and Hellmer, H. H.: Modeling the spreading of glacial meltwater from the Amundsen and Bellinghausen Seas, Geophys. Res. Lett., 41, 7942–7949, doi:10.1002/2014GL061600, 2014b.

Nghiem, S. V., Steffen, K., Neumann, G., and Huff, R.: Mapping of ice layer extent and snow accumulation in the percolation zone of the Greenland ice sheet, J. Geophys. Res., 110, F02017, doi:10.1029/2004JF000234, 2005.

Nitsche, F. O., Jacobs, S., Larter, R., and Gohl, K.: Bathymetry of the Amundsen Sea continental shelf: implications for geology, oceanography, and glaciology, Geochem. Geophy. Geosy., 8, Q10009, doi:10.1029/2007GC001694, 2007.

Nøst, O. A.: Measurements of ice thickness and seabed topography at Fimbul Ice Shelf, Dronning Maud Land, Antarctica, J. Geophys. Res., 109, C10010, doi:10.1029/2004JC002277, 2004.

Paul, S., Willmes, S., and Heinemann, G.: Daily MODIS composites of thin-ice thickness and ice-surface temperatures for the Southern Weddell Sea. doi:10.1594/PANGAEA.848612, Supplement to: Paul, S et al. (2015): Long-term coastal-polynya dynamics in the Southern Weddell Sea from MODIS thermal-infrared imagery, The Cryosphere, 9, 2027–2041, doi:10.5194/tc-9-2027-2015, 2015.

Pritchard, H., Ligtenberg, S., Fricker, H., Vaughan, D., van den Broeke, M., and Padman, L.: Antarctic ice-sheet loss driven by basal melting of ice shelves, Nature, 484, 502–505, doi:10.1038/nature10968, 2012.

Rignot, E., Mouginot, J., and Scheuchl, B.: Antarctic Grounding Line Mapping from Differential Satellite Radar Interferometry, Geophys. Res. Lett., 38, L10504, doi:10.1029/2011GL047109, 2011.

Sandwell, D. T. and Smith, W. H. F.: Global marine gravity from retracked Geosat and ERS-1 altimetry: Ridge segmentation versus spreading rate, J. Geophys. Res., 114, B01411, doi:10.1029/2008JB006008, 2009.

Schaffer, J. and Timmermann, R.: Greenland and Antarctic ice sheet topography, cavity geometry, and global bathymetry (RTopo-2), links to NetCDF files, doi:10.1594/PANGAEA.856844, 2016.

Seale, A., Christoffersen, P., Mugford, R. I., and O'Leary, M.: Ocean forcing of the Greenland Ice Sheet: Calving fronts and patterns of retreat identified by automatic satellite monitoring of eastern outlet glaciers, J. Geophys. Res., 116, F03013, doi:10.1029/2010JF001847, 2011.

Seroussi, H., Morlighem, M., Rignot, E., Larour, E., Aubry, D., Dhia, H. B., and Kristensen, S. S.: Ice flux divergence anomalies on 79north Glacier, Greenland, Geophys. Res. Lett., 38, L09501, doi:10.1029/2011GL047338, 2011.

Smedsrud, L. H., Jenkins, A., Holland, D. M., and Nøst, O. A.: Modeling ocean processes below Fimbulisen, Antarctica, J. Geophys. Res., 111, C01007, doi:10.1029/2005JC002915, 2006.

Straneo, F. and Heimbach, P.: North Atlantic warming and the retreat of Greenland's outlet glaciers, Nature, 504, 36–43, doi:10.1038/nature12854, 2013.

Straneo, F., Sutherland, D. A., Holland, D., Gladish, C., Hamilton, G. S., Johnson, H. L., Rignot, E., Xu, Y., and Koppes, M.: Characteristics of ocean waters reaching Greenland's glaciers, Ann. Glaciol., 53, 202–210, doi:10.3189/2012AoG60A059, 2012.

Timmermann, R. and Hellmer, H. H.: Southern Ocean warming and increased ice shelf basal melting in the twenty-first and twenty-second centuries based on coupled ice-ocean finite-element modelling, Ocean Dynam., 63, 1011–1026, doi:10.1007/s10236-013-0642-0, 2013.

Timmermann, R., Le Brocq, A., Deen, T., Domack, E., Dutrieux, P., Galton-Fenzi, B., Hellmer, H., Humbert, A., Jansen, D., Jenkins, A., Lambrecht, A., Makinson, K., Niederjasper, F., Nitsche, F., Nøst, O. A., Smedsrud, L. H., and Smith, W. H. F.: A consistent data set of Antarctic ice sheet topography, cavity geometry, and global bathymetry, Earth Syst. Sci. Data, 2, 261–273, doi:10.5194/essd-2-261-2010, 2010.

Weatherall, P. K., Marks, K., Jakobsson, M., Schmitt, T., Tani, S., Arndt, J. E., Rovere, M., Chayes, D., Ferrini, V., and Wigley, R.: A new digital bathymetric model of the world's oceans, Earth and Space Science, 2, 331–345, doi:10.1002/2015EA000107, 2015.

Wilson, N. J. and Straneo, F.: Water exchange between the continental shelf and the cavity beneath Nioghalvfjerdsbræ (79 North Glacier), Geophys. Res. Lett., 42, 7648–7654, doi:10.1002/2015GL064944, 2015.

7

Precipitation at Dumont d'Urville, Adélie Land, East Antarctica: the APRES3 field campaigns dataset

Christophe Genthon[1], Alexis Berne[2], Jacopo Grazioli[3], Claudio Durán Alarcón[1], Christophe Praz[2], and Brice Boudevillain[1]

[1]Univ. Grenoble Alpes, CNRS, IRD, Grenoble INP, IGE, 38000 Grenoble, France
[2]Environmental Remote Sensing Laboratory, Environmental Engineering Institute, School of Architecture, Civil and Environmental Engineering, École Polytechnique Fédérale de Lausanne, 1015 Lausanne, Switzerland
[3]Federal Office of Meteorology and Climatology, MeteoSwiss, Locarno-Monti, Switzerland

Correspondence: Christophe Genthon (christophe.genthon@cnrs.fr)

Abstract. Compared to the other continents and lands, Antarctica suffers from a severe shortage of in situ observations of precipitation. APRES3 (Antarctic Precipitation, Remote Sensing from Surface and Space) is a program dedicated to improving the observation of Antarctic precipitation, both from the surface and from space, to assess climatologies and evaluate and ameliorate meteorological and climate models. A field measurement campaign was deployed at Dumont d'Urville station at the coast of Adélie Land in Antarctica, with an intensive observation period from November 2015 to February 2016 using X-band and K-band radars, a snow gauge, snowflake cameras and a disdrometer, followed by continuous radar monitoring through 2016 and beyond. Among other results, the observations show that a significant fraction of precipitation sublimates in a dry surface katabatic layer before it reaches and accumulates at the surface, a result derived from profiling radar measurements. While the bulk of the data analyses and scientific results are published in specialized journals, this paper provides a compact description of the dataset now archived in the PANGAEA data repository (https://www.pangaea.de, https://doi.org/10.1594/PANGAEA.883562) and made open to the scientific community to further its exploitation for Antarctic meteorology and climate research purposes.

1 Introduction

The Antarctic ice sheet is a huge continental storage of water which, if altered through climate change, has the potential to significantly affect global sea level. While climate models consistently predict an increase in precipitation in the future in Antarctica (e.g. Palerme et al., 2016), most of which falls in the form of snow that will not melt and thus will accumulate further ice, observational data to raise confidence in the current precipitation in the models are still in demand. Antarctica is the poor cousin of global continental precipitation observation and climatology building efforts: citing Schneider et al. (2014) of the Global Precipitation Climatology Center (GPCC), "The GPCC refrains from providing a (precipitation) analysis over Antarctica" because of poor data coverage. The GPCC's global maps of continental precipita-

tion from in situ observations are left blank only over Antarctica. Satellites offer rising prospects to monitor remote, difficult and/or uninhabited regions, but even then Antarctica tends to be excluded from comprehensive and/or global studies (e.g. Funk et al., 2015). Only those studies that specifically focus on the polar regions and Antarctica have presented and discussed aspects of the Antarctic precipitation by satellite (Palerme et al., 2014, 2016, 2017; Behrangi et al., 2016). However, ground-based observations are still lacking to suitably calibrate and validate the satellite products.

The measurement of solid precipitation is notoriously difficult (Goodison et al., 1998; Nilu, 2013). Difficulties are exacerbated in Antarctica because access and operations are logistically difficult and environmental conditions are extreme. Antarctica is the driest continent on Earth in terms of precipitation: satellite data estimate the mean precipitation at 171 mm yr^{-1} of water equivalent north of 81° S, the latitude reached by the polar orbiting satellites (Palerme et al., 2014). Low precipitation is supported by net accumulation measurements at the surface using glaciological methods (Eisen et al., 2007) which yield equally low numbers (Arthern et al., 2006). On the high Antarctic plateau, the accumulation is only a few cm yr^{-1} annually (e.g. Genthon et al., 2015). Such a low precipitation rate would be very hard to monitor even in more hospitable environments. It is not possible with conventional instruments in Antarctica. Satellite data and glaciological reconstructions, as well as models and meteorological analyses, support a dry interior but indicate that precipitation is much larger at the peripheries of the Antarctic ice sheet, yearly reaching several tens of centimetres, or even metres, locally (Palerme et al. 2014). However, strong katabatic winds frequently blow at the peripheries, which adversely affect the conventional precipitation measurement methods. Collecting instruments (bucket-style instruments that capture and collect to measure snowfall, typically by weighing or tipping bucket counts) actually undercatch or overcatch because of air deflection and turbulence caused by the instruments themselves. In addition, they catch not only fresh falling snow, but also drifting/blowing snow which was previously deposited at the surface, and then eroded and remobilized by the strong winds. Non-catching instruments, including in situ (disdrometer) and remote (radar, lidar) sensing instruments, offer interesting prospects. Radars are particularly attractive because they can profile through the air layers. They can sense both horizontally to expand the spatial significance of the measurement and vertically to scan the origin and fate of precipitation since condensation in the atmospheric column, from the clouds (see Witze, 2016, for an application in Antarctica) and above to the surface, and separate blowing snow in the lower layers from precipitation higher up.

However, while radars are customarily used in other regions to monitor liquid precipitation (e.g. Krajewski and Smith, 2002; Fabry, 2015), and many campaigns have also been conducted in high-latitude and high-altitude regions to study snowfall (e.g. Schneebeli et al., 2013; Grazioli et al., 2015; Medina and Houze, 2015; Moisseev et al., 2015; Kneifel et al., 2015), experience is still limited in the Antarctic environment (Gorodetskaya et al., 2015). Because such instruments do not collect and directly measure the mass of falling precipitation, but rather measure the fraction of an emitted radiation which is reflected back by the hydrometeors, quantification in terms of precipitation involves both physically based (electromagnetic laws of diffusion, diffraction and propagation) and hypothesis-based (particle population size and shape, habits) post-processing. The hypothesis-based part requires calibration and validation using various sources of in situ measurements (e.g. Souverijns et al., 2017).

As part of the APRES3 project (Antarctic Precipitation, Remote Sensing from Surface and Space, http://apres3.osug.

fr, last access: 1 August 2018), starting in November 2015 until February 2016 for the intense observing period but still ongoing for some observations, an unprecedentedly comprehensive field campaign was launched at the French Dumont d'Urville Antarctic scientific station at the coast of Adélie Land. The objective was to measure and monitor precipitation not only in terms of quantity, but also of falling snow particle characteristics and microphysics. The range of instruments included a profiling K-band and a polarimetric scanning X-band radar, a multi-angle snowflake camera (MASC), an OTT Pluvio2 weighing gauge, and a Biral VPF-730 disdrometer. A weather station reporting temperature, moisture and wind conditions near the instruments was also deployed. Finally, a depolarization lidar was tentatively operated but had problems and is not further mentioned here. All instruments were removed at the end of January 2016 except the K-band radar, which remained in operation throughout 2016 and beyond. Grazioli et al. (2017a) provide a comprehensive description of the data and analysis techniques and discuss scientific outcomes. Further work is ongoing to address the calibration and validation of meteorological and climate models and of satellite remote sensing techniques with the data (snowfall occurrences and rates, but also vertical profiles). Meanwhile, because this is a unique dataset, dissemination to the wider community for similar use with other models and remote sensing processing approaches or other research purposes is considered timely. This paper provides a compact description of the dataset and dissemination.

2   Dataset description

Grazioli et al. (2017a) provide ample information on the observation site, most instruments and methods. A summary and complementary information are provided below.

2.1   Site description

The main APRES3 (austral) summer field campaign took place at French Antarctic scientific station Dumont d'Urville (DDU) in Adélie Land (66.6628° S, 140.0014° E; 41 m a.s.l. on average). The station is on Petrel Island located only ∼ 5 km off the continent and the ice sheet proper: the observations are thus representative of the very coast of the Antarctic ice sheet. Because the station was operated for more than 60 years uninterruptedly, the means and statistics of meteorology and climate are documented (König-Langlo et al., 1998; Grazioli et al., 2017a). A main meteorological feature is the strong katabatic winds that frequently blow in the area. Adélie Land was coined "the home of the blizzard" by Mawson (1915) after the first Australian Antarctic winter over in this region. However, much of the coasts of Antarctica are affected by the katabatic winds (Parish and Bromwich, 1987). DDU is a perfect place to sample their consequences, including in relation to precipitation.

2.2   Observations and instruments

Standard measurements of atmospheric variables (temperature, wind speed, wind direction, relative and specific humidity, atmospheric pressure) are collected regularly all year long by the French meteorological service (Météo France), and a radiosounding is made daily at 00:00 UTC. The routine program does not involve any instrumental measurement of precipitation. There are reports of visual estimation of the occurrence and type in the METAR (METeorological Airport Report) convention, but no quantification. For the APRES3 campaign, several instruments were deployed from the beginning of November 2015 to the end of January 2016 to objectively characterize and quantify the occurrences and amounts of precipitation, as described below.

2.2.1   Surface-based remote sensing instruments

As reported in the introduction, traditional collecting precipitation gauges are unreliable in Antarctica in general, and in particular in the coastal regions strongly affected by katabatic winds. Radars are the core instruments of the APRES3 campaign. Radars remotely sense the hydrometeors, estimate quantities and speed, and from this derive precipitation rates. Radars can scan and profile through atmospheric and hydrometeor layers and look beyond blowing snow near the surface. Two radars were deployed: a K-band frequency-modulated continuous-wave vertically staring profiler and an X-band pulsed dual-polarization scanning Doppler radar. The first instrument, a Metek micro-rain radar (MRR), is designed to measure rainfall rather than snowfall using the backscattering and vertical velocity information. However, the raw Doppler spectra can be reprocessed using Maahn and Kollias (2012)'s improved and innovative processing chain for data collected in snow to retrieve Doppler radar moments such as reflectivity Z and Doppler velocity. Most Z–S relations for radars have been derived for 10, 35 or 94 GHz and therefore the measured equivalent radar reflectivity at 24 GHz is first converted to X-band. Once mapped to X-band reflectivity this can be converted to snowfall rate S by means of a Z–S power law fitted to the local conditions using the weighing gauge information or parameterizations from the existing literature (for more details, see Grazioli et al., 2017a). The MRR was used with a 100 m vertical resolution. The second instrument, a mobile X-band polarimetric radar (MXPol), for which extensive experience with the measurement of snow is available (Schneebeli et al., 2013; Scipión et al., 2013; Grazioli et al., 2015), provided more detailed information and served as a control and reference for the calibration of the method to use the MRR data. It was used with a 75 m radial resolution, maximum radial distance 30 km, and different types of scans within a repeating scanning sequence of 5 min (plan position indicator (PPI), range height indicator (RHI), vertical profiles). While the X-band radar could only be deployed during the summer campaign and had to be

shipped back after completion in February 2016, the K-band radar sheltered by a radome could remain on site after the summer campaign. The radome significantly attenuates the signal (6.14 dB, Fig. 4 and Eq. 1 of Grazioli et al., 2017a), but it is necessary to protect the radar against the fierce winter winds in Adélie Land.

2.2.2   Disdrometer and MASC

The Biral VPF 730 disdrometer is also a non-capture instrument, which estimates the size and speed of airborne particles from the diffusion and diffraction of an infrared light beam within a 400 cm^3 air volume. The volumetric sampling of the VPF730 presents an advantage over 2-D sampling instruments, which is that it does not miss particles with a much larger horizontal (due to strong wind) than vertical (falling) speed. The downside is that the instrument does not straightforwardly distinguish between falling and blowing snow (Bellot et al., 2011). A Biral proprietary algorithm directly provides precipitation rates from the size–speed matrix. Because this is based on various assumptions, including on the phase, shape and density of the particles, particularly unwarranted in the atypical Antarctic environment, the database described here presents the matrices rather than the estimated precipitation.

A MASC was deployed next to the disdrometer. This instrument collects high-resolution stereoscopic photographs of snowflakes in free fall while they cross the sampling area (Garrett et al., 2012), thus providing information about snowfall microphysics and particle fall velocity. The MASC uses three identical 2448 × 2048 pixels cameras (with a common focal point) with apertures and exposure times adjusted to trade off between the contrast on snowflake photographs and motion blur effects. The resolution is about 33 μm per pixel. The cameras are triggered when a falling particle crosses two series of near-infrared sensors. A detailed description of the system and its calibration can be found in Garrett et al. (2012). Information from disdrometers (Souverijns et al., 2017) and more particularly from MASC images, after image processing, provides characterizations and classification of snow particles (Praz et al., 2017) that can be used to better process radar data.

2.2.3   Precipitation gauge, meteorology, and setting of the instruments

What fraction of snowfall a traditional precipitation gauge captures is unwarranted. On the other hand, unlike remote sensing instruments, the mass quantification of any captured snow is direct and straightforward. An OTT Pluvio2 precipitation gauge was deployed for the duration of the summer campaign. Snow falling in the instrument is definitely captured and weighted. The instrument used here was equipped with a manufacturer-design wind shield meant to limit wind impacts on capture efficiency. Further, the instru-

Table 1. Summary of data from the APRES3 observation campaigns available by download from the PANGAEA repository (Berne et al., 2017) or by request to the authors. MASC data are provided for each picture taken, the taking of which varies with the occurrence of particle detection.

Instrument	Variables	Format/source	Time period	Sampling	
				Time	Space
Weather station	Temperature, moisture, wind	ASCII + NetCDF/PANGAEA	21 Nov 2015–6 Feb 2016	30 s	local
K-band MRR radar	Precipitation profiles (28 levels)	NetCDF/PANGAEA	21 Nov 2015–11 Nov 2016	1 h	Vertical: 100 m
Pluvio2 weighing gauge	Surface precipitation	ASCII + NetCDF/PANGAEA	17 Nov 2015–21 Jan 2016	1 min	local
Biral VPF730 disdrometer	Size/speed matrices	ASCII + NetCDF/PANGAEA	2 Dec 2015–23 Feb 2016	10 min	local
MASC	Snow particle classification and microphysics	ASCII + NetCDF/PANGAEA	11 Nov 2015–21 Jan 2016	Variable	local
X-band MXPol radar	Polarimetric radar variables and hydrometeor types	Request to authors	21 Nov 2015 to 1 Feb 2016	5 min	3-D, radial: 75 m

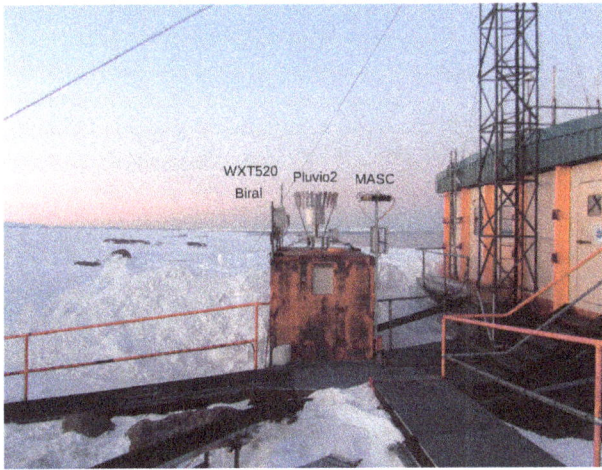

Figure 1. Setting of the in situ sensors (weather station WXT520, disdrometer Biral, snow gauge Pluvio 2 and MASC) on the roof of a small shelter close to other buildings.

ment was relatively shielded from the strongest wind due to its location, on the roof of a container but on the side of a building. The MASC and disdrometer were deployed at the same partially sheltered site (Fig. 1), the local meteorology of which was sampled locally by a Vaisala WXT520 weather transmitter, the principles, instrumental accuracy and performance of which can be found in the manufacturer's User's Guide (https://www.vaisala.com/sites/default/files/documents/M210906EN-C.pdf, last access: 1 August 2018). Note that this station integrates an acoustic rain gauge not appropriate for measuring snowfall; thus, the deployment of the Pluvio2. The radars were closely located, within at most 200 m of the other instruments. A composite picture of the various instruments and instrument settings is provided by Fig. 2 of Grazioli et al. (2017a).

3 Data, samples and conclusions

Table 1 summarizes the data streams from the APRES3 measurement campaign. Grazioli et al. (2017a) extensively process and discuss the data from the different instruments. Further analyses and presentation are beyond the scope of this data paper, and only a few snapshots are provided to illustrate the content of the database. Figure 2 shows the cumulative precipitation during the intensive summer campaign, as yielded by the Pluvio2 snow gauge and the processed MRR at the lowest useful level and at 741 m above sea level. Only 28 out of 31 MRR levels are provided in the database. This is because several simplifications necessary for a tractable quantitative interpretation of radar signal power do not apply in the two lowest levels. Data processing is based on assumptions that are not valid as it may lead to overestimation of reflectivity (Peters et al., 2005). In the uppermost level, the data become noisy, as according to Kneifel et al. (2011), the detectability is highest close to the ground, at -2 dBZ (35 GHz equivalent) at 500 m, but decreases with height to 3 dBz at 3000 m. Precipitation rates were retrieved from MRR data following Grazioli et al. (2017a): the reflectivity was converted into liquid water equivalent rate by fitting the prefactor and exponent of a Z–S relationship using carefully filtered nearby Pluvio2 data. Censoring the Pluvio2 data for wind-induced biases such as vibrations and turbulence effects by cross-referencing with the MRR data removes up to 30 % of the quantities (Grazioli et al., 2017a), as visible by the accumulation of the Pluvio2 in the time periods between snowfall events. As the Pluvio2 is a standard instrument but there is no standard correction method for wind effects, others might want to test other approaches, and the primary rather than the censored data are shown here and distributed in the database.

The MRR precipitation at the lowest useful level (341 m a.s.l.) is significantly less than that at 741 m a.s.l., showing that a significant fraction of the precipitation formed above sublimates in the dry katabatic air layer near the sur-

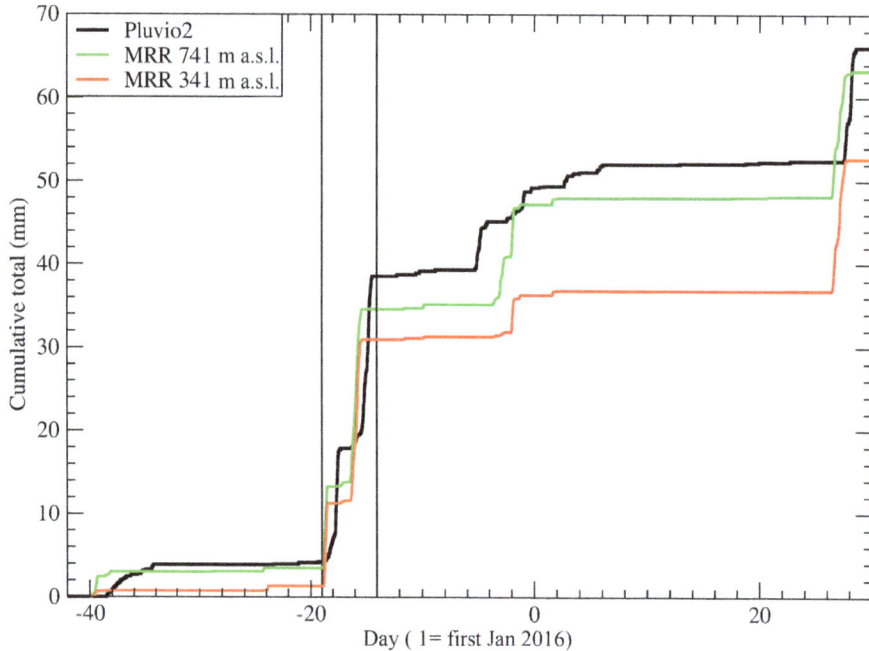

Figure 2. Cumulative precipitation during the APRES3 summer campaign, from the Pluvio2 and MRR instruments. Thin black vertical lines bracket the largest precipitation event in the period, from 12 to 17 December 2015. Precipitation from the MRR is reported for two levels above sea level, 341 and 741 m.

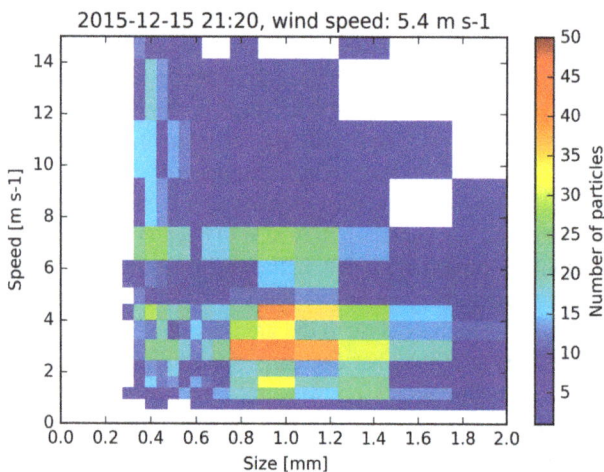

Figure 3. An example of the 21 × 16 fixed-level size–speed matrix of particle density distribution from the Biral disdrometer during the large precipitation event shown in Fig. 2. The date and time (15 December, 21:20 local time) and local wind speed ($5.4\,\mathrm{m\,s^{-1}}$) are printed at the top of the graph.

face. Further observations show that this frequently occurs in all seasons of the year (see below). Meteorological and climate models suggest that at the full scale of the Antarctic ice sheet up to 17 % of the precipitation evaporates in a dry surface layer before reaching the ground, and thus does not contribute to feeding the ice sheet (Grazioli et al., 2017b). Al-

together, the 2015–2016 summer was relatively dry and few strong precipitation events occurred. One such event happened from 12 to 17 December 2015 (delineated by thin vertical black lines in Fig. 2), during which the largest part of the total cumulative precipitation this summer was recorded. Figure 3 shows an example of the Biral disdrometer size–speed matrix during this event. The local wind was relatively strong ($5.4\,\mathrm{m\,s^{-1}}$ averaged on the same 10 min as the matrix Fig. 3, with significant gusts in the period). Considering that the anemometer is set at a relatively sheltered place and thus underestimates the large-scale wind, a contribution of blowing snow to the disdrometer report is likely. However, a significant fraction of the density number of particles detected is associated with moderate speed below $4\,\mathrm{m\,s^{-1}}$. Large particles (0.8–1.2 mm) are detected, the fall speed of which may indeed be over $1\,\mathrm{m\,s^{-1}}$, as reported by the instrument.

Figure 4 shows the probability distribution function of the degree of riming of the snowfall particles as obtained by processing the MASC photographs. No less than 426 229 photographs of falling snow particles were collected during the season. Each picture is processed as described in Praz et al. (2017). The database offers the processed results in the form of a classification, rather than the photographs themselves. Figure 4 cumulates all single estimates of the degree of riming in the database. The degree of riming is defined in this context as a continuous index between 0 (no riming on the particle detected) and 1 (fully rimed, graupel-like particle). Almost half of the particles are close to fully rimed, in-

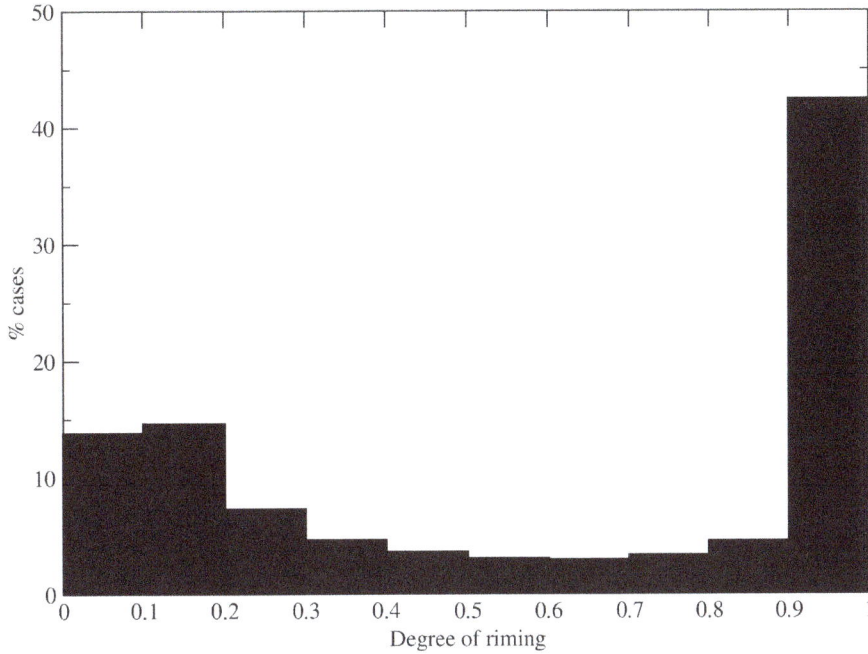

Figure 4. PDF of snow particle riming from the MASC data over the observation period.

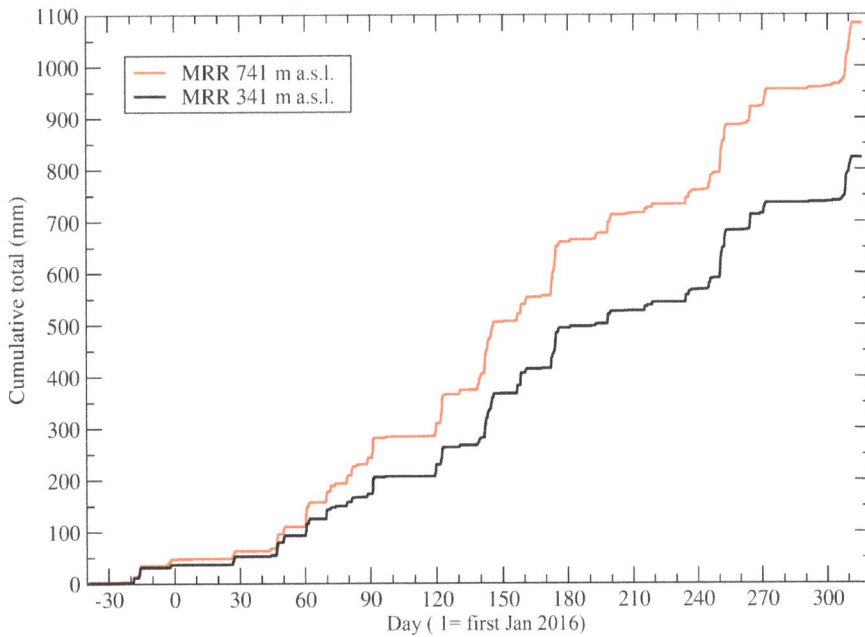

Figure 5. One year (November 2015–November 2016) of cumulative precipitation from MRR backscattering at 341 and 741 m above the surface.

dicating that cloud liquid water is very frequent in summer. Finally, Fig. 5 shows precipitation from the MRR dataset over the full record in the database, for more than a year from 21 November 2015 to 11 December 2016. Again, re-ports from two elevations, 341 and 741 m a.s.l., are displayed. This shows that at DDU, cumulated over a full year, $\sim 25\,\%$ of the precipitation formed in the atmosphere sublimates before reaching the surface.

4 Data availability

The APRES3 field campaign database is available in open access on PANGAEA.

5 Conclusions

In conclusion, observations at DDU carried out as part of the APRS3 project provide an unprecedented dataset of precipitation at the coast of Antarctica, complementing existing documentation efforts (Gorodetskaya et al., 2015) in a region which otherwise suffers from a severe shortage of such data. Our analysis of the data yields new insights into the characteristics and particularities of Antarctic snowfall, in particular that a large fraction of the precipitation formed in the atmosphere sublimates before reaching the surface. This information could only be obtained with instruments that can profile through the atmospheric layers, like radars here. However, the dataset goes beyond radar data and provides extensive complementary characterization of snow particle geometry and cumulative quantities of snowfall at the surface. Except for the dataset from the MXPol dual-polarization scanning radar during the summer campaign, the size of which (about 4 TB) is too large to be shared online but can be obtained by direct request to the authors, all data are now distributed (Berne et al., 2017) and can be freely accessed from the PANGAEA repository (https://doi.org/10.1594/PANGAEA.883562). Table 1 provides a summary of the the variables and periods covered and distributed online for each instrument. At the time of writing this paper, the project carries on with continuous collection of precipitation profiles with the MRR, and a planned contribution to the Year Of Polar Prediction (YOPP, http://www.polarprediction.net/, last access: 1 August 2018) international austral special coordinated observation period from November 2018 to February 2019, the data from which will also be made available to the community. Because of a significant weather service's (Météo France) involvement including additional radiosoundings, in addition to the planned APRES3 contribution, DDU is identified as one of the YOPP observation hotspots for the southern special observing period.

Author contributions. CG, AB and JG organized and ran the field campaigns. JG, AB, CDA and BB developed radar data processing methods and produced data series. CG processed Biral disdrometer data and CP processed MASC data. All authors contributed data analysis, discussion and conclusions.

Competing interests. The authors declare that they have no conflict of interest.

Acknowledgements. We thank the French Polar Institute, which logistically supports the APRES3 measurement campaigns. We particularly acknowledge the support of the French National Research Agency (ANR) to the APRES3 project. The Expecting Earth-Care, Learning from the ATrain (EECLAT) project funded by the Centre National d'Etudes Spatiales (CNES) also contributed support to this work. The Swiss National Science foundation (SNF) is acknowledged for grant 200021_163287, financing the Swiss participation in the project. PANGAEA is gratefully acknowledged for hosting and distributing the APRES3 data. Steve Colwell, of the British Antarctic Survey, and two anonymous reviewers provided thoughtful comments and suggestions that helped correct and clarify a number of issues in the preliminary manuscript.

Edited by: David Carlson

References

Arthern, R. J., Winebrenner, D. P., and Vaughan, D. G.: Antarctic snow accumulation mapped using polarization of 4.3-cm wavelength microwave emission, J. Geophys. Res., 111, D06107, https://doi.org/10.1029/2004JD005667, 2006.

Behrangi, A., Christensen, M., Richardson, M., Lebsock, M., Stephens, G., Huffman, G. J., Bolvin, D., Adler, R. F., Gardner, A., Lambrigtsen, B., and Fetzer, E.: Status of high-latitude precipitation estimates from observations and reanalyses. J. Geophys. Res.-Atmos. 121, 4468–4486, 2016.

Bellot, H., Trouvilliez, A., Naaim-Bouvet, F., Genthon, C., and Gallée, H.: Present weather sensors tests for measuring drifting snow, Ann. Glaciol., 52, 176–184, 2011.

Berne, A., Grazioli, J., and Genthon, C.: Precipitation observations at the Dumont d'Urville Station, Adélie Land, East Antarctica, PANGAEA, 5 datasets, https://doi.org/10.1594/PANGAEA.883562, 2017.

Eisen O., Frezzotti, M., Genthon, C., Isaksson, E., Magand, O., van den Broeke, M. R., Dixon, D. A., Ekaykin, A., Holmlund, P., Kameda, T., Karlöf, L., Kaspari, S., Lipenkov, V. Y., Oerter, H., Takahashi, S., and Vaughan, G.: Snow accumulation in East Antarctica, Rev. Geophys. 46, RG2001, https://doi.org/10.1029/2006RG000218., 2007.

Fabry, F.: Radar meteorology: principles and practice, Cambridge University Press, 2015.

Funk, C., Verdin, A., Michaelsen, J., Peterson, P., Pedreros, D., and Husak, G.: A global satellite-assisted precipitation climatology, Earth Syst. Sci. Data, 7, 275–287, https://doi.org/10.5194/essd-7-275-2015, 2015.

Garrett, T. J., Fallgatter, C., Shkurko, K., and Howlett, D.: Fall speed measurement and high-resolution multi-angle photography of hydrometeors in free fall, Atmos. Meas. Tech., 5, 2625–2633, https://doi.org/10.5194/amt-5-2625-2012, 2012.

Genthon, C., Six, D., Scarchilli, C., Giardini, V., and Frezzotti, M.: Meteorological and snow accumulation gradients across dome C, east Antarctic plateau, Int. J. Clim., 36, 455-466, https://doi.org/10.1002/joc.4362, 2015.

Goodison B. E., Louie, P. Y. T., and Yang, D.: Solid precipitation measurement intercomparison, WMO/TD – No. 872, WMO, Instrument and Observing Methods report No. 67, WMO, Geneva, 1998.

Gorodetskaya, I. V., Kneifel, S., Maahn, M., Van Tricht, K., Thiery, W., Schween, J. H., Mangold, A., Crewell, S., and Van Lipzig, N. P. M.: Cloud and precipitation properties from ground-based remote-sensing instruments in East Antarctica, The Cryosphere, 9, 285–304, https://doi.org/10.5194/tc-9-285-2015, 2015.

Grazioli, J., Lloyd, G., Panziera, L., Hoyle, C. R., Connolly, P. J., Henneberger, J., and Berne, A.: Polarimetric radar and in situ observations of riming and snowfall microphysics during CLACE 2014, Atmos. Chem. Phys., 15, 13787–13802, https://doi.org/10.5194/acp-15-13787-2015, 2015.

Grazioli, J., Genthon, C., Boudevillain, B., Duran-Alarcon, C., Del Guasta, M., Madeleine, J.-B., and Berne, A.: Measurements of precipitation in Dumont d'Urville, Adélie Land, East Antarctica, The Cryosphere, 11, 1797–1811, https://doi.org/10.5194/tc-11-1797-2017, 2017a.

Grazioli, J., Madeleine, J.-B., Gallée, H., Forbes, R. M., Genthon, C., Krinner, G., and Berne, A.: Katabatic winds diminish precipitation contribution to the Antarctic ice mass balance, P. Natl. Acad. Sci. USA, 114, 10858–10863, https://doi.org/10.1073/pnas.1707633114, 2017b.

Kneifel, S., Maahn, M., Peters, G., and Simmer, C.: Observation of snowfall with a low-power FM-CW K-band radar (Micro Rain Radar), Meteorol. Atmos. Phys., 113, 75–87, 2011.

Kneifel, S., Lerber, A., Tiira, J., Moisseev, D., Kollias, P., and Leinonen, J.: Observed relations between snowfall microphysics and triple-frequency radar measurements, J. Geophys. Res.-Atmos., 120, 6034–6055, 2015.

König-Langlo, G., King, J. C., and Pettré, P.: Climatology of the three coastal Antarctic stations Dumont d'Urville, Neumayer, and Halley, J. Geophys. Res., 103, 10935–10946, https://doi.org/10.1029/97JD00527, 1998.

Krajewski, W. F. and Smith, J. A.: Radar hydrology: rainfall estimation, Adv. Water Resour., 25, 1387–1394, 2002.

Maahn, M. and Kollias, P.: Improved Micro Rain Radar snow measurements using Doppler spectra post-processing, Atmos. Meas. Tech., 5, 2661–2673, https://doi.org/10.5194/amt-5-2661-2012, 2012.

Mawson, D.: The home of the blizzard, William Heinemann, London, 1915.

Medina, S. and Houze Jr., R. A.: Small-scale precipitation elements in midlatitude cyclones crossing the California Sierra Nevada, Mon. Weather Rev., 143, 2842–2870, 2015.

Moisseev, D. N., Lautaportti, S., Tyynela, J., and Lim, S.: Dual-polarization radar signatures in snowstorms: Role of snowflake

aggregation, J. Geophys. Res.-Atmos., 120, 12644–12655, 2015.

Nilu, R.: Cold as SPICE, Meteorol. Tech. Int., April 2013, 148–150, 2013.

Palerme, C., Kay, J. E., Genthon, C., L'Ecuyer, T., Wood, N. B., and Claud, C.: How much snow falls on the Antarctic ice sheet?, The Cryosphere, 8, 1577–1587, https://doi.org/10.5194/tc-8-1577-2014, 2014.

Palerme, C., Genthon, C., Claud, C., Kay, J. E., Wood, N. B., and L'Ecuyer, T.: Evaluation of Antarctic precipitation in CMIP5 models, current climate and projections, Clim. Dynam., 48, 225–239, https://doi.org/10.1007/s00382-016-3071-1, 2016.

Palerme, C., Claud, C., Dufour, A., Genthon, C., Kay, J. E., Wood, N. B., and L'Ecuyer, T.: Evaluation of Antarctic snowfall in global meteorological reanalyses, Atmos. Res., 48, 225–239, https://doi.org/10.1007/s00382-016-3071-1, 2017.

Parish, T. R. and Bromwich, D. H.: The surface wind field over the Antarctic ice sheets, Nature, 328, 51–54, 1987.

Peters, G., Fischer, B., Münster, H., Clemens, M., and Wagner, A.: Profiles of raindrop size distributions as retrieved by Microrain Radars, J. Appl. Meteorol., 44, 1930–1949, https://doi.org/10.1175/JAM2316.1, 2005.

Praz, C., Roulet, Y.-A., and Berne, A.: Solid hydrometeor classification and riming degree estimation from pictures collected with a Multi-Angle Snowflake Camera, Atmos. Meas. Tech., 10, 1335–1357, https://doi.org/10.5194/amt-10-1335-2017, 2017.

Schneebeli, M., Dawes, N., Lehning, M., and Berne, A.: High-resolution vertical profiles of X-band polarimetric radar observables during snowfall in the Swiss Alps, J. Appl. Meteorol. Climatol., 52, 378–394, 2013.

Schneider, U., Becker, U. A., Finger, P., Meyer-Christoffer, A., Ziese, M., and Rudolf, B.: GPCC's new land surface precipitation climatology based on quality-con,trolled in situ data and its role in quantifying the global water cycle, Theor. Appl. Climatol., 115, 15–40, https://doi.org/10.1007/s00704-013-0860-x, 2014.

Scipión, D. E., Mott, R., Lehning, M., Schneebeli, M., and Berne, A.: Seasonal small-scale spatial variability in alpine snowfall and snow accumulation, Water Resour. Res., 49, 1446–1457, 2013.

Souverijns, N., Gossart, A., Lhermitte, S., Gorodetskaya, I. V., Kneifel, S., Maahn, M., and van Lipzig, N. F. L.: Estimating radar reflectivity – snow fall rate relationships and their uncertainties over Antarctica by combining disdrometer and radar observations, Atmos. Res., 196, 211–223, https://doi.org/10.1016/J.atmosres.2017.06.001, 2017.

Witze, A.: Climate science Antarctic cloud study takes off, Nature, News in Focus, 529, 12 pp., 2016.

8

A new bed elevation model for the Weddell Sea sector of the West Antarctic Ice Sheet

Hafeez Jeofry[1,2], Neil Ross[3], Hugh F. J. Corr[4], Jilu Li[5], Mathieu Morlighem[6], Prasad Gogineni[7], and Martin J. Siegert[1]

[1]Grantham Institute and Department of Earth Science and Engineering, Imperial College London, South Kensington, London, UK

[2]School of Marine Science and Environment, Universiti Malaysia Terengganu, Kuala Terengganu, Terengganu, Malaysia

[3]School of Geography, Politics and Sociology, Newcastle University, Claremont Road, Newcastle Upon Tyne, UK

[4]British Antarctic Survey, Natural Environment Research Council, Cambridge, UK

[5]Center for the Remote Sensing of Ice Sheets, University of Kansas, Lawrence, Kansas, USA

[6]Department of Earth System Science, University of California, Irvine, Irvine, California, USA

[7]Department of Electrical and Computer Engineering, The University of Alabama, Tuscaloosa, Alabama 35487, USA

Correspondence: Hafeez Jeofry (h.jeofry15@imperial.ac.uk) and Martin J. Siegert (m.siegert@imperial.ac.uk)

Abstract. We present a new digital elevation model (DEM) of the bed, with a 1 km gridding, of the Weddell Sea (WS) sector of the West Antarctic Ice Sheet (WAIS). The DEM has a total area of $\sim 125\,000\,\mathrm{km}^2$ covering the Institute, Möller and Foundation ice streams, as well as the Bungenstock ice rise. In comparison with the Bedmap2 product, our DEM includes new aerogeophysical datasets acquired by the Center for Remote Sensing of Ice Sheets (CReSIS) through the NASA Operation IceBridge (OIB) program in 2012, 2014 and 2016. We also improve bed elevation information from the single largest existing dataset in the region, collected by the British Antarctic Survey (BAS) Polarimetric radar Airborne Science Instrument (PASIN) in 2010–2011, from the relatively crude measurements determined in the field for quality control purposes used in Bedmap2. While the gross form of the new DEM is similar to Bedmap2, there are some notable differences. For example, the position and size of a deep subglacial trough ($\sim 2\,\mathrm{km}$ below sea level) between the ice-sheet interior and the grounding line of the Foundation Ice Stream have been redefined. From the revised DEM, we are able to better derive the expected routing of basal water and, by comparison with that calculated using Bedmap2, we are able to assess regions where hydraulic flow is sensitive to change. Given the potential vulnerability of this sector to ocean-induced melting at the grounding line, especially in light of the improved definition of the Foundation Ice Stream trough, our revised DEM will be of value to ice-sheet modelling in efforts to quantify future glaciological changes in the region and, from this, the potential impact on global sea level. The new 1 km bed elevation product of the WS sector can be found at https://doi.org/10.5281/zenodo.1035488.

1 Introduction

The Intergovernmental Panel on Climate Change (IPCC) concluded that global sea level rise may range from 0.26 to 0.82 m by the end of the 21st century (Stocker, 2014). The rising oceans pose a threat to the socio-economic activities of hundreds of millions of people, mostly in Asia, living at and close to the coastal environment. Several processes drive sea level rise (e.g. thermal expansion of the oceans), but the largest potential factor comes from the ice sheets in Antarctica. The West Antarctic Ice Sheet (WAIS), which if melted would raise sea level by around 3.5 m, is grounded on a bed which is in places more than 2 km below sea level (Bamber et al., 2009a; Ross et al., 2012; Fretwell et al., 2013), allowing the ice margin to have direct contact with ocean water. One of the most sensitive regions of the WAIS to potential ocean warming is the Weddell Sea (WS) sector (Ross et al., 2012; Wright et al., 2014). Ocean modelling studies show that changes in present ocean circulation could bring warm ocean water into direct contact with the grounding lines at the base of the Filchner–Ronne Ice Shelf (FRIS) (Hellmer et al., 2012; Wright et al., 2014; Martin et al., 2015; Ritz et al., 2015; Thoma et al., 2015), which would act in a manner similar to the ocean-induced basal melting under the Pine Island Glacier ice shelf (Jacobs et al., 2011). Enhanced melting of the FRIS could lead to a decrease in the buttressing support to the upstream grounded ice, causing enhanced flow to the ocean. A recent modelling study, using a general ocean circulation model coupled with a 3-D thermodynamic ice-sheet model, simulated the inflow of warm ocean water into the Filchner–Ronne Ice Shelf cavity on a 1000-year timescale (Thoma et al., 2015). A second modelling study, this time using an ice-sheet model only, indicated that the Institute and Möller ice streams are highly sensitive to melting at the grounding lines, with grounding-line retreat up to 180 km possible across the Institute and Möller ice streams (Wright et al., 2014). While the Foundation Ice Stream was shown to be relatively resistant to ocean-induced change (Wright et al., 2014), a dearth of geophysical measurements of ice thickness across the ice stream at the time means the result may be inaccurate.

The primary tool for measurements of subglacial topography and basal ice-sheet conditions is radio-echo sounding (RES) (Dowdeswell and Evans, 2004; Bingham and Siegert, 2007). The first topographic representation of the land surface beneath the Antarctic ice sheet (Drewry, 1983) was published by the Scott Polar Research Institute (SPRI), University of Cambridge, in collaboration with the US National Science Foundation Office of Polar Programs (NSF OPP) and the Technical University of Denmark (TUD), following multiple field seasons of RES surveying in the late 1960s and 1970s (Drewry and Meldrum, 1978; Drewry et al., 1980; Jankowski and Drewry, 1981; Drewry, 1983). The compilation included folio maps of bed topography, ice-sheet surface elevation and ice thickness. The bed was digitized on a 20 km grid for use in ice-sheet modelling (Budd et al., 1984). However, only around one-third of the continent was measured at a line spacing of less than ~ 100 km, making the elevation product erroneous in many places, with obvious knock-on consequences for modelling. Several RES campaigns were thenceforth conducted, and data from them were compiled into a single new Antarctic bed elevation product, named Bedmap (Lythe et al., 2001). The Bedmap digital elevation model (DEM) was gridded on 5 km cells and included an over 1.4 million km and 250 000 km line track of airborne and ground-based radio-echo sounding data, respectively. Subglacial topography was extended north to 60° S, for purposes of ice-sheet modelling and determination of ice–ocean interactions. Since its release, Bedmap has proved to be highly useful for a wide range of research, yet inherent errors within it (e.g. inaccuracies in the DEM and conflicting grounding lines compared with satellite-derived observations) restricted its effectiveness (Le Brocq et al., 2010). After 2001, several new RES surveys were conducted to fill data gaps revealed by Bedmap, especially during and after the fourth International Polar Year (2007–2009). These new data led to the most recent Antarctic bed compilation, named Bedmap2 (Fretwell et al., 2013). Despite significant improvements in the resolution and accuracy of Bedmap2 compared with Bedmap, a number of inaccuracies and poorly sampled areas persist (Fretwell et al., 2013; Pritchard, 2014), preventing a comprehensive appreciation of the complex relation between the topography and internal ice-sheet processes and indeed a full appreciation of the sensitivity of the Antarctic ice sheet to ocean and atmospheric warming.

The WS sector was the subject of a major aerogeophysical survey in 2010–2011 (Ross et al., 2012), revealing the ~ 2 km deep Robin Subglacial Basin immediately upstream of present-day grounding lines, from which confirmation of the ice-sheet sensitivity from ice-sheet modelling was determined (Wright et al., 2014). Further geophysical surveying of the region has been undertaken since Bedmap2 (Ross et al., 2012), which has provided an enhanced appreciation of the importance of basal hydrology to ice flow (Siegert et al., 2016b) and complexities associated with the interaction of basal water flow, bed topography and ice-surface elevation (Lindbäck et al., 2014; Siegert et al., 2014; Graham et al., 2017), emphasizing the importance of developing accurate and high-resolution DEMs, both for the bed and the surface, in glaciology.

In this paper, we present an improved bed DEM for the WS sector, based on a compilation of new airborne radar surveys. The DEM has a total area of $\sim 125\,000$ km^2 and is gridded to 1 km cells. From this dataset, we reveal changes for the routing of subglacial melt water and discuss the differences between the new DEM and Bedmap2. Our new bed DEM can also be easily combined with updated surface DEMs to improve ice-sheet modelling and subglacial water pathways.

2 Study area

The WS sector covers the Institute, Möller and Foundation ice streams, as well as the Bungenstock ice rise (Fig. 1). The region covered by the DEM extends 135 km south of the Bungenstock ice rise, 195 km east of the Foundation Ice Stream, over the Pensacola Mountains and 185 km west of the Institute Ice Stream. In comparison with Bedmap2, our new DEM benefits from several new airborne geophysical datasets (e.g. NASA Operation IceBridge, OIB, 2012, 2014 and 2016). In addition, the new DEM is improved by the inclusion of ice thickness picks derived from synthetic aperture radar (SAR)-processed RES data from the British Antarctic Survey (BAS) aerogeophysical survey of the Institute and Möller ice streams conducted in 2010–2011. This has improved the accuracy of the determination of the ice–bed interface in comparison to Bedmap2. We used the Differential Interferometry Synthetic Aperture Radar (DInSAR) grounding line (Rignot et al., 2011c) to delimit the ice shelf-facing margin of our grid.

3 Data and methods

The RES data used in this study were compiled from four main sources (Fig. 1a): first, SPRI data collected during six survey campaigns between 1969 and 1979 (Drewry, 1983); second, the BAS airborne radar survey accomplished during the austral summer 2006–2007 (known as GRADES/IMAGE: Glacial Retreat of Antarctica and Deglaciation of the Earth System/Inverse Modelling of Antarctica and Global Eustasy) (Ashmore et al., 2014); third, the BAS survey of the Institute and Möller ice streams, undertaken in 2010–2011 (Institute–Möller Antarctic Funding Initiative, IMAFI) (Ross et al., 2012); fourth, flights conducted by the Center for Remote Sensing of Ice Sheets (CReSIS) during the NASA OIB programme in 2012, 2014 and 2016 (Gogineni, 2012) supplement the data previously used in the Bedmap2 bed elevation product to accurately characterize the subglacial topography of this part of the WS sector.

3.1 Scott Polar Research Institute survey

The SPRI surveys covered a total area of 6.96 million km^2 ($\sim 40\%$ of continental area) across West and East Antarctica (Drewry et al., 1980). The data were collected using a pulsed radar system operating at centre frequencies of 60 and 300 MHz (Christensen, 1970; Skou and Søndergaard, 1976) equipped on an NSF LC-130 Hercules aircraft (Drewry and Meldrum, 1978; Drewry et al., 1980; Jankowski and Drewry, 1981; Drewry, 1983). The 60 MHz antennas, built by the Technical University of Denmark, comprised an array of four half-wave dipoles, which were mounted in neutral aerofoil architecture of insulating components beneath the starboard wing. The 300 MHz antennas were composed of four dipoles attached underneath a reflector panel below the port

wing. The purpose of this unique design was to improve the backscatter acquisition and directivity. The returned signals were archived on 35 mm film and dry-silver paper by a fibre optic oscillograph. Aircraft navigation was assisted using an LTN-51 inertial navigation system giving a horizontal positional error of around 3 km. Navigation and other flight data were stored on magnetic and analogue tape by an Airborne Research Data System (ARDS) constructed by the US Naval Weapons Center. The system recorded up to 100 channels of six-digit data with a sampling rate of 303 Hz per channel. Navigation, ice thickness and ice-surface elevation records were recorded every 20 s, corresponding to around 1.6 km between each data point included on the Bedmap2 product. The data were initially recorded on a 35 mm photographic film (i.e. Z-scope radargrams) and were later scanned and digitized, as part of a NERC Centre for Polar Observation and Modelling (CPOM) project, in 2004. Each film record was scanned separately and reformatted to form a single electronic image of a RES transect. The scanned image was loaded into an image analysis package (i.e. ERDAS Imagine) to trace the internal and the ice–bed interface which were then digitized. The digitized dataset was later standardized with respect to the ice surface.

3.2 BAS GRADES/IMAGE surveys

The GRADES/IMAGE project was conducted during the austral summer of 2006–2007 and acquired $\sim 27\,550$ km of airborne RES data across the Antarctic Peninsula, Ellsworth Mountains and Filchner–Ronne Ice Shelf. The BAS Polarimetric radar Airborne Science Instrument (PASIN) operates at a centre frequency of 150 MHz, has a 10 MHz bandwidth and a pulse-coded waveform acquisition rate of 312.5 Hz (Corr et al., 2007; Ashmore et al., 2014). The PASIN system interleaves a pulse and chirp signal to acquire two datasets simultaneously. Pulse data are used for imaging layering in the upper half of the ice column, whilst the more powerful chirp is used for imaging the deep ice and sounding the ice-sheet bed. The peak transmitted power of the system is 4 kW. The spatial sampling interval of ~ 20 m resulted in $\sim 50\,000$ traces of data for a typical 4.5 h flight. The radar system consisted of eight folded dipole elements: four transmitters on the port side and four receivers on the starboard side. The receiving backscatter signal was digitized and sampled using a sub-Nyquist sampling technique. The pulses are compressed using a matched filter, and side lobes are minimized using a Blackman window. Aircraft position was recorded by an onboard carrier-wave global positioning system (GPS). The absolute horizontal positional accuracy for GRADES/IMAGE was 0.1 m (Corr et al., 2007). Synthetic aperture radar processing was not applied to the data.

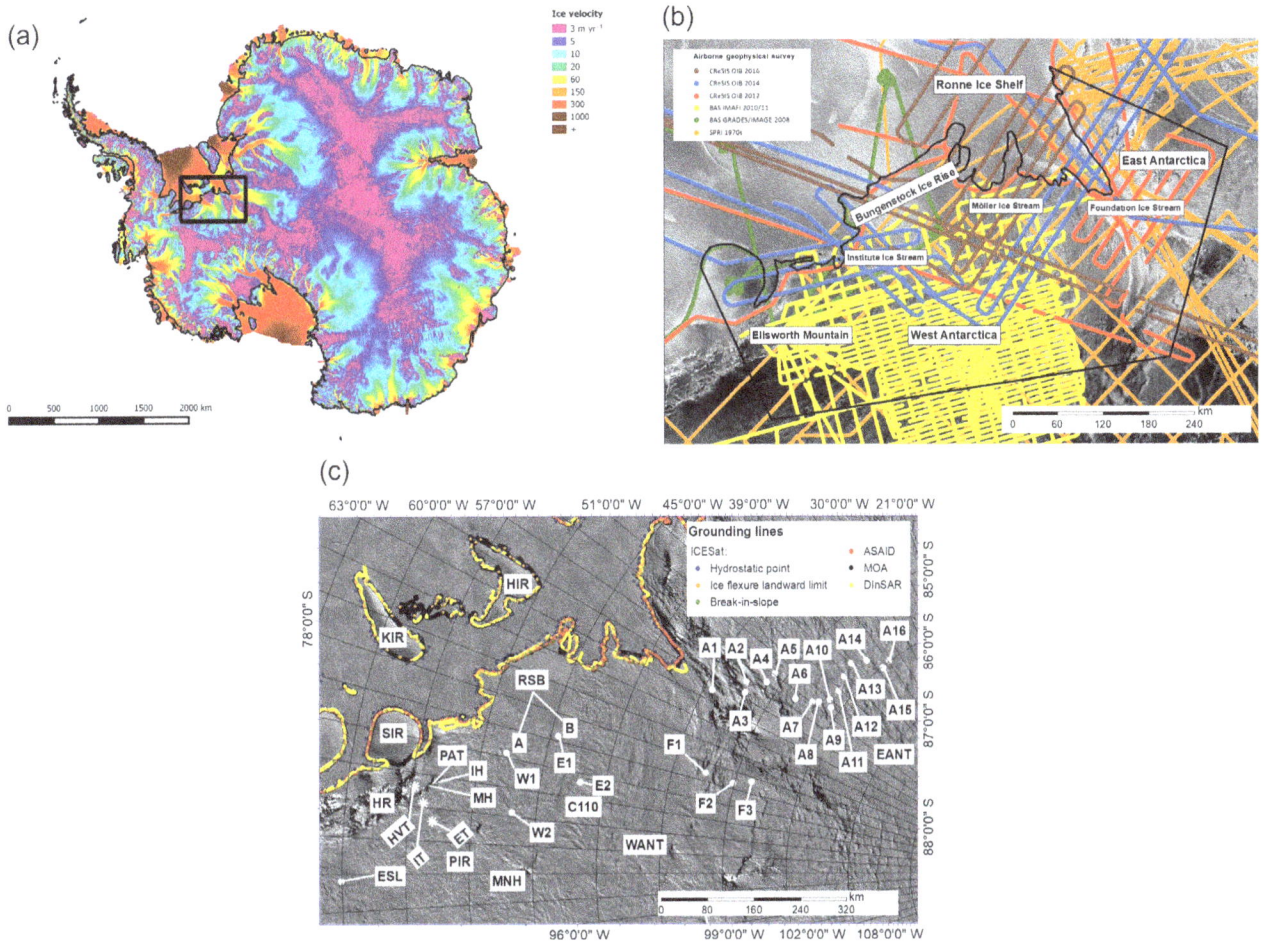

Figure 1. (a) Location of the study area overlain with InSAR-derived ice-surface velocities (Rignot et al., 2011d). (b) Aerogeophysical flight lines across the WS sector superimposed over RADARSAT (25 m) satellite imagery mosaic (Jezek, 2002); SPRI airborne survey 1969–1979 (orange); BAS GRADES/IMAGE survey 2008 (green); BAS Institute Ice Stream survey in 2010–2011 (IMAFI) (yellow); OIB 2012 (red); OIB 2014 (blue); and OIB 2016 (brown). (c) Map of the WS sector based on MODIS imagery (Haran et al., 2014) with grounding lines superimposed and denoted as follows: ICESat laser altimetry (blue: hydrostatic point; orange: ice flexure landward limit; green: break-in-slope) grounding line (Brunt et al., 2010), Antarctic Surface Accumulation and Ice Discharge (ASAID) grounding line (red line) (Bindschadler et al., 2011), Mosaic of Antarctica (MOA) grounding line (black line) (Bohlander and Scambos, 2007), and Differential Interferometry Synthetic Aperture Radar (DInSAR) grounding line (yellow line) (Rignot et al., 2011a). Annotations are as follows: KIR – Korff ice rise; HIR – Henry ice rise; SIR – Skytrain ice rise; HR – Heritage Range; HVT – Horseshoe Valley Trough; IT – Independence Trough; ET – Ellsworth Trough; ESL – Ellsworth Subglacial Lake; E1 – Institute E1 Subglacial Lake; E2 – Institute E2 Subglacial Lake; A1 – Academy 1 Subglacial Lake; A2 – Academy 2 Subglacial Lake; A3 – Academy 3 Subglacial Lake; A4 – Academy 4 Subglacial Lake; A5 – Academy 5 Subglacial Lake; A6 – Academy 6 Subglacial Lake; A7 – Academy 7 Subglacial Lake; A8 – Academy 8 Subglacial Lake; A9 – Academy 9 Subglacial Lake; A10 – Academy 10 Subglacial Lake; A11 – Academy 11 Subglacial Lake; A12 – Academy 12 Subglacial Lake; A13 – Academy 13 Subglacial Lake; A14 – Academy 14 Subglacial Lake; A15 – Academy 15 Subglacial Lake; A16 – Academy 16 Subglacial Lake; A17 – Academy 17 Subglacial Lake; W1 – Institute W1 Subglacial Lake; W2 Institute W2 Subglacial Lake; F1 – Foundation 1 Subglacial Lake; F2 – Foundation 2 Subglacial Lake; F3 – Foundation 3 Subglacial Lake; PAT – Patriot Hills; IH – Independence Hills; MH – Marble Hills; PIR – Pirrit Hills; MNH – Martin–Nash Hills; RSB – Robin Subglacial Basin; WANT – West Antarctica; EANT – East Antarctica.

3.3 BAS Institute–Möller Antarctic Funding Initiative surveys

The BAS data acquired during the IMAFI project consist of ~ 25 000 km of aerogeophysical data collected during 27 flights from two field camps. A total of 17 flights were flown from C110, which is located close to Institute E2 Subglacial

Lake, and the remaining 10 flights were flown from Patriot Hills (Fig. 1b). Data were acquired with the same PASIN radar used for the GRADES/IMAGE survey. The data rate of 13 Hz gave a spatial sampling interval of ~ 10 m for IMAFI. The system was installed on the BAS de Havilland Twin Otter aircraft with a four-element folded dipole array mounted below the starboard wing used for reception and the identical array attached below the port wing for transmission. The flights were flown in a stepped pattern during the IMAFI survey to optimize potential field data (gravity and magnetics) acquisition (Fig. 1b). Leica 500 and Novatel DL-V3 GPS receivers were installed in the aircraft, corrected with two Leica 500 GPS base stations which were operated throughout the survey to calculate the position of the aircraft (Jordan et al., 2013). The positional data were referenced to the WGS 84 ellipsoid. The absolute positional accuracy for IMAFI (the standard deviation for the GPS positional error) was calculated to be 7 and 20 cm in the horizontal and vertical dimensions (Jordan et al., 2013). Two-dimensional focused SAR processing (Hélière et al., 2007) (see Sect. 4.1) was applied to the IMAFI data.

3.4   NASA OIB CReSIS surveys

The OIB project surveyed a total distance of $\sim 32\,693$, $\sim 52\,460$ and $\sim 53\,672$ km in Antarctica in 2012, 2014 and 2016, respectively, using the Multichannel Coherent Radar Depth Sounder (MCoRDS) system developed at the University of Kansas (Gogineni, 2012). The system was operated with a carrier frequency of 195 MHz and a bandwidth of 10 and 50 MHz in 2012 (Rodriguez-Morales et al., 2014) and 2014 onwards (Siegert et al., 2016b). The radar consisted of a five-element antenna array housed in a customized antenna fairing which is attached beneath the NASA DC–8 aircraft fuselage (Rodriguez-Morales et al., 2014). The five antennas were operated from a multichannel digital direct synthesis (DDS) controlled waveform generator enabling the user to adjust the frequency, timing, amplitude and phase of each transmitted waveform (Shi et al., 2010). The radar employs an eight-channel waveform generator to emit eight independent transmit chirp pulses. The system is capable of supporting five receiver channels with an Analog Devices AD9640 14 bit analogue-to-digital converter (ADC) acquiring the waveform at a rate of 111 MHz in 2012 (Gogineni, 2012). The system was upgraded in 2014 and 2016 utilizing six-channel chirp generation and supports six receiver channels with a waveform acquisition rate at 150 MHz. Multiple receivers allow array processing to suppress surface clutter in the cross-track direction which could potentially conceal weak echoes from the ice–bed interface (Rodriguez-Morales et al., 2014). The radar data are synchronized with the GPS and inertial navigation system (INS) using the GPS timestamp to determine the location of data acquisition.

4   Data processing

We assume rain waves propagate through ice at a constant wave speed of $0.168\,\mathrm{m\,ns^{-1}}$, which presupposes the ice to be homogenous (Gogineni et al., 2001, 2014; Lythe and Vaughan, 2001; Plewes and Hubbard, 2001; Dowdeswell and Evans, 2004). Low-density firn, ice chemistry and/or ice anisotropy (Diez et al., 2014; Fujita et al., 2014; Picotti et

al., 2015; Shafique et al., 2016) violate this assumption, typically resulting in a depth bias of the order of $\sim 10\,\mathrm{m}$. The radar pulse travels through a medium until it meets a boundary of differing dielectric constant, which causes some of the radio wave to be reflected and subsequently captured by the receiver antenna. The time travelled by the radar pulse between the upper and lower reflecting surface is measured and converted to ice thickness with reference to WGS 84 (Fig. 2). The digitized SPRI–NSF–TUD bed picks data are available through the BAS web page (https://data.bas.ac.uk/ metadata.php?id=GB/NERC/BAS/AEDC/00326). The two-dimensional SAR-processed radargrams in SEG-Y format for the IMAFI survey are provided at https://doi.org/10.5285/ 8a975b9e-f18c-4c51-9bdb-b00b82da52b8, whereas the ice thickness datasets in comma-separated value (CSV) format for both GRADES/IMAGE and IMAFI are available via the BAS aerogeophysical processing portal (https://secure. antarctica.ac.uk/data/aerogeo/). The ice thickness data for IMAFI are provided in two folders: (1) the region of thinner ice ($< 200\,\mathrm{m}$) picked from the pulse dataset and (2) the overall ice thickness data, derived from picking of SAR-processed chirp radargrams. The data are arranged according to the latitude, longitude, ice thickness values and the pulse repetition interval radar shot number that is used to index the raw data. The OIB SAR images (level 1B) in MAT (binary MATLAB) format and the radar depth sounder level 2 (L2) data in CSV format are available via the CReSIS website (https://data.cresis.ku.edu/). The L2 data include measurements for GPS time during data collection, latitude, longitude, elevation, surface, bottom and thickness. For more information on these data, we refer the reader to the appropriate CReSIS guidance notes for each field season (i.e. https://data.cresis.ku.edu/data/rds/rds_readme.pdf).

4.1   GRADES/IMAGE and Institute–Möller Antarctic Funding Initiative data processing

The waveform was retrieved and sequenced according to its respective transmit pulse type. The modified data were then collated using MATLAB data binary files. Doppler filtering (Hélière et al., 2007) was used to remove the backscattering hyperbola in the along-track direction (Corr et al., 2007; Ross et al., 2012). Chirp compression was then applied to the along-track data. Unfocused synthetic aperture (SAR) processing was used for the GRADES/IMAGE survey by applying a moving average of 33 data points (Corr et al., 2007), whereas two-dimensional SAR (i.e. focused) processing based on the Omega-K algorithm was used to process the IMAFI data (Hélière et al., 2007; Winter et al., 2015) to enhance both along-track resolution and echo signal noise. The bed echo was depicted in a semi-automatic manner using ProMAX seismic processing software. All picking for IMAFI was undertaken by a single operator (Neil Ross). A nominal value of 10 m is used to correct for the firn layer during the processing of ice thickness, which introduces an

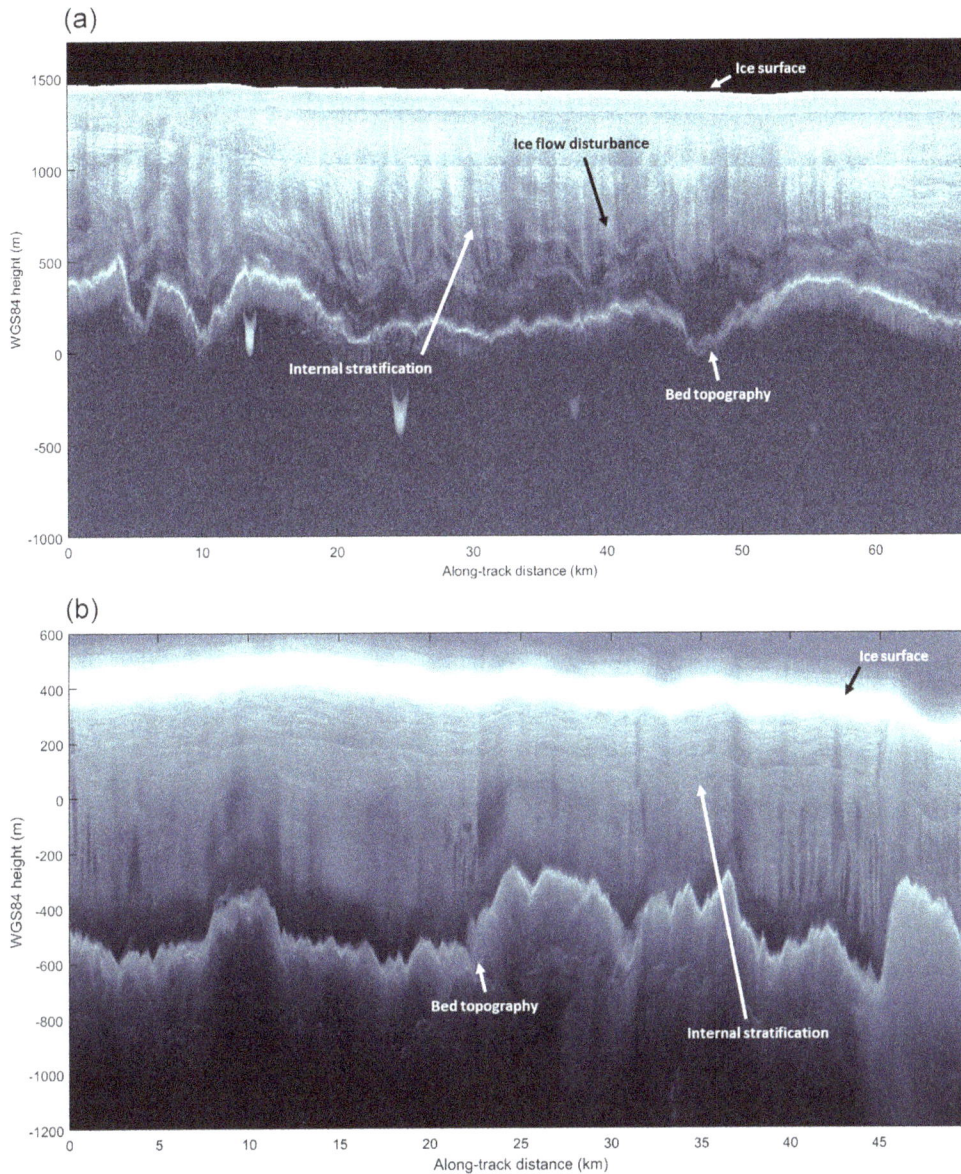

Figure 2. Two-dimensional synthetic aperture radar (SAR)-processed radargrams of **(a)** BAS IMAFI data and **(b)** OIB data.

error of the order of ±3 m across the survey field (Ross et al., 2012). This is small relative to the total error budget of the order of ±1 %. Finally, the GPS and RES data were combined to determine the ice thickness, ice-surface and bed elevation datasets. Elevations are measured with reference to WGS 84. The ice-surface elevation was calculated by subtracting terrain clearance from the height of the aircraft, whereas the bed elevation was computed by subtracting the ice thickness from the ice-surface elevation.

4.2 OIB data processing

The OIB radar adopts SAR processing in the along-track di-

rection to provide higher-resolution images of the subglacial profile. The data were processed in three steps to improve the signal-to-noise ratio and increase the along-track resolution (Gogineni et al., 2014). The raw data were first converted from a digital quantization level to a receiver voltage level. The surface was captured using the low-gain data, microwave radar or laser altimeter. A normalized matched filter with frequency-domain windowing was then used for pulse compression. Two-dimensional SAR processing was used after conditioning the data, which is based on the frequency-wavenumber (F-K) algorithm. The F-K SAR processing requires straight and uniformly sampled data, however, which in the strictest sense are not usually met in the raw data since the aircraft's speed is not consistent and its trajectory is not straight. The raw data were thus spatially resampled along track using a sinc kernel to approximate a uniformly sampled dataset. The vertical deviation in aircraft trajectory from the horizontal flight path was compensated for in the frequency domain with a time-delay phase shift. The phase shift was later removed for array processing as it is able to account for the non-uniform sampling; the purpose is to maintain the original geometry for the array processing. Array processing was performed in the cross-track flight path to reduce surface clutter as well as to improve the signal-to-noise ratio. Both the delay-and-sum and minimum variance distortionless response (MVDR) beamformers were used to combine the multichannel data, and for regions with significant surface clutter the MVDR beamformer could effectively minimize the clutter power and pass the desired signal with optimum weights (Harry and Trees, 2002).

4.3 Quantifying ice thickness, bed topography and subglacial water flow

The new ice thickness DEM was formed from the RES data using the "Topo to Raster" function in ArcGIS, based on the Australian National University DEM (ANUDEM) elevation gridding algorithm (Hutchinson, 1988). This is the same algorithm employed by Bedmap2. The ice thickness DEM was then subtracted from the ice-sheet surface elevation DEM (from European Remote Sensing Satellite-1 (ERS-1) radar and Ice, Cloud and land Elevation Satellite (ICESat) laser satellite altimetry datasets, Bamber et al., 2009b) to derive the bed topography. The ice thickness, ice-sheet surface and bed elevations were then gridded at a uniform 1 km spacing and referenced to the polar stereographic projection (Snyder, 1987) to form the new DEMs. The Bedmap2 bed elevation product (Fretwell et al., 2013) was transformed from the gl04c geoid projection to the WGS 1984 Antarctic Polar Stereographic projection for comparison purposes. A difference map between the new DEM and the Bedmap2 product was computed by subtracting the Bedmap2 bed elevation DEM from the new bed elevation DEM. Crossover analysis for the 2006–2007 data onwards (including data acquired on flight lines beyond the extent of our DEM) shows the RMS errors of 9.1 m (GRADES/IMAGE), 15.8 m (IMAFI), 45.9 m (CRESIS 2012), 23.7 m (CRESIS 2014) and 20.3 m (CRESIS 2016).

Subglacial water flow paths were calculated based on the hydraulic potentiometric surface principle, in which basal water pressure is balanced by the ice overburden pressure as follows:

$$\varphi = g(\rho_w y + \rho_i h), \tag{1}$$

where φ is the theoretical hydropotential surface; y is the bed elevation; h is the ice thickness; ρ_w and ρ_i are the density of water (1000 kg m^{-3}) and ice (920 kg m^{-3}), assuming ice to be homogenous, respectively; and g is the gravitational constant (9.81 m s^{-2}) (Shreve, 1972). Sinks in the hydrostatic pressure field raster were filled to produce realistic hydrologic flow paths. The flow direction of the raster was then defined by assigning each cell a direction to the steepest downslope neighbouring cell. Sub-basins less than 200 km^2 were removed due to the coarse input of bed topography and ice thickness DEMs.

4.4 Inferring bed elevation using mass conservation (MC) and kriging

Relying on the conservation of mass (MC) to infer the bed between flight lines (Morlighem et al., 2011), we were able to investigate how the bed can be developed further in fast-flowing regions, using a new interpolation technique. To perform the MC procedure, we used InSAR-derived surface velocities (Fig. 1a) from Rignot et al. (2011b), surface mass balance from RACMO 2.3 (Regional Atmospheric Climate Model, Van Wessem et al., 2014), and assumed that the ice thinning–thickening rate and basal melt are negligible. We constrained the optimization with ground-penetrating radar from CReSIS, GRADES, IMAFI and SPRI, described above, and used a mesh horizontal resolution of 500 m.

5 Results

5.1 A new 1 km DEM of the WS sector

We present a new DEM of the WS sector of West Antarctica (Fig. 3a). The Bedmap2 bed elevation and the difference map are shown in Fig. 3b and c, respectively. The new DEM contains substantial changes in certain regions compared with Bedmap2, whereas in others there are consistencies between the two DEMs, for example across the Bungenstock ice rise, where there are little new data. The mean error between the two DEMs is −86.45 m, indicating a slightly lower bed elevation in the new DEM data compared to Bedmap2, which is likely the result of deep parts of the topography (i.e. valley bottoms) not being visible in the fieldwork non-SAR-processed quality-control (QC) radargrams from the IMAFI project (e.g. Horseshoe Valley, near Patriot Hills in the Ellsworth Mountains; Winter, 2016). The bed elevation upstream of the Bungenstock ice rise and across the Robin Subglacial Basin shows a generally good agreement with Bedmap2, with only small areas across the Möller Ice Stream and Pirrit Hills significantly different from Bedmap2, with differences in bed elevation typically ranging between −109 and 172 m (Fig. 3c). There is, however, large disagreement between the two DEMs in the western region of Institute Ice Stream, across the Ellsworth Mountains (e.g. in the Horseshoe valley), the Foundation Ice Stream and towards East Antarctica where topography is more rugged. It is also worth noting the significant depth of the bed topogra-

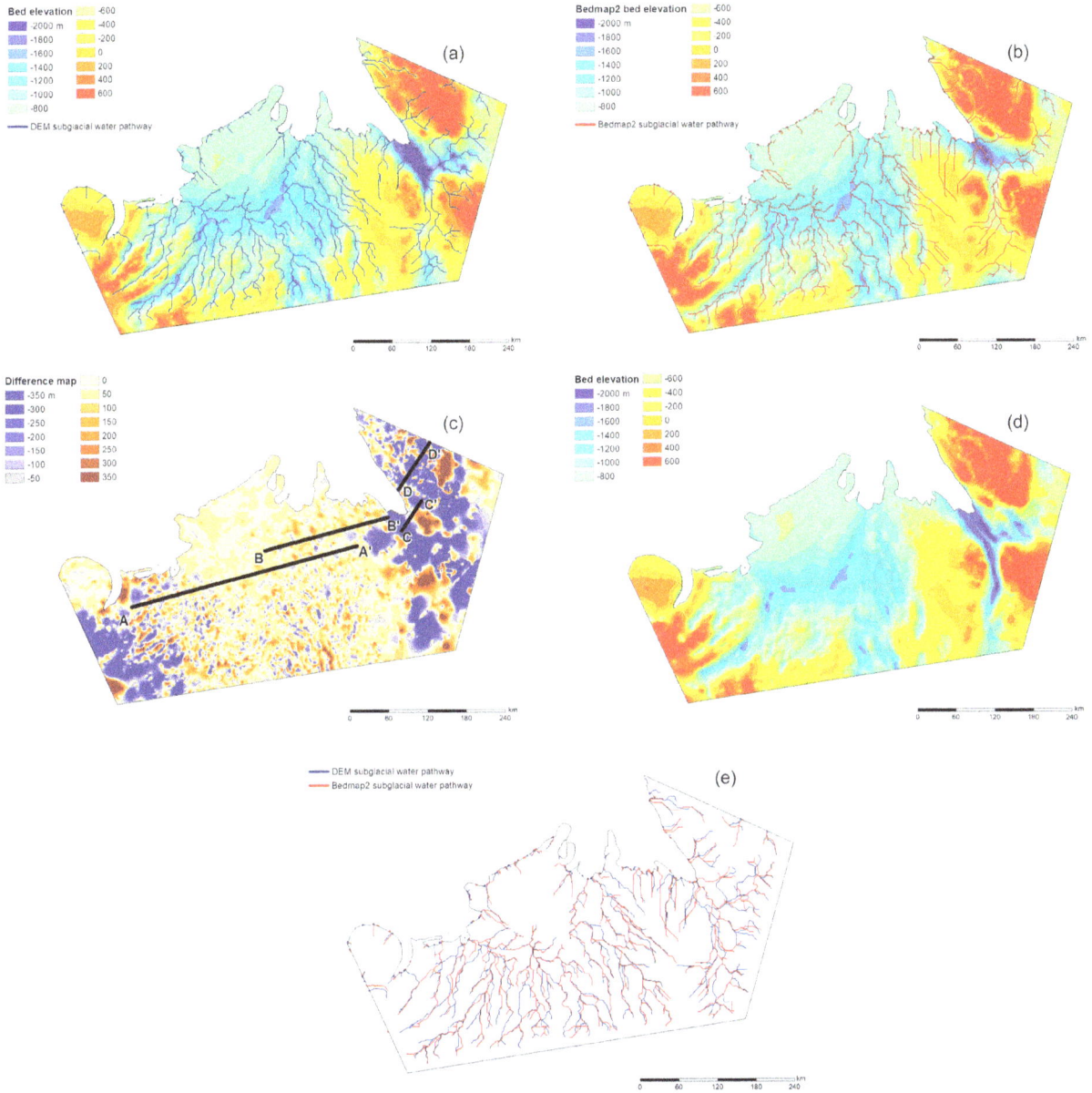

Figure 3. (a) The new bed DEM for the WS sector; **(b)** Bedmap2 bed elevation product (Fretwell et al., 2013); pathways of subglacial water are superimposed in **(a)** and **(b)**. **(c)** Profiles A–A′, B–B′, C–C′ and D–D′ overlain by a map showing differences in bed elevation between the new DEM and Bedmap2; **(d)** bed DEM inferred using mass conservation and kriging; and **(e)** subglacial water pathways calculated with the new DEM (blue) and Bedmap2 (red).

phy beneath the trunk of the Foundation Ice Stream, where a trough more than ∼2 km deep is located and delineated (Fig. 3a). The trough is ∼38 km wide and ∼80 km in length, with the deepest section ∼2.3 km below sea level. The new DEM shows a significant change in the depiction of Foundation Trough; we have measured it to be ∼1 km deeper and far more extensive relative to the Bedmap2 product.

In order to further quantify the differences between Bedmap2 and our new DEM, we present terrain profiles of both DEMs relative to four RES flight lines (Fig. 3c). The new DEM is consistent with the bed elevations from the RES data picks compared to Bedmap2 (Fig. 4a and b). The new DEMs show a correlation coefficient of 0.96 and 0.92 for Profile A and B, respectively. This is higher compared with Bedmap2 which is 0.94 (Profile A) and 0.91 (Profile B), with relative errors of 2 and 1 % for Profile A and B, respectively. Although inaccuracies of the bed elevation persist across the Foundation Ice Stream for both DEMs, the gross pattern of the bed elevation for the new DEM is more consistent with the RES transects relative to Bedmap2 (Fig. 4c and d), with correlation coefficients of 0.97 and 0.94 for Profile C and D, respectively. These values contrast with correlation coefficients from Bedmap2 of 0.87 for Profile C and 0.83 for Profile D, with relative errors of 12 and 13 %, respectively.

The new DEM can be refined further to deal with bumps and irregularities associated with interpolation effects from along-track data in otherwise data-sparse regions. We used the (MC) technique to infer the bed elevation (Fig. 3d) beneath the fast-moving ice, similar to that employed in Greenland (Morlighem et al., 2017). In general, the bed elevation derived from MC and kriging is consistent with our new DEM. However, using MC, significant changes in the bed morphology beneath fast-flowing ice occur. For example, using MC, the tributary of the Foundation Ice Stream has been extended for ~ 100 km further inland relative to our new DEM.

5.2 Hydrology

Computing the passageway of subglacial water beneath the ice sheet is critical for comprehending ice-sheet dynamics (Bell, 2008; Stearns et al., 2008; Siegert et al., 2016a). The development of subglacial hydrology pathways is highly sensitive to ice-surface elevation and, to a lesser degree, to bed morphology (Wright et al., 2008; Horgan et al., 2013). Figure 3e shows a comparison of subglacial hydrology pathways between our new bed DEM and the Bedmap2 DEM. The gross patterns of water flow are largely unchanged between the two DEMs, especially across and upstream of Institute and Möller ice streams. The similar water pathway pattern between both DEMs in these regions is also consistent with the small errors in bed topography (Fig. 3c). Despite large differences in bed topography across the Foundation Ice Stream and the Ellsworth Mountains region, the large-scale patterns of water flow are also similar between both DEMs, due to the dominance of the ice-surface slope in driving basal water flow in these regions (Shreve, 1972). Nonetheless, there are several local small-scale differences in the water pathways (Fig. 3e), which highlight hydraulic sensitivity. The subglacial water network observed in the new DEM across the Foundation Ice Stream appears to be more arborescent than that derived from Bedmap2. This is due to the introduction of new data, resulting in a better-defined bed across the Foundation Trough (Fig. 3a). The subglacial water pathway observed in the new DEM adjacent to the grounding line across the Möller Ice Stream is in good agreement with the position of sub-ice-shelf channels, which have been delineated from a combination of satellite images and RES data (Le Brocq et al., 2013).

Subglacial lakes discovered across the WS sector form an obvious component of the basal hydrological system (Wright and Siegert, 2012). These lakes exist due to sufficient amount of geothermal heating (50–70 m W m^{-2}), which allows the base of the ice sheet to melt especially in areas of thick ice. In addition, the pressure exerted on the bed by the over-lying ice causes the melting point to be lowered. West of the Ellsworth Mountains lies a body of the subglacial water known as Ellsworth Subglacial Lake. The lake measures 28.9 km^2 with a depth ranging between 52 and 156 m capable of carrying a water body volume of 1.37 km^3 (Woodward et al., 2010). In some cases, ice-surface elevation changes have been linked with subglacial lake hydrological change (Siegert et al., 2016a) – referred to as "active" subglacial lakes. There are four known active subglacial lakes distributed across the Institute Ice Stream (Wright and Siegert, 2012): Institute W1 is located close to the Robin Subglacial Basin A whereas Institute W2 is located to the northeast of Pirrit Hills; Institute E1 and E2 are located to the southwest of Robin Subglacial Basin B and near the field camp of C110, respectively. There are three active subglacial lakes in the Foundation Ice Stream catchment: Foundation 1, Foundation 2 and Foundation 3 (Wright and Siegert, 2012). East of the Foundation Ice Stream, there are 16 active subglacial lakes distributed along the main trunk of Academy Glacier (Wright and Siegert, 2012).

5.3 Geomorphological description of the bed topography

The WS sector of WAIS is composed of three major ice-sheet outlets (Fig. 1a and b): the Institute, Möller and Foundation ice streams, feeding ice to the FRIS, the second largest ice shelf in Antarctica. Geophysical data in this area reveal features such as steep reverse bed slopes, similar in scale to that measured for upstream Thwaites Glacier, close to the Institute and Möller Ice Stream grounding lines. The bed slopes inland to a ~ 1.8 km deep basin (the Robin Subglacial Basin), which is divided into two sections with few obvious significant ice-sheet pinning points (Ross et al., 2012). Elevated beds in other parts of the WS sector allow the ice shelf to ground, causing ice-surface features known as ice rises and rumples (Matsuoka et al., 2015).

The Institute Ice Stream has three tributaries, within our survey grid, to the south and west of the Ellsworth Mountains, occupying the Horseshoe Valley, Independence and Ellsworth troughs (Fig. 1c) (Winter et al., 2015). The Horseshoe Valley Trough, around 20 km wide and 1.3 km below sea level at its deepest point, is located downstream of the steep mountains of the Heritage Range. A subglacial ridge is located between the mouth of the Horseshoe Valley Trough and the main trunk of the Institute Ice Stream (Winter et al., 2015). The Independence Trough is located subparallel to the Horseshoe Valley Trough, separated by the 1.4 km high Independence Hills. The trough is ~ 22 km wide and is 1.1 km below sea level at its deepest point. It is characterized by two distinctive plateaus (~ 6 km wide each) on each side of the trough, aligned alongside the main trough axis. Ice flows eastward through the Independence Trough for ~ 54 km before it shifts to a northward direction where the trough widens to 50 km and connects with the main Institute Ice

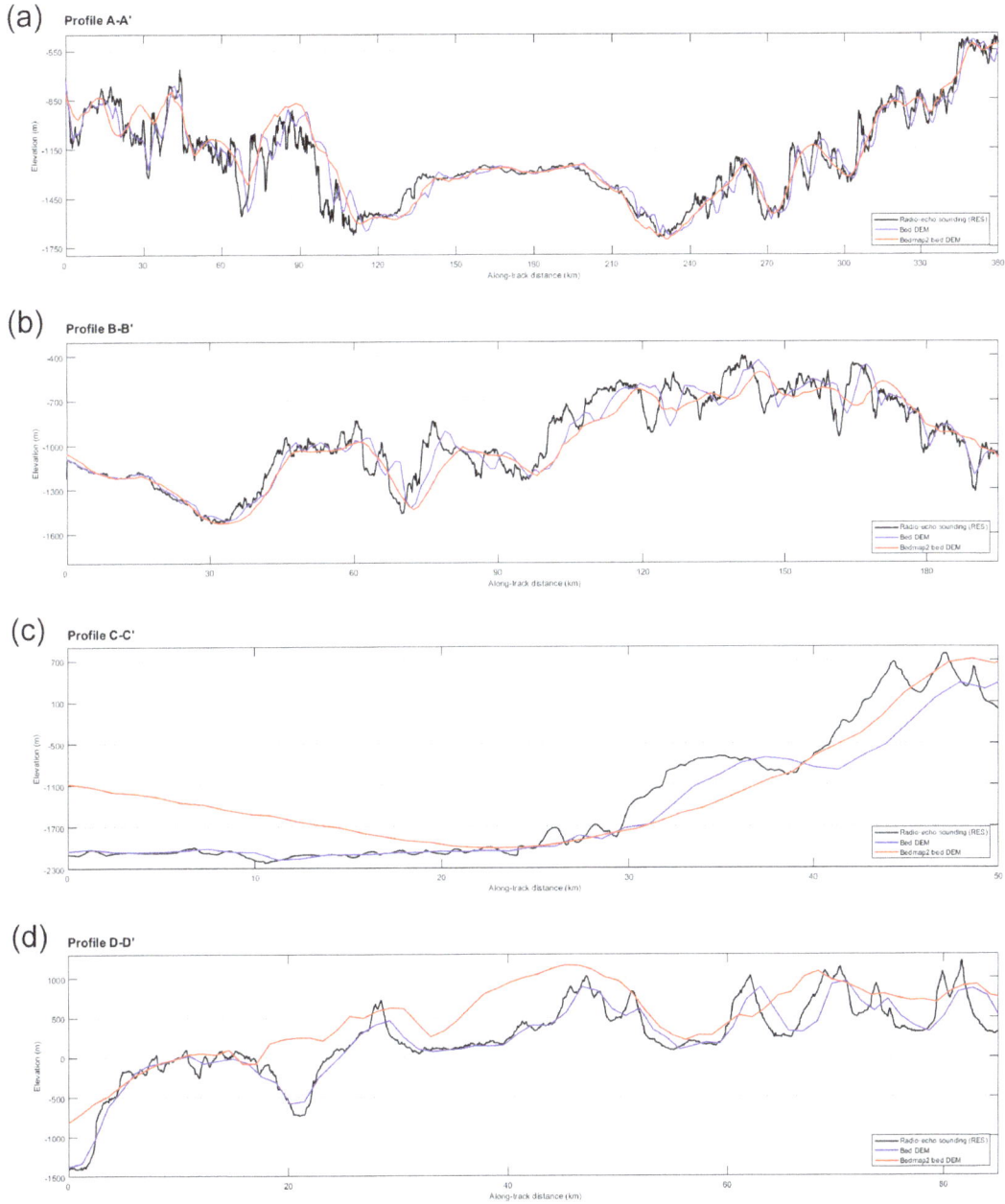

Figure 4. Bed elevations for RES transects (black), DEM (blue) and Bedmap2 (red) for **(a)** Profile A–A′, **(b)** Profile B–B′, **(c)** Profile C–C′ and **(d)** Profile D–D′.

Stream. The Ellsworth Trough is aligned with the Independence Trough, and both are orthogonal to the orientation of the Amundsen–Weddell ice divide, dissecting the Ellsworth Subglacial Highlands northwest to southeast. The Ellsworth Trough measures ∼ 34 km in width, and is ∼ 2 km below sea level at its deepest point and is ∼ 260 km in length. It is considered to be the largest and deepest trough-controlled tributary in this region (Winter et al., 2015). The Ellsworth Trough is intersected by several smaller valleys aligned perpendicularly to the main axis, which are relic landforms from a previous small dynamic ice mass (predating the WAIS in its present configuration) (Ross et al., 2014). The Ellsworth Trough contains the ∼ 15–20 km long Ellsworth Subglacial Lake (Siegert et al., 2004a, 2012; Woodward et al., 2010; Wright and Siegert, 2012). It is worth noting that the lake is outside the grid of our DEM.

Satellite altimetry and imagery are able to estimate the grounding line that separates the grounded ice sheet from the floating ice shelf, based on surface changes due to tidal oscillations and the subtle ice-surface features. Such analysis is

Table 1. Data files and locations.

Products	Files	Location	DOI/URL
1 km bed elevation DEM	1 km bed elevation DEM	Zenodo Data Repository	https://doi.org/10.5281/zenodo.1035488
1 km ice thickness DEM	Digitized SPRI–NSF–TUD bed picks data	UK Polar Data Centre (UKPDC)	https://data.bas.ac.uk/metadata.php?id=GB/NERC/BAS/AEDC/00326
	BAS GRADES/IMAGE ice thickness data	UK Polar Data Centre	https://secure.antarctica.ac.uk/data/aerogeo/
	BAS IMAFI ice thickness data	UK Polar Data Centre	https://secure.antarctica.ac.uk/data/aerogeo/
	NASA Operation Ice-Bridge radar depth sounder level 2 (L2) data	Center for Remote Sensing of Ice Sheet (CReSIS)	https://data.cresis.ku.edu/
1 km ice-sheet-surface DEM	ERS-1 radar and ICESat laser satellite altimetry	National Snow and Ice Data Center (NSIDC)	https://nsidc.org/data/docs/daac/nsidc0422antarctic1kmdem/
Two-dimensional synthetic aperture radar (SAR)-processed radargrams	BAS IMAFI airborne survey	UK Polar Data Centre	https://doi.org/10.5285/8a975b9e-f18c-4c51-9bdb-b00b82da52b8
	NASA Operation Ice-Bridge airborne survey	Center for Remote Sensing of Ice Sheet	https://data.cresis.ku.edu/
Ice velocity map of central Antarctica	MEaSUREs InSAR-based ice velocity	National Snow and Ice Data Center	https://doi.org/10.5067/MEASURES/CRYOSPHERE/nsidc-0484.001
Ice-sheet-surface satellite imagery	MODIS Mosaic of Antarctica (2008–2009) (MOA2009)	National Snow and Ice Data Center	https://doi.org/10.7265/N5KP8037
	RADARSAT (25 m) satellite imagery	Byrd Polar and Climate Research Center	https://research.bpcrc.osu.edu/rsl/radarsat/data/

prone to uncertainty, however. There are currently four proposed grounding-line locations, based on different satellite datasets and/or methods of analysis (Bohlander and Scambos, 2007; Bindschadler et al., 2011; Brunt et al., 2011; Rignot et al., 2011c). Each of the grounding lines were delineated from satellite images but without direct measurement of the subglacial environment. This results in ambiguities for the grounding-line location (Jeofry et al., 2017a). In addition, RES data have demonstrated clear errors in the position of the grounding line with a large, hitherto unknown, subglacial embayment near the Institute Ice Stream grounding line. The subglacial embayment is ~ 1 km deep and is potentially open to the ice shelf cavity, causing the inland ice sheet to have a direct contact with ocean water. Our RES analysis also reveals a better-defined Foundation Trough, in which the grounding line is perched on very deep topography around 2 km below sea level.

A previous study revealed a series of ancient large sub-parallel subglacial bed channels between Möller Ice Stream and Foundation Ice Stream, adjacent to the marginal basins (Fig. 1b) (Rose et al., 2014). While these subglacial channels are likely to have been formed by the flow of basal water, they are presently located beneath slow-moving and cold-based ice. It is thought, therefore, that the channels are ancient and were formed at a time when surface melting was prevalent in West Antarctica (e.g. the Pliocene).

The bed topography of the WS appears both rough (over the mountains and exposed bedrock) and smooth (across the sediment-filled regions) (Bingham and Siegert, 2007). Studies of bed roughness calculated using the fast Fourier transform (FFT) technique based on the relative measurement of bed obstacle amplitude and frequency of the roughness obstacles have indicated that the Institute Ice Stream and Möller Ice Stream are dominated by relatively low roughness values, less than 0.1 (Bingham and Siegert, 2007; Rippin et al., 2014), which was suggested as being the result of the emplacement of marine sediments as in the Siple Coast region (Siegert et al., 2004b; Peters et al., 2005). Radar-derived roughness analysis has evidenced a smooth bed across the Robin Subglacial Basin where sediments may exist (Rippin et al., 2014). The deepest parts of the Robin Subglacial Basin are anomalously rough, marking the edge of a sedimentary drape where the highest ice flow velocities are generated (Siegert et al., 2016b). As such, the smooth basal topography of the Institute and Möller Ice Stream catchments is less extensive than proposed by Bingham and Siegert (2007). The subglacial topography of the region between the Robin Subglacial Basin and the Pirrit and Martin–Nash Hills is relatively flat, smooth, and gently sloping and has been interpreted as a bedrock planation surface (Rose et al., 2015). Although the exact formation process of the planation surface is unknown, it is thought that this geomorphological feature formed due to marine and/or fluvial erosion (Rose et al., 2015).

6 Data availability

The new 1 km bed elevation product of the WS sector can be found at https://doi.org/10.5281/zenodo.1035488. We used four radar datasets to construct the 1 km ice thickness DEM, as follows: (1) digitized radar data from the 1970s SPRI–NSF–TUD surveys, in which the bed was picked every 15–20 s (1–2 km), recorded here in an Excel 97–2003 Worksheet (XLS), which can be obtained from the UK Polar Data Centre (UKPDC) website at https://data.bas.ac.uk/metadata.php?id=GB/NERC/BAS/AEDC/00326; (2) BAS GRADES/IMAGE and (3) BAS IMAFI airborne surveys, both available from the UKPDC Polar Airborne Geophysics Data Portal at https://secure.antarctica.ac.uk/data/aerogeo/; and (4) NASA Operation IceBridge radar depth sounder level 2 (L2) data, available from the Center for Remote Sensing of Ice Sheet (CReSIS) website at https://data.cresis.ku.edu/.

The 1 km ice-sheet surface elevation DEM was derived from a combination of ERS-1 surface radar and ICESat laser altimetry, which is downloadable from the National Snow and Ice Data Center (NSIDC) website at https://nsidc.org/data/docs/daac/nsidc0422_antarctic_1km_dem/. Newer surface elevation models (i.e. Helm et al., 2014) can easily be combined with our improved bed DEM.

Two-dimensional SAR-processed radargrams in SEG-Y format for the BAS IMAFI airborne survey and the NASA Operation IceBridge SAR images (level 1B) in MAT (binary MATLAB) format are provided at https://doi:10.5285/8a975b9e-f18c-4c51-9bdb-b00b82da52b8 and https://data.cresis.ku.edu/, respectively.

Ancillary information for the MEaSUREs InSAR-based ice velocity map of central Antarctica can be found at https://doi:10.5067/MEASURES/CRYOSPHERE/nsidc-0484.001 and the MODIS Mosaic of Antarctica 2008–2009 (MOA 2009) ice-sheet-surface image map is available at https://doi.org/10.7265/N5KP8037. The RADARSAT (25 m) ice-sheet-surface satellite imagery is accessible from the Byrd Polar and Climate Research Center website at https://research.bpcrc.osu.edu/rsl/radarsat/data/ (Byrd Polar and Climate Research Center, 2012). A summary of the data used in this paper and their availability is provided in the Table 1.

7 Conclusions

We have compiled airborne radar data from a number of geophysical surveys, including the SPRI–NSF–TUD surveys of the 1970s; the GRADES/IMAGE and IMAFI surveys acquired by BAS in 2006–2007 and 2010–2011, respectively; and new geophysical datasets collected by CReSIS from the NASA OIB project in 2012, 2014 and 2016. From these data, we produce a bed topography DEM which is gridded to 1 km postings. The DEM covers an area of $\sim 125\,000\,\text{km}^2$ of the WS sector including the Institute, Möller and Foundation ice streams, as well as the Bungenstock ice rise. Large differences can be observed between the new and previous DEMs (i.e. Bedmap2), most notably across the Foundation Ice Stream where we reveal the grounding line to be resting on a bed $\sim 2\,\text{km}$ below sea level, with a deep trough immediately upstream as deep as 2.3 km below sea level. In addition, improved processing of existing data better resolves deep regions of bed compared to Bedmap2. Our new DEM also revises the pattern of potential basal water flow across the Foundation Ice Stream and towards East Antarctica in comparison to Bedmap2. Our new DEM and the data used to compile it are available to download and will be of value to ice-sheet modelling experiments in which the accuracy of the DEM is important to ice flow processes in this particularly sensitive region of the WAIS.

Author contributions. HJ carried out the analysis, created the figures and compiled the database. HJ, NR and MS wrote the paper. All authors contributed to the database compilation, analysis and writing of the paper.

Competing interests. The authors declare that they have no conflict of interest.

Acknowledgements. The data used in this project are available at the Center for the Remote Sensing of Ice Sheets data portal https://data.cresis.ku.edu/ and at the UK Airborne Geophysics Data Portal https://secure.antarctica.ac.uk/data/aerogeo/. Prasad Gogineni and Jilu Li acknowledge funding by NASA for CReSIS data collection and development of radars (NNX10AT68G); Martin J. Siegert and Neil Ross acknowledge funding from the NERC Antarctic Funding Initiative (NE/G013071/1); and the IMAFI data collection team consist of Hugh F. J. Corr, Fausto Ferraccioli, Rob Bingham, Anne Le Brocq, David Rippin, Tom Jordan, Carl Robinson, Doug Cochrane, Ian Potten and Mark Oostlander. Hafeez Jeofry acknowledges funding from the Ministry of Higher Education Malaysia and the Norwegian Polar Institute for the Quantarctica GIS package. All authors gratefully acknowledge the anonymous reviewers for their insightful suggestions to this manuscript.

Edited by: Reinhard Drews

References

Ashmore, D. W., Bingham, R. G., Hindmarsh, R. C., Corr, H. F., and Joughin, I. R.: The relationship between sticky spots and radar reflectivity beneath an active West Antarctic ice stream, Ann. Glaciol., 55, 29–38, 2014.

Bamber, J. L., Riva, R. E., Vermeersen, B. L., and LeBrocq, A. M.: Reassessment of the potential sea-level rise from a collapse of the West Antarctic Ice Sheet, Science, 324, 901–903, 2009a.

Bamber, J. L., Gomez-Dans, J., and Griggs, J.: Antarctic 1 km digital elevation model (DEM) from combined ERS-1 radar and ICE-

Sat laser satellite altimetry, National Snow and Ice Data Center, Boulder, 2009b.

Bell, R. E.: The role of subglacial water in ice-sheet mass balance, Nat. Geosci., 1, 297–304, 2008.

Bindschadler, R., Choi, H., Wichlacz, A., Bingham, R., Bohlander, J., Brunt, K., Corr, H., Drews, R., Fricker, H., Hall, M., Hindmarsh, R., Kohler, J., Padman, L., Rack, W., Rotschky, G., Urbini, S., Vornberger, P., and Young, N.: Getting around Antarctica: new high-resolution mappings of the grounded and freely-floating boundaries of the Antarctic ice sheet created for the International Polar Year, The Cryosphere, 5, 569–588, https://doi.org/10.5194/tc-5-569-2011, 2011.

Bingham, R. G. and Siegert, M. J.: Radar-derived bed roughness characterization of Institute and Möller ice streams, West Antarctica, and comparison with Siple Coast ice streams, Geophys. Res. Lett., 34, https://doi.org/10.1029/2007GL031483, 2007.

Bohlander, J. and Scambos, T.: Antarctic coastlines and grounding line derived from MODIS Mosaic of Antarctica (MOA), National Snow and Ice Data Center, Boulder, CO, USA, 2007.

Brunt, K. M., Fricker, H. A., Padman, L., Scambos, T. A., and O'Neel, S.: Mapping the grounding zone of the Ross Ice Shelf, Antarctica, using ICESat laser altimetry, Ann. Glaciol., 51, 71–79, 2010.

Brunt, K. M., Fricker, H. A., and Padman, L.: Analysis of ice plains of the Filchner–Ronne Ice Shelf, Antarctica, using ICESat laser altimetry, J. Glaciol., 57, 965–975, 2011.

Budd, W., Jenssen, D., and Smith, I.: A three-dimensional time-dependent model of the West Antarctic ice sheet, Ann. Glaciol., 5, 29–36, 1984.

Byrd Polar and Climate Research Center: RADARSAT-1 Antarctic Mapping Project (RAMP) Data, available at: https://research.bpcrc.osu.edu/rsl/radarsat/data/ (last access: 28 January 2018), 2012.

Christensen, E. L.: Radioglaciology 300 MHz radar, Technical University of Denmark, Electromagnetics Institute, 1970.

Corr, H. F., Ferraccioli, F., Frearson, N., Jordan, T., Robinson, C., Armadillo, E., Caneva, G., Bozzo, E., and Tabacco, I.: Airborne radio-echo sounding of the Wilkes Subglacial Basin, the Transantarctic Mountains and the Dome C region, Terra Ant. Reports, 13, 55–63, 2007.

Diez, A., Eisen, O., Weikusat, I., Eichler, J., Hofstede, C., Bohleber, P., Bohlen, T., and Polom, U.: Influence of ice crystal anisotropy on seismic velocity analysis, Ann. Glaciol., 55, 97–106, 2014.

Dowdeswell, J. A. and Evans, S.: Investigations of the form and flow of ice sheets and glaciers using radio-echo sounding, Rep. Prog. Phys., 67, 1821, https://doi.org/10.1088/0034-4885/67/10/R03, 2004.

Drewry, D. and Meldrum, D.: Antarctic airborne radio echo sounding, 1977–78, Polar Rec., 19, 267–273, 1978.

Drewry, D., Meldrum, D., and Jankowski, E.: Radio echo and magnetic sounding of the Antarctic ice sheet, 1978–79, Polar Rec., 20, 43–51, 1980.

Drewry, D. J.: Antarctica, Glaciological and Geophysical Folio, University of Cambridge, Scott Polar Research Institute, 1983.

Fretwell, P., Pritchard, H. D., Vaughan, D. G., Bamber, J. L., Barrand, N. E., Bell, R., Bianchi, C., Bingham, R. G., Blankenship, D. D., Casassa, G., Catania, G., Callens, D., Conway, H., Cook, A. J., Corr, H. F. J., Damaske, D., Damm, V., Ferraccioli, F., Forsberg, R., Fujita, S., Gim, Y., Gogineni, P., Griggs,

J. A., Hindmarsh, R. C. A., Holmlund, P., Holt, J. W., Jacobel, R. W., Jenkins, A., Jokat, W., Jordan, T., King, E. C., Kohler, J., Krabill, W., Riger-Kusk, M., Langley, K. A., Leitchenkov, G., Leuschen, C., Luyendyk, B. P., Matsuoka, K., Mouginot, J., Nitsche, F. O., Nogi, Y., Nost, O. A., Popov, S. V., Rignot, E., Rippin, D. M., Rivera, A., Roberts, J., Ross, N., Siegert, M. J., Smith, A. M., Steinhage, D., Studinger, M., Sun, B., Tinto, B. K., Welch, B. C., Wilson, D., Young, D. A., Xiangbin, C., and Zirizzotti, A.: Bedmap2: improved ice bed, surface and thickness datasets for Antarctica, The Cryosphere, 7, 375–393, https://doi.org/10.5194/tc-7-375-2013, 2013.

Fujita, S., Hirabayashi, M., Goto-Azuma, K., Dallmayr, R., Satow, K., Zheng, J., and Dahl-Jensen, D.: Densification of layered firn of the ice sheet at NEEM, Greenland, J. Glaciol., 60, 905–921, 2014.

Gogineni, P.: CReSIS Radar Depth Sounder Data, Lawrence, Kansas, USA, Digital Media, available at: http://data.cresis.ku.edu/ (last access: 15 December 2017), 2012.

Gogineni, S., Tammana, D., Braaten, D., Leuschen, C., Akins, T., Legarsky, J., Kanagaratnam, P., Stiles, J., Allen, C., and Jezek, K.: Coherent radar ice thickness measurements over the Greenland ice sheet, J. Geophys. Res.-Atmos., 106, 33761–33772, 2001.

Gogineni, S., Yan, J.-B., Paden, J., Leuschen, C., Li, J., Rodriguez-Morales, F., Braaten, D., Purdon, K., Wang, Z., and Liu, W.: Bed topography of Jakobshavn Isbræ, Greenland, and Byrd Glacier, Antarctica, J. Glaciol., 60, 813–833, 2014.

Graham, F. S., Roberts, J. L., Galton-Fenzi, B. K., Young, D., Blankenship, D., and Siegert, M. J.: A high-resolution synthetic bed elevation grid of the Antarctic continent, Earth Syst. Sci. Data, 9, 267–279, https://doi.org/10.5194/essd-9-267-2017, 2017.

Haran, T., Bohlander, J., Scambos, T., Painter, T., and Fahnestock, M.: MODIS Mosaic of Antarctica 2008–2009 (MOA2009) Image Map, Boulder, Colorado USA, National Snow and Ice Data Center, 10, N5KP8037, 2014.

Harry, L. and Trees, V.: Optimum array processing: part IV of detection, estimation, and modulation theory, John Wiley and Sons Inc, 2002.

Hélière, F., Lin, C.-C., Corr, H., and Vaughan, D.: Radio echo sounding of Pine Island Glacier, West Antarctica: Aperture synthesis processing and analysis of feasibility from space, IEEE T. Geosci. Remote, 45, 2573–2582, 2007.

Hellmer, H. H., Kauker, F., Timmermann, R., Determann, J., and Rae, J.: Twenty-first-century warming of a large Antarctic ice-shelf cavity by a redirected coastal current, Nature, 485, 225–228, 2012.

Helm, V., Humbert, A., and Miller, H.: Elevation and elevation change of Greenland and Antarctica derived from CryoSat-2, The Cryosphere, 8, 1539–1559, https://doi.org/10.5194/tc-8-1539-2014, 2014.

Horgan, H. J., Alley, R. B., Christianson, K., Jacobel, R. W., Anandakrishnan, S., Muto, A., Beem, L. H., and Siegfried, M. R.: Estuaries beneath ice sheets, Geology, 41, 1159–1162, 2013.

Hutchinson, M. F.: Calculation of hydrologically sound digital elevation models, in Proceedings: Third International Symposium on Spatial Data Handling, 117–133, International Geographic Union, Commission on Geographic Data Sensings and Processing, Columbus, Ohio, 17–19 August 1988.

Jacobs, S. S., Jenkins, A., Giulivi, C. F., and Dutrieux, P.: Stronger ocean circulation and increased melting under Pine Island Glacier ice shelf, Nat. Geosci., 4, 519–523, 2011.

Jankowski, E. J. and Drewry, D.: The structure of West Antarctica from geophysical studies, Nature, 291, 17–21, 1981.

Jeofry, H., Ross, N., Corr, H. F., Li, J., Gogineni, P., and Siegert, M. J.: A deep subglacial embayment adjacent to the grounding line of Institute Ice Stream, West Antarctica, Geological Society, London, Special Publications, 461, 411 pp., 2017a.

Jeofry, H., Ross, N., Corr, H. F., Li, J., Gogineni, P., and Siegert, M. J.: 1-km bed topography digital elevation model (DEM) of the Weddell Sea sector, West Antarctica, Polar Data Centre, Natural Environment Research Council, UK, https://doi.org/10.5281/zenodo.1035488 (last access: 28 January 2018), 2017b.

Jezek, K. C.: RADARSAT-1 Antarctic Mapping Project: change-detection and surface velocity campaign, Ann. Glaciol., 34, 263–268, 2002.

Jordan, T. A., Ferraccioli, F., Ross, N., Corr, H. F. J., Leat, P. T., Bingham, R. G., Rippin, D. M., le Brocq, A., and Siegert, M. J.: Inland extent of the Weddell Sea Rift imaged by new aerogeophysical data, Tectonophysics, 585, 137–160, 2013.

Le Brocq, A. M., Payne, A. J., and Vieli, A.: An improved Antarctic dataset for high resolution numerical ice sheet models (ALBMAP v1), Earth Syst. Sci. Data, 2, 247–260, https://doi.org/10.5194/essd-2-247-2010, 2010.

Le Brocq, A. M., Ross, N., Griggs, J. A., Bingham, R. G., Corr, H. F., Ferraccioli, F., Jenkins, A., Jordan, T. A., Payne, A. J., Rippin, D. M., and Siegert, M. J.: Evidence from ice shelves for channelized meltwater flow beneath the Antarctic Ice Sheet, Nat. Geosci., 6, 945–948, 2013.

Lindbäck, K., Pettersson, R., Doyle, S. H., Helanow, C., Jansson, P., Kristensen, S. S., Stenseng, L., Forsberg, R., and Hubbard, A. L.: High-resolution ice thickness and bed topography of a land-terminating section of the Greenland Ice Sheet, Earth Syst. Sci. Data, 6, 331–338, https://doi.org/10.5194/essd-6-331-2014, 2014.

Lythe, M. B., Vaughan, D. G., and the Bedmap Consortium: BEDMAP: A new ice thickness and subglacial topographic model of Antarctica, J. Geophys. Res.-Sol. Ea., 106, 11335–11351, 2001.

Martin, M. A., Levermann, A., and Winkelmann, R.: Comparing ice discharge through West Antarctic Gateways: Weddell vs. Amundsen Sea warming, The Cryosphere Discuss., https://doi.org/10.5194/tcd-9-1705-2015, 2015.

Matsuoka, K., Hindmarsh, R. C. A., Moholdt, G., Bentley, M. J., Pritchard, H. D., Brown, J., Conway, H., Drews, R., Durand, G., Goldberg, D., Hattermann, T., Kingslake, J., Lenaerts, J. T. M., Martín, C., Mulvaney, R., Nicholls, K. W., Pattyn, F., Ross, N., Scambos, T., and Whitehouse, P. L.: Antarctic ice rises and rumples: Their properties and significance for ice-sheet dynamics and evolution, Earth-Sci. Rev., 150, 724–745, 2015.

Morlighem, M., Rignot, E., Seroussi, H. L., Larour, E., Ben Dhia, H., and Aubry, D.: A mass conservation approach for mapping glacier ice thickness, Geophys. Res. Lett., 38, https://doi.org/10.1029/2011GL048659, 2011.

Morlighem, M., Williams, C., Rignot, E., An, L., Arndt, J. E., Bamber, J. L., Catania, G., Chauché, N., Dowdeswell, J. A., and Dorschel, B.: BedMachine v3: Complete bed topography and ocean bathymetry mapping of Greenland from multibeam echo sounding combined with mass conservation, Geophys. Res. Lett., 44, 11051–11061, https://doi.org/10.1002/2017GL074954, 2017.

Peters, M. E., Blankenship, D. D., and Morse, D. L.: Analysis techniques for coherent airborne radar sounding: Application to West Antarctic ice streams, J. Geophys. Res.-Sol. Ea., 110, https://doi.org/10.1029/2004JB003222, 2005.

Picotti, S., Vuan, A., Carcione, J. M., Horgan, H. J., and Anandakrishnan, S.: Anisotropy and crystalline fabric of Whillans Ice Stream (West Antarctica) inferred from multicomponent seismic data, J. Geophys. Res.-Sol. Ea., 120, 4237–4262, 2015.

Plewes, L. A. and Hubbard, B.: A review of the use of radio-echo sounding in glaciology, Prog. Phys. Geog., 25, 203–236, 2001.

Pritchard, H. D.: Bedgap: where next for Antarctic subglacial mapping?, Antarct. Sci., 26, 742–757, https://doi.org/S095410201400025X, 2014.

Rignot, E., Mouginot, J., and Scheuchl, B.: Antarctic grounding line mapping from differential satellite radar interferometry, Geophys. Res. Lett., 38, https://doi.org/10.1029/2011GL047109, 2011a.

Rignot, E., Mouginot, J., and Scheuchl, B.: Ice flow of the Antarctic ice sheet, Science, 333, 1427–1430, 2011b.

Rignot, E., Mouginot, J., and Scheuchl, B.: MEaSUREs Antarctic Grounding Line from Differential Satellite Radar Interferometry, National Snow and Ice Data Center, Boulder, CO, USA, 2011c.

Rignot, E., Mouginot, J., and Scheuchl, B.: MEaSUREs InSAR-based Antarctica ice velocity map, National Snow and Ice Data Center, Boulder, CO, USA, 2011d.

Rippin, D., Bingham, R., Jordan, T., Wright, A., Ross, N., Corr, H., Ferraccioli, F., Le Brocq, A., Rose, K., and Siegert, M.: Basal roughness of the Institute and Möller Ice Streams, West Antarctica: Process determination and landscape interpretation, Geomorphology, 214, 139–147, 2014.

Ritz, C., Edwards, T. L., Durand, G., Payne, A. J., Peyaud, V., and Hindmarsh, R. C.: Potential sea-level rise from Antarctic ice-sheet instability constrained by observations, Nature, 528, 115–118, 2015.

Rodriguez-Morales, F., Gogineni, S., Leuschen, C. J., Paden, J. D., Li, J., Lewis, C. C., Panzer, B., Alvestegui, D. G.-G., Patel, A., and Byers, K.: Advanced multifrequency radar instrumentation for polar research, IEEE T. Geosci. Remote, 52, 2824–2842, 2014.

Rose, K. C., Ross, N., Bingham, R. G., Corr, H. F. J., Ferraccioli, F., Jordan, T. A., Le Brocq, A. M., Rippin, D. M., and Siegert, M. J.: A temperate former West Antarctic ice sheet suggested by an extensive zone of subglacial meltwater channels, Geology, 42, 971–974, 2014.

Rose, K. C., Ross, N., Jordan, T. A., Bingham, R. G., Corr, H. F. J., Ferraccioli, F., Le Brocq, A. M., Rippin, D. M., and Siegert, M. J.: Ancient pre-glacial erosion surfaces preserved beneath the West Antarctic Ice Sheet, Earth Surf. Dynam., 3, 139–152, https://doi.org/10.5194/esurf-3-139-2015, 2015.

Ross, N., Bingham, R. G., Corr, H. F. J., Ferraccioli, F., Jordan, T. A., Le Brocq, A., Rippin, D. M., Young, D., Blankenship, D. D., and Siegert, M. J.: Steep reverse bed slope at the grounding line of the Weddell Sea sector in West Antarctica, Nat. Geosci., 5, 393–396, 2012.

Ross, N., Jordan, T. A., Bingham, R. G., Corr, H. F., Ferraccioli, F., Le Brocq, A., Rippin, D. M., Wright, A. P., and Siegert, M. J.: The Ellsworth subglacial highlands: inception and retreat of the West Antarctic Ice Sheet, Geol. Soc. Am. Bull., 126, 3–15, 2014.

Shafique, U., Anwar, J., Munawar, M. A., Zaman, W.-U., Rehman, R., Dar, A., Salman, M., Saleem, M., Shahid, N., and Akram, M.: Chemistry of ice: Migration of ions and gases by directional freezing of water, Arab. J. Chem., 9, S47–S53, 2016.

Shi, L., Allen, C. T., Ledford, J. R., Rodriguez-Morales, F., Blake, W. A., Panzer, B. G., Prokopiack, S. C., Leuschen, C. J., and Gogineni, S.: Multichannel coherent radar depth sounder for NASA operation ice bridge, Paper presented at the Geoscience and Remote Sensing Symposium (IGARSS), 2010 IEEE International, 1729–1732, IEEE, Honolulu, HI, USA, 25–30 July 2010.

Shreve, R.: Movement of water in glaciers, J. Glaciol., 11, 205–214, 1972.

Siegert, M. J., Hindmarsh, R., Corr, H., Smith, A., Woodward, J., King, E. C., Payne, A. J., and Joughin, I.: Subglacial Lake Ellsworth: A candidate for in situ exploration in West Antarctica, Geophys. Res. Lett., 31, https://doi.org/10.1029/2004GL021477, 2004a.

Siegert, M. J., Taylor, J., Payne, A. J., and Hubbard, B.: Macro-scale bed roughness of the siple coast ice streams in West Antarctica, Earth Surf. Processes, 29, 1591–1596, 2004b.

Siegert, M. J., Clarke, R. J., Mowlem, M., Ross, N., Hill, C. S., Tait, A., Hodgson, D., Parnell, J., Tranter, M., and Pearce, D.: Clean access, measurement, and sampling of Ellsworth Subglacial Lake: a method for exploring deep Antarctic subglacial lake environments, Rev. Geophys., 50, https://doi.org/10.1029/2011RG000361, 2012.

Siegert, M. J., Ross, N., Corr, H., Smith, B., Jordan, T., Bingham, R. G., Ferraccioli, F., Rippin, D. M., and Brocq, A. L.: Boundary conditions of an active West Antarctic subglacial lake: implications for storage of water beneath the ice sheet, The Cryosphere, 8, 15–24, https://doi.org/10.5194/tc-8-15-2014, 2014.

Siegert, M. J., Ross, N., and Le Brocq, A. M.: Recent advances in understanding Antarctic subglacial lakes and hydrology, Philos. T. Roy. Soc. A, 374, https://doi.org/10.1098/rsta.2014.0306, 2016a.

Siegert, M. J., Ross, N., Li, J., Schroeder, D. M., Rippin, D., Ashmore, D., Bingham, R., and Gogineni, P.: Subglacial controls on the flow of Institute Ice Stream, West Antarctica, Ann. Glaciol., 57, 19–24, https://doi.org/10.1017/aog.2016.17, 2016b.

Siegert, M. J., Bingham, R., Corr, H. F., Ferraccioli, F., Le Brocq, A. M., Jeofry, H., Rippin, D., Ross, N., Jordan, T., and Robinson, C.: Synthetic-aperture radar (SAR) processed airborne radio-echo sounding data from the Institute and Möller ice streams, West Antarctica, 2010-11; Polar Data Centre, Natural Environment

Research Council, UK, https://doi.org/10.5285/8a975b9e-f18c-4c51-9bdb-b00b82da52b8, 2017.

Skou, N. and Søndergaard, F.: Radioglaciology, A 60 MHz ice sounder system, Technical University of Denmark, 1976.

Snyder, J. P.: Map projections–A working manual, US Government Printing Office, Washington, D.C., 1987.

Stearns, L. A., Smith, B. E., and Hamilton, G. S.: Increased flow speed on a large East Antarctic outlet glacier caused by subglacial floods, Nat. Geosci., 1, 827–831, 2008.

Stocker, T.: Climate change 2013: the physical science basis: Working Group I contribution to the Fifth assessment report of the Intergovernmental Panel on Climate Change, Cambridge University Press, 2014.

Thoma, M., Determann, J., Grosfeld, K., Goeller, S., and Hellmer, H. H.: Future sea-level rise due to projected ocean warming beneath the Filchner Ronne Ice Shelf: A coupled model study, Earth Planet. Sc. Lett., 431, 217–224, 2015.

Van Wessem, J., Reijmer, C., Morlighem, M., Mouginot, J., Rignot, E., Medley, B., Joughin, I., Wouters, B., Depoorter, M., and Bamber, J.: Improved representation of East Antarctic surface mass balance in a regional atmospheric climate model, J. Glaciol., 60, 761–770, 2014.

Winter, K.: Englacial stratigraphy, debris entrainment and ice sheet stability of Horseshoe Valley, West Antarctica, 2016, Doctoral thesis, University of Northumbria, Newcastle upon Tyne, England, UK, 263 pp., 2016.

Winter, K., Woodward, J., Ross, N., Dunning, S. A., Bingham, R. G., Corr, H. F. J., and Siegert, M. J.: Airborne radar evidence for tributary flow switching in Institute Ice Stream, West Antarctica: Implications for ice sheet configuration and dynamics, J. Geophys. Res.-Earth, 120, 1611–1625, 2015.

Woodward, J., Smith, A. M., Ross, N., Thoma, M., Corr, H., King, E. C., King, M., Grosfeld, K., Tranter, M., and Siegert, M.: Location for direct access to subglacial Lake Ellsworth: An assessment of geophysical data and modeling, Geophys. Res. Lett., 37, https://doi.org/10.1029/2010GL042884, 2010.

Wright, A. and Siegert, M.: A fourth inventory of Antarctic subglacial lakes, Antarct. Sci., 24, 659–664, 2012.

Wright, A., Siegert, M., Le Brocq, A., and Gore, D.: High sensitivity of subglacial hydrological pathways in Antarctica to small ice-sheet changes, Geophys. Res. Lett., 35, https://doi.org/10.1029/2008GL034937, 2008.

Wright, A. P., Le Brocq, A. M., Cornford, S. L., Bingham, R. G., Corr, H. F. J., Ferraccioli, F., Jordan, T. A., Payne, A. J., Rippin, D. M., Ross, N., and Siegert, M. J.: Sensitivity of the Weddell Sea sector ice streams to sub-shelf melting and surface accumulation, The Cryosphere, 8, 2119–2134, https://doi.org/10.5194/tc-8-2119-2014, 2014.

9

Subglacial topography, ice thickness, and bathymetry of Kongsfjorden, northwestern Svalbard

Katrin Lindbäck[1], **Jack Kohler**[1], **Rickard Pettersson**[2], **Christopher Nuth**[3], **Kirsty Langley**[4], **Alexandra Messerli**[1], **Dorothée Vallot**[2], **Kenichi Matsuoka**[1], **and Ola Brandt**[5]

[1]Norwegian Polar Insitute, Framsentret, Postboks 6606, Langnes, 9296 Tromsø, Norway
[2]Department of Earth Sciences, Uppsala University, Villavägen 16, 752 36 Uppsala, Sweden
[3]University of Oslo, Postboks 1047 Blindern, 0316 Oslo, Norway
[4]Asiaq Greenland Survey, Postboks 1003, 3900 Nuuk, Greenland
[5]Norwegian Coastal Administration, Kystveien 30, 4841 Arendal, Norway

Correspondence: Katrin Lindbäck (katrin.lindback@npolar.no)

Abstract. Svalbard tidewater glaciers are retreating, which will affect fjord circulation and ecosystems when glacier fronts become land-terminating. Knowledge of the subglacial topography and bathymetry under retreating glaciers is important to modelling future scenarios of fjord circulation and glacier dynamics. We present high-resolution (150 m gridded) digital elevation models of subglacial topography, ice thickness, and ice surface elevation of five tidewater glaciers in Kongsfjorden (1100 km^2), northwestern Spitsbergen, based on $\sim$ 1700 km airborne and ground-based ice-penetrating radar profiles. The digital elevation models (DEMs) cover the tidewater glaciers Blomstrandbreen, Conwaybreen, Kongsbreen, Kronebreen, and Kongsvegen and are merged with bathymetric and land DEMs for the non-glaciated areas. The large-scale subglacial topography of the study area is characterized by a series of troughs and highs. The minimum subglacial elevation is -180 m above sea level (a.s.l.), the maximum subglacial elevation is 1400 m a.s.l., and the maximum ice thickness is 740 m. Three of the glaciers, Kongsbreen, Kronebreen, and Kongsvegen, have the potential to retreat by $\sim$ 10 km before they become land-terminating. The compiled data set covers one of the most studied regions in Svalbard and is valuable for future studies of glacier dynamics, geology, hydrology, and fjord circulation.

1 Introduction

Ocean waters around Svalbard are warming, which in combination with the overall atmospheric warming has made Svalbard's tidewater glaciers particularly vulnerable to climate change (Nuth et al., 2013). Air temperatures have increased steadily over the last 4 decades, similar to the rest of the Arctic (Overland et al., 2004). Summer temperatures have the strongest influence on Svalbard glacier mass balance (van Pelt et al., 2012), and the recent summer warming has led to increasing rates of mass loss (Kohler et al., 2007). The current overall mass balance for Svalbard glaciers is negative (Moholdt et al., 2010; Nuth et al., 2010; Wouters et al., 2008), with tidewater glaciers having the greatest retreat

rates overall (Nuth et al., 2013). More than half of Svalbard's total land area of $\sim$ 60 000 km is covered by glaciers (König et al., 2014). Over 1100 glaciers are larger than 1 km^2, and of these, 163 (15 %) are tidewater glaciers. In terms of area, more than 60 % of all glacier fronts terminate at sea, and the total length of calving ice-cliffs around Svalbard is estimated to be $\sim$ 860 km (Błaszczyk et al., 2009). A significant portion of the meltwater is delivered to the ocean at calving glacier fronts. With further warming in the Arctic, we expect the Svalbard glaciers to continue to retreat, and concomitant declines in the number of tidewater calving glaciers and total length of calving fronts around Svalbard, providing a contribution to rising global sea levels.

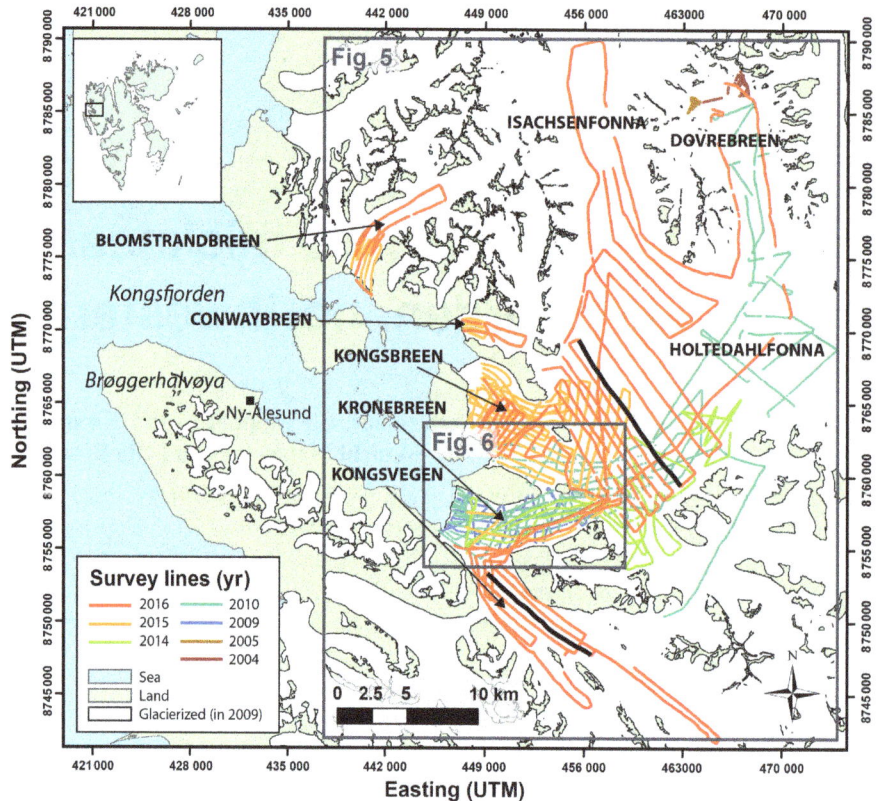

Figure 1. Radar survey lines collected between 2004 and 2016. Kongsfjorden, tidewater glaciers, icefields, the peninsula Brøggerhalvøya, and the research town Ny-Ålesund are marked in the map. The small inset map shows the location of Kongsfjorden in Svalbard. The blue areas are sea, green areas are land, and white areas are glacierized regions (in 2009). The black lines indicate the location of the profiles in Fig. 4 and the boxes show the coverage of Figs. 5 and 6. Grid projection is Universal Transverse Mercator Zone 33W.

Glacier front areas are important feeding areas for seabirds and marine mammals (Kovacs et al., 2011; Kovacs and Michel, 2011; Loeng et al., 2005; Lydersen et al., 2014). In summer, glacier meltwater flows on, in, and under the glacier towards the front. This meltwater is typically discharged below the seawater surface, often at the base of the calving front. The relatively low density of these fresh waters forces them to rise rapidly, entraining large volumes of ambient fjord water. These meltwater plumes can breach the surface, then flow outward towards the mouth of the fjord, further entraining subsurface water. In Svalbard, several bird species can be found in large numbers, up to thousands of individuals, at tidewater glacier fronts. The birds are often found in the so-called "brown zone", the meltwater plume, which is ice-free and muddy due to upwelling suspended sediments and currents. These brown zones are also foraging hotspots for Svalbard's ringed seals and white whales (Lydersen et al., 2014). When the tidewater glaciers retreat so much that they become land-terminating, outflow into the fjord will only occur via surface drainage, just as with any un-glaciated fjord, with a cap of fresh river water flowing over the denser ocean water. This will lead to fewer nutrients and plankton being brought to the surface from the fjord bottom, which is

likely to affect fjord ecosystems. Changes in freshwater flux from western Svalbard glaciers may also, in extreme climate warming scenarios, disturb deep-water production on the Svalbard shelf (Hagen et al., 2003). The amplified climatic warming at northern high latitudes (Serreze and Barry, 2011) makes Svalbard glaciers prime targets for understanding not only glacial dynamics but also the effects of ongoing climate change on glaciers, oceans, and ecosystems. To model future scenarios of fjord circulation and glacier dynamics, knowledge on the subglacial topography and bathymetry under the retreating glaciers is vital. Here, we present high-resolution (150 m gridded) digital elevation models (DEMs) of the subglacial topography, ice thickness, and elevation of five tidewater glaciers in Kongsfjorden, northwestern Spitsbergen, near Ny-Ålesund (78.9° N, 12.4° E).

2 Study area

Kongsfjorden is the southern branch of the Kongsfjorden–Krossfjorden system that merges towards the open sea, in a large submarine trough, Kongsfjordrenna, which channelled a fast-flowing ice stream during the last glacial maximum

(Ingólfsson and Landvik, 2013; Ottesen et al., 2005). Kongsfjorden is ~ 20 km long and between 4 and 10 km wide and covers an area of ~ 200 km^2 and a water volume of ~ 30 km^3 (Ito and Kudoh, 1997). The maximum depth in the outer part of the fjord is 350 m and 100 m in the inner fjord. The mouth of the fjord lacks a well-defined sill and is therefore interconnected with neighbouring water masses on the West Spitsbergen Shelf, including Atlantic Water (Svendsen et al., 2002). Five tidewater glaciers terminate in Kongsfjorden (Fig. 1): Blomstrandbreen, Conwaybreen, Kongsbreen (with a north and south branch around Ossian Sarsfjellet), Kronebreen, and Kongsvegen. Kronebreen is among the fastest-flowing glaciers in Svalbard, with speeds up to ~ 3 m d^{-1} (Schellenberger et al., 2015). Upglacier from Kongsbreen and Kronebreen are two large icefields, Holtedahlfonna (named Dovrebreen in the upper part) and Isachsenfonna. In the following section, a short summary of the glacio-geomorphological setting of the study area is presented.

Glacio-geomorphological setting

The youngest deposits in Kongsfjorden are of Quaternary age and these landforms around Kongsfjorden are shaped by glacial activity (Ingólfsson and Landvik, 2013). Brøggerhalvøya and the areas to the north were probably completely ice-covered during the last Weichselian ice age. Compared to most other places in Svalbard, the Kongsfjorden area shows a more complete glacial sedimentary record dating back to before the last interglacial, the Eemian (Landvik et al., 2005). Together with Bellsund and Isfjorden, Kongsfjorden was one of the largest outlets for palaeo-ice streams in western Svalbard. The glaciers along the west coast of Svalbard had a complicated topographically controlled configuration during the Weichselian (Howe et al., 2003). The ice stream in Kongsfjordrenna was fed by ice draining through the deep fjord systems of Kongsfjorden and Krossfjorden, which drained a large section of the ice fields over northwestern Spitsbergen. Adjacent to the ice stream there were sharp boundaries to dynamically less active ice.

The glaciers started to retreat during the early Holocene and the region was likely largely ice-free until the neoglacial advance ~ 4.5 thousand years ago. The bedrock has a relict subglacial, ice-scoured topography from the glacial re-advances of the Weichselian glaciation, with drumlins and glacial flutes common across the sea floor (Howe et al., 2003). Brøggerhalvøya shows four isostatically induced cycles of emergence out of the sea during the Weichselian glaciation (Miller et al., 1989), with beach ridges up to the marine limit at ~ 80 m above sea level (a.s.l.). The current isostatic uplift rate is 8 mm y^{-1} (Kierulf et al., 2009).

The bottom of inner Kongsfjorden has waveform morphology interpreted as moraines, partly originating from surges (Howe et al., 2003). Three glaciers are documented as surge type glaciers: Kronebreen–Kongsvegen surged around 1869 and 1897 (Bennett et al., 1999), Kongsvegen around 1948

(Liestøl, 1988; Woodward et al., 2002), and Blomstrandbreen, possibly between 1911 and 1928 (Burton et al., 2016), around 1960 (Hagen et al., 1993), and recently in 2010 (Mansell et al., 2012). Glacier surges lead to short-term reworking of sediments and deposition of sediment lobes containing massive glaciomarine muds with sedimentation accumulation rates up to 30 cm y^{-1}. Suspension settling from meltwater plumes and ice rafting are the dominant sedimentary processes, leading to the deposition of stratified glaciomarine muds with clasts from melting icebergs. The fjord topography has been smoothed by bottom currents.

3 Data and methods

We map the glacier beds with ice-penetrating radar (Dowdeswell and Evans, 2004). Our analysis builds on extensive radar campaigns conducted in the area from 2004 to 2016 (Fig. 1). Earlier campaigns (1988 to 2005) covered the upper parts of the glaciers, but the airborne radar failed to detect the bed in the lower reaches. This was caused by a too-high radar frequency (dictated by limitations on antenna size on an airplane) and too-high travel speed with respect to the data acquisition rate, as well as radar clutter from the rough surface, crevasses, and water within the glacier (Hagen and Sætrang, 1991). In recent years, radar surveys have successfully detected the bed in the lower parts of the glaciers using a lower frequency set-up mounted on a helicopter frame (Fig. 2). In the following sections, we describe the methods used to collect, process, and interpolate the radar data sets into the final products of gridded subglacial elevation and ice thickness. Surveys of crevassed glaciers from helicopters are less common (e.g. Blindow et al., 2012; Kennett et al., 1993; Langhammer et al., 2017; Rutishauser et al., 2016) than surveys from fixed-wing airplanes (e.g. Bamber et al., 2013; Fretwell et al., 2013; Morlighem et al., 2017). Therefore, we describe the set-up in detail.

3.1 Radar systems and uncertainties

3.1.1 Radar data collected 2014 to 2016

During early spring (April to May) in 2014, 2015, and 2016 we collected ~ 1300 km of common-offset radar profiles with an impulse radar system that was either suspended under a helicopter (for crevassed areas; Fig. 2a) or towed behind a snowmobile. The system is based on radar developed for surveying ice thickness on the Greenland ice sheet (Lindbäck et al., 2014). The radar system consisted of resistively loaded half-wavelength dipole antennas of 10 MHz centre frequency. We used a commercial off-the-shelf Kentech impulse transmitter with an average output power of 35 W and a pulse repetition frequency of 1 kHz. The trace acquisition was triggered by the direct wave pulse between transmitter and receiver. The 14-bit A/D converter sampled two channels at 125 MHz sampling frequency, with different sensitiv-

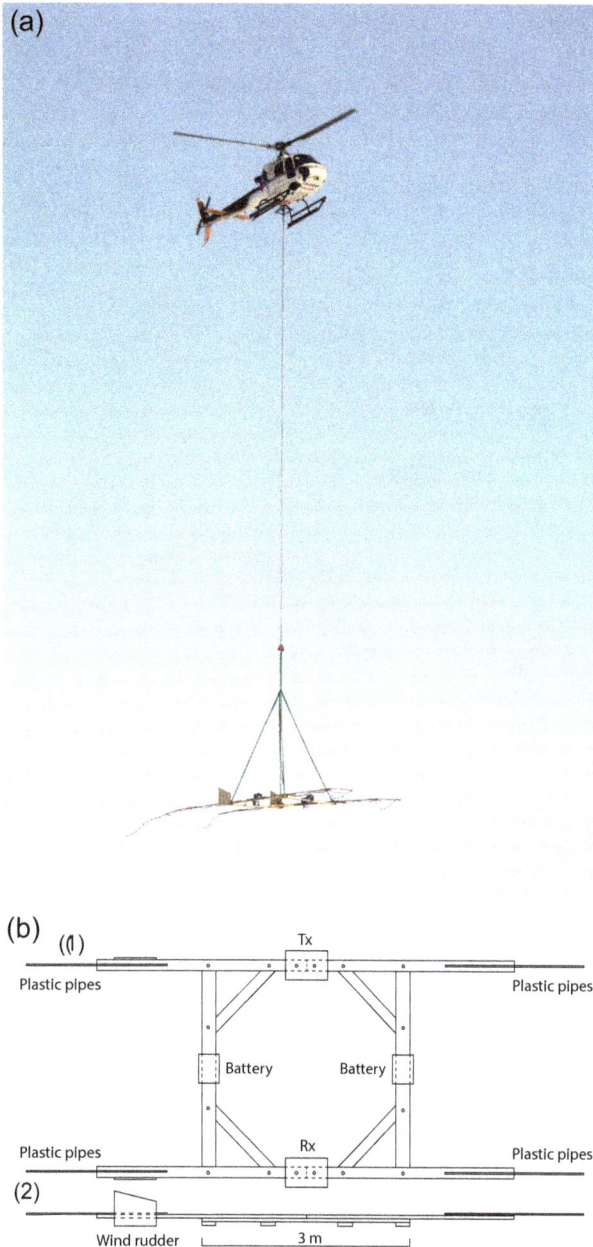

Figure 2. (a) Helicopter with radar frame. Photo: Nick Hulton. (b) Helicopter wooden frame (1) from above, with transmitter (Tx), receiver (Rx), two batteries and four plastic pipes for holding the antennas, and (2) from the side, with two fins on one side that function as wind rudders to prevent the frame from spinning. The antennas are connected to the Tx and Rx and fixed to the frame extending out on the plastic rods.

ity ranges. One channel was attenuated with 20 dB to record both the surface and the bed return.

Using the helicopter-based system, we surveyed the glaciers at a nominal speed of $\sim 40\,\mathrm{km\,h^{-1}}$, along tracks separated by ~ 0.5 to 1 km. By stacking 125 traces, a mean

trace spacing of 4 m was achieved. We mounted the system on a $3 \times 3\,\mathrm{m}$ wooden frame, with extended wooden arms and plastic rods for the antenna (Fig. 2b). The frame was suspended 20 m below the helicopter. The radar was controlled by wireless connection to a PC inside the helicopter. We used the ground-based system to survey the snowmobile-accessible Kongsvegen glacier. The system was mounted on two sleds and towed behind the snowmobile at a speed of $\sim 20\,\mathrm{km\,h^{-1}}$. Stacking 125 traces resulted in a mean trace spacing of 2 m. We positioned the traces by using data from a code-phase global positioning system (GPS) receiver in 2014 and 2015 and a carrier-phase dual-frequency GPS receiver in 2016, mounted on the radar receiver box 1.5 m in front of the common mid-point along the travelled trajectory on the helicopter frame and 15 m from the common mid-point on the snowmobile. For the dual-frequency receiver we processed the data kinematically using the Canadian Spatial Reference System precise point positioning service (Natural Resources Canada, 2017).

We applied several corrections and filters to the radar data: (1) a Butterworth bandpass filter, with cut-off frequencies of 2 and 50 MHz, to remove unwanted frequency components in the data; (2) normal move-out correction to correct for antenna separation (including adjusted travel times for the trigger delay); (3) rubber-band correction to re-sample the data to a uniform trace spacing; and (4) two-dimensional frequency wave-number migration (Stolt, 1978) to collapse hyperbolic reflectors back to their original positions in the profile direction. On the high gain channel, we applied a spreading and exponential compensation (SEC) gain to amplify bed returns. The surface and bed returns were digitized semi-automatically with a cross-correlation picker (Irving et al., 2007) at the first break of the bed reflection. We calculated ice thickness from the picked travel times of the bed return using a constant radio-wave velocity of $169\,\mathrm{m\,\mu s^{-1}}$ for ice. For the airborne profiles, we removed the travel times to the surface return using a constant radio-wave velocity of $300\,\mathrm{\mu m\,s^{-1}}$ for air. We converted the GRS80 ellipsoidal heights to heights above sea level with a geoid model developed by the Norwegian Polar Institute, where the average geoid height is $\sim 35 \pm 0.5\,\mathrm{m}$ in the study area relative to the ellipsoid. Figure 3 shows examples of processed radar images.

3.1.2   Radar data collected 2004 to 2010

In addition to the data collected in this study, we used two additional data sets of unpublished radar data collected earlier by the Norwegian Polar Institute on (1) Dovrebreen in 2004 and 2005, and (2) Kronebreen and Holtedahlfonna in 2009 and 2010. We did not use additional data sets collected in Kongsvegen in 1988 (ice thickness; Hagen and Sætrang, 1991) and 1995 (subglacial elevation and ice thickness; Melvold and Hagen, 1998), because these data sets had large (> 50 m) differences in subglacial elevation. This is be-

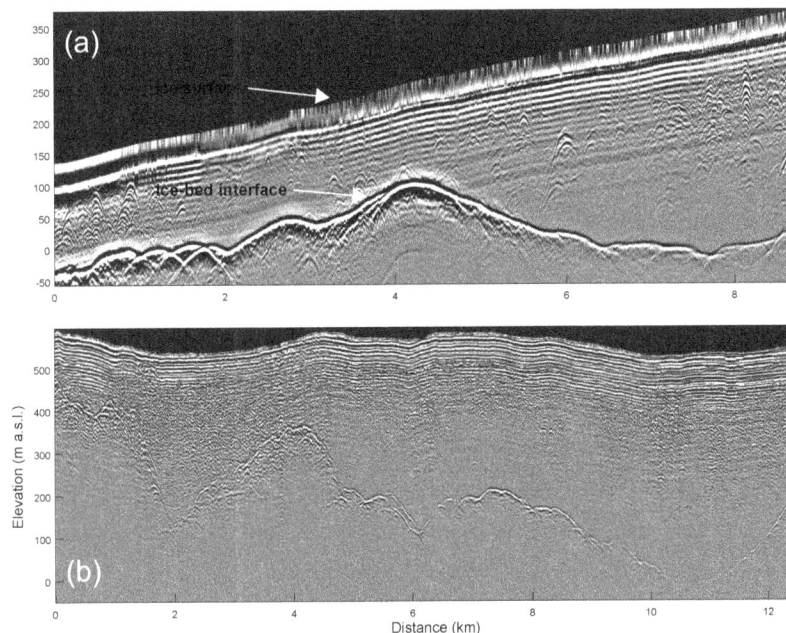

Figure 3. Examples of processed radar images collected on **(a)** Kongsvegen by snowmobile and **(b)** Holtedahlfonna by helicopter. Locations of the profiles are marked in Fig. 1 with black lines.

cause the data were collected over 20 years ago and possibly significant changes in glacier surface and subglacial sediment may hinder accurate estimates of subglacial elevations from these old ice-thickness data. Here follows a short summary of the two included data sets:

The Dovrebreen campaign. The data set consists of $\sim$ 22 km of radar profiles collected on the upper parts of Dovrebreen, during an ice coring campaign (Beaudon et al., 2013; Sjögren et al., 2007). The subglacial elevation and ice thickness were measured in April 2004 and 2005, with a 10 MHz centre frequency impulse radar. A single channel impulse radar based on a Narod transmitter (Narod and Clarke, 1994) and a 12-bit A/D converter in the receiver were used with restively loaded dipoles as antennas. The radar was operated at both 100 MHz sampling frequency and at 200, 300, and 500 MHz sampling frequency using repetitive sampling. The repetitive sampling gave a non-uniform sampling frequency in the scan, and the data had therefore been resampled to an equal time base between the samples with linear interpolation. An antenna separation of 20 m was used. The antennas were configured with the transmitter in the back and the receiver approximately 25 m behind a snowmobile. The profiles were positioned with a code-phase GPS receiver attached to the radar receiver and the position was recorded each second.

The Kronebreen and Holtedahlfonna campaign. The data consist of $\sim$ 340 km of radar profiles collected in 2009 to 2010, where both helicopter and snowmobiles were used. Data were collected with an impulse dipole radar comprising a Kentech pulser (average output power of 35 W), 10 MHz

resistively loaded wire dipole antennas, and a 12-bit A/D converter. The helicopter and ground-based system was similar to the one previously described (see Sect. 3.1.1). The A/D converter sampled with two channels at 50 MHz sampling frequency. Five traces were stacked in flight, and further stacking was done during post-processing. Positioning was made with a code-phase GPS receiver attached to the radar receiver.

3.1.3 Radar system errors and uncertainty

We used standard analytical error propagation methods (Lapazaran et al., 2016; Taylor, 1996) to calculate the error in subglacial elevation for each data point:

$$\varepsilon_{\text{bed data}} = \sqrt{\varepsilon_{\text{radar}}^2 + \varepsilon_{xy}^2}, \tag{1}$$

where $\varepsilon_{\text{radar}}$ was the error in the radar acquisition and ε_{xy} was the positioning error. The error in radar acquisition was calculated by the following:

$$\varepsilon_{\text{radar}} = \frac{1}{2}\sqrt{v^2 \cdot \varepsilon_t^2 + t^2 \cdot \varepsilon_v^2}, \tag{2}$$

where v was the radio-wave velocity used for time-to-depth conversation, t was the two-way-travel time of the radio wave and ε_t and ε_v were the errors in t and v respectively. We used a constant wave-propagation speed for the ground-based and airborne surveys (169 m µs^{-1}). Wave velocity can vary spatially, depending mostly on density. Profiles were collected in the ablation and accumulation zone with a snow and firn

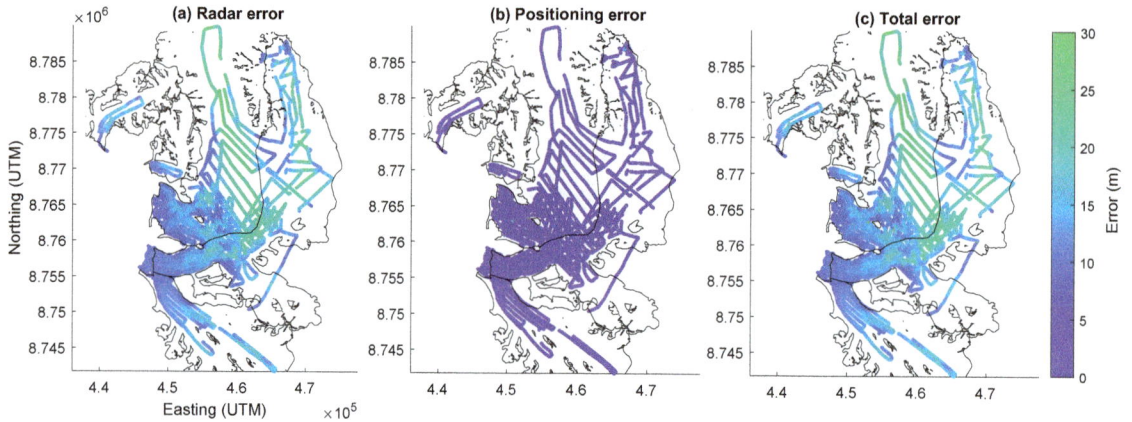

Figure 4. Errors in the subglacial elevation for each data point: **(a)** radar error consisting of the technical and theoretical capacity of the radar systems, **(b)** positioning error, and **(c)** the total error when combining radar and positioning errors. Grid projection is Universal Transverse Mercator Zone 33W.

cover of up to 20 m thick in the upper parts of Dovrebreen (Beaudon et al., 2013; Woodward et al., 2003). We used a typical variation of 4 % of glacier ice density for the calculation of ε_v (Seligman, 1936). Variations in the wave velocity can also occur because of varying ice temperature and the presence of inhomogeneities and liquid water in the ice (Drewry, 1975). These effects are expected to have a small impact on the average velocity for the whole ice column, while water content in the ice can influence the velocity in a substantial way. In most parts of the study area the ice is cold (Beaudon et al., 2013; Woodward et al., 2003) and there are limited amounts of liquid water. We therefore neglect variations of velocity due to water content. The upper parts of Holtedahlfonna contain a firn aquifer (Christianson et al., 2015), but it comprises a small part of the total glacierized area, and is not accounted for. For ε_t we calculated the range resolution, which is the accuracy of the measurement of distance between the antenna and the bed and can be determined from the characteristics of the source pulse (i.e. bandwidth) and the digitization frequency. The range resolution for the data collected in this study was estimated at 8.5 m. We also included the vertical resolution, by taking the inverse of the radar frequency (Lapazaran et al., 2016). This results in values of $\varepsilon_{\text{radar}}$ between 8.5 m (thin ice) and 30.5 m (thick ice) with a mean value of 14.3 m and standard deviation of 4.2 m (Fig. 4a). We calculated the positioning error ε_{xy} at each data point depending on the bed slope angle along the profile, following the method by Lapazaran et al. (2016). We assumed a helicopter-travel speed of 40 km h^{-1}, snowmobile-travel speed of 20 km h^{-1}, and $T_{\text{GPS}} \leq T_{\text{GPR}}$ (case a' in Appendix B of Lapazaran et al. 2016). This produced values of ε_{xy} between 0 (flat bed) and 76.0 m (steep bed $\sim$ 90°), with a mean value of 1.4 m and standard deviation of 2.1 m (Fig. 4b). The total error in subglacial elevation along the profiles (Eq. 2) varied between 8.5 and 78.0 m, with a mean of 14.5 m and standard deviation of 4.3 m (Fig. 4c).

To test the consistency between the data sets we also compared the crossover differences in the subglacial elevation estimates between different profiles and data sets. The data set collected in this study (2014 to 2016) had a median crossover misfit in subglacial elevation of 11.1 m with a standard deviation (σ) of 17.4 m based on 208 crossing points. The Dovrebreen campaign data set (2004 to 2005) had a median crossover misfit of 13.1 m ($\sigma = 12.7$ m) based on 85 crossing points. The Kronebreen and Holtedahlfonna campaign data set (2009 to 2010) had a median crossover misfit of 3.9 m ($\sigma = 13.0$ m) based on 136 crossing points. As the crossover analysis within the same data set does not capture systematic errors between the different data sets, we also did a comparison between the data sets. When we ran a crossover analysis between all the data sets the median misfit was 9.2 m ($\sigma = 15.7$ m).

3.2 Surface and bathymetric elevation data

To obtain surface elevation for the study area that is most temporally consistent with the acquired helicopter thickness measurements, we used TanDEM-X monostatic radar images (Moreira et al., 2004) acquired on 20 December 2014. These images were processed by differential interferometry using Gamma Software (e.g. Neckel et al., 2013; Rankl and Braun, 2016) as precise orbital information is not publicly available. The monoscopic images were first co-registered to each other and to a previous baseline DEM derived from the TanDEM-X intermediate DEM (Wessel, 2016). After phase filtering, the differential phase was unwrapped using a minimum cost flow (MCF) algorithm and triangulation to provide elevation differences in metres between the intermediate DEM and that from the monoscopic images. Unwrapping was successful over the relatively flat terrain, with no apparent blunders even over steeper terrain. These differences were then added back to the TanDEM-X intermediate DEM to provide elevations from 20 December 2014 at a 12 m resolution. The obtained 2014 TanDEM-X elevations were for the underlying ice and ground surface as the X-band satellite radar wave can penetrate through the winter snowpack, providing a reflector from the ground and ice interface. On stable terrain surrounding the glaciers, the 2014 TanDEM-X DEM shows little bias, with a standard deviation of ∼ 4 m compared with the 2009 aerial-photogrammetric DEM from the Norwegian Polar Insitute (2014), which has a 5 m gridded resolution and a stated accuracy of ∼ 2 to 5 m. Elevation changes between 2009 and 2014 have mostly occurred close to the margins of the tidewater glaciers (< 5 km), with up to 5 m of surface lowering (Cesar Deschemps, personal communication, 2017). This TanDEM-X DEM was merged with the NPI (2014) DEM to cover Blomstrandbreen, which is located outside of the TanDEM-X DEM, and downsampled to 150 m.

The offshore bathymetric DEM was compiled by the Norwegian Mapping Authority Hydrographic Service and is a publicly available data product (Kartverket, 2018). Data were acquired in 2000 with an EM 1002 multi-beam echo sounder. In 2007, 2010, and 2011 an EM 3002 multi-beam echo sounder was used. They derived the 50 m grid DEM in 2014 with the software QPS Fledermaus and CARIS HIPS/CARIS BathyDataBASE. The surface and bathymetric DEMs were point sampled and added to the subgrid of the radar data, described in detail below.

3.3 Assimilation of the data sets

We combined the different data sets of subglacial, land, and bathymetry elevation to a final gridded elevation DEM. The measuring interval for the radar data sets are dense along the profiles, with a data point spacing of ∼ 5 m, compared with ∼ 50 to 1000 m spacing between individual profiles. As this non-uniform spacing is not optimal for gridding algorithms, we sub-gridded data sets into a 100 m pseudo-grid to reduce the data density along individual profiles. The subgrid was produced by calculating the median values for the points that fell within the distance of half the grid cell. To prevent steps at the borders between the subglacial and proglacial DEMs we point-sampled the land-topography and bathymetric DEMs outside the glacierized areas and added these points to the subgrid. We used a universal kriging algorithm (e.g. Isaaks and Srivastava, 1989) for the interpolation. To calculate glacier ice thickness in 2014, we subtracted the subglacial DEM from the combined TanDEM-X DEM and NPI DEM. We did this instead of using ice thickness measurement for each radio echo-sounding data set, to make the subglacial DEM more consistent over the area, as the ice thickness has changed since the first data were collected in 2004.

The subglacial DEM agrees well with a borehole study in Kronebreen in 2014 (How et al., 2017), with a measured subglacial elevation of −93 m a.s.l., where the gridded DEM

predicts a depth of 90 m. To assess the error in interpolation we cross-validated the gridded data, which is a common validation technique to see how well an interpolated model is influenced by the observed data. By removing one observation from the data set, the remaining data were used to interpolate a value for the removed observation. This process was continued for 1000 random observations in the data, where the error is the residual between the observed and the interpolated value (Isaaks and Srivastava, 1989). The standard deviation of the residuals was estimated to be 18 m, and increases with distance from the profiles. To summarize, the total accuracy of the gridded subglacial elevation depends on: (1) the technical and theoretical capability of the radar systems, (2) positioning errors, and (3) interpolation errors. By assessing all these potential sources of error, we estimate the maximum vertical root-mean-squared uncertainty in the final subglacial and ice thickness DEMs to be approximately ±24 m.

4 Results

We present DEMs of subglacial topography (Fig. 5a), ice thickness (Fig. 5b), and ice surface elevation (Fig. 5c) of a 1100 km² area of Svalbard on a 150 m grid. The DEMs cover the tidewater glaciers Blomstrandbreen, Conwaybreen, Kongsbreen, Kronebreen, and Kongsvegen and are merged with bathymetric and land DEMs for the non-glaciated areas. The large-scale subglacial topography of the study area is characterized by a series of troughs and highs. The minimum subglacial elevation is −180 m a.s.l., the maximum subglacial elevation is 1400 m a.s.l., and the maximum ice thickness is 740 m. We estimate the maximum vertical root-mean-squared uncertainty in the subglacial and ice thickness DEMs to be approximately ±24 m. The statistics for each glacier are summarized in Table 1. For Kronebreen, which has a dense data coverage (Fig. 1), we also present a 50 m gridded subglacial DEM (Fig. 6). In the following paragraphs, we describe the main troughs and highs in the subglacial topography. Overdeepenings in Blomstrandbreen and Conwaybreen (−110 and −60 m a.s.l., respectively; Figs. 5a, 7a, and b) lie behind sills at the glacier front, which limit the extension of the fjord further upglacier; both overdeepenings will therefore likely be either filled with sediments or freshwater, as the glaciers retreat. Further south, Kongsbreen consists of two tributary outlets around the Ossian Sarsfjellet. Kongsbreen North has a deep trough beneath it with a minimum elevation of −180 m a.s.l (Figs. 5a and 7c). The continuation of the fjord (i.e. elevation beneath current sea level) extends 11 km inland from the current front to north of the nunatak Steindolpen, upglacier from Collethøgda. The fjord may possibly connect with Kronebreen in a 500 m wide embayment, with only 10 m deep waters. Kongsbreen South has a sill at its front, with a minimum elevation of 8 m a.s.l.,

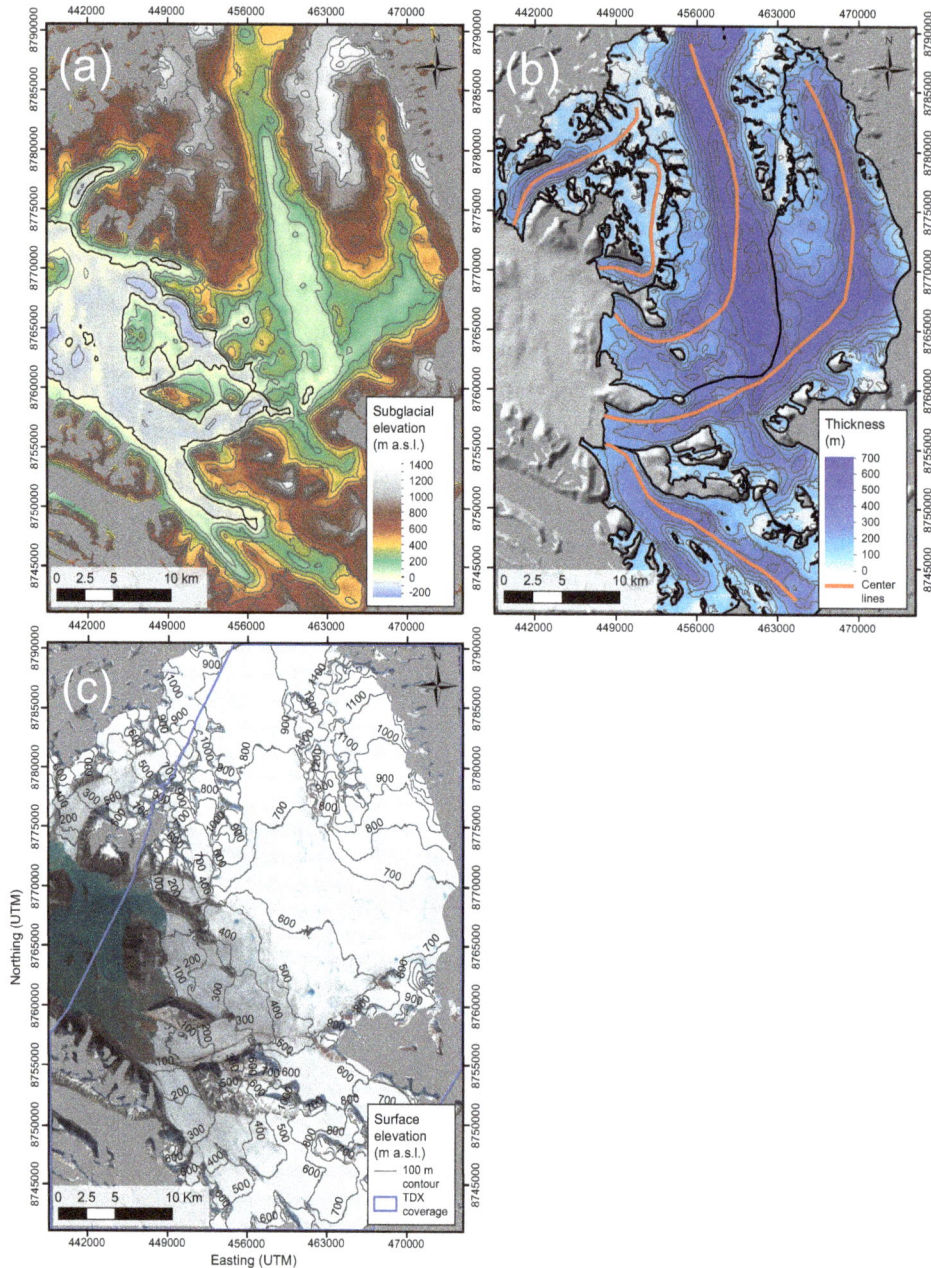

Figure 5. (a) Subglacial, bathymetric, and land elevation with 100 m elevation contours (grey lines). **(b)** Ice thickness in 2014 with 100 m elevation contours (grey lines) and glacier surface elevation catchments (black polygons). Hillshade image in the background from Fig. 5a. Statistics for each glacier are specified in Table 1. **(c)** Surface elevation in 100 m contours showing the extent of the TANDEM-X (TDX) DEM (blue polygon). Background image is Sentinel-2 satellite image taken on 10 July 2016 (Copernicus, 2016). Grey areas in all figures are glaciers not covered in the study. Grid projection is Universal Transverse Mercator Zone 33W.

preventing the embayment in front of the glacier connecting to Kongsbreen North.

Kronebreen, the fastest flowing glacier in the fjord (Schellenberger et al., 2015), has a trough beneath it with a minimum elevation of −130 m a.s.l (Figs. 5a, 6 and 7d). The trough continues 10 km inland, where it ends at a 350 m wide

and 2 km long embayment with 20 m shallow waters. Up-glacier from the embayment there is a steep sill, with a minimum elevation of 130 m a.s.l., which can also be seen at the glacier surface (Fig. 7d), where there is a steep and heavily crevassed ice fall. Further inland, the ice thickens again and there is a small overdeepening with a minimum subglacial el-

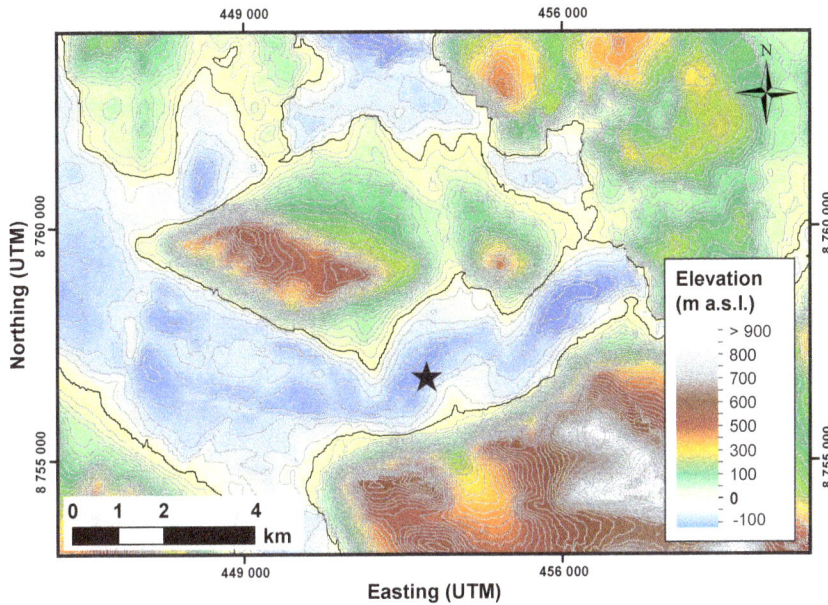

Figure 6. Subglacial, bathymetric, and land elevation of Kronebreen at 50 m gridded resolution with 20 m elevation contours. Location of the borehole study is marked with a black star (How et al., 2017). Grid projection is Universal Transverse Mercator Zone 33W.

Table 1. Statistics for each glacier with subglacial elevation and ice thickness.

Glacier	Subglacial elevation (m a.s.l.)	Ice thickness (m)
Blomstrandbreen	Max: 1190 Min: −110 Mean: 250 SD: 340	Max: 410 Min: 0 Mean: 160 SD: 130
Conwaybreen	Max: 1200 Min: −60 Mean: 340 SD: 400	Max: 320 Min: 0 Mean: 110 SD: 90
Kongsbreen	Max: 1400 Min: −180 Mean: 250 SD: 370	Max: 740 Min: 0 Mean: 330 SD: 190
Kronebreen	Max: 1390 Min: −130 Mean: 160 SD: 290	Max: 580 Min: 0 Mean: 280 SD: 150
Kongsvegen	Max: 1010 Min: −70 Mean: 160 SD: 260	Max: 450 Min: 0 Mean: 190 SD: 120
Total	Max: 1400 Min: −180 Mean: 480 SD: 340	Max: 740 Min: 0 Mean: 280 SD: 180

evation of −30 m a.s.l. The high-resolution subglacial DEM of Kronebreen has so far been used in studies of basal sliding, subglacial hydrology, and calving (How et al., 2017; Vallot et al., 2017, 2018).

The subglacial topography beneath Isachsenfonna consists of a 3 km wide flat valley with a minimum subglacial elevation of 40 m a.s.l (Figs. 5a and 7c). Holtedahlfonna has a 2 km wide valley with higher subglacial elevations, with a minimum elevation of 120 m a.s.l, and gradually higher elevations up on Dovrebreen (Figs. 5a and 7d). Finally, Kongsvegen, flowing in from the southeast towards Kronebreen, has a trough beneath it with a minimum subglacial elevation of −70 m a.s.l., which continues 9 km inland until just north of the nunatak Vorehaugen (Figs. 5a and 7e).

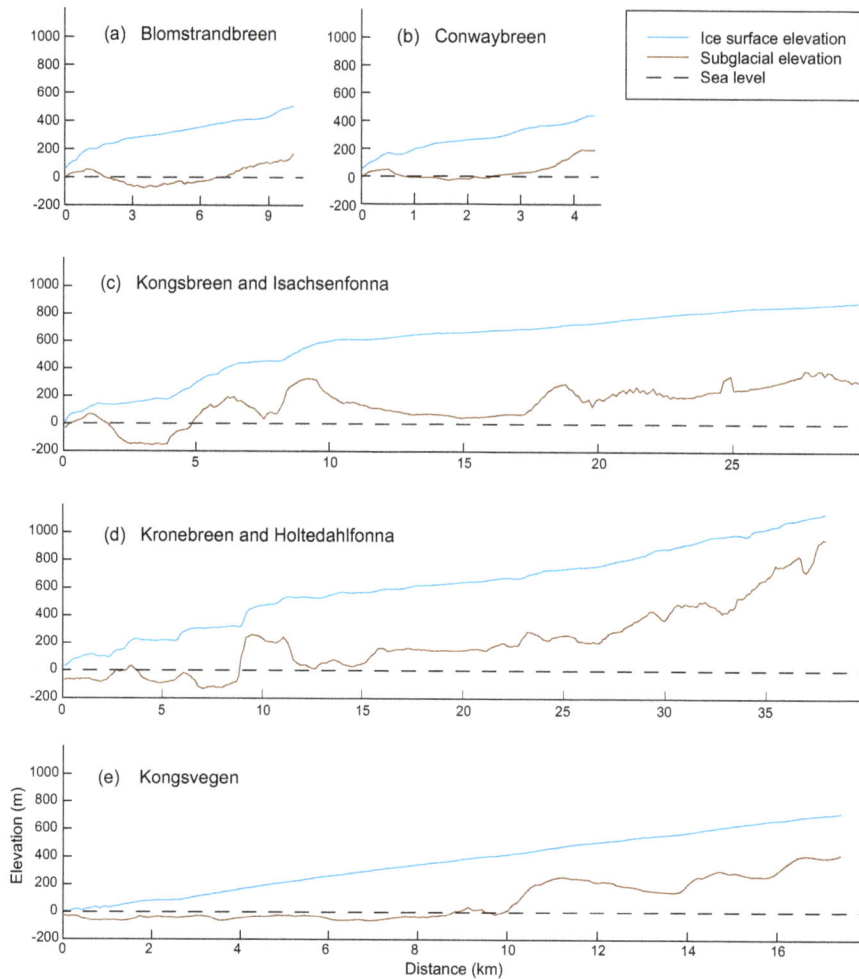

Figure 7. Glacier surface elevation in 2014 (blue line) and subglacial elevation (brown line) along the glacier ice flow centre lines (Fig. 5b) of (**a**) Blomstrandbreen, (**b**) Conwaybreen, (**c**) Kongsbreen and Isachsenfonna, (**d**) Kronebreen and Holtedahlfonna, and (**e**) Kongsvegen. Dashed black line is present-day sea level. Distance is measured from the present-day glacier front. Notice that the *x* axis scale varies between the plots.

Summary

Tidewater glaciers have a major influence on circulation in the water bodies in which they sit, particularly in constricted bays or fjords. In this study, we produced subglacial topography and ice thickness DEMs of five tidewater glaciers in Kongsfjorden on a 150 m grid. We also produced a 50 m gridded resolution DEM for Kronebreen, where the data coverage is dense. The subglacial elevation and ice thickness data consist of ~ 1700 km common-offset radar profiles collected in 2004 to 2016 with an impulse radar system that was either suspended under a helicopter (for crevassed areas) or towed behind a snowmobile. We combined the data sets of subglacial elevation with land elevation and bathymetry elevation DEMs to a final gridded DEM. The large-scale subglacial topography of the study area is characterized by a series of troughs and highs, where the glaciers Kongsbreen, Kronebreen, and Kongsvegen have the potential to retreat by ~ 10 km before they become land-terminating. The compiled data set covers one of the most studied regions in Svalbard and is valuable for future studies of glacier dynamics, geology, hydrology, and fjord circulation.

Author contributions. KaL was primarily responsible for collecting, processing, and analysing the data and prepared the paper with contributions from all co-authors. JK was the project leader and was the main responsible for fieldwork and data management. RP was primarily responsible for the radar system used in 2014 to 2016. KaL, JK, AM, and DV collected the radar data in the field (campaigns 2014 to 2016). CN provided the surface TanDEM-X data set. KiL, KM, and OB provided radar data from the earlier campaigns (2004 to 2010).

Competing interests. The authors declare that they have no conflict of interest.

Acknowledgements. This work was part of the TIGRIF (Tidewater Glacier Retreat Impact on Fjord circulation and ecosystems) project, funded by the Research Council of Norway, Oceans and Coastal Areas Programme (project 243808). Funding has also been provided by the GLAERE project (the Polish-Norwegian Research Programme) and TW-ICE projects (Centre for Ice, Climate, and Ecosystems of the Norwegian Polar Institute). Field support was given from the Swedish Society for Anthropology and Geography (SSAG), Svalbard Science Forum (SSF; RIS no. 6660) and the Nordic Centre of Excellence SVALI. We would also like to thank Geir Gunleiksrud and Boele Kuipers for providing the bathymetric DEM. Christopher Nuth acknowledges funding from European Union, through the ERC (grant no. 320816) and ESA (project Glaciers_CCI, 4000109873/14/I-NB). The TanDEM-X DEM and IDEM were provided by the German Space Agency (DLR) satellites TerraSAR-X and TanDEM-X (proposals XTI_GLAC6716 and IDEM_GLAC0435). We thank Neil Ross, an anonymous referee, and editor Reinhard Drews for reviewing the paper. We also thank Edward King and Bryn Hubbard for reviewing an earlier version of the paper.

Edited by: Reinhard Drews

References

Bamber, J. L., Griggs, J. A., Hurkmans, R. T. W. L., Dowdeswell, J. A., Gogineni, S. P., Howat, I., Mouginot, J., Paden, J., Palmer, S., Rignot, E., and Steinhage, D.: A new bed elevation dataset for Greenland, The Cryosphere, 7, 499–510, https://doi.org/10.5194/tc-7-499-2013, 2013.

Beaudon, E., Moore, J. C., Pohjola, V. A., van de Wal, R. S. W., Kohler, J., and Isaksson, E.: Lomonosovfonna and Holtedahlfonna ice cores reveal east–west disparities of the Spitsbergen environment since AD 1700, J. Glaciol., 59, 1069–1083, https://doi.org/10.3189/2013JoG12J203, 2013.

Bennett, M. R., Hambrey, M. J., Huddart, D., Glasser, N. F., and Crawford, K.: The landform and sediment assemblage produced by a tidewater glacier surge in Kongsfjorden, Svalbard, Quaternary Sci. Rev., 18, 1213–1246, https://doi.org/10.1016/S0277-3791(98)90041-5, 1999.

Błaszczyk, M., Jania, J. A., and Hagen, J. O.: Tidewater glaciers of Svalbard: Recent changes and estimates of calving fluxes, Pol. Polar Res., 30, 85–142, 2009.

Blindow, N., Salat, C., and Casassa, G.: Airborne GPR sounding of deep temperate glaciers – Examples from the Northern Patagonian Icefield, in: 2012 14th International Conference on Ground Penetrating Radar (GPR), Shanghai, China, 4–8 June 2012, IEEE, 664–669, https://doi.org/10.1109/ICGPR.2012.6254945, 2012.

Burton, D. J., Dowdeswell, J. A., Hogan, K. A., and Noormets, R.: Marginal Fluctuations of a Svalbard Surge-Type Tidewater Glacier, Blomstrandbreen, Since the Little Ice Age: A Record of Three Surges, Arct. Antarct. Alp. Res., 48, 411–426, https://doi.org/10.1657/AAAR0014-094, 2016.

Christianson, K., Kohler, J., Alley, R. B., Nuth, C., and Van Pelt, W. J. J.: Dynamic perennial firn aquifer on an Arctic glacier, Geophys. Res. Lett., 42, 1418–1426, https://doi.org/10.1002/2014GL062806, 2015.

Copernicus: Copernicus Sentinel Data, available at: https://scihub.copernicus.eu/, last access: 1 June 2016.

Dowdeswell, J. A. and Evans, S.: Investigations of the form and flow of ice sheets and glaciers using radio-echo sounding, Rep. Prog. Phys., 67, 1821–1861, https://doi.org/10.1088/0034-4885/67/10/R03, 2004.

Drewry, D. J.: Seismic-gravity ice thickness measurements in East Antarctica, J. Glaciol., 15, 137–150, 1975.

Fretwell, P., Pritchard, H. D., Vaughan, D. G., Bamber, J. L., Barrand, N. E., Bell, R., Bianchi, C., Bingham, R. G., Blankenship, D. D., Casassa, G., Catania, G., Callens, D., Conway, H., Cook, A. J., Corr, H. F. J., Damaske, D., Damm, V., Ferraccioli, F., Forsberg, R., Fujita, S., Gim, Y., Gogineni, P., Griggs, J. A., Hindmarsh, R. C. A., Holmlund, P., Holt, J. W., Jacobel, R. W., Jenkins, A., Jokat, W., Jordan, T., King, E. C., Kohler, J., Krabill, W., Riger-Kusk, M., Langley, K. A., Leitchenkov, G., Leuschen, C., Luyendyk, B. P., Matsuoka, K., Mouginot, J., Nitsche, F. O., Nogi, Y., Nost, O. A., Popov, S. V., Rignot, E., Rippin, D. M., Rivera, A., Roberts, J., Ross, N., Siegert, M. J., Smith, A. M., Steinhage, D., Studinger, M., Sun, B., Tinto, B. K., Welch, B. C., Wilson, D., Young, D. A., Xiangbin, C., and Zirizzotti, A.: Bedmap2: improved ice bed, surface and thickness datasets for Antarctica, The Cryosphere, 7, 375–393, https://doi.org/10.5194/tc-7-375-2013, 2013.

Hagen, J. O. and Sætrang, A.: Radio-echo soundings of sub-polar glaciers with low-frequency radar, Polar Res., 9, 99–107, 1991.

Hagen, J. O., Liestøl, O., Roland, E., and Jorgensen, T.: Glacier atlas of Svalbard and Jan Mayen, vol. 129, Norsk Polarinstitutt Middelelser, Oslo, 1993.

Hagen, J. O., Melvold, K., and Dowdeswellt, J. A.: On the Net Mass Balance of the Glaciers and Ice Caps in Svalbard, Norwegian Arctic, Arct. Antarct. Alp. Res., 35, 264–270, 2003.

How, P., Benn, D. I., Hulton, N. R. J., Hubbard, B., Luckman, A., Sevestre, H., van Pelt, W. J. J., Lindbäck, K., Kohler, J., and Boot, W.: Rapidly changing subglacial hydrological pathways at a tidewater glacier revealed through simultaneous observations of water pressure, supraglacial lakes, meltwater plumes and surface velocities, The Cryosphere, 11, 2691–2710, https://doi.org/10.5194/tc-11-2691-2017, 2017.

Howe, J. A., Moreton, S. G., Morri, C., and Morris, P.: Multibeam bathymetry and the depositional environments of Kongsfjorden and Krossfjorden, western Spitsbergen, Svalbard, Polar Res., 22, 301–316, https://doi.org/10.1111/j.1751-8369.2003.tb00114.x, 2003.

Ingólfsson, Ó. and Landvik, J. Y.: The Svalbard-Barents Sea ice-sheet – Historical, current and future perspectives, Quaternary Sci. Rev., 64, 33–60, https://doi.org/10.1016/j.quascirev.2012.11.034, 2013.

Irving, J. D., Knoll, M. D., and Knight, R. J.: Improving cross-hole radar velocity tomograms: A new approach to incorporating high-angle traveltime data, Geophysics, 72, 31–41, https://doi.org/10.1190/1.2742813, 2007.

Isaaks, E. H. and Srivastava, R. M.: An Introduction to Applied Geostatistics, Oxford University Press, New York, New York, 1989.

Ito, H. and Kudoh, S.: Characteristics of water in Kongsfjorden, Svalbard, Proceedings of the NIPR Symposium on Polar Meteorology and Glaciology, 11, 211–232, 1997.

Kartverket: Sjø terrengmodeller DTM 50, available at: https://kartkatalog.geonorge.no/metadata/uuid/ 67a3a191-49cc-45bc-baf0-eaaf7c513549, last access: 1 August 2018.

Kennett, M., Laumann, T., and Lund, C.: Helicopter-borne radio-echo sounding of Svartisen, Norway, Ann. Glaciol., 17, 23–26, 1993.

Kierulf, H. P., Plag, H. P., and Kohler, J.: Surface deformation induced by present-day ice melting in Svalbard, Geophys. J. Int., 179, 1–13, https://doi.org/10.1111/j.1365-246X.2009.04322.x, 2009.

Kohler, J., James, T. D., Murray, T., Nuth, C., Brandt, O., Barrand, N. E., and Aas, H. F.: Acceleration in thinning rate on western Svalbard glaciers, Geophys. Res. Lett., 34, L18502, https://doi.org/10.1029/2007GL030681, 2007.

König, M., Nuth, C., Kohler, J., Moholdt, G., and Pettersen, R.: A Digital Glacier Database for Svalbard, in: Global Land Ice Measurements from Space, edited by: Kargel, J., Leonard, G., Bishop, M., Kääb, A., and Raup, B., Springer Praxis Books. Springer, Berlin, Heidelberg, 229–239, 2014.

Kovacs, K. M. and Michel, C.: Biological impacts of changes in sea ice in the Arctic, in Snow, Water, Ice and Permafrost in the Arctic (SWIPA): Climate Change and the Cryosphere, AMAP, Oslo, Norway, chap. 9.3, 32–51, 2011.

Kovacs, K. M., Lydersen, C., Overland, J. E., and Moore, S. E.: Impacts of changing sea-ice conditions on Arctic marine mammals, Mar. Biodivers., 41, 181–194, https://doi.org/10.1007/s12526-010-0061-0, 2011.

Landvik, J. Y., Ingólfsson, Ó., Mienert, J., Lehman, S. J., Solheim, A., Elverhøi, A., and Ottesen, D.: Rethinking Late Weichselian ice-sheet dynamics in coastal NW Svalbard, Boreas, 34, 7–24, https://doi.org/10.1111/j.1502-3885.2005.tb01001.x, 2005.

Langhammer, L., Rabenstein, L., Bauder, A., and Maurer, H.: Ground-penetrating radar antenna orientation effects on temperate mountain glaciers, Geophysics, 82, H15–H24, https://doi.org/10.1190/geo2016-0341.1, 2017.

Lapazaran, J. J., Otero, J., and Navarro, F. J.: On the errors involved in ice-thickness estimates I: ground-penetrating radar measurement errors, J. Glaciol., 62, 1008–1020, https://doi.org/10.1017/jog.2016.93, 2016.

Liestøl, O.: The glaciers in the Kongsfjorden area, Spitsbergen, Norsk Geogr. Tidsskr., 42, 231–238, https://doi.org/10.1080/00291958808552205, 1988.

Lindbäck, K., Pettersson, R., Doyle, S. H., Helanow, C., Jansson, P., Kristensen, S. S., Stenseng, L., Forsberg, R., and Hubbard, A. L.: High-resolution ice thickness and bed topography of a land-terminating section of the Greenland Ice Sheet, Earth Syst. Sci. Data, 6, 331–338, https://doi.org/10.5194/essd-6-331-2014, 2014.

Lindbäck, K., Kohler, J., Pettersson, R., Nuth, C., Langley, K., Messerli, A., Vallot, D., Matsuoka, K., and Brandt, O.: Subglacial topography, ice thickness, and bathymetry of Kongsfjor-

den, northwestern Svalbard, [Data set], Norwegian Polar Institute, https://doi.org/10.21334/npolar.2017.702ca4a7, 2018.

Loeng, H., Brander, K., Carmack, E., Denisenko, S., Drinkwater, K., Hansen, B., Kovacs, K., Livingston, P., McLaughlin, F., Sakshaug, E., Bellerby, R., Browman, H., Furevik, T., Grebmeier, J. M., Jansen, E., Jónsson, S., Lindal Jørgensen, L., Malmberg, S.-A., Østerhus, S., Ottersen, G., and Shimada, K.: Marine Systems, in: Arctic Climate Impact Assessment, Cambridge University Press, Cambridge, UK, chap. 9, 453–538, 2005.

Lydersen, C., Assmy, P., Falk-Petersen, S., Kohler, J., Kovacs, K. M., Reigstad, M., Steen, H., Strøm, H., Sundfjord, A., Varpe, Ø., Walczowski, W., Marcin, J., and Zajaczkowski, M.: The importance of tidewater glaciers for marine mammals and seabirds in Svalbard, Norway, J. Marine Syst., 129, 452–471, https://doi.org/10.1016/j.jmarsys.2013.09.006, 2014.

Mansell, D., Luckman, A., and Murray, T.: Dynamics of tidewater surge-type glaciers in northwest Svalbard, J. Glaciol., 58, 110–118, https://doi.org/10.3189/2012JoG11J058, 2012.

Melvold, K. and Hagen, J. O.: Evolution of a surge-type glacier in its quiescent phase: Kongsvegen, Spitsbergen, 1964–95, J. Glaciol., 44, 394–404, 1998.

Miller, G. H., Sejrup, H. P., Lehman, S. J., and Forman, S. L.: Glacial history and marine environmental change during the last interglacial-glacial cycle, western Spitsbergen, Svalbard, Boreas, 18, 273–296, https://doi.org/10.1111/j.1502-3885.1989.tb00403.x, 1989.

Moholdt, G., Nuth, C., Ove, J., and Kohler, J.: Recent elevation changes of Svalbard glaciers derived from ICE-Sat laser altimetry, Remote Sens. Environ., 114, 2756–2767, https://doi.org/10.1016/j.rse.2010.06.008, 2010.

Moreira, A., Krieger, G., Hajnsek, I., Hounam, D., Werner, M., Riegger, S. and Settelmeyer, E.: TanDEM-X: a TerraSAR-X add-on satellite for single-pass SAR interferometry, in: IGARSS 2004. 2004 IEEE International Geoscience and Remote Sensing Symposium, Anchorage, AK, USA, 20–24 September 2004, IEEE, 2, 1000–1003, 2004.

Morlighem, M., Williams, C. N., Rignot, E., An, L., Arndt, J. E., Bamber, J. L., Catania, G., Chauché, N., Dowdeswell, J. A., Dorschel, B., Fenty, I., Hogan, K., Howat, I., Hubbard, A., Jakobsson, M., Jordan, T. M., Kjeldsen, K. K., Millan, R., Mayer, L., Mouginot, J., Noël, B. P. Y., O'Cofaigh, C., Palmer, S., Rysgaard, S., Seroussi, H., Siegert, M. J., Slabon, P., Straneo, F., van den Broeke, M. R., Weinrebe, W., Wood, M., and Zinglersen, K. B.: BedMachine v3: Complete Bed Topography and Ocean Bathymetry Mapping of Greenland From Multibeam Echo Sounding Combined With Mass Conservation, Geophys. Res. Lett., 44, 11051–11061, https://doi.org/10.1002/2017GL074954, 2017.

Narod, B. B. and Clarke, G. K. C.: Instruments and Methods. Miniature high-power impulse transmitter for radio-echo sounding, J. Glaciol., 40, 190–194, 1994.

Natural Resources Canada: CSRS-PPP: On-Line GNSS PPP Post-Processing Service, available at: http://webapp.geod.nrcan.gc.ca/ geod/tools-outils/ppp.php (last access: 1 August 2018), 2017.

Neckel, N., Braun, A., Kropácek, J., and Hochschild, V.: Recent mass balance of the Purogangri Ice Cap, central Tibetan Plateau, by means of differential X-band SAR interferometry, The Cryosphere, 7, 1623–1633, https://doi.org/10.5194/tc-7-1623-2013, 2013.

Norwegian Polar Institute: Terrengmodell Svalbard (S0 Terreng-modell), https://doi.org/10.21334/npolar.2014.dce53a47, 2014.

Nuth, C., Moholdt, G., Kohler, J., Hagen, J. O., and Kääb, A.: Svalbard glacier elevation changes and contribution to sea level rise, J. Geophys. Res., 115, F01008, https://doi.org/10.1029/2008JF001223, 2010.

Nuth, C., Kohler, J., König, M., von Deschwanden, A., Hagen, J. O., Kääb, A., Moholdt, G., and Pettersson, R.: Decadal changes from a multi-temporal glacier inventory of Svalbard, The Cryosphere, 7, 1603–1621, https://doi.org/10.5194/tc-7-1603-2013, 2013.

Ottesen, D., Dowdeswell, J. A., and Rise, L.: Submarine landforms and the reconstruction of fast-flowing ice streams within a large Quaternary ice sheet: The 2500-km-long Norwegian-Svalbard margin (57°–80° N), Geol. Soc. Am. Bull., 117, 1033–1050, 2005.

Overland, J. E., Spillane, M. C., Percival, D. B., Wang, M., and Mofjeld, H. O.: Seasonal and Regional Variation of Pan-Arctic Surface Air Temperature over the Instrumental Record, J. Climate, 17, 3263–3282, 2004.

Rankl, M. and Braun, M.: Glacier elevation and mass changes over the central Karakoram region estimated from TanDEM-X and SRTM/X-SAR digital elevation models, Ann. Glaciol., 57, 273–281, https://doi.org/10.3189/2016AoG71A024, 2016.

Rutishauser, A., Maurer, H., and Bauder, A.: Helicopter-borne ground-penetrating radar investigations on temperate alpine glaciers: A comparison of different systems and their abilities for bedrock mapping, Geophysics, 81, WA119–WA129, https://doi.org/10.1190/geo2015-0144.1, 2016.

Schellenberger, T., Dunse, T., Kääb, A., Kohler, J., and Reijmer, C. H.: Surface speed and frontal ablation of Kronebreen and Kongsbreen, NW Svalbard, from SAR offset tracking, The Cryosphere, 9, 2339–2355, https://doi.org/10.5194/tc-9-2339-2015, 2015.

Seligman, G.: Snow structure and ski fields: being an account of snow and ice forms met with in nature and a study on avalanches & snowcraft, Macmillan, London, 1936.

Serreze, M. C. and Barry, R. G.: Processes and impacts of Arctic amplification: A research synthesis, Global Planet. Change, 77, 85–96, https://doi.org/10.1016/j.gloplacha.2011.03.004, 2011.

Sjögren, B., Brandt, O., Nuth, C., Isaksson, E., Pohjola, V., Kohler, J., and van de Wal, R. S. W.: Instruments and Methods Determination of firn density in ice cores using image analysis, J. Glaciol., 53, 1–7, 2007.

Stolt, R. H.: Migration by Fourier Transform, Geophysics, 43, 23–48, 1978.

Svendsen, H., Beszczynska-møller, A., Hagen, J. O., Lefauconnier, B., Tverberg, V., Gerland, S., Ørbæk, J. B., Bischof, K., Papucci, C., Zajaczkowski, M., Azzolini, R., Bruland, O., Wiencke, C., Winther, J., and Dallmann, W.: The physical environment of Kongsfjorden–Krossfjorden, an Arctic fjord system in Svalbard, Polar Res., 21, 133–166, 2002.

Taylor, J. R.: An Introduction to Error Analysis: The Study of Uncertainties in Physical Measurements, 2nd edn., University Science Books, 1996.

Vallot, D., Pettersson, R., Luckman, A., Benn, D. I., Zwinger, T., Van Pelt, W. J. J., Kohler, J., Schäfer, M., Claremar, B., and Hulton, N. R. J.: Basal dynamics of Kronebreen, a fast-flowing tidewater glacier in Svalbard: Non-local spatio-temporal response to water input, J. Glaciol., 63, 1012–1024, https://doi.org/10.1017/jog.2017.69, 2017.

Vallot, D., Åström, J., Zwinger, T., Pettersson, R., Everett, A., Benn, D. I., Luckman, A., van Pelt, W. J. J., Nick, F., and Kohler, J.: Effects of undercutting and sliding on calving: a global approach applied to Kronebreen, Svalbard, The Cryosphere, 12, 609–625, https://doi.org/10.5194/tc-12-609-2018, 2018.

van Pelt, W. J. J., Oerlemans, J., Reijmer, C. H., Pohjola, V. A., Pettersson, R., and van Angelen, J. H.: Simulating melt, runoff and refreezing on Nordenskiöldbreen, Svalbard, using a coupled snow and energy balance model, The Cryosphere, 6, 641–659, https://doi.org/10.5194/tc-6-641-2012, 2012.

Wessel, B.: TanDEM-X Ground Segment – DEM Products Specification Document, Oberpfaffenhofen, Germany, available at: https://elib.dlr.de/108014/1/TD-GS-PS-0021_DEM-Product-Specification_v3.1.pdf (last access: 1 August 2018), 2016.

Woodward, J., Murray, T., and McCaig, A.: Formation and reorientation of structure in the surge-type glacier Kongsvegen, Svalbard, J. Quaternary Sci., 17, 201–209, https://doi.org/10.1002/jqs.673, 2002.

Woodward, J., Murray, T., Clark, R. A., and Stuart, G. W.: Glacier surge mechanisms inferred from ground-penetrating radar: Kongsvegen, Svalbard, J. Glaciol., 49, 473–480, https://doi.org/10.3189/172756503781830458, 2003.

Wouters, B., Chambers, D., and Schrama, E. J. O.: GRACE observes small-scale mass loss in Greenland, Geophys. Res. Lett., 35, L20501, https://doi.org/10.1029/2008GL034816, 2008.

10

Meteorological and snow distribution data in the Izas Experimental Catchment (Spanish Pyrenees) from 2011 to 2017

Jesús Revuelto[1,2], Cesar Azorin-Molina[1,3], Esteban Alonso-González[1], Alba Sanmiguel-Vallelado[1], Francisco Navarro-Serrano[1], Ibai Rico[1,4], and Juan Ignacio López-Moreno[1]

[1]Pyrenean Institute of Ecology, CSIC, Zaragoza, Spain
[2]Météo-France – CNRS, CNRM (UMR3589), Centre d'Etudes de la Neige, Grenoble, France
[3]Regional Climate Group, Department of Earth Sciences, University of Gothenburg, Gothenburg, Sweden
[4]University of the Basque Country. Department of Geography, Prehistory and Archaeology, Vitoria, Spain

Correspondence: Jesús Revuelto (jesus.revuelto@meteo.fr)

Abstract. This work describes the snow and meteorological data set available for the Izas Experimental Catchment in the Central Spanish Pyrenees, from the 2011 to 2017 snow seasons. The experimental site is located on the southern side of the Pyrenees between 2000 and 2300 m above sea level, covering an area of 55 ha. The site is a good example of a subalpine environment in which the evolution of snow accumulation and melt are of major importance in many mountain processes. The climatic data set consists of (i) continuous meteorological variables acquired from an automatic weather station (AWS), (ii) detailed information on snow depth distribution collected with a terrestrial laser scanner (TLS, lidar technology) for certain dates across the snow season (between three and six TLS surveys per snow season) and (iii) time-lapse images showing the evolution of the snow-covered area (SCA). The meteorological variables acquired at the AWS are precipitation, air temperature, incoming and reflected solar radiation, infrared surface temperature, relative humidity, wind speed and direction, atmospheric air pressure, surface temperature (snow or soil surface), and soil temperature; all were taken at 10 min intervals. Snow depth distribution was measured during 23 field campaigns using a TLS, and daily information on the SCA was also retrieved from time-lapse photography. The data set (https://doi.org/10.5281/zenodo.848277) is valuable since it provides high-spatial-resolution information on the snow depth and snow cover, which is particularly useful when combined with meteorological variables to simulate snow energy and mass balance. This information has already been analyzed in various scientific studies on snow pack dynamics and its interaction with the local climatology or topographical characteristics. However, the database generated has great potential for understanding other environmental processes from a hydrometeorological or ecological perspective in which snow dynamics play a determinant role.

1 Introduction

Snowpack distribution and its temporal evolution have a marked influence on many mountain processes. These include erosion rates and sediment transport (Colbeck et al., 1979; Lana-Renault et al., 2011) and geomorphological and glaciological processes (López-Moreno et al., 2017; Serrano et al., 2001). Phenological cycles (Liston, 1999; Wipf et al., 2009) are directly controlled by the evolution of snow cover over time. Conversely, snowmelt dynamics are also of major importance from a hydrological perspective since one-sixth of the Earth's total population depends on the water storage in mountain river headwaters (Barnett et al., 2005). In downstream areas exposed to extreme climatic conditions, the snowmelt runoff from mountain areas becomes a key element (Viviroli et al., 2007), especially in zones affected by water shortages. This is the case of semiarid regions, like the Mediterranean area, which are characterized by an irregular climate with long drought periods (Vicente-Serrano, 2006), and therefore are highly dependent on water stored in mountain areas, such as the Pyrenees (López-Moreno, 2005; López-Moreno et al., 2008).

The Pyrenees are a midlatitude mountain range, with significant snowfalls in the high-elevation areas throughout the year. During the spring, Pyrenean river discharges depend on the snowmelt timing, with approximately 40 % of spring runoff being directly attributable to snow (López-Moreno and García-Ruiz, 2004). Thus, snow accumulation has a heavy influence on Pyrenean headwaters. This dependence is mostly due to the generally continuous snow cover from November to April above 2000 m above sea level (a.s.l.) (Alvera and Garcia-Ruiz, 2000; García-Ruiz et al., 1986; López-Moreno et al., 2002) and, therefore, the study of the snowpack at high elevations in the Pyrenees is crucial for understanding and managing mountain river discharges (López-Moreno, 2005), especially in the scenario of global climate change (García-Ruiz et al., 2011). However, continuous snow observations above 2000 m a.s.l. are scarce in this mountain range since most only have information from 1600 to 2000 m a.s.l. and those that are available only cover short time spans. Therefore, well-established study areas at high elevations with continuous measurements of meteorological variables and snowpack distribution are required in the Pyrenees.

This paper presents the recently acquired data set of meteorological and snowpack variables obtained from a small experimental catchment on the southern face of the Pyrenees. Although meteorological and hydrological data are available from previous years (some variables have been measured since the late 1980s; Alvera and Garcia-Ruiz, 2000), we present data from the 2011/12 to 2016/17 snow seasons, as data series provide higher quality and continuity, and they also match in situ observations of snow depth and snow cover. The data set consists of (i) continuous meteorological variables acquired from an automatic weather station (AWS), (ii) detailed information on snow depth distribution collected with a terrestrial laser scanner (TLS, lidar technology) for certain dates across the snow season (between two and six TLS surveys per snow season) and (iii) time-lapse images showing the snow-covered area (SCA) evolution. Some years of this data set have already been used to study the topographic control on snow depth distribution (Revuelto et al., 2014b), the spatial variability in snowpack at different distances (López-Moreno et al., 2012) or to investigate how detailed snowpack simulation could be improved by including snow distribution information (Revuelto et al., 2016a, b).

The paper is structured as follows: Sect. 2 describes the study area characteristics; Sect. 3 presents meteorological data acquired from the AWS with a general description of the observed climatology; Sect. 4 describes the distributed measurements on snow depth distribution from the TLS and

the SCA derived from time-lapse images; Sect. 5 concludes with information for downloading the database; and finally Sect. 6 summarizes all information available and the potential application of the database.

2 Study area characteristics and climatology

2.1 The Pyrenees

The Pyrenees lie on the northeastern border of the Iberian Peninsula (Fig. 1) and form an orographic barrier between the north and south faces. Due to this, progressively higher aridity is found toward the south as the mountain range blocks humid air masses from the Atlantic (López-Moreno and Vicente-Serrano, 2007; Vicente-Serrano, 2005). Thus, the natural barrier directly influences precipitation, leading to areas above 2000 m a.s.l. receiving about 2000 mm year^{-1}, increasing to 2500 mm year^{-1} in the highest divides of the mountain range and rapidly decreasing to 600–800 mm year^{-1} in low-elevation areas on the southern side (García-Ruiz, et al., 2001).

Another distinct feature of the Pyrenees is their location between two water masses with contrasting conditions, i.e., the Atlantic Ocean is on the west side, while the Mediterranean Sea lies in the east. This position between both water masses causes a climatic transition from oceanic to mediterranean conditions in the east. During autumn, fronts approaching from the Atlantic bring the highest monthly averages of precipitation in the western observatories, with their total contribution accounting for 40 % of total annual precipitation in this area (Creus-Novau, 1983). Conversely, spring and summer storms mostly affect the eastern areas of the Pyrenees, promoted by the development of zones in which sea breezes and local winds converge to initiate deep moist convection along the eastern fringe of the Iberian Mediterranean area (Azorin-Molina et al., 2015). Therefore, Pyrenean observatories in the east record a large number of convective events, i.e up to 32 % of total annual precipitation in eastern valleys, but dropping below 16 % of annual precipitation in western valleys (Cuadrat et al., 2007). In early winter, the arrival of fronts from the northwest and west are the most frequent, leading to the highest snow accumulation found in the western Pyrenees (Navarro-Serrano and López-Moreno, 2017). The Azores high, which usually affects the Iberian Peninsula at certain times in the winter, gives rise to relatively long periods with no snow accumulation in this season. Subsequently, in spring, snow accumulation is associated with southwesterly advections, which lead to heavy snow accumulations in the western Pyrenees (Revuelto et al., 2012). Snow remains for long periods above 1600 m a.s.l., between November and April (López-Moreno and Nogués-Bravo, 2006).

Similar to precipitation, air temperature is influenced by the Atlantic–Mediterranean transitions, but elevation plays a major role in its distribution. For instance, the lower annual

Figure 1. The Izas Experimental Catchment study site. Panel **(a)** shows the location of the study site. Panel **(b)** shows an overview of the catchment with marginal snow presence. The right map **(c)** shows the topographic characteristics of the catchment and the location of the TLS scanning positions (Scan stations), the meteorological station and the field of view of the time-lapse camera (continuous lines from Scan station 1).

thermal amplitude observed in the western Pyrenees is because of the proximity of the ocean (Cuadrat et al., 2007). As a general tendency in the Central Pyrenees, the annual 0 °C isotherms lie between 2700 and 2900 m a.s.l. (del Barrio et al., 1990; Chueca, 1993).

Additionally the Pyrenees exhibit a high interannual variability in air temperature and precipitation, which makes the annual snow accumulation very uncertain (López-Moreno, 2005). This variability is influenced by the interannual variability in atmospheric circulation, with a decrease in snow accumulation weather types being identified under positive North Atlantic Oscillation (NAO) phases (López-Moreno and Vicente-Serrano, 2007). As observed with precipitation, snow accumulation correlates to Atlantic–Mediterranean proximity and distance from the main divide of the mountain range (Revuelto et al., 2012), and it is strongly dependent on the fluctuations of the 0 °C isotherm during winter and spring. This high climatic variability is also the cause of large interannual variability in total snow accumulation and its temporal distribution across the snow season (López-Moreno, 2005).

2.2 The Izas Experimental Catchment

The Izas Experimental Catchment (42°44′ N, 0°25′ W) has a surface area of 33 ha, but snow depth information covers a total of 55 ha, with elevations ranging between 2075 and 2325 m a.s.l. This area is close to the main divide of the Pyrenees in the headwaters of the Gállego River, near the

Spain–France border (Fig. 1). The Izas Experimental Catchment exemplifies the general characteristics of subalpine areas of the Pyrenees. In this environment, snowpack dynamics are of major importance throughout the year. Thus, the atmosphere–snowpack interactions observed at this experimental site will enable a better understanding of many processes in subalpine areas.

The mean annual precipitation is 2000 mm, and snow accounts for approximately 50 % of total precipitation (Anderton et al., 2004). For an average of 130 days each year the mean daily air temperature is below 0 °C, with a mean annual air temperature of 3 °C (del Barrio et al., 1997). Snow covers a high percentage of the catchment from November to the end of May (López-Moreno et al., 2010). Lithology shows limestones and sandstones of the Cretaceous period, and limestones of the Paleocene, much more resistant to erosion. The zonal vegetation type corresponds to a high mountain steppe, mainly covered by bunch grasses, namely *Festuca eskia*, *Nardus stricta*, *Trifolium alpinum*, *Plantago alpine* and *Carex sempervirens*. Rocky outcrops dominate the upper and steeper slopes (less than 15 % of the study area). There are no trees present in the study area. The catchment is predominantly east-facing, with some areas also facing north or south. The mean slope of the catchment is 16° (López-Moreno et al., 2012), with the topographic characteristics displaying the typical high spatial heterogeneity of subalpine areas, with flat concave and convex areas.

Figure 2. Pictures of the experimental site equipment. (**a**) AWS sensors. 1A: Young wind sensor; 2A: radiation shield with HMP 155 humidity and temperature probe; 3A: BP1 air pressure recorder; 4A: IR100 infrared remote temperature sensor; 5A: CMA6 Kipp & Zonen albedometer; 6A: SR50A range sensor; 7A: Geonor T-200B with wind shield; 8A: CR3000 data logger and modem; 9A: solar panel and battery; 10A: Campbell Scientific 107 ground temperature probes. (**b**) RIEGL LPM-321 TLS mounted on the tripod during an acquisition campaign. The upper-right part shows one of the 12 fixed reflective targets fixed on the terrain. (**c**) Campbell CC640 camera mounted in the metal structure with 1C: digital camera inside the enclosure house; 2C: modem; 3C: protection glass of the digital camera; and 4C: frontal view of the camera and its structure.

3 Meteorological data

The study site is equipped with an AWS located in the lower elevation of the catchment (42°44′33.65″ N, 0°25′8.83″ W, 2113 m a.s.l.; Fig. 1), located in a flat open area with sparse vegetation (mountain pastures). The AWS measures wind speed and direction, atmospheric air temperature, relative humidity and air pressure, soil temperature for 0, 5, 10, and 20 cm, temperature of the surface close to the AWS (snow or soil, depending on whether snow is present or not), global and reflected solar irradiance, snow depth, and precipitation (the precipitation gauge is located at 15 m from the AWS tower) (see Fig. 2). Information on the main atmospheric variables has been recorded since the end of 2011 (AWS installed in November 2011). Therefore, data availability covers five complete snow seasons. Since the station is located in the lower elevation of the catchment and despite air temperature lapse rate with elevation, the AWS records serve to describe the evolution of atmospheric variables occurring at the Izas Experimental Catchment.

The data acquisition system consists of a Campbell Scientific CR3000 data logger that samples each instrument and stores data at 10 min time intervals. All data are transmitted via modem to the Pyrenean Institute of Ecology where automatic quality-control checks are applied to remove outliers. Data gaps are rare for almost all variables and, therefore, instead of gap filling with interpolation methods, only measured data are available. However, some variables had long data gaps and certain periods have been discarded from further analysis. This is the case of precipitation for the first three snow seasons, which were useless because of the length of data gaps.

Since the main application of the data collected by the AWS is to assess the evolution of snow cover in the study area, in the following subsections we focus our analyses on the accumulation and melt periods, i.e., accumulation (January, February and March; JFM) and melt (April, May and June; AMJ). Annual values observed during a whole snow season are also presented for each subsection.

3.1 Wind speed and direction

The AWS is equipped with a Young wind monitor – Alpine model (Young Company, 2010), placed at the highest point of the meteorological tower (8 m above the ground). The Pyrenees are commonly affected by strong westerly to northerly winds as shown in the wind roses displayed in Fig. 3. With the exception of south winds that mainly occur during the melt period, westerly to northerly winds dominate. Additionally, the most frequently moderate to strong winds come from the northwest.

3.2 Air temperature, relative humidity and atmospheric air pressure

Air temperature and relative humidity were measured with the HMP155 Vaisala sensor (Vaisala Company, 2012), and atmospheric air pressure was recorded with the BP1 sensor from ADCON Telemetry (ADCON Telemetry Company, 2015). The HMP 155 humidity and temperature probe was placed inside a standard radiation shield at 3.2 m from the ground in order to prevent the snowpack from eventually covering the sensors.

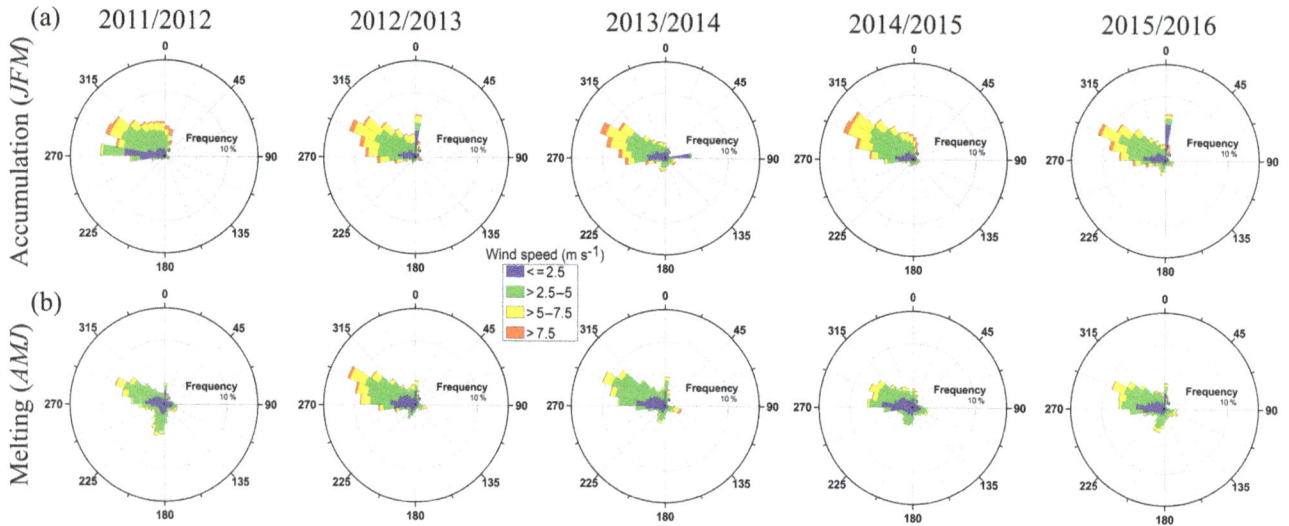

Figure 3. Wind roses showing the frequency (%) of wind speed and direction observed in the AWS for accumulation **(a)** and melt **(b)** snow seasons.

Table 1. Mean and standard deviation of air temperature for the five snow seasons for the annual, accumulation and melt periods. Also shown are maximum and minimum air temperatures for each period of the snow seasons.

		Air temperature (°C)					
		2011/12	2012/13	2013/14	2014/15	2015/16	2016/17
Mean	Annual	5.13 ± 7.73	3.50 ± 6.88	4.17 ± 6.11	5.26 ± 7.02	5.08 ± 6.69	Nan
	Accumulation	−1.15 ± 5.69	−2.78 ± 4.57	−1.71 ± 3.44	−1.65 ± 4.87	−1.66 ± 3.69	−0.56 ± 4.20
	Melting	5.80 ± 6.60	2.79 ± 4.79	5.51 ± 4.07	7.23 ± 4.86	4.45 ± 5.12	7.58 ± 6.00
Max	Annual	25.87	20.85	21.42	24.07	24.23	Nan
	Accumulation	7.89	10.69	10.20	10.98	11.62	11.39
	Melting	18.29	17.13	18.32	23.07	19.26	22.51
Min	Annual	−18.51	−15.26	−11.35	−15.24	−11.78	Nan
	Accumulation	−18.51	−15.26	−11.35	−15.24	−11.78	−14.97
	Melting	−9.33	−9.04	−3.71	−4.76	−8.20	−8.33

Nan: no data observed during the period.

Over the six snow seasons analyzed, the mean annual air temperature ranged between 5.26 °C (2014/15) and 3.51 °C (2012/13), with an average value of 4.59 °C. The mean air temperature in the accumulation period ranged from −2.78 °C (2012/13) to −0.56 °C (2016/17), with an average value of −1.59 °C for the whole study period. Finally, the melt period returned a mean value of 5.56 °C ranging from 2.79 °C (2012/13) to 7.58 °C (2016/17). Table 1 shows that the 2012/13 snow season was the coldest in the study period. Figure 4 depicts the temporal evolution of air temperature and other variables observed in the AWS from 2011 to 2016. Thus, this figure shows the control points for air temperature on the ground and the surface temperature.

The relative air humidity and the atmospheric air pressure are shown in Tables 2 and 3, respectively. The mean annual value of the relative humidity for the five seasons is 65%, with 67% during the accumulation period and 66% during the melt. Similarly, atmospheric air pressure has a mean annual value of 791 mbar, with 787 mbar for the accumulation period and 792 mbar for the melt.

3.3 Ground temperature

On 22 November 2012 four Campbell Scientific 107 temperature probes (Campbell Scientific Ltd, 2012) were installed in the AWS to measure ground temperature at different depths. One sensor was located in the atmosphere–ground interface (slightly buried, 0 cm depth), while the other three were placed at depths of 5, 10 and 20 cm. Table 4 and Fig. 4 show the average values of ground temperatures and the temporal evolution of ground temperature. Data are lacking from August 2016 onwards because tem-

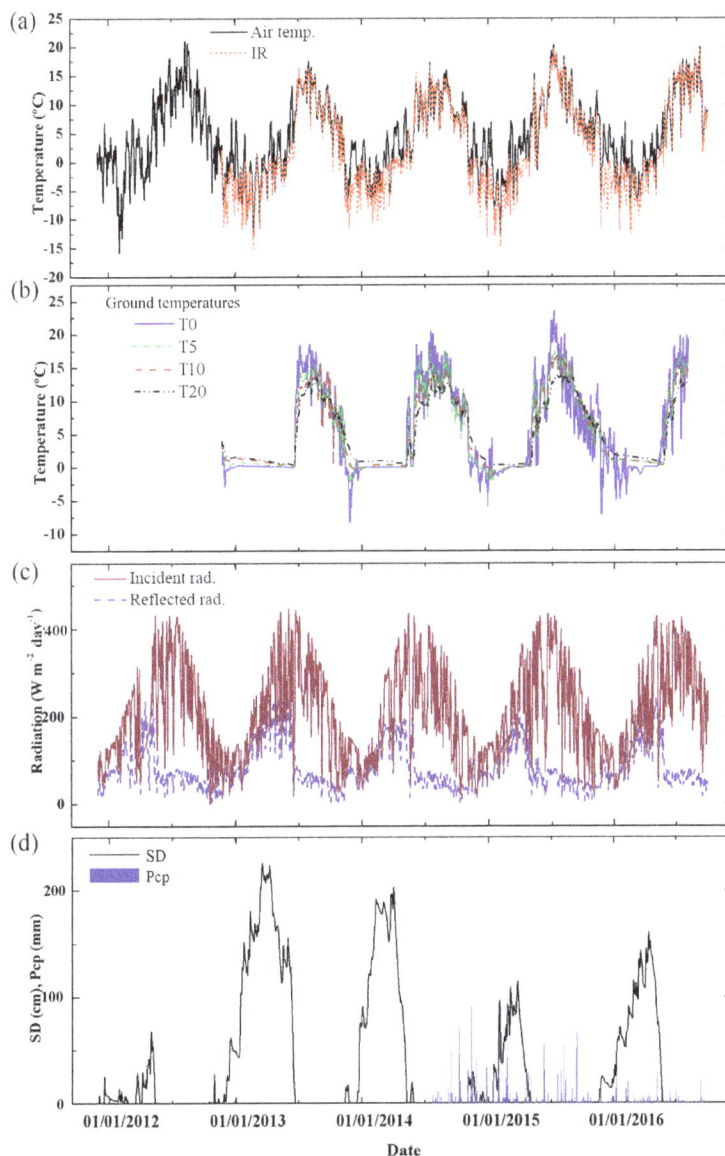

Figure 4. Temporal evolution of meteorological variables from 2011 to 2016. From top to bottom are air temperature and surface temperature (from the IR sensor); ground temperature for the four depths; global (Incident) and reflected solar irradiance; punctual snow depth (SD); and daily accumulated precipitation (Pcp) (sum of solid and liquid).

Table 2. Mean and standard deviation of relative humidity for the five snow seasons for the annual, accumulation and melt periods.

	Relative air humidity (%)					
	2011/12	2012/13	2013/14	2014/15	2015/16	2016/17
Annual	59.9 ± 18.9	70.1 ± 17.1	68.8 ± 17.3	64.8 ± 19.2	65.9 ± 18.5	Nan
Accumulation	67.1 ± 18.1	70.5 ± 19.3	72.7 ± 15.8	62.8 ± 22.2	71.3 ± 18.3	61.0 ± 20.8
Melting	57.1 ± 15.2	74.4 ± 14.5	68.7 ± 15.9	63.9 ± 15.8	69.9 ± 14.1	62.9 ± 15.65

Nan: no data observed during the period.

Table 3. Mean and standard deviation of atmospheric air pressure for the five snow seasons for the annual, accumulation and melt periods.

	Atmospheric air pressure (mbar)					
	2011/12	2012/13	2013/14	2014/15	2015/16	2016/17
Annual	794.5 ± 5.9	790.7 ± 7.7	791.3 ± 6.5	792.4 ± 6.9	791.8 ± 7.1	Nan
Accumulation	790.9 ± 7.2	784.7 ± 8.3	786.4 ± 6.9	789.7 ± 9.3	786.8 ± 7.9	788.5 ± 5.5
Melting	797.1 ± 3.6	790.9 ± 6.6	791.8 ± 4.6	794.2 ± 4.4	788.9 ± 5.4	791.5 ± 5.0

Nan: no data observed during the period.

Table 4. Mean and standard deviation ground temperature for different depths for the five snow seasons for the annual, accumulation and melt periods.

	Depth (cm)	Ground temperatures (°C)					
		2011/12	2012/13	2013/14	2014/15	2015/16	2016/17
Annual	0	Nan	4.60 ± 6.71	5.13 ± 6.45	5.98 ± 7.02	4.24 ± 6.02	Nan
	5	Nan	4.35 ± 5.67	5.61 ± 6.52	6.06 ± 5.52	4.66 ± 5.12	Nan
	10	Nan	4.38 ± 5.19	5.07 ± 5.46	5.99 ± 6.09	4.55 ± 4.87	Nan
	20	Nan	4.26 ± 4.66	5.01 ± 4.62	5.08 ± 3.26	4.51 ± 3.88	Nan
Acc.	0	Nan	0,22 ± 0,05	0.03 ± 0.04	−0.26 ± 0.87	−0.66 ± 1.13	Nan
	5	Nan	0,69 ± 0.12	0.11 ± 0.08	−0.39 ± 0.54	0.99 ± 0.10	Nan
	10	Nan	1.10 ± 0.16	0.31 ± 0.18	−0.27 ± 0.23	0.98 ± 0.11	Nan
	20	Nan	1.34 ± 0.19	0.94 ± 0.06	0.39 ± 0.08	1.57 ± 0.17	Nan
Melting	0	Nan	1.21 ± 3.49	5.53 ± 6.41	7.87 ± 6.41	4.57 ± 5.46	Nan
	5	Nan	1.04 ± 2.45	5.19 ± 6.08	7.03 ± 5.71	4.43 ± 5.09	Nan
	10	Nan	1.06 ± 1.78	4.15 ± 4.68	6.46 ± 5.32	4.15 ± 4.79	Nan
	20	Nan	1.04 ± 1.36	3.46 ± 3.49	5.35 ± 4.15	3.50 ± 3.47	Nan

Nan: no data observed during the period.

perature probes were damaged by cows. The average values during the period with information for the 0, 5, 10 and 20 cm depths are, respectively, 5.26 ± 6.22, 4.97, 4.93 ± 6.17 and 4.89 ± 4.56 °C.

The temporal evolution of air and ground temperatures depicts the impact of the snowpack on ground energy dynamics. The snowpack shelters the ground from the high temporal variability in air temperature. Therefore, the daily variability in ground temperatures is significantly lower. Furthermore, the different ground temperatures tend to reach 0 °C while snow covers the ground, i.e., the typical soil–snow interface temperature.

3.4 Surface temperature

Together with the installation of the ground temperature sensors, an IR100 infrared remote temperature sensor (Campbell Scientific Ltd, 2015) was also set up to measure surface temperature of near-target ground or snow. Table 5 shows the average land surface temperatures. The mean annual surface temperature is 2.56 °C, with a mean value of −4.58 °C during the accumulation period and 3.94 °C during the melt period.

The infrared remote sensor shows the tendency of the snow surface to cool faster than soil. During winter and spring, while snow is present on the ground, the differences between air and surface temperature are more marked, with surface temperatures always observed to be lower (see the occurrence of snow below the AWS when lower surface temperatures are observed in Fig. 4). This plainly exemplifies the higher energy irradiance of snow when compared to snow-free soils.

3.5 Global and reflected solar irradiance

The AWS also obtains information on the global and reflected solar irradiance with a CMA 6 Kipp & Zonen albedometer (Kipp & Zonen B. V., 2016) placed at 3.4 m height. Figure 4 shows the daily evolution of the values recorded and how these are interrelated, with the reflected radiation increasing at the same time as the incident. The average values of these variables are presented in Table 6. For the whole period, the average values of the incident radiation are 207.97 W m^{-2} day, taking complete snow seasons into account, 164.73 m^{-2} day for accumulation and 280.95 m^{-2} day

Table 5. Mean surface temperature from the infrared sensor for the five snow seasons for the annual, accumulation and melt periods.

			Surface temperature (°C)			
	2011/12	2012/13	2013/14	2014/15	2015/16	2016/17
Annual	Nan	1.29 ± 7.83	2.44 ± 7.06	3.26 ± 8.14	3.26 ± 7.71	Nan
Accumulation	Nan	-5.38 ± 3.58	-4.18 ± 2.65	-5.36 ± 3.61	-4.32 ± 2.99	-3.68 ± 3.58
Melting	Nan	-0.09 ± 3.44	3.75 ± 5.16	5.95 ± 6.02	3.47 ± 5.96	6.64 ± 6.67

Nan: no data observed during the period.

Table 6. Mean global and reflected radiation for the five snow seasons for the annual, accumulation and melt periods.

				Radiation (W m^{-2} day)			
		2011/12	2012/13	2013/14	2014/15	2015/16	2016/17
Annual	Global	219.48 ± 110.60	205.36 ± 114.50	196.64 ± 110.49	207.63 ± 116.50	211.03 ± 113.95	Nan
	Reflected	82.87 ± 49.60	96.20 ± 64.92	79.35 ± 52.78	76.34 ± 64.90	83.61 ± 53.76	Nan
Acc.	Global	181.09 ± 68.18	154.83 ± 67.30	150.04 ± 84.02	166.97 ± 65.80	152.83 ± 83.18	182.64 ± 85.36
	Reflected	99.14 ± 40.34	117.04 ± 44.35	108.50 ± 47.43	114.24 ± 44.35	108.94 ± 48.59	110.29 ± 41.13
Melting	Global	245.37 ± 120.56	289.59 ± 114.10	283.33 ± 102.80	287.65 ± 117.15	278.71 ± 114.37	301.07 ± 107.57
	Reflected	103.11 ± 67.15	169.56 ± 60.28	114.83 ± 61.10	90.51 ± 60.28	120.06 ± 67.30	104.28 ± 66.7

Nan: no data observed during the period.

Table 7. Accumulated precipitation (liquid and solid) for snow seasons with observations available. Average snow depth values for accumulation and melt periods for the five snow seasons.

				Accumulated total precipitation (mm)			
		2011/12	2012/13	2013/14	2014/15	2015/16	2016/17
Ann	Pcp (mm)	Nan	Nan	Nan	1572	411	Nan
Acc.	SD (cm)	14.74 ± 14.60	145.61 ± 52.3	148.54 ± 41.60	55.90 ± 36.50	81.42 ± 31.67	114.44 ± 73.80
	Pcp (mm)	Nan	Nan	Nan	454.35	147.22	82.38
Mlt.	SD (cm)	1.60 ± 1.57	131.42 ± 64.64	51.57 ± 64.95	12.70 ± 22.24	64.30 ± 59.85	25.14 ± 32.22
	Pcp (mm)	Nan	Nan	Nan	249.61	121.05	162.51

Nan: no data observed during the period.

for all melt periods. Similarly, the reflected radiation average values are 83.67 m^{-2} day for entire snow seasons, 109.69 m^{-2} day for the accumulation and 117.06 m^{-2} day for melt periods.

Similar to ground and surface temperatures, the radiation reflected is heavily influenced by the presence of snow. When snow covers the ground, the sensor shows higher values of reflected radiation in comparison with snow-free periods (Fig. 4).

3.6   Snow depth and precipitation

The AWS is also equipped with a Campbell SR50A sonic ranging sensor (Campbell Scientific Ltd, 2011). For the sake of simplicity we will refer to it as a snow depth sensor since it is used for measuring the changing distance between the surface and the sensor (the sensor is placed 2.64 m from the ground, and the snow depth is obtained by subtracting this value from the observed distance). This sensor worked uninterruptedly during the study period and provided a good record of the snow depth evolution in the Izas Experimental Catchment. Therefore, the information on the snow depth can be used as a reference for other observations of snowpack evolution. The average values for the whole study period are 93.4 cm for the accumulation period and 47.8 cm for the melt period (Table 7 shows the seasonal values). The temporal evolution of the snow depth is shown in Fig. 4.

In addition, Fig. 4 shows the precipitation values for the period with data in the precipitation gauge (from the end of July 2014). The sensor installed is a Geonor T-200B with a wind shield (Geonor A/S, 2010), which continuously weighs the accumulated precipitation (liquid and solid). The height of the gauge orifice is 3.25 m (2.5 m metal pedestal plus the height of the T-200B inlet). The precipitation accumulated over a certain period was calculated by subtracting final and initial weighted values. Table 7 includes the accumulated precipitation for the whole snow year and also during the accumulation and melt periods.

4   Information on snow distribution

4.1   TLS acquisitions of snow depth distribution

During the five snow seasons presented here, three to six TLS surveys were carried out each year in the Izas Experimental Catchment. TLSs are devices using lidar technology, a remote sensing method to obtain the distance between a target area and the device. During a TLS data acquisition, the device measures the distance of some hundreds of thousands of points within the area defined by the operator, creating a cloud of data points representing the topography of the target surface. The device used in this study is a long-range TLS (RIEGL LPM-321 (Fig. 2), RIEGL Laser Measurements, 2010). The technical characteristics of this model include (i) light pulses of 905 nm wavelength (near infrared), appropriate for acquiring data from snow cover (Prokop, 2008); (ii) a minimum angular step width of 0.018°; (iii) a laser beam divergence of 0.046°; and (iv) a maximum working distance of 6000 m. In order to reduce topographic shadowing (note that terrain topography limits the line of sight of the TLS), two scanning positions (Scan station in Fig. 1) were established within the study site (Fig. 1). Also fixed on the terrain, were 12 reflected targets (Fig. 2). The location of these targets was acquired on each TLS acquisition date since this information is used in the post-processing phase for comparing the point clouds acquired on different dates. The protocol for obtaining the information in the field and the methodology for generating the snow depth distribution maps for the different TLS survey dates is fully explained in Revuelto et al. (2014a). The method is mainly based on calculating the elevation difference between the point clouds obtained on different dates with and without snow cover across the study area. The final products are snow depth distribution maps with a grid size of 1 m × 1 m, with a mean absolute error of 0.07 m in the snow depth values (Revuelto et al., 2014a).

Figure 5 shows the snow depth maps obtained for the 2012/13 snow season. The information for this snow season is presented because six TLS surveys were completed. Furthermore, the accumulated snow depths were significant and thus provide an interesting example of snowpack evolution over time. These maps show the high spatial variability in the snowpack within the study area, with marked changes in the snow depth distribution within short distances. It was also observed how high-accumulation areas had large accumulations during the whole snow season, with a thick snowpack

Table 8. Observed mean and maximum snow depth values, snow-covered area (SCA, % of the total area covered by the TLS), and coefficient of variation for the observed snow distribution on the TLS survey dates.

Snow season	Date	Mean SD (m)	Max SD (m)	SCA (%)	CV
2011/12	22 Feb	0.46	5.53	67.2	1.35
	2 Apr	0.17	3.86	33.5	2.23
	17 Apr	0.56	5.34	94.1	1.07
	2 May	0.90	6.11	98.8	0.74
	14 May	0.21	4.47	30.9	1.90
	24 May	0.09	4.32	18.9	1.29
2012/13	17 Feb	2.91	10.89	98.8	0.63
	3 Apr	3.19	11.20	100	0.56
	25 Apr	2.42	10.10	96.3	0.76
	6 Jun	1.98	9.64	86.4	0.86
	12 Jun	1.69	8.90	77.1	0.90
	20 Jun	0.76	7.97	67.0	1.35
2013/14	3 Feb	2.16	10.20	96.0	0.59
	22 Feb	2.56	10.47	98.6	0.57
	9 Apr	2.54	9.72	89.0	0.65
	5 May	1.67	9.02	75.2	0.87
2014/15	6 Nov	0.22	2.78	85.0	0.81
	26 Jan	0.74	4.88	89.3	0.85
	6 Mar	2.13	11.55	94.0	0.69
	12 May	0.67	7.75	56.0	1.21
2015/16	4 Feb	0.82	6.20	91.1	0.63
	25 Apr	1.86	10.82	97.0	0.50
	26 May	1.16	7.81	74.8	0.70
2016/17	20 Jan	1.26	6.33	93	0.72
	8 May	0.77	7.25	57.2	0.81

for dates on which the snow cover had already completely melted over wide areas of the catchment.

Table 8 presents the average snow depth and the maximum snow depth value observed for each TLS acquisition. This table also shows the coefficient of variation on each snow distribution map and the fraction of the SCA. The values obtained depict the heavy accumulation of snow in some areas of the catchment, while the average snow depth is lower.

4.2   Snow-covered area from time-lapse photographs

The Izas Experimental Catchment is also equipped with a Campbell CC640 digital camera (Campbell Scientific Ltd, 2010). This camera was mounted on a solid metal structure set into the ground with concrete (Fig. 2), which ensured a constant position to obtain consistent information. The digital camera has a resolution of 640 × 480 pixels with a focal length of 6–12 mm. The field of view of the photographs obtained with the camera mounted on the metal structure covers approximately 30 ha (Fig. 1), which repre-

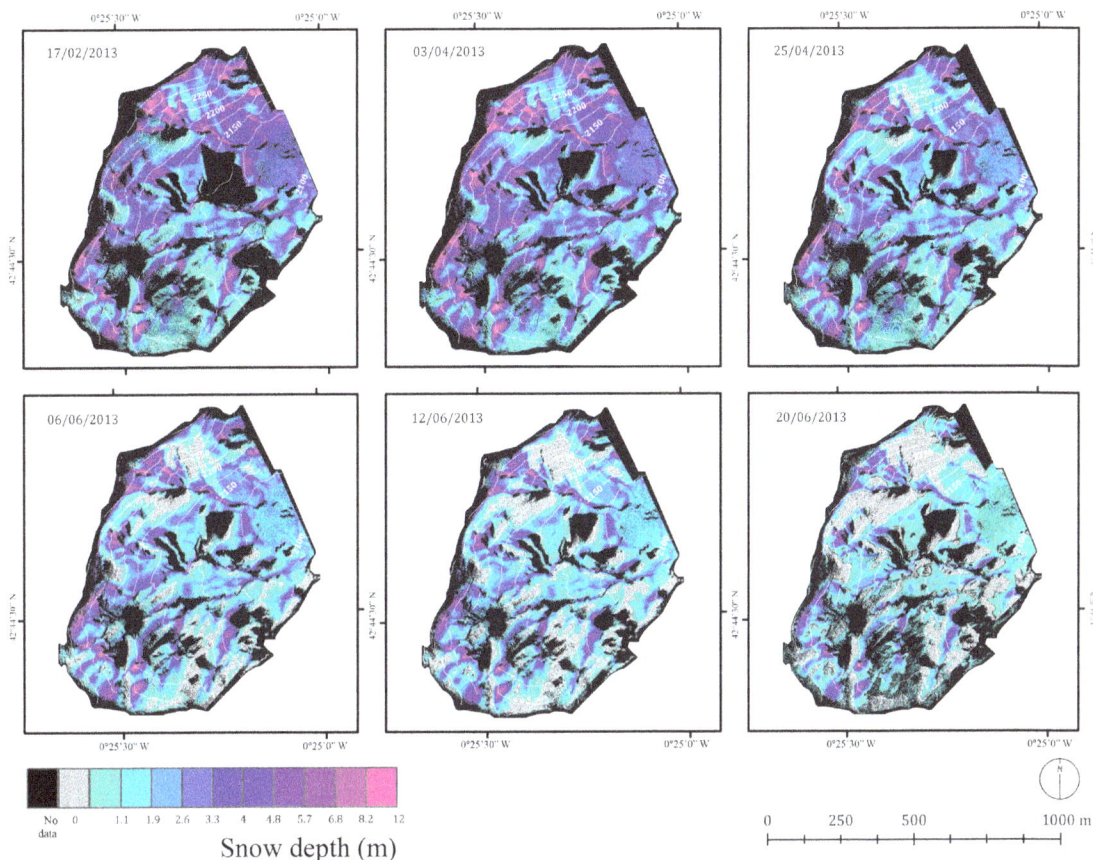

Figure 5. Snow depth distribution maps obtained for the six TLS acquisition dates of the 2012/13 snow season.

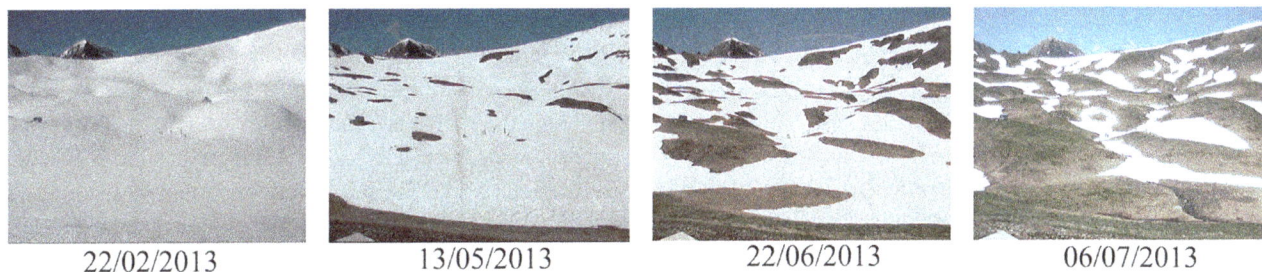

Figure 6. Example of a sequence of four photographs for the 2012/13 snow season, showing the evolution of the snow cover.

sents about 52 % of the total surface covered by the TLS. The camera obtained three pictures per day (time-lapse photography) at 10:00, 11:00 and 12:00 UTC, ensuring good illumination of the area. Figure 6 contains four photographs from the 2012/13 snow season, showing how the SCA evolved in time.

The pictures can be projected into a digital elevation model (DEM) of the study site. Projecting the pictures into the 1 m × 1 m DEM for an entire snow season provides distributed information on the evolution of the SCA in the same reference system as snow depth maps. The approach for projecting the pictures into the DEM is described by Corri-

pio (2004) and the specific features of the methodology applied in the Izas Experimental Catchment are fully described in Revuelto et al. (2016a). The routines applied first make a viewing transformation allowing for the optics of the camera and, second, a perspective projection, providing a virtual image of the DEM. Therefore, in the second step, the correspondence of ground control points with the pixels of the photograph must be established. Since this stage is quite sensitive, the coordinates of ground control points were acquired with a differential GPS. With this process, images projected into the DEM had a 3.3 pixel performance in the calibration of the transformation. Finally, the daily series of the projected im-

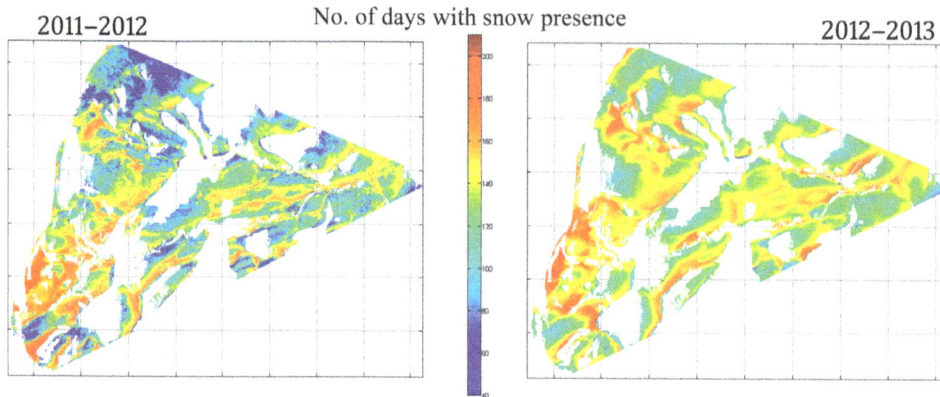

Figure 7. Number of days with snow present for each pixel for 2011/12 and 2012/13 snow seasons.

ages can be binarized to create daily snow presence–absence maps. This information can also be used for other applications, such as to observe the growth timing of plant species.

Since the binarized snow presence–absence maps were recorded on almost a daily frequency (note that about 20 % of all photographs from the camera had to be discarded because cloud or snow obscured the camera lens), many parameters can be derived from this information, including the SCA temporal evolution, the numbers of days with snow presence or the melt-out date (MOD) on each pixel. Figure 7 shows an example of the number of days with snow presence for the 2011/12 and 2012/13 snow seasons.

5   Data availability

The database presented and described in this article is available for download at Zenodo (Revuelto et al., 2017; https://doi.org/10.5281/zenodo.848277). Meteorological data of the AWS are given in .csv format. The meteorological data set includes observations at 10 min intervals. The TLS survey snow depth distribution maps are available online (one file for each TLS acquisition). These files are in ASCII format in the UTM 30T north coordinate system with a 1 m × 1 m spatial resolution. The DEM of the study area is also provided in same coordinate system. Photographs of cloud-free days from the time-lapse camera are available in the online repository, with the correspondence of pixel ground control points to GPS coordinates. Information on the optics and chip size of the camera is also provided. Additionally, all available MOD distribution maps (last Julian day with snow presence on each pixel) are included in the database.

6   Summary

The Izas Experimental Catchment is a well-established study area on the south face of the Pyrenees, in which different meteorological and snow variables are automatically acquired.

Additionally, great effort has been made on field data acquisition with TLS over the last five snow seasons and is ongoing. The data set described here is novel in the Pyrenees because, for the first time, it represents high-spatial-resolution information on the snowpack distribution and its evolution in time, as well as making continuous information available on meteorological variables. The high quality of the information obtained has already been exploited for different studies on the understanding of snowpack dynamics and the improvement of simulation approaches to snowpack evolution in mountain areas (López-Moreno et al., 2012, 2014, Revuelto et al., 2014b, 2016a, 2016b). However, many scientific questions still go unanswered, such as the long-term influence of topography on snow dynamics, and the spatial distribution of snow during precipitation and strong wind events. Also, the high interannual variability in snow accumulation in the Pyrenees has serious consequences for water management, especially in the Mediterranean area (García-Ruiz et al., 2011). Thus, it is very important to continue obtaining information on snowpack evolution and the meteorological variables controlling snow dynamics. This information will allow the scientific community to better understand processes involved in snow dynamics and make for better adaptation to climate change scenarios. Moreover, offering the possibility of exploiting the information to other fields provides, as INARCH does, the opportunity of establishing new collaboration networks to push forward the frontiers of science in mountain areas.

Competing interests. The authors declare that they have no conflict of interest.

Special issue statement. This article is part of the special issue "Hydrometeorological data from mountain and alpine research catchments". It is not associated with a conference.

Acknowledgements. This study was funded by the research projects CGL2014-52599-P "Estudio del manto de nieve en la montaña española y su respuesta a la variabilidad y cambio climatico" (Ministry of Economy and Development, MINECO) and CLIMPY "Characterization of the evolution of climate and provision of information for adaptation in the Pyrenees" (FEDER-POCTEFA). The authors are thankful for this unique opportunity to share information through the International Network for Alpine Research Catchment Hydrology (INARCH). Jesús Revuelto is supported by a postdoctoral Fellowship from the AXA research fund 2016 (le Post-Doctorant Jesús Revuelto est bénéficiaire d'une bourse postdoctorale du Fonds AXA pour la Recherche). Cesar Azorin-Molina is supported by the Marie Skłodowska-Curie Individual Fellowship (STILLING project – 703733) funded by the European Commission.

Edited by: John Pomeroy

References

ADCON Telemetry Company: Adcon BP1 Barometric Pressure Sensor, Copyright © 2017 Adcon Telemetry, OTT Hydromet GmbH, Klosterneuburg, Austria, 2017.

Alvera, B. and Garcia-Ruiz, J. M.: Variability of Sediment Yield from a High Mountain Catchment, Central Spanish Pyrenees, Arct. Antarct. Alp. Res., 32, 478–484, 2000.

Anderton, S. P., White, S. M., and Alvera, B.: Evaluation of spatial variability in snow water equivalent for a high mountain catchment, Hydrol. Process., 18, 435–453, 2004.

Azorin-Molina, C., Tijm, S., Ebert, E. E., Vicente-Serrano, S.-M., and Estrela, M.-J.: High Resolution HIRLAM Simulations of the Role of Low-Level Sea-Breeze Convergence in Initiating Deep Moist Convection in the Eastern Iberian Peninsula, Bound.-Lay. Meteorol., 154, 81–100, 2015.

Barnett, T. P., Adam, J. C., and Lettenmaier, D. P.: Potential impacts of a warming climate on water availability in snow-dominated regions, Nature, 438, 303–309, 2005.

Campbell Scientific Ltd: CC640 Digital Camera, Instruction Manual, Copyright © 2005–2010 Campbell Scientific Ltd, Campbell Park, Shepshed, Loughborough, UK, 2010.

Campbell Scientific Ltd: SR50A Sonic Ranging Sensor, User Guide, Copyright © 2007–2011 Campbell Scientific Ltd, Campbell Park, Shepshed, Loughborough, UK, 2011.

Campbell Scientific Ltd: 107 temperature Probe, User Guide, Copyright © 2003–2012 Campbell Scientific Ltd, Campbell Park, Shepshed, Loughborough, UK, 2012.

Campbell Scientific Ltd: IR100/IR120 Infra-red remote temperature sensor, User Guide, Copyright © 2007–2015 Campbell Scientific Ltd., Campbell Park, Shepshed, Loughborough, UK, 2015.

Chueca, J.: Geomorfología de la Alta Ribagorza: análisis de la dinámica de procesos en el ámbito superficial, Asoc. Guayente, Benasque, Spain, 1993.

Colbeck, S. C., Anderson, E. A., Bissell, V. C., Crock, A. G., Male, D. H., Slaughter, C. W., and Wiesnet, D. R.: Snow accumulation, distribution, melt, and runoff, Eos T. Am. Geophys. Un., 60, 465–468, 1979.

Corripio, J. G.: Snow surface albedo estimation using terrestrial photography, Int. J. Remote Sens. 25, 5705–5729, 2004.

Creus-Novau, J.: El clima del alto Aragón occidental, CSIC – Instituto de Estudios Pirenaicos, Publicada con los Patrocinios de la Excma, Diputación Provincial de Huesca y la General de Aragón, 109, 233 pp., 1983.

Cuadrat, J. M., Saz, M. A., and Vicente-Serrano, S. M.: Atlas climático de Aragón, Gobierno de Aragón, 2007.

del Barrio, G., Creus, J., and Puigdefabregas, J.: Thermal Seasonality of the High Mountain Belts of the Pyrenees, Mt. Res. Dev. 10, 227–233, 1990.

del Barrio, G., Alvera, B., Puigdefabregas, J., and Diez, C.: Response of high mountain landscape to topographic variables: Central pyrenees, Landscape Ecol., 12, 95–115, 1997.

García-Ruiz, J. M., Puigdefabregas-Tomas, J., and Creus-Novau, J.: Snowpack accumulation in the Central Pyrenees, and its hydrological effects, Pirineos, 127, 27–72, 1986.

García-Ruiz, J. M., Beguería, S., López-Moreno, J. I., Lorente, A., and Seeger, M.: Los recursos hídricos superficiales del Pirineo aragonés y su evolución reciente, Geofroma Logroño, 2001.

García-Ruiz, J. M., López-Moreno, J. I., Vicente-Serrano, S. M., Lasanta-Martínez, T., and Beguería, S.: Mediterranean water resources in a global change scenario, Earth-Sci. Rev., 105, 121–139, 2011.

Geonor A/S: Geonor T-200B series, All-weather precipitation gauges specifications, © 2010 Geonor A/S, Oslo, Norway, 2010.

Kipp & Zonen B.V.: CMP/CMA series pyranometer and albedometer, Instruction Manual, Copyright © 2016 Kipp & Zonen B.V., Delft, the Netherlands, 2016.

Lana-Renault, N., Alvera, B., and García-Ruiz, J. M.: Runoff and Sediment Transport during the Snowmelt Period in a Mediterranean High-Mountain Catchment, Arct. Antarct. Alp. Res., 43, 213–222, 2011.

Liston, G. E.: Interrelationships among Snow Distribution, Snowmelt, and Snow Cover Depletion: Implications for Atmospheric, Hydrologic, and Ecologic Modeling, J. Appl. Meteorol., 38, 1474–1487, 1999.

López-Moreno, J. I.: Cambio ambiental y gestión de los embalses en el Pirineo Central Español, Consejo de Protección de la Naturaleza de Aragón, 2005.

López-Moreno, J. I. and García-Ruiz, J. M.: Influence of snow accumulation and snowmelt on streamflow in the central Spanish Pyrenees/Influence de l'accumulation et de la fonte de la neige sur les écoulements dans les Pyrénées centrales espagnoles, Hydrolog. Sci. J., 49, 787–802, 2004.

López-Moreno, J. I. and Nogués-Bravo, D.: Interpolating local snow depth data: an evaluation of methods, Hydrol. Process., 20, 2217–2232, 2006.

López-Moreno, J. I. and Vicente-Serrano, S. M.: Atmospheric circulation influence on the interannual variability of snow pack in the Spanish Pyrenees during the second half of the 20th century, Nord. Hydrol., 38, 33–44, 2007.

López-Moreno, J. I., Beguería, S., and García-Ruiz, J. M.: Influence of the Yesa reservoir on floods of the Aragón River, central Spanish Pyrenees, Hydrol. Earth Syst. Sci., 6, 753–762, https://doi.org/10.5194/hess-6-753-2002, 2002.

López-Moreno, J. I., Beniston, M., and García-Ruiz, J. M.: Environmental change and water management in the Pyrenees: Facts and future perspectives for Mediterranean mountains, Global Planet. Change, 61, 300–312, 2008.

López-Moreno, J. I., Latron, J., and Lehmann, A.: Effects of sample and grid size on the accuracy and stability of regression-based snow interpolation methods, Hydrol. Process., 24, 1914–1928, 2010.

López-Moreno, J. I., Pomeroy, J. W., Revuelto, J., and Vicente-Serrano, S. M.: Response of snow processes to climate change: spatial variability in a small basin in the Spanish Pyrenees, Hydrol. Process., 27, 2637–2650, 2012.

López-Moreno, J. I., Revuelto, J., Fassnacht, S. R., Azorín-Molina, C., Vicente-Serrano, S. M., Morán-Tejeda, E., and Sexstone, G. A.: Snowpack variability across various spatio-temporal resolutions, Hydrol. Process., 29, 1213–1224, 2014.

López-Moreno, J. I., Revuelto, J., Rico, I., Chueca-Cía, J., Julián, A., Serreta, A., Serrano, E., Vicente-Serrano, S. M., Azorin-Molina, C., Alonso-González, E., and García-Ruiz, J. M.: Thinning of the Monte Perdido Glacier in the Spanish Pyrenees since 1981, The Cryosphere, 10, 681–694, https://doi.org/10.5194/tc-10-681-2016, 2016.

Navarro-Serrano, F. M. and López-Moreno, J. I.: Spatio-temporal analysis of snowfall events in the spanish Pyrenees and their relationship to atmospheric circulation, Cuad. Investig. Geográfica, 43, 233–254, 2017.

Prokop, A.: Assessing the applicability of terrestrial laser scanning for spatial snow depth measurements, Cold Reg. Sci. Technol., 54, 155–163, 2008.

Revuelto, J., López-Moreno, J. I., Morán-Tejeda, E., Fassnacht, S.R., and Vicente-Serrano, S. M.: Variabilidad interanual del manto de nieve en el Pirineo: Tendencias observadas y su relación con índices de telconexión durante el periodo 1985–2011, Universidad de Salamanca, Salamanca, Spain, 613–621, 2012.

Revuelto, J., López-Moreno, J. I., Azorin-Molina, C., Zabalza, J., Arguedas, G., and Vicente-Serrano, S. M.: Mapping the annual evolution of snow depth in a small catchment in the Pyrenees using the long-range terrestrial laser scanning, J. Maps, 10, 1–15, 2014a.

Revuelto, J., López-Moreno, J. I., Azorin-Molina, C., and Vicente-Serrano, S. M.: Topographic control of snowpack distribution in a small catchment in the central Spanish Pyrenees: intra- and inter-annual persistence, The Cryosphere, 8, 1989–2006,

https://doi.org/10.5194/tc-8-1989-2014, 2014b.

Revuelto, J., Jonas, T., and López-Moreno, J.-I.: Backward snow depth reconstruction at high spatial resolution based on time-lapse photography, Hydrol. Process., 30, 2976–2990, 2016a.

Revuelto, J., Vionnet, V., López-Moreno, J. I., Lafaysse, M., and Morin, S.: Combining snowpack modeling and terrestrial laser scanner observations improves the simulation of small scale snow dynamics, J. Hydrol., 291–307, 2016b.

Revuelto, J., Azorin-Molina, C., Alonso-González, E., Sanmiguel-Vallelado, A., Navarro-Serrano, F., Rico, I., and López-Moreno, J. I.: Observations of snowpack distribution and meteorological variables at the Izas Experimental Catchment (Spanish Pyrenees) from 2011 to 2017 [Data set], Zenodo, https://doi.org/10.5281/zenodo.848277, 2017.

RIEGL Laser Measurements: LPM-321, Long Range Laser Profil Measrurement System, © RIEGL, Horn, Austria, 2010.

Serrano, E., Agudo, C., Delaloyé, R., and González-Trueba, J. J.: Permafrost distribution in the Posets massif, Central Pyrenees, Norsk Geogr. Tidsskr., 55, 245–252, 2001.

Vaisala Company: HMP155 Humidity and Temperature Probe, specifications, © Vaisala 2012, Ref. B21072EN-E, 2012.

Vicente-Serrano, S. M.: Las sequías climáticas en el valle medio del Ebro: Factores atmosféricos, evolución temporal y variabilidad espacial, Consejo de Protección de la Naturaleza de Aragón, 2005.

Vicente-Serrano, S. M.: Spatial and temporal analysis of droughts in the Iberian Peninsula (1910–2000), Hydrolog. Sci. J., 51, 83–97, 2006.

Viviroli, D., Dürr, H. H., Messerli, B., Meybeck, M., and Weingartner, R.: Mountains of the world, water towers for humanity: Typology, mapping, and global significance, Water Resour. Res., 43, W07447, https://doi.org/10.1029/2006WR005653, 2007.

Wipf, S., Stoeckli, V., and Bebi, P.: Winter climate change in alpine tundra: plant responses to changes in snow depth and snowmelt timing, Climatic Change, 94, 105–121, 2009.

Young Company: Model 05103-45-5, Wind Monitor – Alpine Model, specifications, Copyright © 2010 R.M. Young Company, R.M. Young Company, Traverse City, Michigan, USA, 2010.

11

Geomatic methods applied to the study of the front position changes of Johnsons and Hurd Glaciers, Livingston Island, Antarctica, between 1957 and 2013

Ricardo Rodríguez Cielos[1], Julián Aguirre de Mata[2], Andrés Díez Galilea[2], Marina Álvarez Alonso[3], Pedro Rodríguez Cielos[4], and Francisco Navarro Valero[5]

[1]Departamento de Señales, Sistemas y Radiocomunicaciones, ETSI de Telecomunicación, Universidad Politécnica de Madrid, Madrid, Spain

[2]Departamento de Ingeniería Topográfica y Cartografía, ETSI en Topografía, Geodesia y Cartografía, Universidad Politécnica de Madrid, Madrid, Spain

[3]Departamento de Lenguajes y Sistemas Informáticos e Ingeniería de Software, ETS de Ingenieros Informáticos, Universidad Politécnica de Madrid, Madrid, Spain

[4]Departamento de Matemática Aplicada, ETSI de Telecomunicación, Universidad de Málaga, Málaga, Spain

[5]Departamento de Matemática Aplicada a las Tecnologías de la Información y las Comunicaciones, ETSI de Telecomunicación, Universidad Politécnica de Madrid, Madrid, Spain

Correspondence to: Ricardo Rodríguez Cielos (ricardo.rodriguez@upm.es)

Abstract. Various geomatic measurement techniques can be efficiently combined for surveying glacier fronts. Aerial photographs and satellite images can be used to determine the position of the glacier terminus. If the glacier front is easily accessible, the classic surveys using theodolite or total station, GNSS (Global Navigation Satellite System) techniques, laser-scanner or close-range photogrammetry are possible. When the accessibility to the glacier front is difficult or impossible, close-range photogrammetry proves to be useful, inexpensive and fast. In this paper, a methodology combining photogrammetric methods and other techniques is applied to determine the calving front position of Johnsons Glacier. Images taken in 2013 with an inexpensive nonmetric digital camera are georeferenced to a global coordinate system by measuring, using GNSS techniques, support points in accessible areas close to the glacier front, from which control points in inaccessible points on the glacier surface near its calving front are determined with theodolite using the direct intersection method. The front position changes of Johnsons Glacier during the period 1957–2013, as well as those of the land-terminating fronts of Argentina, Las Palmas and Sally Rocks lobes of Hurd glacier, are determined from different geomatic techniques such as surface-based GNSS measurements, aerial photogrammetry and satellite optical imagery. This provides a set of frontal positions useful, e.g., for glacier dynamics modeling and mass balance studies.

1   Introduction: study area and background

Hurd and Johnsons glaciers are located in Hurd Peninsula, Livingston Island, the second largest island of the South Shetland Islands (SSI) archipelago (Fig. 1). Johnsons is a tidewater glacier, calving small icebergs into the proglacial bay known as Johnsons Dock, while the fronts of the various tongues of Hurd Glacier (Argentina, Las Palmas and Sally Rocks) are land-terminating. The three unnamed small sea-terminating glacier basins draining to False Bay, to the south-east of Hurd Peninsula (U1, U2, U3 in Fig. 1), which are very steep and heavily crevassed, are not covered in the current study. Johnsons and Hurd glaciers are polythermal, though Johnsons, as compared with Hurd, has a higher proportion of temperate ice (Navarro et al., 2009). Typical velocities near the calving front of Johnsons Glacier are about $50\,\mathrm{m\,yr^{-1}}$ (Rodríguez, 2014), whereas maximum velocities in Hurd Glacier are reached in its central zone, and are typically around $5\,\mathrm{m\,yr^{-1}}$ (Molina, 2014), decreasing towards the terminal zones, which have been suggested to be frozen to bed on the basis of ground-penetrating radar studies, glacier velocities and geomorphological evidences (Navarro et al., 2009; Molina et al., 2007). The annual average temperature at Juan Carlos I Station (12 m a.s.l., in Hurd Peninsula) between 1988 and 2011 was $-0.9\,°\mathrm{C}$, with average summer (DJF) and winter (JJA) temperatures of 2.4 and $-4.4\,°\mathrm{C}$, respectively (Osmanoglu et al., 2014). The main glaciological studies in Hurd Peninsula include cartography of volcanic ash layers (Palà et al., 1999; Ximenis, 2001; Molina, 2014), shallow ice coring (Furdàda et al., 1999), numerical modeling of glaciers dynamics (Martín et al., 2004; Otero et al., 2010), analysis of glacier volume changes 1957–2000 (Molina et al., 2007), seismic and ground-penetrating radar surveys (Benjumea et al., 1999, 2001, 2003; Navarro et al., 2005, 2009), modeling of melting (Jonsell et al., 2012), mass-balance observations (Ximenis et al., 1999; Ximenis, 2001; Navarro et al., 2013) and geomorphological and glacier dynamics studies (Ximenis et al., 2000; Molina, 2014). Glaciological studies covering the whole Livingston Island include the analysis of Livingston ice cap front position changes 1956–1996 (Calvet et al., 1999), ground-penetrating radar surveys (Macheret et al., 2009) and estimates of ice discharge to the ocean (Osmanoglu et al., 2014). The latter study also includes ice-cap-wide mass balance estimates and ice velocity fields determined from satellite synthetic aperture radar measurements.

Focusing on the studies dealing with geomatic techniques, Palà et al. (1999) did photogrammetric work in February 1999 focused on the terminal zone of Johnsons Glacier. However, their study only dealt with cartography of ash layers which originated from the eruptions in the neighboring Deception Island (see inset of Fig. 1). They did not determine Johnsons calving front position, since, due to the location of the observation points on the top of Johnsons/Charrúa Peak (340 m a.s.l., see upper right corner of Fig. 1), only a small fraction of the south-western part of the calving front was visible. Calvet et al. (1999) did an interesting study of the front position changes of the main basins of Livingston Island during the period 1956–1996, based on the Directorate of Overseas Surveys (DOS, 1968a, b) 1 : 200 000 maps (based, in turn, on the 1956–1957 British aerial photographs), a 1962 satellite photograph (declassified intelligence satellite photography, project code "Argon", 18 May 1962) and LANDSAT (1986, 1988, 1989) and SPOT (1991, 1996) optical images. However, their study did not cover Hurd Peninsula glacier fronts. Calvet et al. (1999) estimated a reduction of the glacier-covered area of Livingston Island by 4.3 % during the period 1956–1996 (from 734 to $703\,\mathrm{km^2}$). More recently, Osmanoglu et al. (2014), using unpublished outlines for 2004 from Jaume Calvet and David García-Sellés, estimated the glacierized area to be $697\,\mathrm{km^2}$. Molina et al. (2007) analyzed the volume changes 1957–2000 of Hurd Peninsula glaciers comparing digital elevation models (DEMs) for 1957, obtained by photogrammetric restitution from the British Antarctic Survey (BAS) aerial photographs of 1957 and theodolite and GNSS surface-based measurements in 1999–2001. We note, however, that their photogrammetric restitution was done from paper-printed photos (the only available at that time), while our work is based on a digitization from the original films performed later by the BAS. We also note that Molina et al. (2007) used the 1957 photographs, and not those of 1956 as incorrectly stated in their paper (Rodríguez, 2014). Finally, we note that, although the Randolph Glacier Inventory (Pfeffer et al., 2014), based for the Antarctic peripheral glaciers on the inventory by Bliss et al. (2013), provides outlines for Livingston Island glacier basins, those for Hurd Peninsula are outdated. Bliss' inventory is mainly based on the Antarctic Digital Database (ADD consortium, 2000), digitized from paper maps, aerial photos and LANDSAT, MODIS and ASTER imagery with a range of dates from 1957 to 2005. In the particular case of Hurd Peninsula, at least some of the data clearly correspond to 1957, because Sally Rocks, Las Palmas and Argentina are classified as sea-terminating (in 1957 their fronts were indeed very close to the coastline), while they are clearly land-terminating since several decades ago.

The current study aims to make available to the scientific community the whole set of front positions of Johnsons Glacier, and Argentina, las Palmas and Sally Rocks fronts of Hurd Glacier, between 1957 and 2013. As discussed above, only some front positions for 1957 and 2000 have been published before (Molina et al., 2007; Navarro et al., 2013) while all others were unpublished so far. There are two main reasons for the interest of these front position changes: (1) the surface mass balance of both Hurd and Johnsons glaciers has been monitored since 2001 as part of the World Glacier Monitoring Service database (http://wgms.ch), and the correct estimate of the surface mass balance requires one to know the evolution of the glacier outlines; and (2) the modeling of glacier dynamics requires a precise knowledge of the glacier geometry, including the glacier boundaries, and their evolution in case the model applied is time-dependent. The first subject is particularly relevant, since the glaciers in the Antarctic Peninsula region are an exception among the WGMS-monitored glaciers, in the sense that they have experienced sustained positive surface mass balances starting around 2007–2008. The exclusion of the unnamed glacier basins U1, U2, U3 (Fig. 1) from our study is not relevant, since Hurd and Johnsons glaciers are the only Hurd Peninsula glacier basins included in the WGMS database.

2 Data and methods

2.1 Summary of data

The aerial and satellite photographs, and surface-based photogrammetric and GNSS measurements which form the basis of the present study are summarized below (Rodríguez et al., 2015):

- DOCU 1: flight made by the British Antarctic Survey in December 1957. We have selected a total of 5 frames (X26FID0052130, X26FID0052131, X26FID0052132, X26FID0052160 and X26FID0052161) to study Hurd Peninsula glacier fronts.

- DOCU 2: flight made by the United Kingdom Hydrographic Office (UKHO) in January 1990. We have selected a total of 3 frames (0097, 0098 and 0099) for our study.

- DOCU 3: photogrammetric survey, using metric camera, performed by Palà et al. (1999) from the top of Johnsons Peak in 1999.

- DOCU 4: satellite image obtained by the Quickbird system in January 2010 for Hurd Peninsula.

- DOCU 5: satellite image obtained by the Quickbird system in February 2007 for Hurd Peninsula.

- DOCU 6: inventory of data (2000–2012) by the Group of Numerical Simulation in Science and Engineering (GSNCI, in its Spanish acronym) of Universidad Politécnica de Madrid (several authors are affiliated to this group). These observations are made with GNSS techniques and theodolite and are focused on Sally Rocks, Las Palmas and Argentina land-terminating fronts, excluding the calving front of Johnsons Glacier.

- DOCU 7: photogrammetric survey (using non-metric camera) of the calving front of Johnsons Glacier conducted in February 2013.

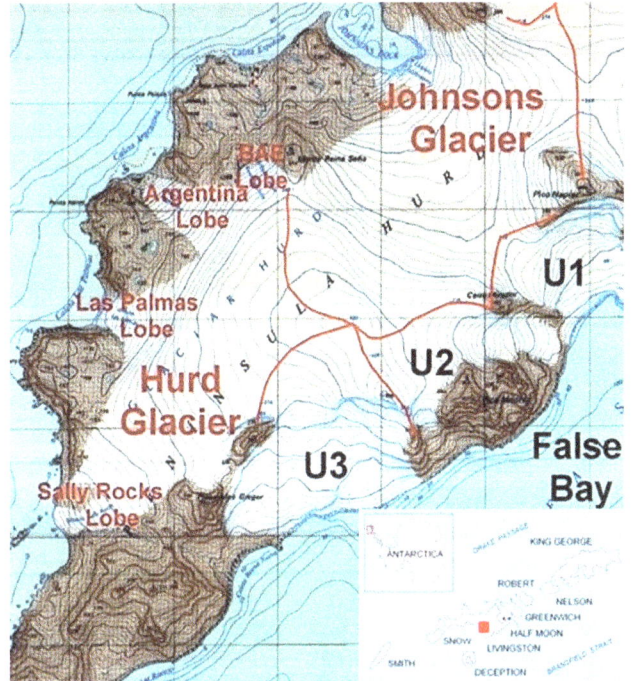

Figure 1. Location of Johnsons and Hurd glaciers in Hurd Peninsula. Base map: 1 : 25 000 map from Geographic Service of the Army (Spain), 1991. The inset shows the location of the South Shetland Islands, and Livingston Island in particular. The red rectangle in the inset indicates the location of Hurd Peninsula.

2.2 Photogrammetry: fundamentals

Photogrammetry is a well-known technique allowing one to obtain three-dimensional information from photographs using stereoscopic vision provided by two different points of view (Wolf, 1983). Its fundamental principle is triangulation. By taking photographs from at least two different locations, so-called "lines of sight" can be traced from each camera to points on the object. These lines of sight (sometimes called rays due to their optical nature) are mathematically intersected to produce the three-dimensional coordinates of the points of interest. This requires a precise knowledge of the position and orientation of the cameras. Resection is the procedure used to determine the position and orientation (also called aiming direction) of the camera, using ground control points appearing on the images that have known coordinates. For a good resection, at least more than 10 well-distributed points in each photograph are needed (Kraus, 1993). Each camera's position is defined by three coordinates, while three angles are needed to define its orientation. The theoretical central projection can be deformed by lens and film distortion. These influences can be accounted for in a bundle block adjustment by introducing correction polynomials in the observation equations, whose coefficients are determined in the adjustment (Kraus, 2007). The distortion varies with the distance from each point to the center of the optical axis. Dis-

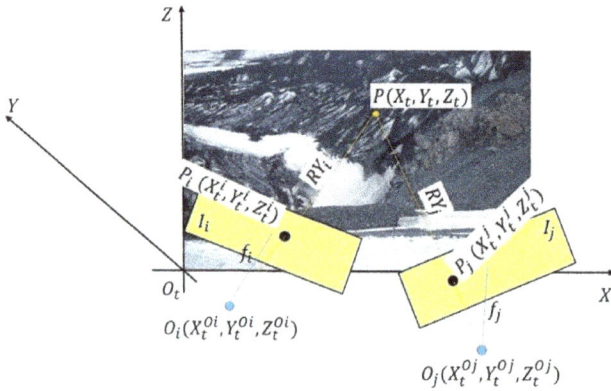

Figure 2. Photogrammetric restitution. By applying spatial similarity transformations (Kraus, 2007), one can calculate the unknown coordinates (X_t, Y_t, Z_t) of a point P from its coordinates P_i and P_j on flat photographs taken from two different points of view O_i and O_j (locations of each camera) with known coordinates.

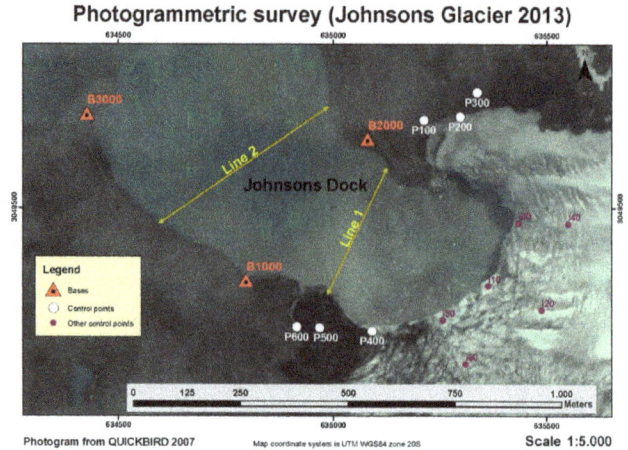

Figure 3. Location of the various kinds of control points. Red triangles: location of the bases, measured using GNSS techniques, where theodolite was positioned (B1000, B2000, B3000). White circles: control points measured using GNSS techniques (P100, P200, P300, P400, P500, P600). Magenta points: other control points, measured using direct intersection method (I10, I20, I30, I40, I50, I60 points). Line 1 and Line 2 represent the zodiac boat tracks from which the photographs were taken. The scale of the figure does not match with the numerical map scale indicated. The latter corresponds to the full-size printed version of the figure. This comment also applies to Figs. 8–12.

tortion is often decomposed into radial and tangential components (Brown, 1971), of which the radial is much larger and the tangential is customarily ignored in practice. The radial distortion Δr for a point with image coordinates (x, y) can be expressed as

$$\begin{aligned}
\Delta r &= k_1 r^3 + k_2 r^5 + k_3 r^7 \\
r &= \sqrt{(x - x_0)^2 + (y - y_0)^2},
\end{aligned} \qquad (1)$$

where k_i are the coefficients of radial distortion and (x_0, y_0) are the coordinates of the principal point of symmetry (PPS) in the image plane. Once these corrections have been applied, the photogrammetric part of the solution consists of obtaining three-dimensional information from a pair of two-dimensional photographs providing stereoscopic vision. The procedure is illustrated in Fig. 2.

2.3 Photogrammetry of Johnsons Glacier calving front using non-metric camera

We took (February 2013) the photographs of Johnsons Glacier calving front using a non-metric DSLR (Digital Single-Lens Reflex) camera Nikon D60. This is a typical, inexpensive 10 MP digital camera without excessive loss of accuracy. Obviously, its use for photogrammetric purposes requires a photogrammetric calibration, aimed to determine with sufficient accuracy the internal geometry (internal orientation) of the camera. This calibration process involves the use of Eq. (1), to calculate the k_i coefficients and the calibrated focal length. We detail below the main steps of the whole process.

a. Photogrammetric survey

The fieldwork took place in February 2013, under local conditions of few clouds and high visibility (more than 500 m). Several control points were established making

a network with permanent base stations near Johnsons Glacier front (Fig. 3, B1000, B2000 and B3000). These bases were measured using GNSS techniques, in particular, a Trimble 5700 GPS with horizontal and vertical accuracies of $\sigma_{xy} = 0.005$ m, $\sigma_z = 0.008$ m respectively. In addition, six control points were measured by the same technique at the sides of the calving front, on lateral moraines (points P100, P200, P300, P400, P500 and P600 in Fig. 3), with accuracies $\sigma_{xy} = 0.011$ m, $\sigma_z = 0.015$ m. These points were marked with red flags, to make them easy to recognize on the photographs. Finally, using a Wild Heerbrugg T1 theodolite, placed at B1000, B2000 and B3000, we measured the coordinates of the control points I10, I20, I30, I40, I50, I60 (Fig. 3) on the terminal part of the glacier (inaccessible because of high crevassing), with accuracies $\sigma_{xz} = 0.17$ m, $\sigma_z = 0.30$ m. We used the direct intersection method by resection from the three bases, allowing for some redundancy in the observations (Domínguez, 1993). T1 theodolite does not allow the measurement of distances, so only angles were measured.

b. Photograph shooting

As mentioned, the camera used is a Nikon D60 DSLR 10 MP camera, with lens 55–200 mm AF-S DX. In our case, the only possibility for taking pictures approximately perpendicular to the glacier calving front (normal photogrammetry) is using a boat. We used a zodiac boat, taking photos from lines 1 and 2 in Fig. 3. Line 1 is at approximately 400 m distance from the glacier front, and we used a focal length of 95 mm and set the focus to infinity. We used the same focal length and focus to infinity for the photos taken from Line 2, located at an approximate distance of 700 m from the glacier front. The overlap was higher than 95 %. To increase the number of control points over the glacier, we took more pictures from a location near P500. These photographs were taken with a focal length of 130 mm and focus to infinity. In this case we mounted the camera on a tripod over the ground, and took the photos using convergent photogrammetry. We thus obtained coordinates for additional control points (about 20 new points). In these photographs we could also observe three stakes from the net of velocity and mass balance measurements, which are customarily measured using GNSS techniques (Navarro et al., 2013).

c. Camera calibration

Calibration is made both before and after the photographic fieldwork, using the same settings. We chose a building named "Mirador" for calibration, which is located at Princesa de Eboli street in Madrid, which has an ideal configuration for this project (Fig. 4). This building has a large open space (square) at its front, allowing shot distances similar to those used during fieldwork at Johnsons Glacier (400 m). This distance also makes it easy to measure the corners of the windows to be used as calibration grid, using a total station (theodolite plus laser distance meter). For establishing the corners of the reticle, windows situated in a lower horizontal line, an upper horizontal line, two central horizontal lines and three vertical lines, defining the dot pattern shown in Fig. 4, were chosen. Measuring the calibration points using total station was motivated by the fact that we had no access to the architectural plans of the building. On the other hand, there is no guarantee that the real elements of the building coincide with those in the plans. We had to determine the coordinates of the calibration points using angular measurements because the total station did not allow measuring distances exceeding 100 m. We applied the direct intersection method (Domínguez, 1993). Observations were made from two stations. One of them had arbitrary local coordinates, and provided those for the other station and for all calibration points. The coordinates of the second station were obtained with an error of 12 mm. In this process we used Leica Geo Office software to obtain the coordinates of all calibration points with a root mean square (RMS) error of 53 mm, eliminating the points with residuals exceeding this quantity. We did the camera calibration for three focal lengths (85, 95 and 130 mm). We used the coordinates of the calibration points to obtain the k_i coefficients in Eq. (1) and

the position of the PPS, and finally calibrated the focal length using 102 points (Table 1).

d. Image preparation

One of the problems of fieldwork in extreme environments is that they often do not allow the repetition of field observations. Consequently, it is extremely important to collect as much data as possible during the observations (in our case, taking a large number of photographs), so that errors which are detected back in the office can be corrected or, at least, minimized. To minimize errors, we did a selection of photographic peer models with an overlap of 80 %. We also corrected distortion of all photographs using internal orientation parameters and generated a new set of distortion-corrected images (Fig. 5), which can be used by any photogrammetric software.

e. Calculation of ground coordinates

Taking the radial distortion-free photographs as a starting point, we applied collinearity conditions using the known control points to obtain the parameters of different transformations (Helmert 3D transformations), with which the ground coordinates of any point on the photographs can be retrieved. We used a total of 10 photos to make 9 models, with estimated errors $\sigma_{xy} = 0.70$ m, $\sigma_z = 0.55$ m. Note that in this case the altimetry error is lower than the planimetry error because the X-Z plane is parallel to the glacier calving front (Fig. 2). Once all parameters are calculated from different transformations, the photographs are introduced, together with these parameters, into our in-house developed software that allows the photogrammetric restitution, without the need of artificial stereoscopic vision (Fig. 6), using semi-automatic correlation (Luhmann et al., 2006).

Johnsons Glacier front has two main lines requiring photogrammetric restitution. The upper one is the top of the cliff of the calving front and the lower one is the waterline on the calving front. The calving front, as usual, is heavily crevassed (Fig. 7b), which facilitates the use of automatic correlation. An orthophoto corresponding to a total of 180 000 points produced by automatic correlation is shown in Fig. 7a, and the restitution of the upper and lower lines of the glacier front, as well as the main crevasses and/or fractures, are shown in Fig. 7b.

2.4 Surface-based GNSS measurements of the land-terminating fronts of Hurd Glacier

In addition to the positions of the land-terminating fronts of Sally Rocks, Las Palmas and Argentina lobes of Hurd Glacier determined from aerial photos (BAS flight of 1957 and UKHO flight of 1990) or satellite images (Quickbird images of 2007 and 2010), researchers from GSNCI have

Table 1. Coefficients of radial distortion and position for PPS. The two last columns show the RMS error and the maximum error for radial distortion respectively. We considered 102 points for the polynomial adjustment. The first column shows the values for the calibrated focal lengths (mm).

Focal mm	Points	x_0	y_0	k_1	k_2	k_3	σ	σ_{max}
85.23	102	1975 px	608 px	1.25213889313851E-08	−4.53330480188616E-15	4.48262617928856E-22	1 px	15 px
94.28	102	2043 px	602 px	1.18819659796645E-08	−4.07129365804265E-15	3.78566248003642E-22	1 px	13 px
131.50	102	2017 px	544 px	1.20656410994392E-08	−4.04403355146344E-15	3.68227991661486E-22	2 px	20 px

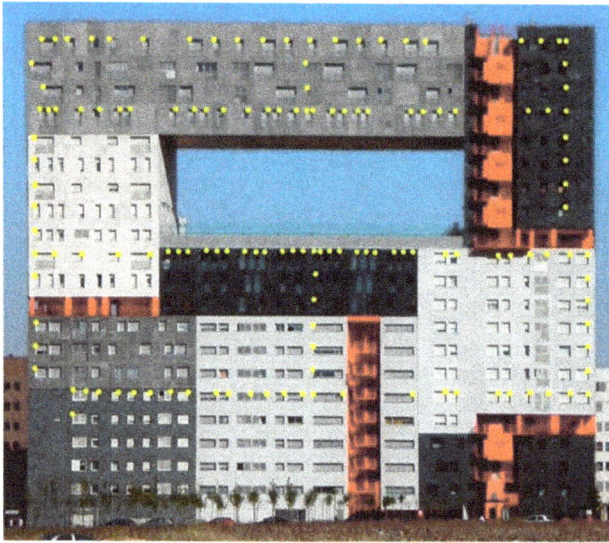

Figure 4. View of the facade of the building "Mirador", where we did the calibration of the non-metric camera used for photographic shots of Johnsons Glacier front. In yellow, the points measured using a total station.

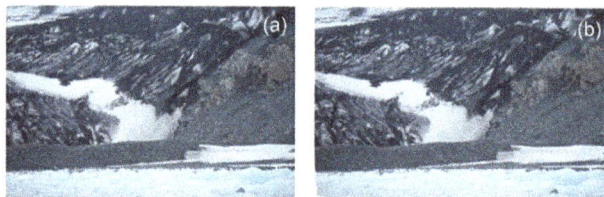

Figure 5. **(a)** Original image obtained with non-metric camera. **(b)** The rectified image, after applying the distortion function. At the lower left corner of the image an area affected by radial distortion can be seen.

measured the positions of these fronts several times during the period 2000–2012. The measurements were done using GNSS techniques (with a Trimble 5700 system, with Data Controller TSC2), with estimated horizontal accuracy between 0.07 and 0.60 m, depending on the campaign. The measurements were done either in real-time kinematics or in fast static (post-processed) mode. In all cases the GNSS base station was located at the neighboring Juan Carlos I station.

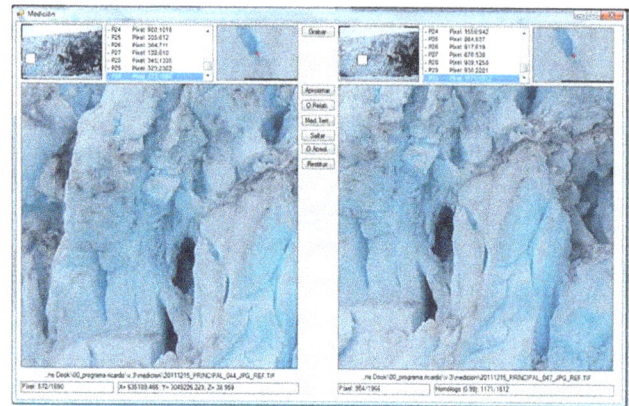

Figure 6. A snapshot of the main screen of our in-house developed software for photogrammetric restitution. This software does not require artificial stereoscopic vision for the restitution. By clicking on an item in the frame to the left, it locates the corresponding point in the right frame using automatic correlation.

2.5 Processing of data from various sources and metadata compilation

Here we further discuss the compilation and processing of the data from the various sources, with an emphasis on those not discussed so far (aerial photos and satellite images), and we also summarize the errors for each one.

The first set of data (DOCU 1) corresponds to the British photogrammetric flight of 26 December 1957, performed at a flight altitude of 13 500 feet, using a metric camera IX Eagle Mk I and a nominal focal length of 153.19 mm. Once restored using Digi3D software, it allowed us to obtain the position of the different glacier fronts, including Johnsons Glacier (Table 2). The estimated horizontal accuracies range within 0.60–1.0 m. The earlier work by Molina et al. (2007) also used these same photographs for 3-D restitution, but, as mentioned in the Introduction, they used paper-printed photos (implying additional distortion) while we used digitized versions from the original films. Moreover, in the current study we use the images only to get the planimetry at sea level, so we reach a better accuracy. Using the certificate of calibration for IX Eagle Mk I, we have rectified the photos and then georeferenced the photograms X26FID0052160 and X26FID0052131 using ARCGIS software with an 8-parameter transformation.

Table 2. ARCGIS shape files for the BAS photogrammetric flight of 1957. In the third column the RMS error is shown.

Filename	Year	σ_{xy} (m)	Geomatic acquisition method
CNDP-ESP_SIMRAD_FRONT_JOHNSON_1957.shp	1957	±0.60	Photogrammetric restitution, DOCU 1.
CNDP-ESP_SIMRAD_FRONT_SALLY_1957.shp	1957	±1.20	Photogrammetric restitution, DOCU 1.
CNDP-ESP_SIMRAD_FRONT_LAS_PALMAS_1957.shp	1957	±1.20	Photogrammetric restitution, DOCU 1.
CNDP-ESP_SIMRAD_FRONT_ARGENTINA _1957.shp	1957	±1.20	Photogrammetric restitution, DOCU 1.

Table 3. ARCGIS shape files for the UKHO photogrammetric flight of 1990. In the third column the RMS error is shown.

Filename	Year	σ_{xy} (m)	Geomatic acquisition method
CNDP-ESP_SIMRAD_FRONT_JOHNSON_1990.shp	1990	±2.00	Photogrammetric restitution, DOCU 2.
CNDP-ESP_SIMRAD_FRONT_SALLY_1990.shp	1990	±2.00	Photogrammetric restitution, DOCU 2.
CNDP-ESP_SIMRAD_FRONT_LAS_PALMAS_1990.shp	1990	±2.00	Photogrammetric restitution, DOCU 2.
CNDP-ESP_SIMRAD_FRONT_ARGENTINA _1990.shp	1990	±2.00	Photogrammetric restitution, DOCU 2.

Table 4. ARCGIS shape file corresponding to the satellite image of QUICKBIRD system program (2010). The image is corrected to sea level to obtain the correct planimetric position of Johnsons Glacier calving front. In the third column the RMS error is shown.

Name	Year	σ_{xy} (m)	Geomatic acquisition method
CNDP-ESP_SIMRAD_FRONT_JOHNSON_2010.shp	2010	±0.60	Aerial photo, DOCU 4.

The second set of data (DOCU 2) corresponds to another British photogrammetric flight, done in January 1990. In this case, a helicopter based on the ship HMS Endurance was used as platform, at a flight altitude of 10 000 feet, using a metric camera RMK A 15/23 with a nominal focal length of 153 mm. Once restored the data using the same Digi3D software as before, we obtained the position of the different glacier fronts, including Johnsons Glacier (Table 3). The estimated horizontal accuracy is 2.0 m.

DOCU 3 corresponds to the photogrammetric survey of Johnsons Glacier front made in 1999 by researchers from the University of Barcelona using a metric camera. However, this has not been considered in the current study because they only determined the top of the calving front cliff and not the position of the waterline, which is the line of interest to us (it is what we are comparing for the various images of Johnsons Glacier calving front).

DOCU 4 corresponds to an image captured in January 2010 by the Quickbird image system, which covers the entire work area. It is a raster file format GEOTIFF UTM 20S on the ellipsoid WGS84. Its original name was 10JAN29132854-P2AS052832138010_01_P001.TIF and it was obtained from http://www.euspaceimaging.com with reference ID 101001000B044C00. It is a black and white image with 16 bit digital values, but actually quantization levels are reduced to 11 bits. We restored this image using ARCGIS software to get a shape file with the position of Johnsons Glacier front (Table 4), upon rectification of the image to a horizontal plane at sea level using a projective transformation (Shan, 1999). The estimated horizontal accuracy is 0.60 m.

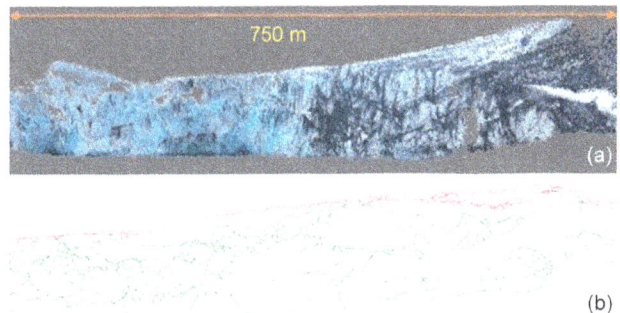

Figure 7. (a) Orthophoto of Johnsons Glacier calving front. This point cloud is made up of 180 000 points obtained by automatic correlation. (b) Johnsons Glacier calving front: in the upper part, in red, the uppermost line of the calving front, obtained by photogrammetric restitution; in the lower part, in blue, the waterline.

DOCU 5 is another Quickbird system image, taken in February 2007, which also covers the entire work area (with the exception of an insignificant rock outcrop to the northwest). It is a raster file format GEOTIFF UTM 20S on the ellipsoid WGS84. Its original name was 07FEB03135449-M2AS-052422572010_01_P001.TIF and it was obtained from http://www.euspaceimaging.com. It is an RGB color image with 16 bit digital values and an extra layer of near-infrared at 16 bits, but actually quantization levels are reduced to 11 bits. We restored this image using ARCGIS software to get a shape file with the positions of Johnsons Glacier and Las Palmas Glacier fronts (Table 5). This image was also rectified to a horizontal plane at sea level, using a projective transformation. The estimated horizontal accuracy is 2.30 m.

DOCU 6 corresponds to the surface-based GNSS measurements of the land-terminating fronts of Hurd Glacier discussed in Sect. 2.3 (Blewitt et al., 1997). The data compilation, with their estimated accuracies (ranging between 0.07 and 0.60 m), is shown in Table 6.

Finally, DOCU 7 corresponds to the photogrammetric survey of Johnsons Glacier front done by the authors in February 2013 with a non-metric camera, discussed in Sect. 2.2. The corresponding metadata are given in Table 7. The estimated horizontal accuracy is 0.70 m.

3   Results and discussion: evolution of the glacier fronts during recent decades

3.1   Johnsons Glacier

The position of Johnsons Glacier calving front (the waterline position) at the beginning and end of 1957 is shown in Fig. 8, while its position for various years within the period 1957–2013 is shown in Fig. 9. As Johnsons is a tidewater glacier, its front position changes, though influenced by climate, are mostly driven by its internal dynamics, including feedback mechanisms involving the balance of forces, the flow and calving. We observe that the calving front advanced 74 m in its central part (segment A) between 1957 and 1990. Then the glacier front retreated 171 m (sum of the segments A, L and F) between 1990 and 2007, to remain stable until 2010 and then re-advance by 31 m in its central part (segment L) between 2010 and 2013. In spite of this latter re-advance of the central part, the southern part of the front retreated ca. 57 m (segment J) during the same period 2010–2013. Also note that, in the northern part, the glacier front has retreated over 97 m between 1957 and 2013 (E segment). The position of the front in the neighbourhood of segment E was estimated using the photo X26FID0093015 obtained at the beginning of 1957 (Rodríguez, 2014) from an incomplete BAS photogrammetric flight, shown in Fig. 8a.

The darker area seen on Fig. 8, in the southwestern area of the terminal part of the glacier, corresponds to an accumulation, due to intense folding and faulting in this part of Johnsons Glacier, of volcanic ash from tephra layers stemming from the volcanic eruptions at the neighboring Deception Island (Ximenis et al., 2000).

3.2   Sally Rocks front of Hurd Glacier

In this case, the front position changes are shown in Fig. 10. Hurd is a land-terminating glacier, which implies that the front position changes are mainly driven by climate-induced variations (e.g., increased melting), coupled with the (slow) dynamic response to the climate variations (due, e.g., to geometry changes associated to accumulation-ablation changes or variation in basal lubrication because of changes in the

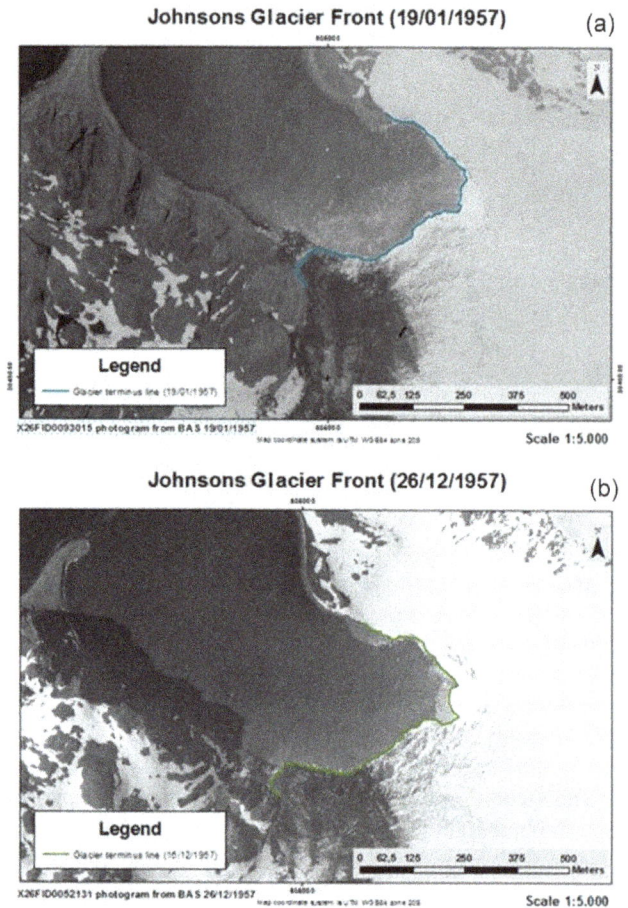

Figure 8. (a) Position of Johnsons Glacier calving front at the beginning of 1957 (blue line). The base map corresponds to the BAS photogrammetric flight of date 19 January 1957. This flight was not completed, and it can only be used for planimetric measurements at sea level. **(b)** Position of Johnsons Glacier calving front at the end of 1957 (green line). The base map corresponds to the BAS photogrammetric flight of date 26 December 1957. The scale of the figure does not match with the numerical map scale indicated.

amount of meltwater reaching the glacier bed). We observe that the glacier front glacier retreated by 116 m in its central area (segment A) between 1957 and 1990. This was followed by a further retreat by 60 m (segment B) between 1990 and 2000, and yet further retreats by 47 m (C segment) during 2000–2006, and by 36 m (D segment) during 2006–2009. From 2009, the front position has remained stable until 2012. These changes are consistent with the overall warming trend in the South Shetland Islands, where the decade 1995–2006 has been the warmest since the 1960s, while the decade 2006–2015 has been colder than the previous one (by 0.5 °C in the summer and 1.0 °C in the winter; Oliva et al., 2016), and also with the mass balance record, which shows a shift to predominantly positive (i.e., mass gains) surface mass balances in Hurd and Johnsons glaciers starting around 2007/2008 (Navarro et al., 2013).

Table 5. ARCGIS shape file corresponding to the satellite image of QUICKBIRD system program (2007). The image is corrected to sea level to obtain the correct planimetric position of the glacier fronts. In the third column the RMS error is shown.

Name	Year	σ_{xy} (m)	Geomatic acquisition method
CNDP-ESP_SIMRAD_FRONT_JOHNSON_2007.shp	2007	±2.30	Aereal photo, DOCU 5.
CNDP-ESP_SIMRAD_FRONT_LAS_PALMAS_2007.shp	2007	±2.30	Aereal photo, DOCU 5.

Table 6. ARCGIS shape files corresponding to the GSNCI data inventory for Argentina, Las Palmas and Sally Rocks fronts of Hurd Glacier, determined using GNSS techniques. In the third column the RMS error is shown.

Name	Year	σ_{xy} (m)	Geomatic acquisition method
CNDP-ESP_SIMRAD_FRONT_SALLY_2000_2001.shp	2000	±0.07	GNSS pole, DOCU 6.
CNDP-ESP_SIMRAD_FRONT_SALLY_2004_2005.shp	2005	±0.60	GNSS backpack, DOCU 6.
CNDP-ESP_SIMRAD_FRONT_SALLY_2005_2006.shp	2006	±0.60	GNSS backpack, DOCU 6.
CNDP-ESP_SIMRAD_FRONT_SALLY_2007_2008.shp	2008	±0.07	GNSS pole, DOCU 6.
CNDP-ESP_SIMRAD_FRONT_SALLY_2008_2009.shp	2009	±0.60	GNSS backpack, DOCU 6.
CNDP-ESP_SIMRAD_FRONT_SALLY_2009_2010.shp	2010	±0.07	GNSS pole, DOCU 6.
CNDP-ESP_SIMRAD_FRONT_SALLY_2010_2011.shp	2011	±0.07	GNSS pole, DOCU 6.
CNDP-ESP_SIMRAD_FRONT_SALLY_2011_2012.shp	2012	±0.07	GNSS pole, DOCU 6.
CNDP-ESP_SIMRAD_FRONT_LAS_PALMAS_2000_2001.shp	2000	±0.07	GNSS pole, DOCU 6.
CNDP-ESP_SIMRAD_FRONT_LAS_PALMAS_2004_2005.shp	2005	±0.60	GNSS backpack, DOCU 6.
CNDP-ESP_SIMRAD_FRONT_LAS_PALMAS_2005_2006.shp	2006	±0.60	GNSS backpack, DOCU 6.
CNDP-ESP_SIMRAD_FRONT_LAS_PALMAS_2007_2008.shp	2008	±0.07	GNSS pole, DOCU 6.
CNDP-ESP_SIMRAD_FRONT_LAS_PALMAS_2008_2009.shp	2009	±0.60	GNSS backpack, DOCU 6.
CNDP-ESP_SIMRAD_FRONT_LAS_PALMAS_2009_2010.shp	2010	±0.07	GNSS pole, DOCU 6.
CNDP-ESP_SIMRAD_FRONT_LAS_PALMAS_2010_2011.shp	2011	±0.07	GNSS pole, DOCU 6.
CNDP-ESP_SIMRAD_FRONT_LAS_PALMAS_2011_2012.shp	2012	±0.07	GNSS pole, DOCU 6.
CNDP-ESP_SIMRAD_FRONT_ARGENTINA _2000_2001.shp	2000	±0.07	GNSS pole, DOCU 6.
CNDP-ESP_SIMRAD_FRONT_ARGENTINA _2004_2005.shp	2005	±0.60	GNSS backpack, DOCU 6.
CNDP-ESP_SIMRAD_FRONT_ARGENTINA _2005_2006.shp	2006	±0.60	GNSS backpack, DOCU 6.
CNDP-ESP_SIMRAD_FRONT_ARGENTINA _2007_2008.shp	2008	±0.07	GNSS pole, DOCU 6.
CNDP-ESP_SIMRAD_FRONT_ARGENTINA _2008_2009.shp	2009	±0.60	GNSS backpack, DOCU 6.
CNDP-ESP_SIMRAD_FRONT_ARGENTINA _2009_2010.shp	2010	±0.07	GNSS pole, DOCU 6.
CNDP-ESP_SIMRAD_FRONT_ARGENTINA _2010_2011.shp	2011	±0.07	GNSS pole, DOCU 6.
CNDP-ESP_SIMRAD_FRONT_ARGENTINA_2011_2012.shp	2012	±0.07	GNSS pole, DOCU 6.

The black spots on the glacier surface seen in Figure 10 are mostly ash from tephra layers, but these are often mixed with subglacial sediments taken to the glacier surface through fractures associated to thrust faults in the glacier snout resulting from the compressional regime. The compressional regime is mainly a consequence of fact that the glacier snout is frozen to bed (Molina et al., 2007; Navarro et al., 2009; Molina, 2014). This mixture of volcanic ash and subglacial sediments often accumulates at the surface in the form of pinnacles. According to Molina (2014, p. 126–127), most tephra shown in Fig. 10 likely correspond to eruptions well before 1829 (what is denoted as "oldest layers" in Ximenis et al., 2000, for Johnsons Glacier). We say likely because Molina did not analyze 1957 photos but images taken during 2004–2009, and did the geochemical analysis (X-ray fluorescence) with samples taken within this period. In any case, as Molina (2014) has pointed out, the tephra deposits in these terminal zones should not be used for purposes of dating

structures, because these spots in the terminal part of Hurd Glacier fronts are the result of accumulation of tephra flushed out, upon melting, from tephra layers corresponding to various eruptions. Consequently, the ash layers in this zone cannot be used to reliably infer past mass balance rates.

3.3 Las Palmas front of Hurd Glacier

The front position changes for Las Palmas Glacier are shown in Fig. 11. We can see that the glacier front retreated 11 m in its central part (segment A) between 1957 and 1990. Then, the front retreated 24 m (segment F) between 1990 and 2000. This was followed by a further retreat by 17 m (segment B) during 2000–2005, another 14 m (D segment) during 2005–2007, and finally another 10 m during 2007–2009, to remain stable thereafter. For Las Palmas front, the retreat during the period 1957–1990 was significantly lower than that for Sally Rocks front. This is likely due to the fact that, in 1957,

Table 7. ARCGIS shape file corresponding to the photogrammetric restitution of the Johnsons Glacier calving front in February 2013. The photo was obtained with a non-metric camera. In the third column the RMS error is shown.

Name	Year	σ_{xy} (m)	Geomatic acquisition method
CNDP-ESP_SIMRAD_FRONT_JOHNSON_2013.shp	2013	±0.70	Photogrammetric restitution, DOCU 7.

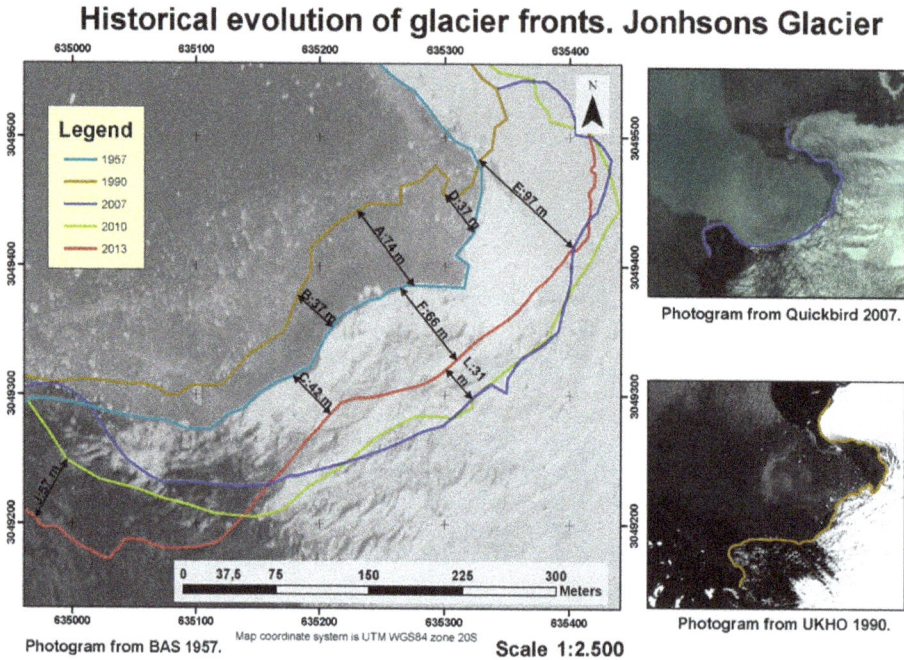

Figure 9. Calving front position changes of Johnsons Glacier. The oldest front position available corresponds to 1957 (in light blue) and the most recent to 2013 (in red). The base map corresponds to the BAS aerial photograph taken on 19 January 1957. The scale of the figure does not match with the numerical map scale indicated.

Las Palmas front ended at sea, and thus had a comparatively larger thickness at its terminal zone as compared with Sally Rocks front.

3.4 Argentina front of Hurd Glacier

In this case, shown in Fig. 12, the glacier front seems to have advanced by a small amount, ca. 5 m, in its central area (segment A) between 1957 and 1990. Then the front retreated by 70 m (segment A + segment B) between 1990 and 2000, to retreat another 15 m (C segment) during 2000–2005, a further 6 m during 2005–2008 (segment D) and yet another 14 m (segment I) until 2009, to remain stable since then. The apparent small advance between 1957 and 1990 does not fit with the overall retreating trend of the land-terminating fronts of Sally Rocks and Las Palmas. In fact, it could be due to a misinterpretation of the front position in 1990, due to remnants of snow cover on the terminal zone; or it could, instead, be real and have been induced by the dynamic response of this particular side lobe to climate changes, since each glacier reacts with different response times according

to its size and geometry (the observed change position is, in fact, very small).

4 Conclusions

The following conclusions may be drawn from our study:

1. Close-range photogrammetry with non-metric cameras is very suitable for determining the position of glacier fronts and, in particular, the calving fronts of tidewater glaciers, for which other techniques, such as surface-based GNSS measurements, are not possible. Moreover, it compares favorably in terms of costs with other methods such as photogrammetry with metric cameras or laser scanner systems. For the measurement of terminus position of land-terminating glaciers, surface-based GNSS techniques are fast and reliable. These surface-based techniques are effectively complemented, whenever available, with remote-sensing images, from either aerial photogrammetric flights or satellite optical imagery.

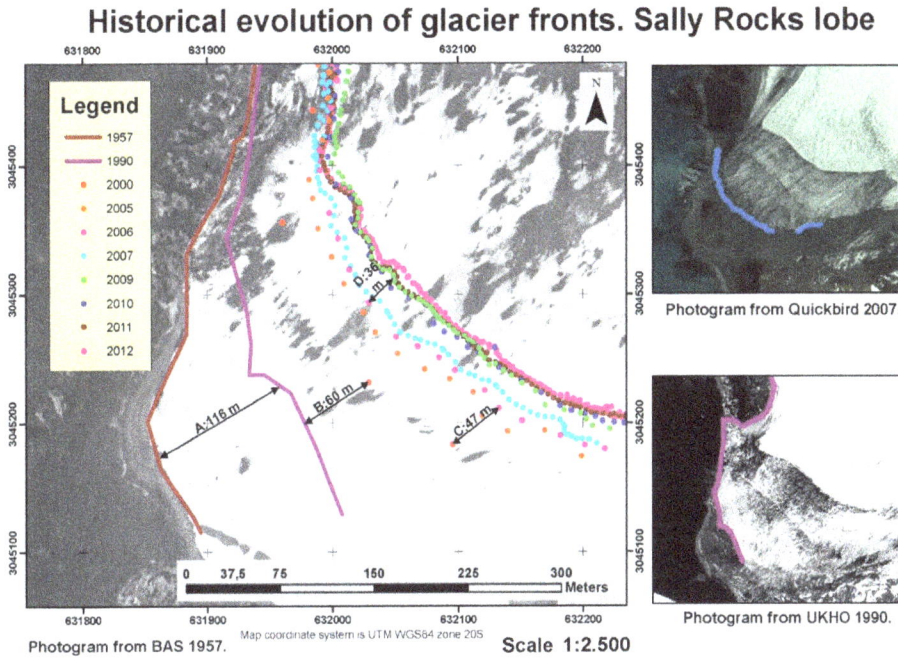

Figure 10. Terminus position changes of Sally Rocks front of Hurd Glacier. The oldest front position available corresponds to 1957 (in purple) and the latest to 2012 (in cyan). The base map corresponds to the BAS aerial photograph of 1957. The scale bar shown in the figure does not match with the map scale indicated.

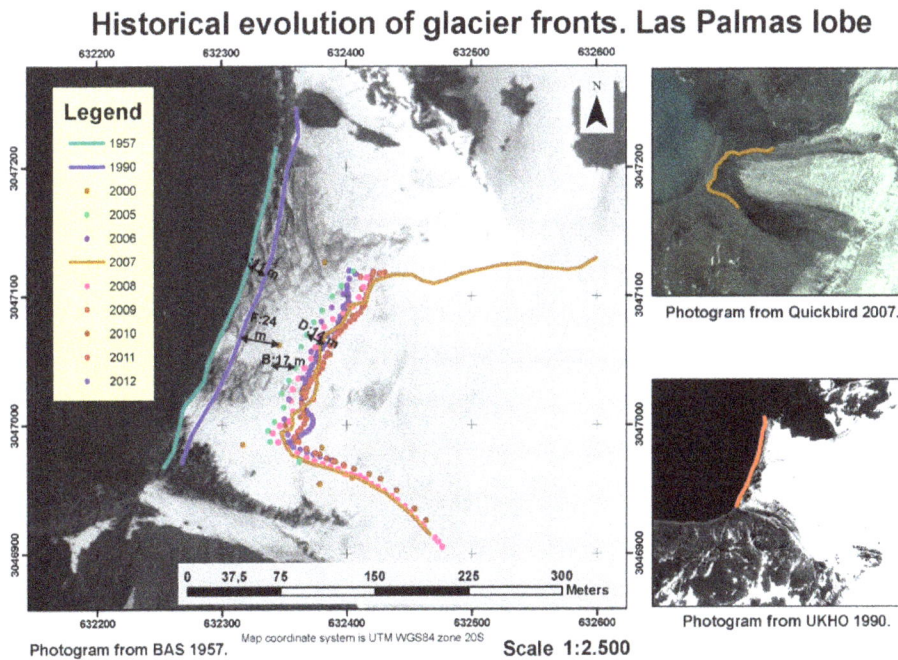

Figure 11. Terminus position changes of the front of Las Palmas side lobe of Hurd Glacier. The oldest front position available corresponds to 1957 (in green) and the latest to 2012 (blue points). The base map corresponds to the BAS aerial photograph of 1957. The scale bar shown in the figure does not match with the map scale indicated.

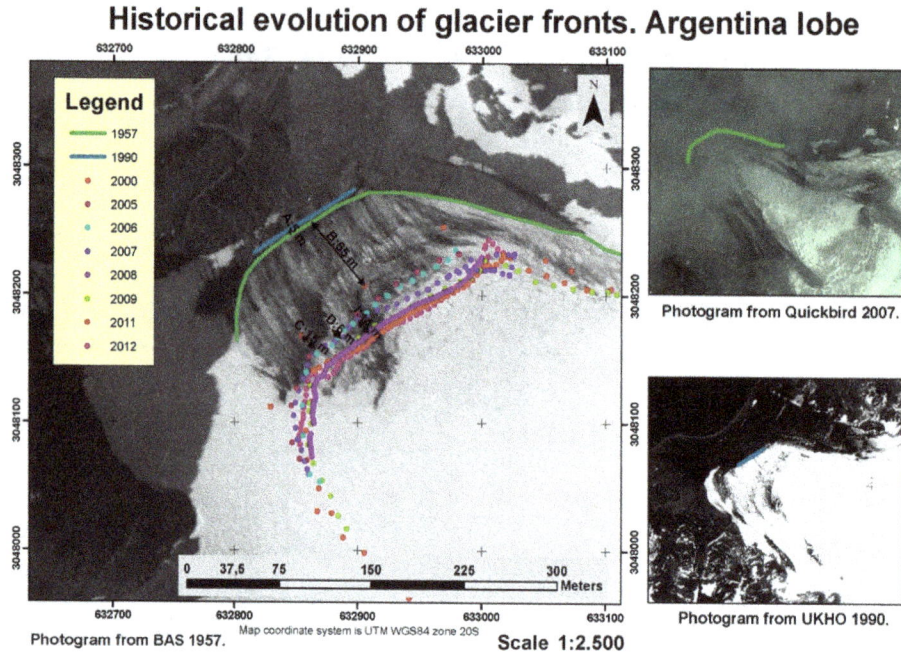

Figure 12. Terminus position changes of the front of Argentina side lobe of Hurd Glacier. The oldest front position available corresponds to 1957 (in purple) and the latest for the year 2012 (in cyan). The base map corresponds to the BAS aerial photograph of 1957. The outline marked on the "diffuse" 2007 Quickbird image corresponds to ground-based GNSS measurements. The scale bar shown in the figure does not match with the map scale indicated.

2. The calving front of Johnsons Glacier has shown, during the period 1957–2013, several episodes of advance and retreat (and likely many more happened during long intermediate periods lacking observations). This is a tidewater glacier, so its front position changes, though can be influenced by climate, are mostly driven by the internal glacier dynamics, including various feedback mechanisms.

3. The land-terminating fronts of Hurd Glacier (Sally Rocks, Las Palmas and Argentina) have experienced during the period 1957–2009 an overall retreating trend, remaining nearly stationary during 2009–2012. Being land-terminating, the front position changes are mostly driven by climate-related processes such as accumulation and melting variations, and their associated glacier dynamic response. The observed front variations are consistent with the regional climate changes. In particular, with an extremely warm 1995–2006 decade, followed by a colder 2006–2015 decade.

Acknowledgements. This work was supported by grant CTM2014-56473-R from the Spanish National Plan of R&D.

Edited by: D. Carlson

References

ADD Consortium: Antarctic Digital Database, Version 3.0, database, manual and bibliography, Scientific Committee on Antarctic Research, Cambridge, 2000.

Benjumea, B. and Teixidó, T.: Seismic reflection constrains on the glacial dynamics of Johnsons Glacier, Antarctica, J. Appl. Geophys., 46, 31–44, 2001.

Benjumea, B., Teixidó, T., Ximenis. L., and Furdàda, G.: Prospección sísmica en el Glacier Johnsons, Isla Livingston (Antártida). (Campañas antárticas 1996–1997 y 1997–1998), Acta Geológica Hispánica, 34, 375–389, 1999.

Benjumea, B., Macheret, Y. Y., Navarro F., and Teixidó, T.: Estimation of water content in a temperate glacier from radar and seismic sounding data, Ann. Glaciol., 37, 317–324, 2003.

Blewitt, G.: Basics of the GPS Technique: Observation Equations, Department of Geomatics, University of Newcastle, United Kingdom, 1997.

Bliss, A., Hock, R., and Cogley, J. G.: A new inventory of mountain glaciers and ice caps for the Antarctic periphery, Ann. Glaciol., 54, 191–199, 2013.

Brown, D. C.: Close-range camera calibration, PE&RS, 37, 855–866, 1971.

Calvet, J., García Sellés, D., and Corbera, J.: Fluctuaciones de la extensión del casquete glacial de la isla Livingston (Shetland del Sur) desde 1956 hasta 1996, Acta Geológica Hispánica, 34, 365–374, 1999.

Domínguez, F.: Topografía general y aplicada, Ediciones Mundi-Prensa, Madrid, 1993.

Directorate of Overseas Surveys (DOS): Sheet W62 58, British Antarctic Territory, South Shetland Island, Scale 1 : 200 000, 1968a.

Directorate of Overseas Surveys (DOS): Sheet W62 60, British Antarctic Territory, South Shetland Island, Scale 1 : 200 000, 1968b.

Furdàda, G., Pourchet, M., and Vilaplana, J. M.: Characterization of Johnsons Glacier (Livingston Island, Antarctica) by means of shallow ice cores and their tephra and ^{137}Cs contents. Acta Geológica Hispánica, 34, 391–401, 1999.

Jonsell, U. Y., Navarro, F. J., Bañón, M., Lapazaran, J. J., and Otero, J.: Sensitivity of a distributed temperature–radiation index melt model based on AWS observations and surface energy balance fluxes, Hurd Peninsula glaciers, Livingston Island, Antarctica, The Cryosphere, 6, 539–552, doi:10.5194/tc-6-539-2012, 2012.

Kraus, K.: Photogrammetry, Vol. 1: Fundamentals and Standard Processes, Dümmlers, 1993.

Kraus, K.: Photogrammetry. Geometry from Images and Laser Scans, 2nd Edn., de Gruyter, 2007.

Luhmann, T., Robson, S., Kyle, S., and Harley, I.: Close Range Photogrammetry: Principles, Techniques and Applications, Whittles Publishing, 2006.

Macheret, Y. Y., Otero, J., Navarro, F. J., Vasilenko, E. V., Corcuera, M. I., Cuadrado, M. L., and Glazovsky, A. F.: Ice thickness, internal structure and subglacial topography of Bowles Plateau ice cap and the main ice divides of Livingston Island, Antarctica, by ground-based radio-echo sounding, Ann. Glaciol., 50, 49–56. 2009.

Martín, C., Navarro, F. J., Otero, J., Cuadrado, M. L., and Corcuera, M. I.: Three-dimensional modelling of the dynamics of Johnsons glacier (Livingston Island, Antarctica), Ann. Glaciol., 39, 1–8, 2004.

Molina, C.: Caracterización dinámica del glaciar Hurd combinando observaciones de campo y simulaciones numérica, PhD Thesis, Universidad Politécnica de Madrid, 2014.

Molina, C., Navarro, F., Calvet, J., García-Selles, D., and Lapazaran, J.: Hurd Peninsula glaciers, Livingston Island, Antarctica, as indicators of regional warming: ice-volume changes during period 1956–2000, Ann. Glaciol., 46, 43–49, 2007.

Navarro, F. J., Macheret, Y. Y., and Benjumea, B.: Application of radar and seismic methods for the investigation of temperate glaciers, J. Appl. Geophys., 57, 193–211, 2005.

Navarro, F. J., Otero, J., Macheret, Y. Y., Vasilenko, E. V., Lapazaran, J. J., Ahlstrøm, A. P., and Machío, F.: Radioglaciological studies on Hurd Peninsula glaciers, Livingston Island, Antarctica, Ann. Glaciol., 50, 17–24, 2009.

Navarro, F., Jonsell, U., Corcuera, M., and Martín-Español, A.: Decelerated mass loss of Hurd and Johnsons glaciers, Livingston Island, Antarctic Peninsula, J. Glaciol., 59, 115–128, 2013.

Oliva, M., Navarro, F., Hrbáček, F., Hernández, A., Nývlt, D., Pereira, P., Ruiz-Fernández, J., and Trigo, R.: Recent regional cooling of the Antarctic Peninsula and its impacts on the cryosphere, Nature Communications, submitted, 2016.

Osmanoglu, B., Navarro, F. J., Hock, R., Braun, M., and Corcuera, M. I.: Surface velocity and mass balance of Livingston Island ice cap, Antarctica, The Cryosphere, 8, 1807–1823, doi:10.5194/tc-8-1807-2014, 2014.

Otero, J., Navarro, F. J., Martín, C., Cuadrado M. L., and Corcuera, M. I.: A three-dimensional calving model: numerical experiments on Johnsons Glacier, Livingston Island, Antarctica, J. Glaciol., 56, 200–214, 2010.

Palà, V., Calvet, J., García-Sellés, D., and Ximenis, L.: Fotogrametría terrestre en el Glaciar Johnsons, Isla Livingston, Antártida, Acta Geológica Hispánica, 34, 427–445, 1999.

Pfeffer, T., Arendt, A., Bliss, A., Bolch, T., Cogley, J. G., Gardner, A. S., Hagen, J. O., Hock, R., Kaser, G., Kienholz, C., Miles, E. S. Moholdt, G., Moelg, N., Paul, F., Radić, V., Rastner, P., Raup, B. H., Rich, J., Sharp, M. J., and the Randolph Consortium: The Randolph Glacier Inventory: a globally complete inventory of glaciers, J. Glaciol., 60, 537–552, 2014.

Rodríguez, R.: Integración de modelos numéricos de glaciares y procesado de datos de georradar en un sistema de información geográfica, PhD Thesis, Universidad Politécnica de Madrid, 2014.

Rodríguez, R. and Navarro, F.: Study of the fronts of Johnsons and Hurd Glaciers (Livingston Island, Antarctica) from 1957 to 2013, Universidad Politécnica de Madrid, https://doi.pangaea.de/10.1594/PANGAEA.845379, 2015.

Shan, J. J.: A Photogrammetric Solution for Proyective Reconstruction, Department of Geomatics Engineering, Purdue University, SPIE Conference on Vision Geometry VIII, Denver, Colorado, 296–304, 1999.

Wolf, P. R.: Elements of Photogrammetry with Air Photo Interpretation and Remote Sensing, Second Edition, McGraw-Hill Book Company, New York, 1983.

Ximenis, L.: Dinàmica de la Glacera Johnsons (Livingston, Shetland del Sud, Antàrtida), PhD Thesis, University of Barcelona, 2001.

Ximenis, L., Calvet, J., Enrique, J., Corbera, J., Fernández de Gamboa C., and Furdàda, G.: The measurement of ice velocity, mass balance and thinning-rate on Johnsons Glacier, Livingston Island, South Shetland Islands, Antarctica, Acta Geológica Hispánica, 34, 403–409, 1999.

Ximenis, L., Calvet, J., García, D., Casas, J. M., and Sàbat, F.: Folding in the Johnsons glacier, Livingston Island, Antarctica, in: Deformation of Glacial Materials, edited by: Maltman, A. J., Hubbard, B. A., and Hambrey, M. J., The Geological Society London, special publ., 176, 147–157, 2000.

12

A consistent glacier inventory for Karakoram and Pamir derived from Landsat data: distribution of debris cover and mapping challenges

Nico Mölg[1], Tobias Bolch[1], Philipp Rastner[1], Tazio Strozzi[2], and Frank Paul[1]

[1]Department of Geography, University of Zurich, Winterthurerstr. 190, 8057 Zurich, Switzerland, Switzerland
[2]Gamma Remote Sensing, Worbstr. 225, 3073 Gümligen, Switzerland

Correspondence: Nico Mölg (nico.moelg@geo.uzh.ch)

Abstract. Knowledge about the coverage and characteristics of glaciers in High Mountain Asia (HMA) is still incomplete and heterogeneous. However, several applications, such as modelling of past or future glacier development, run-off, or glacier volume, rely on the existence and accessibility of complete datasets. In particular, precise outlines of glacier extent are required to spatially constrain glacier-specific calculations such as length, area, and volume changes or flow velocities. As a contribution to the Randolph Glacier Inventory (RGI) and the Global Land Ice Measurements from Space (GLIMS) glacier database, we have produced a homogeneous inventory of the Pamir and the Karakoram mountain ranges using 28 Landsat TM and ETM+ scenes acquired around the year 2000. We applied a standardized method of automated digital glacier mapping and manual correction using coherence images from the Advanced Land Observing Satellite 1 (ALOS-1) Phased Array type L-band Synthetic Aperture Radar 1 (PALSAR-1) as an additional source of information; we then (i) separated the glacier complexes into individual glaciers using drainage divides derived by watershed analysis from the ASTER global digital elevation model version 2 (GDEM2) and (ii) separately delineated all debris-covered areas. Assessment of uncertainties was performed for debris-covered and clean-ice glacier parts using the buffer method and independent multiple digitizing of three glaciers representing key challenges such as shadows and debris cover. Indeed, along with seasonal snow at high elevations, shadow and debris cover represent the largest uncertainties in our final dataset. In total, we mapped more than 27 800 glaciers > 0.02 km^2 covering an area of 35 520 ± 1948 km^2 and an elevation range from 2260 to 8600 m. Regional median glacier elevations vary from 4150 m (Pamir Alai) to almost 5400 m (Karakoram), which is largely due to differences in temperature and precipitation. Supraglacial debris covers an area of 3587 ± 662 km^2, i.e. 10 % of the total glacierized area. Larger glaciers have a higher share in debris-covered area (up to > 20 %), making it an important factor to be considered in subsequent applications.

1 Introduction

Glacier outlines and their accompanying attributes as recorded in glacier inventories provide the baseline for climate change impact assessments (Vaughan et al., 2013), numerous hydrology-related calculations that consider water resources and their changes (e.g. drinking water, irrigation, hydropower production, run-off, sea-level rise) (e.g. Kraaijenbrink et al., 2017; Bliss et al., 2014), climatic character-istics (Sakai et al., 2015), or modelling of past and future glacier changes (e.g. Huss and Hock, 2015). All of this applies to most catchments of High Mountain Asia (HMA), although some of their glacier meltwater does not directly contribute to sea-level rise as related rivers end in endorheic basins (e.g. Tarim basin, Aral Sea basin).

When glacier outlines are of poor quality, related hydrologic calculations at the catchment scale (e.g. Immerzeel et al., 2010) have higher uncertainties. This situation has im-

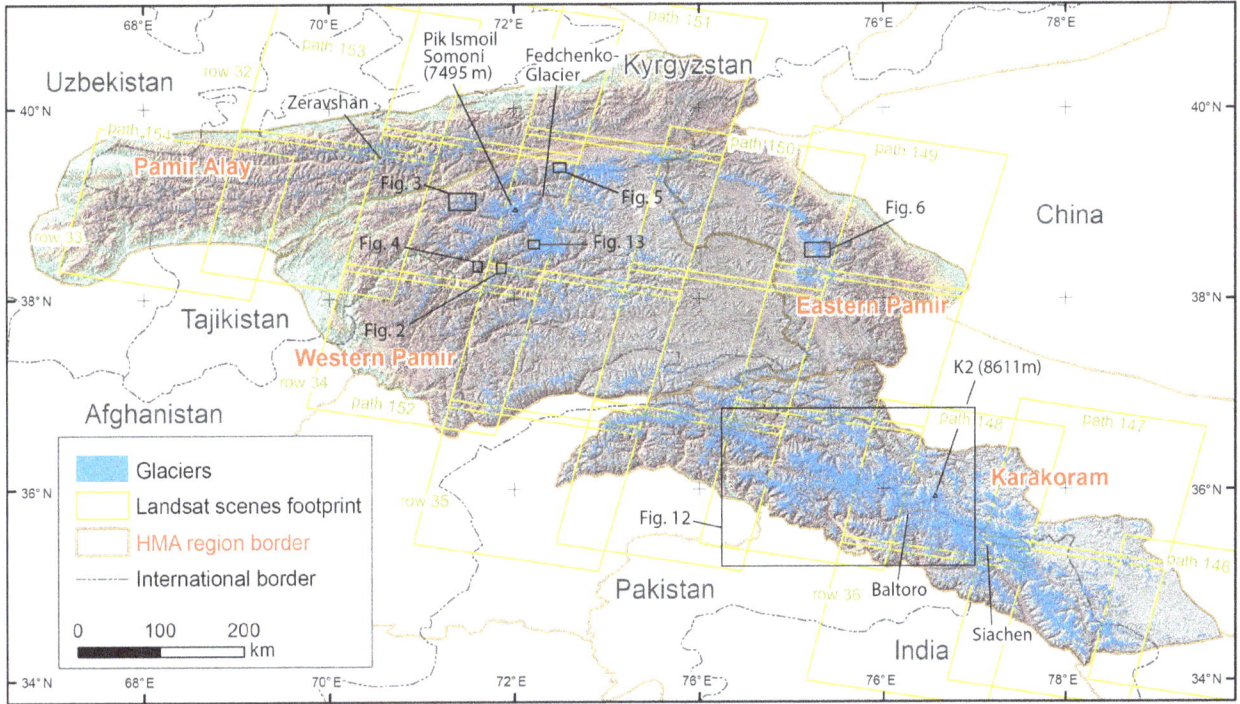

Figure 1. The study region in High Mountain Asia (HMA) covers four mountain ranges. Annotations denote locations of figures in the paper and points of orientation. International borders are tentative only as they are disputed in several regions.

proved significantly during recent years, during which several large-scale glacier inventories for HMA have been published – among others the Glacier Area Mapping for Discharge from the Asian Mountains (GAMDAM) inventory (Nuimura et al., 2015) and the second Chinese Glacier Inventory (CGI, Guo et al., 2015). Because the currently available version of GAMDAM only partially considered ice cover on steep slopes and the CGI only covered Chinese territory, a homogeneous basis for precise calculations covering all relevant catchments at the large scale is still missing. As both inventories have been combined for version 5.0 and transferred unchanged to version 6.0 of the Randolph Glacier Inventory (RGI) (Arendt et al., 2015; RGI Consortium, 2017), the related regional-scale calculations using this version (e.g. Brun et al., 2017; Kraaijenbrink et al., 2017; Dehecq et al., 2015; Kääb et al., 2015) still suffer from uncertainties which stem from outlines of varying quality.

To overcome this situation, several regional-scale studies digitized glacier outlines themselves (e.g. Rankl and Braun, 2016; Minora et al., 2016) to have better control on data quality. But these again applied different criteria to delineate glacier extents and are thus not comparable to the existing datasets, making change assessment difficult. On the other hand, the Karakoram and Pamir regions are characterized by a high number of surge-type glaciers (Bhambri et al., 2017; Copland et al., 2011; Kotlyakov et al., 2008) with often strong geometric changes over a short period of time (Paul, 2015; Quincey et al., 2015). A precise inventory is

key to determine and maybe better understand such changes. Moreover, the large number of debris-covered glaciers in the region (Herreid et al., 2015; Minora et al., 2016) results in interpretation differences and is a large source of uncertainty.

The correct delineation of debris is also important for detecting very subtle past glacier changes (Scherler et al., 2011b) and to correctly model future glacier development (e.g. Kraaijenbrink et al., 2017; Shea et al., 2015), as surface mass balance of ice under a supraglacial debris layer is different from that of clean ice (Ragettli et al., 2015, 2016; Brock et al., 2010; Nicholson and Benn, 2006). The information on debris extent should thus be included in large-scale glacier inventories (Kraaijenbrink et al., 2017).

The main objective of this study is to present a consistent dataset of glacier coverage for the higher and more extensively glacierized mountain ranges – Pamir Alai, the western and eastern Pamir, and the Karakoram of HMA (Fig. 1) – along with the spatial distribution of debris cover for the years around 2000. In addition, we present a structured overview of the difficulties related to glacier mapping in this region as well as an estimate of the respective uncertainties. Key challenges are the entity assignment, as many glaciers in this region are of a surge type, and tributary glaciers can be either connected to or disconnected from a larger main glacier; the already mentioned mapping of debris-covered glacier ice; and the differentiation of debris-covered glaciers from rock glaciers that are increasingly abundant towards the north and the drier east of the study region.

2 Study region

2.1 Location and glacier characteristics

The study area comprises a major share of the western part of HMA (Fig. 1). It stretches over $\sim 300\,000\,\mathrm{km}^2$ and fully covers the mountain ranges of (i) Pamir and its northern neighbour Pamir Alai in Uzbekistan, Turkmenistan, Kyrgyzstan, Tajikistan, Afghanistan and China and (ii) the Karakoram in Pakistan, India, and China.

The mountain ranges reach their highest elevations between 5500 m (Pamir Alai) and more than 8600 m (Karakoram, Hindu Kush), with K2 being the highest peak with 8611 m. Glaciers are found from ~ 2300 m a.s.l. up to the highest peaks. The central Karakoram and inner Pamir are two of the most heavily glacierized mountain regions worldwide and include some extremely large glaciers such as Baltoro, Siachen, and Fedchenko with sizes of about 810, 1094, and $573\,\mathrm{km}^2$, respectively. High peaks and deeply incised valleys create an extreme topographic relief that is also reflected in the geometry of the glaciers, the majority of which are valley glaciers. In the Pamir, numerous cirques are also present, and hanging glaciers can be found at high elevations in all regions. Larger, flat, high-altitude accumulation areas are rare and can only be found for some of the largest glaciers. Due to the steep terrain, most glaciers are partly fed by avalanches from the surrounding steep valley walls (Dobreva et al., 2017; Iturrizaga, 2011; Hewitt, 2011; Scherler et al., 2011a). This also causes an abundance of glaciers with partly or completely debris-covered tongues. Whereas debris cover makes glacier mapping difficult, the strong geometric changes in surging glaciers create additional challenges for glacier inventory compilation. Rock glaciers are present in all periglacial environments and are abundant also in our study region.

2.2 Climate

The climate of the study region can be subdivided into two major and independent regimes that define both the thermal and the moisture conditions. The Pamir and the major part of the Karakoram are predominantly influenced by westerly air flows throughout the year (e.g. Bookhagen and Burbank, 2010; Singh et al., 1995); towards the south-east the influence of the Indian summer monsoon (approx. June–September) becomes continuously stronger (e.g. Archer and Fowler, 2004). In general, outer western areas of the mountain ranges (Pamir Alai, western Pamir, Hindu Kush, southeastern Karakoram) receive more precipitation, whereas further inland and to the east (eastern Pamir, central Karakoram) the climate becomes drier and more continental (Lutz et al., 2014).

For most of the study region, the main share of precipitation falls in winter and spring (Archer and Fowler, 2004; Bookhagen and Burbank, 2010). In regions dominated by westerlies, winter precipitation is mostly advective, whereas convective precipitation plays an important role in drier regions and occurs predominantly in spring and summer (Böhner, 2006). In the eastern Pamir and also the central-eastern Karakoram, the precipitation peak is shifted towards spring, with important annual precipitation shares even in summer (Zech et al., 2005; Aizen et al., 2001, 1997). In monsoon-dominated regions (the border is roughly at 77° E, Bookhagen and Burbank, 2010) a mixture of western disturbances and monsoon dominates in summer (Maussion et al., 2014; Böhner, 2006; Bookhagen and Burbank, 2010; Archer and Fowler, 2004). Little is known about temperature and precipitation at the elevation of glaciers; stations in valley floors yield amounts between 70 and 300 mm yr^{-1} (Seong et al., 2009; Archer and Fowler, 2004). Precipitation amounts along the edge of mountain ranges and in high altitudes are largely unknown, but can be substantially higher ("by a factor of ten": Wake, 1989; Immerzeel et al., 2015), which is also suggested by snow station measurements showing snow accumulations of > 1000 mm w.e. (millimetre water equivalent) around 4000 m in the Hunza basin (Winiger et al., 2005).

2.3 Glacier changes

Glaciers in the Karakoram have gained considerable attention during the last decade. The "Karakoram anomaly" that was first identified by Hewitt (2005), based on the observed unusual behaviour of glacier termini, is now a major research topic, and numerous studies have investigated the recent and longer-term evolution of climate, changes in glacier extent and volume, and glacier dynamics. These studies suggest that since the 1970s the extent and mass of glaciers in the central Karakoram have on average hardly changed (Bolch et al., 2017; Bajracharya et al., 2015; Bhambri et al., 2013), which also applies to the beginning of the 21st century (Lin et al., 2017; Brun et al., 2017; Gardelle et al., 2013; Gardner et al., 2013; Kääb et al., 2012), while glaciers in the mountain ranges of the Hindu Kush and Hindu Raj are mostly retreating (Sarıkaya et al., 2013; Haritashya et al., 2009). However, the patterns of climate-induced glacier change are not to be confounded with the strong geometric changes observed for the abundant surge-type glaciers in the region that might occur independent of climatic forcing (Bhambri et al., 2017; Paul, 2015; Quincey et al., 2015; Rankl et al., 2014; Copland et al., 2011). Glaciers in the eastern Pamir were on average almost in balance, as in the Karakoram (Brun et al., 2017; Holzer et al., 2016). In the western Pamir, glacier volume evolution seems to be more negative, but is for the first decade of the 2000s still relatively modest (Brun et al., 2017); moreover, satellite images of the past two decades reveal that many glaciers in this region have surged (e.g. Wendt et al., 2017). However, many of the non-surge-type glaciers in the Pamir have been continuously retreating and losing area

Table 1. List of Landsat scenes used to compile the inventory.

WRS2 path-row	Date	Scene ID	Sensor	HMA region
146-036	8 Oct 2000	LE71460362000282SGS00	ETM+	Karakoram
147-035	2 Aug 2002	LE71470352002214SGS00	ETM+	Karakoram
147-036	2 Aug 2002	LE71470362002214SGS00	ETM+	Karakoram
147-036	28 Aug 2000	LE71470362000241SGS00	ETM+	Karakoram
148-035	4 Sep 2000	LE71480352000248SGS00	ETM+	Karakoram
148-035	21 Jul 2001	LE71480352001202SGS00	ETM+	Karakoram
148-036	4 Sep 2000	LE71480362000248SGS00	ETM+	Karakoram
149-033	26 Jul 2009	LT51490332009207KHC00	TM	Eastern Pamir
149-034	13 Aug 1998	LT51490341998225XXX01	TM	Eastern Pamir, Karakoram
149-034	11 Sep 2000	LE71490342000255SGS00	ETM+	Eastern Pamir
149-035	13 Aug 1998	LT51490351998225XXX01	TM	Karakoram
149-035	29 Aug 2001	LE71490352001241SGS00	ETM+	Karakoram
150-033	2 Sep 2000	LE71500332000246SGS01	ETM+	Eastern and western Pamir
150-034	20 Aug 1998	LT51500341998232BIK00	TM	Eastern and western Pamir, Karakoram
150-034	2 Sep 2000	LE71500342000246SGS01	ETM+	Western Pamir
150-035	16 Sep 1999	LE71500351999259SGS00	ETM+	Karakoram
151-032	23 Sep 1999	LE71510321999266EDC00	ETM+	Pamir Alai
151-033	24 Aug 2000	LE71510332000237SGS00	ETM+	Western Pamir
151-034	30 Aug 2002	LE71510342002242SGS00	ETM+	Western Pamir
151-034	26 Jul 2001	LE71510342001207SGS00	ETM+	Western Pamir
151-035	30 Aug 2002	LE71510352002242SGS00	ETM+	Karakoram
152-032	16 Sep 2000	LE71520322000260SGS00	ETM+	Pamir Alai
152-033	16 Sep 2000	LE71520332000260SGS00	ETM+	Pamir Alai, western Pamir
152-034	2 Aug 2001	LE71520342001214SGS00	ETM+	Western Pamir
152-034	31 Aug 2000	LE71520342000244SGS00	ETM+	Western Pamir
153-032	15 Sep 2000	LT51530322000259XXX02	TM	Pamir Alai
153-033	15 Sep 2000	LT51530332000259XXX02	TM	Pamir Alai, western Pamir
154-033	29 Aug 2000	LE71540332000242EDC00	ETM+	Pamir Alai

and mass since the Little Ice Age (Holzer et al., 2016; Khromova et al., 2006; Shangguan et al., 2006).

3   Input data

As a mapping basis we have used six Landsat 5 TM and 22 Landsat 7 ETM+ Level 1T scenes, with the latter offering a 15 m panchromatic band for improved mapping quality (Table 1). Additionally, we have also used coherence images derived from the Advanced Land Observing Satellite 1 (ALOS-1) Phased Array type L-band Synthetic Aperture Radar 1 (PALSAR-1) scenes acquired around 2007 to aid in mapping the debris-covered glacier parts (Atwood et al., 2010; Frey et al., 2012) and the global digital elevation model (GDEM) version 2 from ASTER (hereafter referred to as GDEM2; United States Geological Survey, 2018a). The TM and ETM+ scenes served as a basis for glacier mapping while the coherence images were used for corrections of debris-covered glacier areas. Moreover, satellite images available in Google Earth served as a visual control for outline detection, with data originating mainly from very high-resolution optical sensors such as QuickBird, Worldview,

Pléiades 1A and 1B, and SPOT 6 and SPOT 7 (GoogleEarth 2017); unfortunately these were not available for all regions.

Coherence images have been produced from ALOS-1 PALSAR-1 scenes usually separated by 46 days and acquired over summer (Table 2). The processing of the images takes into account a number of effects (e.g. sensor geometry, radiometric calibration, frequency interference) that influence the noise of the radar interferogram. The remaining decorrelation can be ascribed to changes in landscape surface properties, e.g. due to movement of landforms. More details on the processing line can be found in Frey et al. (2012).

A DEM is needed to retrieve drainage divides and topographic information for a glacier inventory. The freely available SRTM DEM (United States Geological Survey, 2018b) and the GDEM2 (both with 30 m cell size) could have been used for this purpose. The optical GDEM2 has a potentially reduced quality in low-contrast regions such as shadow and snow-covered accumulation regions, but it has been averaged from scenes acquired over a 12-year period strongly reducing these factors. On the other hand, the SRTM DEM has a precise acquisition date (February 2000) but suffers from data voids in steep terrain due to radar shadow and layover, which affect the final quality over glacierized areas, in par-

Table 2. List of ALOS-1 PALSAR-1 scenes used to generate the coherence images.

Path	Frames	Date 1	Date 2	Interval (days)	HMA region
533	770–780	22 Jul 2009	6 Sep 2009	46	Pamir Alai
528	720–730	22 Jul 2009	6 Sep 2009	46	Karakoram, western Pamir
528	750–760	22 Jul 2009	6 Sep 2009	46	Western Pamir
522	700	22 Aug 2007	7 Oct 2007	46	Karakoram
523	690	24 Jul 2007	8 Sep 2007	46	Karakoram
524	690	10 Aug 2007	25 Sep 2007	46	Karakoram
524	700–710	10 Aug 2007	25 Sep 2007	46	Karakoram
524	750–760	10 Aug 2007	25 Sep 2007	46	Eastern Pamir
523	690–700	8 Jun 2007	24 Jul 2007	46	Karakoram
525	700–730	12 Jul 2007	27 Aug 2007	46	Karakoram
525	750–770	12 Jul 2007	27 Aug 2007	46	Eastern Pamir
526	710–730	13 Jun 2007	29 Jul 2007	46	Karakoram
526	770	13 Jun 2007	29 Jul 2007	46	Eastern Pamir
527	710–730	15 Aug 2007	30 Sep 2007	46	Karakoram, western Pamir
529	720–750	18 Jun 2007	18 Sep 2007	92	Karakoram, western Pamir
529	760–780	18 Jun 2007	18 Sep 2007	92	Western Pamir, Pamir Alai
530	720–750	5 Jul 2007	20 Aug 2007	46	Karakoram, western Pamir
530	760–770	5 Jul 2007	20 Aug 2007	46	Western Pamir, Pamir Alai
530	780	5 Jul 2007	20 Aug 2007	46	Western Pamir, Pamir Alai
531	730	22 Jul 2007	22 Oct 2007	92	Western Pamir
531	750–770	22 Jul 2007	22 Oct 2007	92	Western Pamir
531	780	22 Jul 2007	22 Oct 2007	92	Western Pamir, Pamir Alai
532	720–750	8 Aug 2007	23 Sep 2007	46	Western Pamir
532	760–770	8 Aug 2007	23 Sep 2007	46	Western Pamir, Pamir Alai
532	780	8 Aug 2007	23 Sep 2007	46	Western Pamir, Pamir Alai
535	770–780	5 Jul 2007	20 Aug 2007	46	Pamir Alai
536	770	22 Jul 2007	22 Oct 2007	92	Pamir Alai
537	770	8 Aug 2007	23 Sep 2007	46	Pamir Alai

ticular when using void-filled versions. A direct comparison (subtraction) of both DEMs as recommended by Frey and Paul (2012) confirmed these differences. We finally decided to work only with the GDEM2 as it had fewer data voids along mountain crests (important to derive correct drainage divides) and because it is spatially consistent; i.e. data voids over glaciers in the SRTM DEM did not have to be filled with some other DEM data (which is beneficial for deriving consistent topographic information and increases traceability). The vertical accuracy was found to be around 9 m (probably higher in steep terrain) and similar for both DEMs (Satgé et al., 2015). For consistency, the glacier separation and all subsequent topographic analysis of glacier elevation, slope, and aspect are thus based on the GDEM2.

4 Methods

4.1 Glacier mapping

We applied the well-established semi-automatic band ratio method (Paul et al., 2002) to classify glaciers (the clean-ice and snow part), taking advantage of the reflection contrast between snow–ice and other land surfaces in the red and short-wave infrared (SWIR) parts of the electromagnetic spectrum, corresponding to Landsat TM or ETM+ bands 3 (red) and 5 (SWIR). An individual, scene-adjusted band ratio threshold between 1.5 and 3.5 is applied to separate glaciers and snow from other terrain and to compute a binary raster image, which is smoothed using a 3 by 3 majority filter and is then converted to a vector file for further editing.

Due to the spectral similarity of debris on and off glaciers, there is so far no method available to automatically map debris cover over a large set of glaciers using optical satellite imagery alone. Hence, several studies have tested combined approaches that generally include topographic information derived from a DEM and other data (Robson et al., 2016; Racoviteanu and Williams, 2012; Rastner et al., 2014; Bolch et al., 2007; Paul et al., 2004). However, all methods require time-consuming manual post-processing, and the quality of the results depends to some extent on the experience of the analyst.

As debris-covered glacier tongues can be difficult to identify visually, even when using high-resolution images (Paul et al., 2013), we have utilized ALOS-1 PALSAR-1 coherence images. Such images have also been used for glacier

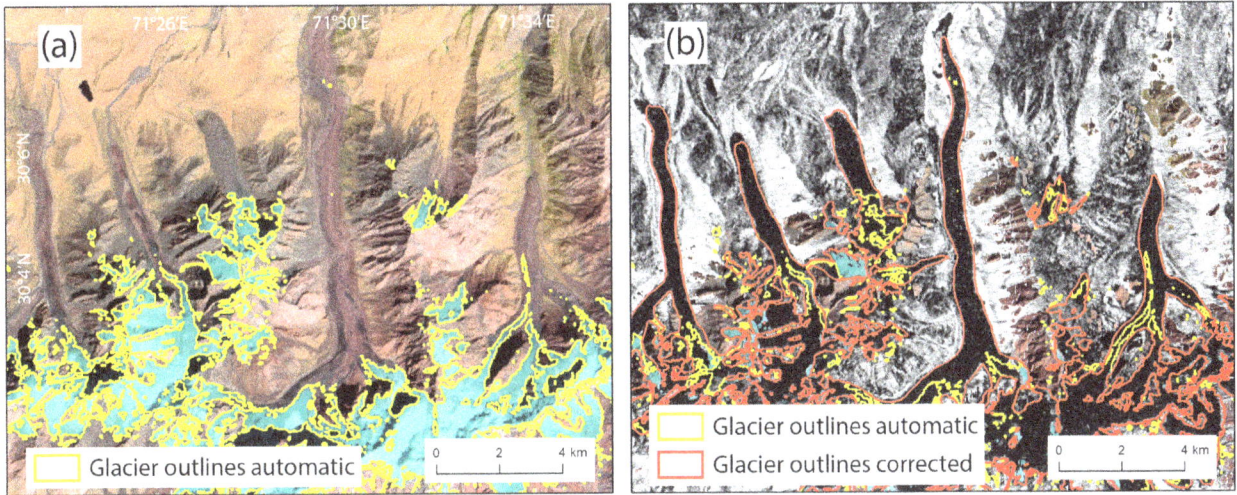

Figure 2. Mapping of heavily debris-covered tongues. **(a)** False-colour image overlaid by raw outlines. **(b)** PALSAR-1 coherence images overlaid by raw (yellow) and corrected (red) outlines. The decorrelated image areas are shown in black.

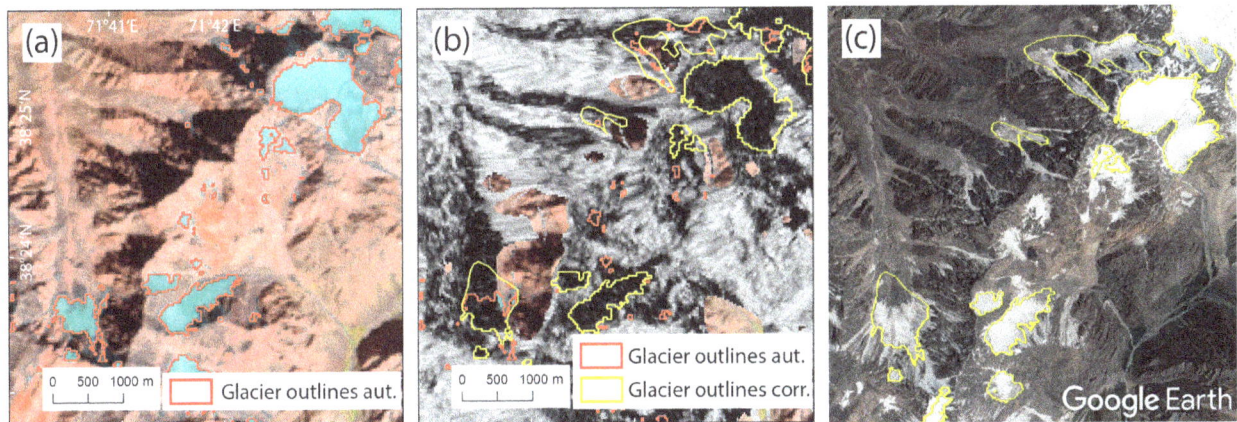

Figure 3. Elongated rock glaciers that are (almost) connected to the active glacier tongue are hard to distinguish. **(a)** False-colour Landsat image overlaid by raw outlines. **(b)** PALSAR-1 coherence images are fuzzy and potentially misleading in this case. **(c)** High-resolution imagery is decisive for a correct decision, because rock glaciers can be visually distinguished from the glacier tongue.

mapping in Alaska (Atwood et al., 2010) or as supportive means for correcting automatically derived glacier outlines in the western Himalaya by Frey et al. (2012). The coherence images were primarily used to detect the existence of debris-covered glacier tongues, while the exact positions of the glacier margin and terminus were detected using the optical Landsat image. The data are combined to account for the time difference of up to 9 years between coherence and Landsat images. Non-surging debris-covered glaciers commonly change little during such periods (Scherler et al., 2011b), especially in this region (e.g. Baltoro glacier has been stable for at least 25 years) (Paul, 2015).

The elevation of a glacier can be described by different elevation parameters. One that is well suited for a comparison between different glacier types and sizes as well as an indication for climatic differences is the median elevation, which is indicative of the equilibrium line altitude (ELA) at a balanced mass budget (Braithwaite and Raper, 2009) and similar to the midpoint elevation (Raper and Braithwaite, 2009), that has been used in several studies to characterize glaciers (e.g. Haeberli and Hoelzle, 1995) and climatic conditions, primarily precipitation amounts (e.g. Sakai et al., 2015; Bolch et al., 2013).

4.2 Mapping challenges and solutions

The main challenges for mapping glaciers in this region are the correct delineation of debris-covered glacier parts (including their separation from rock glaciers), seasonal snow, cast shadow, and orographic clouds. In the following, we shortly describe these challenges and present the techniques applied to overcome them.

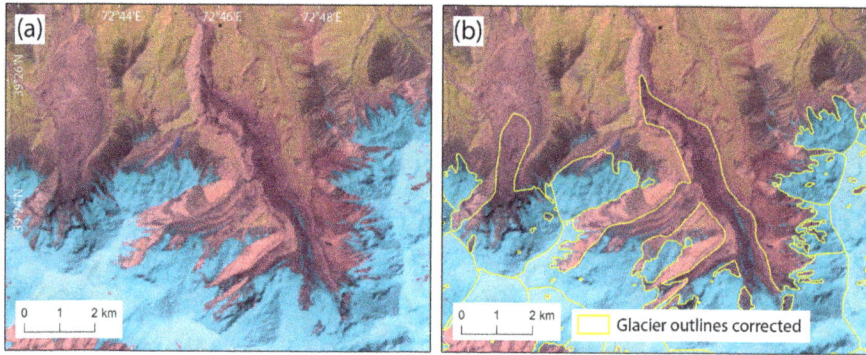

Figure 4. Extensive moraines and large areas of debris can be found on dead ice and active glaciers.

Debris cover. The main reason for extensive debris cover on glaciers is steep/high topography with ice-free rock walls leading to rock falls and avalanches onto the glacier surface (e.g. Herreid et al., 2015; Scherler et al., 2011a; Paul et al., 2004). Apart from the central Karakoram, most regions exhibit glacier recession, which is another factor for increasing debris coverage on the glacier surface (Rowan et al., 2015; Kirkbride and Deline, 2013).

The debris-covered glacier area in this study was mapped manually by editing the automatically derived clean-ice outlines. Key difficulties in identifying these regions are the small solar incidence angle at these latitudes (reducing topographic contrast in the terminus region), unclear boundaries between supra-glacial debris and moraines or rock glaciers, and debris in shadow (e.g. Bishop et al., 2014). Heavily debris-covered glacier tongues are often in contact with lateral or frontal moraines (Figs. 2, 3, and 4) and their composition is very similar, leading to similar spectral properties and the need of applying other measures for identification. Whereas human recognition has the ability to trace very subtle features for identification of debris on glaciers, the 30 m spatial resolution of Landsat images is often too coarse for a clear assignment.

In this study we mostly relied on the ALOS-1 PALSAR-1 coherence images for identifying the margins of debris-covered glaciers (Fig. 2). Their usability decreases with decreasing glacier size as well as when the images become "fuzzy" and glacier margins less clear; in high-alpine terrain this could happen, as also glacier-free terrain can change: permafrost landforms such as rock glaciers, talus ramparts, and moraines complicate the proper identification of glaciers (Figs. 2, 3, and 4). In such cases we also used the very high-resolution imagery available in Google Earth and similar tools for identification. Furthermore, multi-temporal data aided in terminus identification by either providing better contrast or by using them in animations (Paul, 2015). In some cases it was also possible to consider glaciological relations for a first approximation of glacier extent. For example, a tiny accumulation area would likely not support a large glacier tongue (and vice versa).

In contrast to glaciers – massive bodies of ice originating from continuous snow accumulation – rock glaciers probably have a different genesis: they develop in a permafrost environment either from ice-cored moraines or on talus slopes that provide constant debris input, and they commonly have a higher debris content than glaciers (Berthling, 2011; Haeberli et al., 2006; Barsch, 1996). Especially towards the cold–dry regions of central Asia, rock glaciers of both types are increasingly abundant (Bolch and Gorbunov, 2014; Gorbunov and Titkov, 1989). In particular, moraine-derived rock glaciers challenge the analyst as there is often a continuous transition between the glacier and the rock glacier, making it hard to define a divide (cf. Monnier and Kinnard, 2015). A well-developed rock glacier can in principle be distinguished from a debris-covered glacier by characteristic surface patterns such as the arc-shaped transverse ridge and furrow structure instead of the longitudinal debris striations and supra-glacial ponds found on most debris-covered glaciers (Bishop et al., 2014; Bodin et al., 2010). However, identifying such differences using remotely sensed imagery requires a spatial resolution better than 15 m (Paul et al., 2003) and might not work at all when rock glaciers are not well developed. We separated debris-covered glaciers from rock glaciers based on interpreting the above data sources (Google Earth, coherence images) and their known morphological characteristics. In the cases where no clear boundary could be found, we followed a more conservative interpretation that might have resulted in a potential underestimation of the debris-covered glacier area.

Seasonal snow. Seasonal snow can obscure the underlying glacier ice and is included in the automatic classification result due to the similar reflection properties of snow and ice. Seasonal snow and clouds also required consideration of scenes from years other than 2000. For larger glaciers with a low-lying terminus, it would have been possible to adjust the (snow-free) terminus to the year 2000 scene; we have not applied this in favour of temporally consistent glacier outlines. Interestingly, for some regions it was much harder to find satellite scenes with satisfying snow conditions than for others. It was particularly difficult for the eastern Pamir and

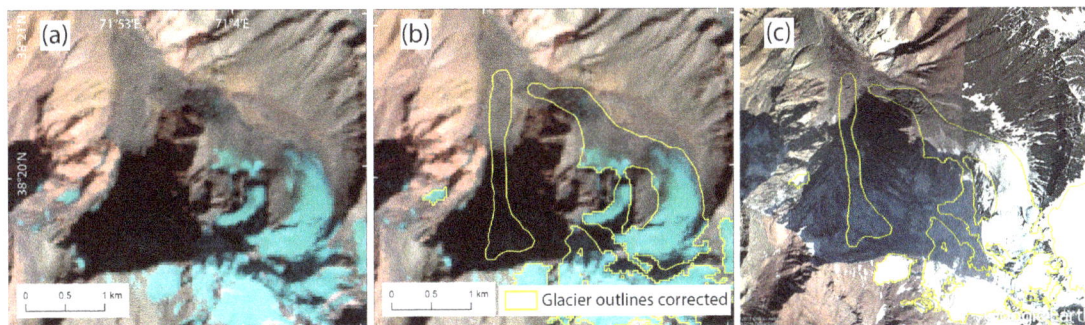

Figure 5. Glacier detection in shadow with the supporting input of high-resolution Google Earth images.

some parts of the northern–central Karakoram, potentially resulting in related higher area uncertainties for the accumulation areas of the glaciers. Our strategy to reduce the impact of wrongly mapped seasonal snow was threefold: we applied a size filter of $0.02\,\text{km}^2$ to remove the smallest snow patches; snow attached to glaciers was manually removed after visual inspection, and in some regions a different scene (with less snow but possibly more clouds) was chosen to improve results (see Table 1). Despite these measures, we assume that glacier area in this inventory is likely overestimated due to the inclusion of seasonal snow.

Shadow. Cast shadows from mountains decrease reflection values, partly down to near-zero. This results in considerable noise in this region for a band ratio using a red (or near infrared) band. As TM band 1 (blue) is strongly influenced by atmospheric scattering, ice and snow in shadow are much more visible and can be distinguished using an additional threshold (e.g. Paul and Kääb, 2005). Although the shadow problem is less pronounced in lower latitudes due to the higher solar elevation angle, it is still a problem in the study region due to the high and steep terrain. We have used the additional blue band to map glaciers in shadow automatically or applied manual corrections on contrast-enhanced true-colour composites in case the automated refinement was not successful. We also analysed scenes from a different date or another sensor (including very high-resolution imagery as available in Google Earth and similar tools) to reveal if glaciers are possibly present (Fig. 5). However, this is time-consuming and in some regions images are not available or do not meet the criteria for glacier identification (e.g. due to snow cover). In the cases where glaciers in shadow could not be identified, a related underestimation of glacier area results.

Cloud coverage. Cloud-free scenes were available for most of the study region. In the few cases when cloud cover prevented glacier mapping, the problem was solved multi-temporally by using additional scenes from years close to the year 2000 (Paul et al., 2017). In some regions, scenes with high cloud coverage and possible precipitation events were followed by scenes with extensive snow coverage, so that we had to use scenes from other years. The entire study region is thus a mosaic of many individual scenes (see Table 1).

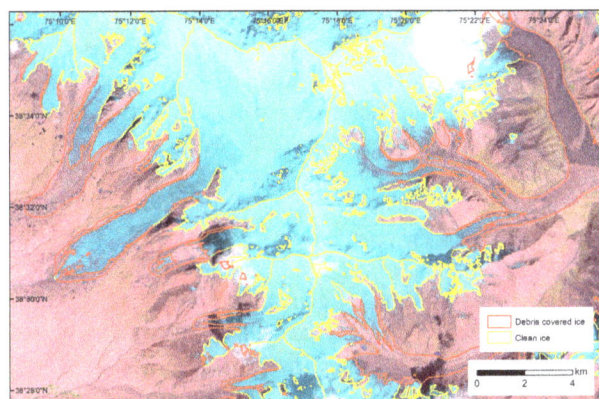

Figure 6. Debris cover classification in the Kongur Shan in the eastern Pamir.

4.3 Calculating the debris-covered area share of glaciers

For calculating the area share of debris cover, we decided to consider only the ablation areas of glaciers (i.e. the region below their median elevation), because debris deposited in the accumulation area should emerge on the glacier surface only below the ELA (Braithwaite and Raper, 2009; Braithwaite and Müller, 1980). We distinguished the debris cover from snow and ice surfaces by applying a constant threshold of 2.0 to all band ratio images from Landsat TM bands 3 and 5 (red and SWIR) and subtracted the resulting clean-ice glacier map from the corrected glacier map. The threshold was found empirically with satisfying results for all scenes from TM and ETM+ sensors. Changing the threshold by ±0.2 changed the result by less than the mapping uncertainties ($\sim 5\,\%$ for debris-covered areas, see Sect. 6).

4.4 Glacier definition and separation using drainage divides

We based the mapping and division of glaciers on the Global Land Ice Measurements from Space (GLIMS) definition of a glacier (Raup and Khalsa, 2007), stating that a glacier includes "all tributaries and connected feeders that contribute ice to the main glacier, plus all debris-covered parts of it". For the sake of consistency with earlier datasets and the GLIMS definition, surge-type tributaries were not separated from the main glacier tongue, even if they contribute only during an active surge phase. The preparation of a dataset where these short-term tributaries are properly separated is worthwhile but a considerable extra effort. Stagnant ice masses (e.g. from a former surge) that were still connected to the glacier tongue were mapped as part of the glacier. In cases where the active glacier has clearly receded away from the stagnant "dead" ice (e.g. after a surge phase), only the active glacier was mapped. In contrast to this definition, the surging Bivachny glacier tongue was separated from Fedchenko glacier in the confluence region although one can argue that Bivachny is a connected feeder (see Fig. 6 in Wendt et al., 2017). A size filter of $0.02\,\mathrm{km}^2$ was applied to remove small seasonal snow-fields and remaining noise. Automatically classified polygons larger than this are considered as glaciers in this inventory, but this does not mean that all seasonal or perennial snowfields have been excluded, nor that some of the mapped glaciers are in fact perennial snow.

Glacier complexes – at least two glaciers connected in their accumulation areas – can be split into single glaciers using drainage divides derived from a DEM. This is performed in two steps. Firstly, raw drainage basins are calculated by watershed analysis using a flow direction grid derived from a sink-filled DEM. Afterwards, overlying raw basins are merged to one basin polygon per glacier considering pour points and a buffer (Falaschi et al., 2017; Kienholz et al., 2013; Bolch et al., 2010). This approach proved to be robust even for the large regions of the Karakoram and Pamir as in general glaciers are divided by very steep mountain crests. Secondly, manual corrections were performed, which took about 90 % of the total processing time. Gross errors were improved using a colour-coded flow direction grid in the background, a hillshade, the original Landsat scenes, and sometimes oblique views in Google Earth. We manually assigned the same ID to separated glacier polygons that were obviously linked by mass transport (e.g. regenerated glaciers).

4.5   Uncertainty estimation

Since there is no ground truth or reference data for any larger set of glaciers in the study region, we calculated uncertainties for the relevant input data rather than accuracy (Paul et al., 2017).

Glacier mapping uncertainties originate from the coverage of glaciers by seasonal snow and/or debris, shadow, and clouds. These need to be corrected manually (on-screen digitizing) by a well-trained analyst. According to the literature, the uncertainty of automatically and manually digitized glacier outlines (clean ice only) ranges between 2 % and 5 % and is dependent on glacier size (Paul et al., 2013, 2011;

Andreassen et al., 2008; Bolch and Kamp, 2006). Paul et al. (2013) estimated uncertainties using a sample of manually and automatically digitized glaciers from a number of experts and found a mean standard deviation of $\sim 5\,\%$. Other studies (Bolch et al., 2010; Granshaw and Fountain, 2006) have used a buffer-based estimate, where the final uncertainty depends on the pixel size of the input image. The study by Paul et al. (2017) suggested a tiered system of uncertainty assessment related to workload. We used three of the methods: (1) fixed uncertainty values applied to all glaciers, (2) the buffer method with different buffer sizes for clean and debris-covered glacier parts, and (3) independent multiple digitization of outlines by all analysts for three difficult debris-covered glaciers.

For (1), we applied an uncertainty of $\pm 2\,\%$ for the clean ice and $\pm 5\,\%$ for the debris-covered ice. This is an upper boundary estimate, because it does not account for the overlapping area of the two surface types. For the buffer method (2) we applied an uncertainty of $\pm 1/2$ pixel for clean-ice parts and ± 1 pixel for debris-covered parts. This also provides an upper-bound estimate and we use the standard deviation of the uncertainty distribution for the estimate, as a normal distribution can be assumed for this type of mapping error. It is applied to glacier complexes excluding overlapping areas as well as the border of clean and debris-covered ice of the same glacier. Due to the abundant debris-covered glaciers in the study region, we also performed method (3) to obtain a more realistic uncertainty estimate for the analysts participating in the outline correction. They manually corrected the outlines of three example glaciers from different regions three times (Glacier 1: $38°34.4'$ N, $72°12.8'$ E; Glacier 2: $39°38.8'$ N, $69°41.9'$ E; Glacier 3: $36°0.3'$ N, $75°14.6'$ E), with differing additional information being considered (e.g. coherence images and Google Earth imagery). The glaciers are of different size and contain a substantial debris-covered part; they also feature difficulties of moraines, glacier confluences, and cast shadow.

As not all satellite scenes used to compile the inventory are from the same year, there is a certain temporal uncertainty introduced. However, glacier changes within the ± 2-year difference to the target year 2000 are likely within the uncertainty of the glacier outlines and should thus not matter. The actual date information is given for each glacier in the attribute table.

5   Results

5.1   Basic statistics

We identified 27 877 glaciers (larger than $0.02\,\mathrm{km}^2$) in the four HMA regions covering $35\,519.7 \pm 1958\,\mathrm{km}^2$; western Pamir and Karakoram each host over 10 000 glaciers, whereas the other regions contain 2000–4000. As in other larger regions where detailed glacier inventories have been

Table 3. The upper table shows the basic inventory statistics for all glaciers, with the lower table only showing the basic inventory statistics for glaciers larger than 5 km^2.

	Glacier area (km^2)	Uncertainty (km^2)	Mean of					
			glacier area (km^2)	median elev. (m)	slope (°)	max. elev. (m)	min elev. (m)	no. of glaciers
All glaciers								
All	35519.7	±1948	1.27	4993	26.4	5251	4739	27 877
Pamir Alai	2078.5	±114	0.57	4147	25.1	4359	3962	3655
Pamir west	9469.4	±519	0.85	4871	25.5	5105	4654	11 098
Pamir east	2277.2	±125	0.98	5050	26.4	5305	4801	2326
Karakoram	21694.6	±1190	2.01	5392	27.7	5693	5076	10 798
Glaciers ≥ 5 km^2								
All	22299.9	±1223	22.5	5173	22.6	6135	4250	990
Pamir Alai	673.1	±37	12.9	4111	19.5	5024	3379	52
Pamir west	4754.5	±261	17.3	4882	21.7	5806	4107	275
Pamir east	1089.7	±60	15.1	5196	25	6342	4204	72
Karakoram	15782.6	±866	26.7	5399	22.9	6362	4399	591

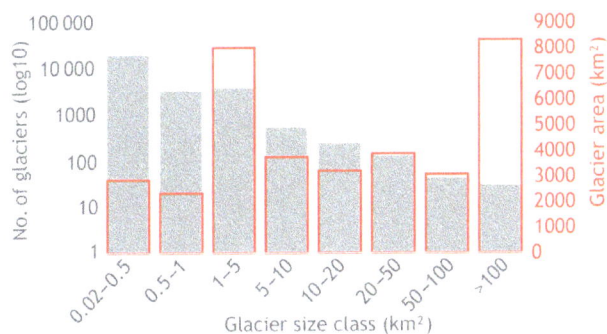

Figure 7. Histogram of all glaciers by number. Please note the logarithmic scale of the left y axis.

compiled (e.g. Kienholz et al., 2015; Guo et al., 2015; Pfeffer et al., 2014; Le Bris et al., 2011; Bolch et al., 2010), the histogram is strongly skewed towards small glaciers (Fig. 7).

Only 3.5 % (985) of all glaciers are larger than 5 km^2, and most of them are located in the Karakoram. In total, they cover over 60 % of the glacierized area. On the other hand, 83 % (23 048) of all glaciers are smaller than 1 km^2 but cover only ∼ 15 % of the total area. The mean glacier size is 1.29 km^2, with large differences between the regions: from 0.57 km^2 in the Pamir Alai to 2.07 km^2 in the Karakoram (Table 3). The average median elevation is 4978 and 5169 m for all glaciers and glaciers larger than 5 km^2, respectively, and differs by only a few metres from the mean elevation.

5.2 Extremes

In the Pamir Alai, the largest glacier is Zeravshan glacier, with an area of 106.3 ± 6.7 km^2, which is 3 times larger than the second largest. Zeravshan glacier stretches over 2600 m from 2800 to 5400 m, close to the highest elevations in the Pamir Alai range. The largest elevation range is covered by Tandykul glacier (39°27′ N, 71°8′ E) 50 km further east, with almost 3000 m (2450–5400 m). Its heavily debris-covered tongue lies in a deep valley that is well shielded from the south. Overall, only a few larger valley glaciers (19 larger than 10 km^2) have several large tributaries.

In the western Pamir, Fedchenko glacier is by far the largest with 573 ± 19.5 km^2 (not including Bivachny glacier with 170 ± 8.5 km^2 since it is in contact but not contributing). Bivachny glacier starts right below the summit of Ismoil Somoni Peak (formerly known as Pik Kommunízma, 7495 m) and terminates at about 3420 m, whereas Fedchenko glacier stretches from Independence Peak, with an elevation of 6940 m, down to below 2900 m; hence, both glaciers are spanning an elevation range of over 4000 m. The region hosts several large glacier systems (13 larger than 50 km^2, 108 larger than 10 km^2) that are arranged in two clusters: one is around Fedchenko glacier in the Yazgulem Range and one around Lenin Peak in the Trans-Alai Range. Also this region has steep topography and several glaciers reach an elevation range of around 4000 m. However, these numbers are a snapshot in time and have to be treated with care, since there are many surge-type glaciers whose current phase state can significantly influence minimum elevation and area (Kotlyakov et al., 2008). We found the lowest-lying terminus at a very small, north-facing and likely avalanche-fed glacier

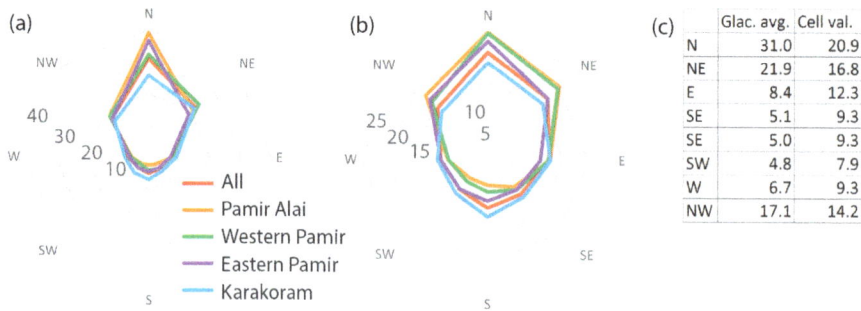

Figure 8. Glacier orientation of the different HMA regions. **(a)** shows the values based on average glacier aspect; **(b)** is based on the 30 m raster cells. Lower elevations tend to have a higher share of the north-facing glacier area. The respective numbers of "All" are given in the table **(c)**.

	Glac. avg.	Cell val.
N	31.0	20.9
NE	21.9	16.8
E	8.4	12.3
SE	5.1	9.3
SE	5.0	9.3
SW	4.8	7.9
W	6.7	9.3
NW	17.1	14.2

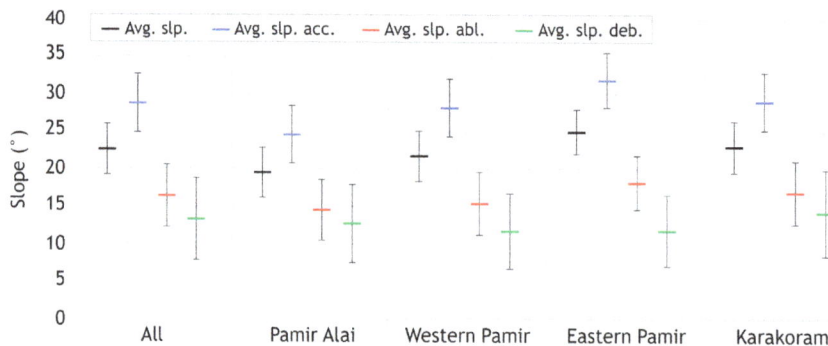

Figure 9. Slope per glacier regions and surface type (avg. slp.: average slope of glacierized area; avg. slp. acc.: average slope in the accumulation area; avg. slp. abl.: average slope in the ablation area; avg. slp. deb.: average slope in the debris-covered area).

(70.65° E/38.99° N) in the Petra Pervogo Range, reaching down to below 2400 m.

The eastern Pamir region has 38 glaciers larger than 10 km², evenly distributed over the individual mountain ranges. The largest glacier (109.4 ± 6.9 km²) is Karayaylak glacier, draining the northern basin of the Kongur Shan. It starts at the top of Kongur Tagh, with the highest mountain in the Pamir at an elevation of 7680 m, and reaches down to 2819 m, spanning an elevation range of over 4800 m, which is by far the largest value in this region. One of its tributaries has reportedly surged in 2015 (Shangguan et al., 2016). The neighbouring Qimgan glacier starts at the same peak facing south-east and reaches down to 3160 m (almost 4500 m elevation range). A smaller, east-facing glacier in the Oytagh glacier park reaches the same low elevation as Karayaylak glacier (2824 m).

Siachen glacier is the largest of its kind in the Karakoram. With an area of 1094.2 ± 31.2 km² (including all of its major tributaries) it is by far the largest glacier in the study area, and with over 70 km in length Siachen and Fedchenko are the longest glaciers in the mid-latitudes. Two more glaciers have an area over 500 km²: Baltoro with 810 ± 36.1 km² and Biafo with 560 ± 23.8 km². Both glaciers reach their lowest elevations in the central part of the Hunza valley (around Gilgit),

with terminus elevations of around 2500 m and below (Hopar glacier: 2260 m). Two large glaciers reach elevation ranges of 5200 m (Batura and Baltoro), but also smaller glaciers like Shishper (45 ± 4.1 km²), Passu (62.2 ± 1.7 km²), and Rakaposhi (14.4 ± 0.9 km²) stretch over an elevation range of 5000 m. Once again, many of these glaciers are of a surge type (e.g. Bhambri et al., 2017), and their minimum elevations and area values after a surge might strongly differ from those at the end of a quiescent phase. The highest glacierized regions in the Karakoram are found around K2 (8611 m; Baltoro glacier) and Distaghil Sar (7885 m; Yazghil glacier, Hispar glacier).

5.3 Glacier aspect analysis

On average, most glaciers are oriented towards the north sector (mean: 71.5 % ± 5.4 %, Fig. 8a). The relative distribution is similar among the regions, and the largest variations occur in the aspects with a small glacier share (SE, S, SW; normalized SD: 0.21, 0.32, 0.31). The distribution of single cells (instead of one mean aspect value per glacier) shows a similar pattern although with less significance of north aspect. Nevertheless, the north sector has the highest share in all re-

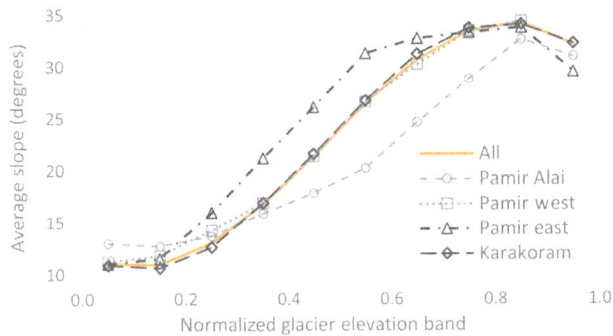

Figure 10. Glacier slope along 10 glacier elevation sections. The glaciers were normalized for elevation to compare high- and low-elevation glaciers.

gions (mean: 56.5 % ± 5.7 %, Fig. 8b), while the south and south-west host the smallest share of glacierized area.

In contrast to other regions, we found no correlation between median elevation and aspect.

5.4 Glacier slope analysis

The mean slope of all glaciers is 26.4°. It decreases to 22.6° for glaciers larger than 5 km^2; hence, mean slope is size dependent. The decrease in mean slope between the sample of all glaciers and glaciers larger than 5 km^2 is relatively large for Pamir Alai (−5.6°) and very small for the eastern Pamir (−1.4°). Mean slope varies between different parts of the glacier, with the accumulation area being the steepest section and the debris-covered areas being far flatter in all regions (Fig. 9).

To determine whether glaciers constantly get steeper from the terminus to the upper reaches of the accumulation area, we normalized the elevation distribution of all glaciers such that each glacier covered the value range from 0 to 1 from the terminus to the upper end, divided into sections of 0.1. The result clearly reveals a mean slope of about 12° in the lower parts and a constant increase to over 30° at the highest elevations of each glacier (Fig. 10). The uppermost band is again somewhat flatter, possibly due to the transition of slope direction at crests. The pattern is similar in all regions, but the slope increase along the glacier is higher than average in the eastern Pamir and lower than average in the Pamir Alai.

5.5 Glacier elevation analysis

The median elevation of glaciers larger than 2 km^2 ranges from 2800 to over 6500 m. There is a statistically significant correlation ($p < 0.001$) between median elevation and latitude ($r^2 = 0.48$) and longitude ($r^2 = 0.66$), which appears as a rise of median elevation from the north-west towards the south-east across the study region (Fig. 11).

This rise becomes even clearer when looking at separated areas along a "fishbone" transverse profile of our study re-

gion (inset). The average values of each segment reveal a rise in median elevation from 3980 m (bin 1) to 5860 m (bin 6), with an average trend of 1.9 m km^{-1} along the profile.

5.6 Hypsography

Plotting the glacier hypsography of the different HMA regions (Fig. 12a) reveals a number of further differences among the regions. Most apparent is the difference in elevation: the median elevation extends from 4141 m (Pamir Alai) to 5419 m (Karakoram), with the western Pamir (4941 m) and eastern Pamir (5119 m) in between the two. Most of the glacierized area is located in the Karakoram (60 %) where the ice is distributed over a large elevation range (Fig. 12b). In contrast, in the Pamir Alai most of the glacier area is situated closely around the median elevation. The large glaciers in the Karakoram reach far down and occupy large areas in lower elevations, further away from the median elevation than in other regions. The eastern Pamir shows a similar drop in the area share of higher elevations, but the curve flattens in elevations over 1000 m above the median elevation. This is related to the shape of topography that is dominated by distinct mountain ranges with large areas above 6500 m (Kongur, Muztag Ata, Kingata Shan). When analysing the hypsography of glaciers with over 10 % debris-covered area compared to the rest of the sample, the insulation effect becomes visible, with debris-covered glaciers occupying considerably more area at lower elevations (Fig. 12c).

5.7 Debris cover

The mapping quality of the debris-covered areas is defined by the corrected outlines as well as by the clean-ice threshold used to differentiate between debris cover and clean-ice surfaces. It contains the same uncertainties and is homogeneous throughout the different Landsat scenes (Fig. 13). The total amount of debris-covered glacier area is 3580 km^2, i.e. 10 % of the total glacierized area with small differences among the four HMA regions. The uncertainty considering a buffer of ±1 pixel along the (manually mapped) debris-covered glacier boundary yields ±662 km^2, including a buffer of ±1/2 pixel along the (automatically mapped) boundary between clean and debris-covered ice, and the uncertainty increases to ±1131 km^2; the latter number is high due to the great number of small debris polygons. The lowest and highest debris-covered area shares are found in the western (8 %) and eastern Pamir (12 %). There is no significant relation between glacier size and debris-covered area share. The distribution in aspect is somewhat skewed towards the north and north-east (12 % and 11 % vs. 8 %–9 % in E, SE, S, SW, W, NW), but this is less of a systematic pattern than for the total glacierized area. The highest values are found in the eastern Pamir where north-facing glaciers are debris covered by over 17 %, whereas Pamir Alai exhibits the largest range (N = 15 % vs. SW = 6 %).

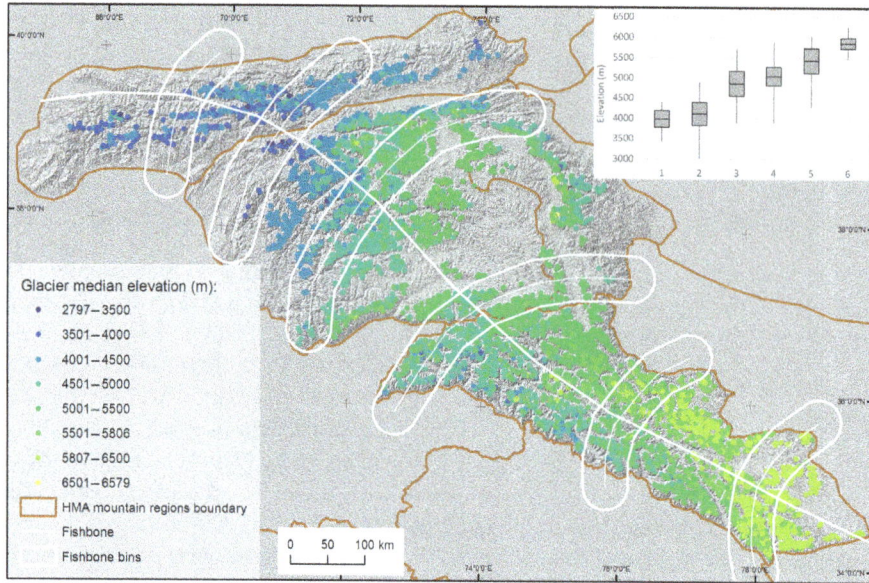

Figure 11. Glacier median elevation over the study area of glaciers larger than 0.5 km². The inset shows median elevation, standard deviations, and minimum and maximum elevations per bin.

Generally, there is no relation between the mean slope of a glacier and the area share of its debris cover. However, the mean slope of the debris-covered part of the glaciers is 16.6° (±5.5), whereas the mean slope of these glaciers is 26.1° (±3.2). This was expected since the debris cover is usually situated at the flatter glacier tongues (Paul et al., 2004). Looking at the ablation area of all glaciers, the mean slope is 25.0° (±4.2). The ablation areas of more strongly debris-covered glaciers are somewhat flatter: glaciers with a debris-covered area of ≥ 10 % on average have a steepness of 22.7° (±4.0); in contrast, glaciers with less than 5 % debris cover have a mean slope of 25.7° (±4.1).

6 Uncertainties and the multiple digitizing experiment

By applying previously assumed area uncertainties (±2.5 % for clean ice, ±5 % for debris-covered ice) to the mapped glacier area, the derived total glacier area is 35 520 ± 1955 km² . With the buffer method (clean ice ±1/2, debris-covered ice ±1 pixel) we obtain a similar uncertainty of ±1948 km². Both methods are applied to glacier complexes to avoid double counting of overlapping areas of adjacent glaciers. Finally, the multiple digitization experiment resulted in a ±13 % standard deviation (averaged over all experiments). This value appears high, but it reflects the mapping reality in challenging situations with debris-covered glacier tongues. For two of the three test glaciers, the difference between the largest and the smallest area mapped was less than 5 % of the mean glacier area. The third example is a small (∼ 2.9 km²) and steep glacier with a high share of

its area hidden in shadow, a large and barely visible debris-covered part and adjacent rock glaciers (see Fig. S2 in the Supplement). Here, the respective uncertainty is ±33 %. Taking this as a worst-case scenario, only few such cases exist in a larger inventory and the high uncertainty has little impact on the overall uncertainty.

Paul et al. (2013, 2015) showed that analyst interpretations for debris-covered glaciers and glacier parts in shadow can differ by up to 50 %. Our experiment showed that, if the glacier is affected by both shadow and debris cover and is additionally small, the differences can be even higher with up to 70 %. The experiment also confirmed that area differences mainly depend on the interpretation of the debris-covered parts. Thereby, using coherence images improved the analyst's interpretation. Although the overall effect was small (on average ∼ 1 %), it reduced the dispersion of the analyst's interpretations considerably (see Fig. 14). The different timing of Landsat (2000) and ALOS-1 PALSAR-1 (2007–2009) imagery had only a small impact, as geometric changes during these 7 years were small. The use of Google Earth imagery did not lead to notable outline modifications as they either had low quality (resolution, snow cover) or provided a mere confirmation of the existing interpretation from Landsat and coherence images. We conclude that the area uncertainty of the debris-covered parts of a glacier is of the order of 10 % to 20 %. However, at least one third of this uncertainty can be disregarded due to direct contact to clean-ice glacier parts (see Fig. 13).

The mapping uncertainty for the clean-ice glacier parts was found to be low, notwithstanding the simple method applied (constant threshold for all scenes). Using different thresholds of 2.0 ± 0.3 yielded results in the range of 5 %

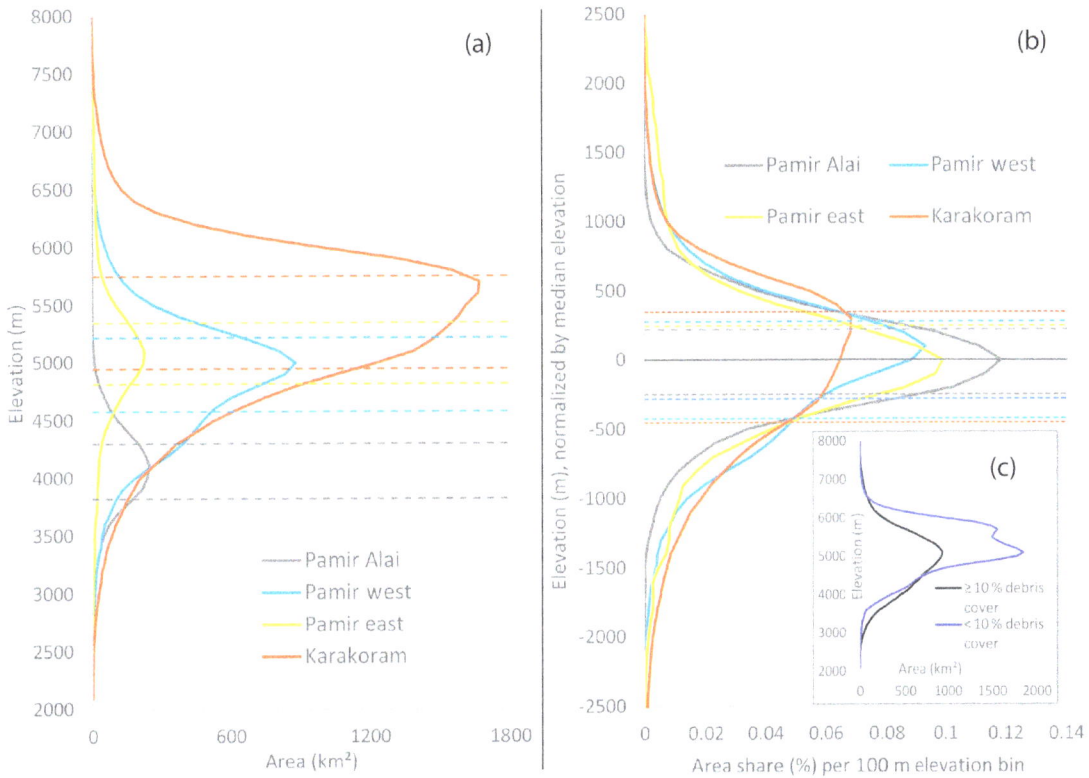

Figure 12. Glacier hypsography of the different regions **(a)**, normalized by the respective median elevation **(b)**. Dashed lines represent the 25 % and 75 % area elevations. **(c)** Hypsography comparison of more and less debris-covered glaciers.

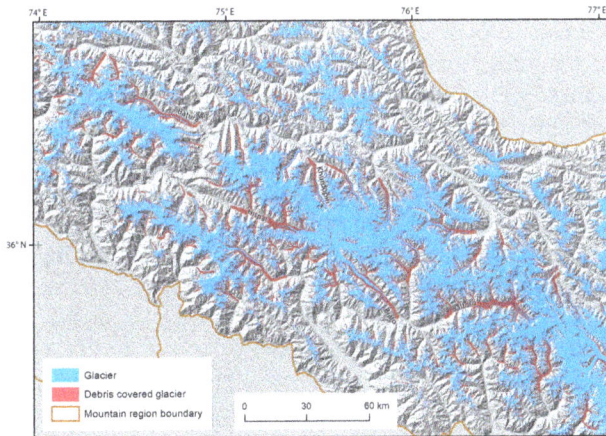

Figure 13. Debris cover on glaciers in the central Karakoram.

of the debris-covered area, which is smaller than the uncertainty from the manual correction of the debris-covered glacier parts.

All uncertainty values have to be seen from the perspective of methodological uncertainties, e.g. the inclusion of possible snowfields at high elevations, which can easily increase the area of a small glacier by 50 % or more. With this in mind, the uncertainties presented above are in general much smaller

and are more of an academic nature. As the uncertainties from the expert round robin are close to those from the buffer method, we use the uncertainty derived by the buffer method as the uncertainties assigned to our results, knowing that they are on the conservative side.

We also performed a comparison in regions where the CGI (Guo et al., 2015), the GAMDAM (Nuimura et al., 2015), and our inventory have mapped the glaciers, to determine major differences among them. Compared to the CGI, our total glacier area is ~ 15 % larger (despite a similar glacier definition) and the CGI overlaps with 82 % of our inventory. Our debris-covered areas are somewhat larger along the margins of the tongues and more of the smaller glaciers at higher elevations are included (Fig. 15). Regions where the CGI area is larger (7 % in total) are related to the inclusion of areas enclosed by different branches of the same glacier, as well as dead ice and rock glaciers in front of a terminus. The GAMDAM inventory covers 13 % less area than ours and also overlaps with 82 % of the area. Here the difference is clearly linked to a diverging glacier mapping definition that mostly excludes headwalls steeper than 40° (Nuimura et al., 2015). Moreover, many debris-covered glacier areas and in some cases entire glaciers have not been mapped. On the other hand, almost all of the areas covered by GAMDAM but not by our inventory are mapped as debris-covered glaciers.

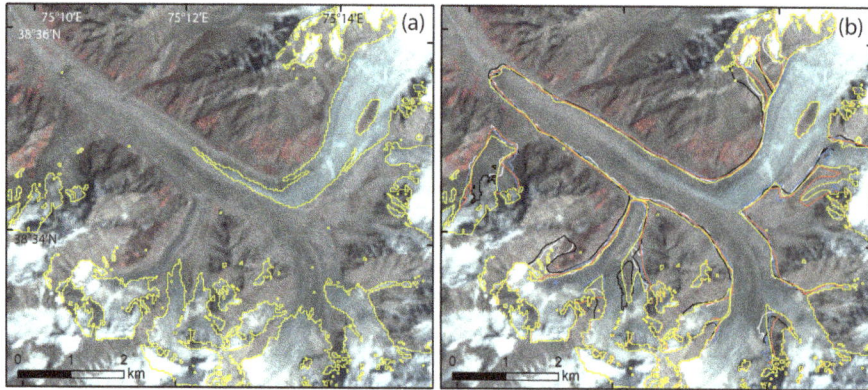

Figure 14. Results of the expert round robin, example Glacier 2. **(a)** shows mapping results solely based on the satellite image, whereas **(b)** shows mapping results after manual corrections using the additional source of coherence images and Google Earth high-resolution imagery.

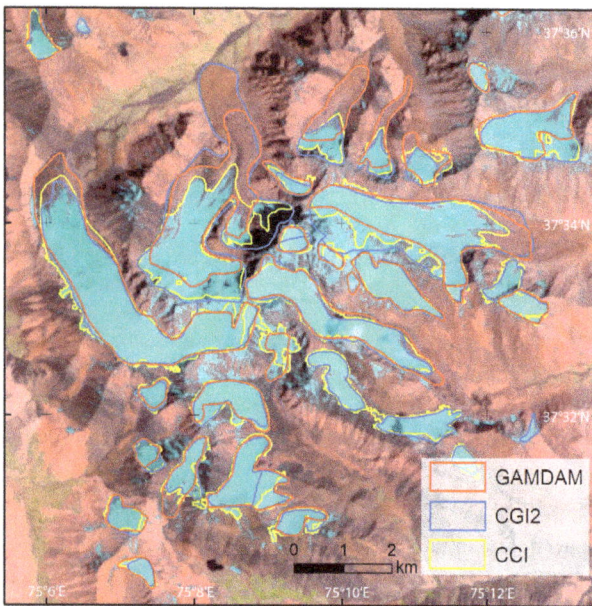

Figure 15. Comparison example of the three inventories.

We think that excluding steep headwalls leads to an incomplete inventory and that the inclusion of rock outcrops in the CGI constitutes a commission error that needs to be corrected for some applications. Overall, the differing interpretation of debris-covered glacier parts and seasonal snow is seemingly the main source of differences in glacier extents for the same region when mapped by different analysts.

7   Discussion

When compiling a large-scale glacier inventory, it is essential to have homogeneous quality of the input data used, at best also in a temporal sense, to ensure high credibility of the resulting outlines and topographic parameters as well as low

uncertainties of the glacierized area. This can be achieved by using globally consistent datasets such as the Landsat images and the GDEM2. However, the latter does not fulfil the criterion of temporal consistency and future work might overcome this issue. There are strategies like DEM fusion to improve DEM quality in regions of very steep terrain or low-contrast glacier surfaces (e.g. Shean et al., 2016; Lee et al., 2015; Tran et al., 2014 and references therein), but the impact of such quality issues is difficult to assess without accurately geo-referenced high-resolution reference data (Kääb et al., 2016; Frey and Paul, 2012). With the semi-automated processing line applied here and the few experts involved in the manual corrections, we assume a homogeneous quality of the glacier outlines throughout the study area has been achieved. However, our glacier extents are likely on the conservative side for debris-covered ablation areas, leading to an underestimation of glacier area, whereas the included perennial ice and snowfields in steep terrain at high elevations are more generous than in other interpretations.

The extraction of debris-covered ice was performed automatically by applying a single threshold value to all scenes and removing the resulting clean-ice areas from the corrected glacier polygons. This is likely the easiest method that still provided good results. Adapting the clean-ice threshold changed the resulting debris-covered area only by $\pm 2.5\%$, indicating that the transition from clean ice to continuous debris cover is relatively sharp. Herreid et al. (2015) used a function applied to Landsat band combinations that was fit to manually derived reference data of a single glacier and adapted it to the various mapping dates (using different Landsat sensors). This method might be superior, but it is more labour-intensive and visual comparison with the figures in Herreid et al. (2015) shows very high agreement. Another approach was applied by Kraaijenbrink et al. (2017), who used the normalized difference snow index (NDSI) together with a composite image of Landsat 8 band 10 (thermal infrared) scenes to detect debris cover based on the RGI 5.0 outlines.

A major prerequisite for all methods is the use of glacier outlines that are well adjusted for debris cover. Glacier retreat was found to correlate with an increase in supraglacial debris cover (e.g. Stokes et al., 2007) and, hence, multi-temporal mapping of debris extent should be applied. As extensive debris cover affects glacier melt and geometry (e.g. Anderson and Anderson, 2016), we recommend including it in the published glacier inventories (GLIMS, RGI), by (a) adding the debris mask as a polygon and (b) including debris cover share in the attribute table. Our results show a total of $\sim 10\%$ debris-covered area, with many of the larger glaciers reaching 20% or more. These numbers complement and confirm existing estimates in HMA that are based on smaller samples. Values reported from the central Karakoram are 20% (Minora et al., 2016) and $\sim 21\%$ (Herreid et al., 2015), Frey et al. (2012) calculated 16% for the western Himalaya, and for the entire Himalaya a $\sim 10\%$ coverage was calculated (Kraaijenbrink et al., 2017; Bolch et al., 2012).

Comparing the freely available results by Kraaijenbrink et al. (2017) to our results, the total difference is $< 1.5\%$ of the total debris-covered area (3365 km^2 vs. 3409 km^2), a value well below the uncertainty of these areas. However, the differences vary from one region to another, from -18.4% in the Pamir west to 5.7% in the Karakoram.

The pattern of glacier median elevations found in our study reflects combinations of climatic and topographic aspects. A similar west–east and north–south gradient was also found in the study by Sakai et al. (2015), who determined median elevations from a glacier inventory (GAMDAM, Nuimura et al., 2015) for all of HMA. Whereas the latitudinal extent of 7° decreases air temperatures and thus median elevations towards the north, the precipitation decrease from west to east due to leeward rain shadow effects increases median elevations in the eastern Pamir and Karakoram. Approximating the balanced-budget ELA (ELA$_0$) with the median elevation has been successfully applied in many mountain ranges and works well for different glacier types (Braithwaite and Raper, 2009). However, this concept likely does not apply to surge-type glaciers and glaciers that are mainly nourished by snow avalanches (Hewitt, 2011). For the latter as well as debris-covered glaciers, ELA$_0$ values are expected to be higher than those calculated here due to the additional accumulation and reduced ablation. This is supported by the fact that we find debris-covered areas also above the median elevation and by a finding from Braithwaite and Raper (2009), who mention possible accumulation-area ratio values below 0.5 for Himalayan glaciers.

We also performed a detailed analysis of uncertainties and analysed the most important sources contributing to uncertainty. It is, however, impossible to perform a rigorous calculation, as this would require a comparison with appropriate reference data. The uncertainties presented here are based on different methods and some of the values are higher than

reported previously. This is mainly because of the high debris coverage and the large number of (very) small glaciers. Under such challenging conditions, area differences among the analysts were as high as uncertainties due to the possible wrong consideration of seasonal snow. Hence, the total area of our inventory will likely be somewhat larger than other inventories for this region as these might have excluded the maybe just snow-covered steep regions at highest elevations. Once scenes without seasonal snow in these regions become available, glacier extents should be revisited and corrected as required.

Conclusion

We derived a new glacier inventory for a substantial part of western High Mountain Asia (Karakoram and Pamir) and have presented in detail the derived characteristics of the glaciers in this region. Special emphasis was given to the description of mapping challenges for debris-covered glaciers (and distinguishing them from rock glaciers), seasonal snow, and shadow, along with the selected solutions. In the absence of appropriate reference datasets, we instead applied various methods for uncertainty assessment and compared our outlines to other existing inventories covering the same region. As an extension to already existing datasets we included outlines and percentages of the debris-covered area for each glacier.

We mapped 27 437 glaciers covering 35 287 ± 1209 km^2, with about 10% of the area being debris covered. The ASTER GDEM2 was found to be superior to the SRTM DEM (1 arcsec) in deriving drainage divides and topographic information for each glacier as the latter suffered from too many (wrongly interpolated) data voids in this region. The application of a constant band ratio threshold to derive clean-ice areas for all scenes to create the debris cover maps was found to be of sufficient quality. Uncertainties derived from three different methods were all in good agreement (3.4%) but the multiple-digitizing experiment also revealed larger deviations among the analysts under challenging conditions (debris, shadow, small glacier). However, the availability of coherence images improved the quality and consistency of the manual corrections for debris-covered glaciers considerably.

The analysis of the topographic information revealed several interesting dependencies among the glaciers and also across the regions. Despite the fact that in the Karakoram the largest glaciers are facing south-east (Siachen, Biafo), east (Batura, Skamri), or west (Baltoro, Hispar), most glacier area (47%) is still exposed to the three northern sectors. Glacier median elevation has little dependence on aspect but a strong one on longitude and latitude (higher towards the drier north and east), indicating a close relation to precipitation amounts. Glacier hypsometry reveals a peak distribution that is highest (~ 5700 m) in the Karakoram, similar but 700 m lower in the eastern and western Pamir, and lowest in Pamir Alai (~ 4200 m). Glaciers in the Karakoram have a comparably higher area share at the lowest elevations and glaciers larger than 5 km^2 or debris-covered glaciers are flatter (22.6 and 16.6°, respectively) than on average (26.4°). By location, glaciers are especially flat ($< 15°$) in their lowest third and progressively steeper ($> 30°$) in the uppermost third, indicating the dominance of large valley glaciers with very flat tongues and steep head walls. Both glacier outlines and the separate outlines of the debris-covered parts will be freely available from the GLIMS database to facilitate applications such as distributed mass balance modelling and albedo calculation; debris-thickness calculation; determination of run-off (with a melt reduction under debris-covered areas); and future geometric evolution, sediment transport, and mountain erosion rates, to name a few.

Author contributions. NM, FP, TB, and PR designed the study. NM and FP wrote the manuscript. NM, PR, TB, and FP generated and edited glacier outlines and basins. TS produced the coherence images. NM extracted debris cover and performed all analysis. All authors contributed to the final form of the manuscript.

Competing interests. The authors declare that they have no conflict of interest.

Acknowledgements. We thank J. Graham Cogley and an anonymous reviewer as well as the journal editor for their thorough reviews and constructive comments that helped improve the clarity of the paper. We acknowledge funding from the ESA projects Glaciers-cci (4000109873/14/I-NB) and Dragon 4 (4000121469/17/I-NB) as well as the Copernicus Climate Change Service (C3S) that is implemented by ECMWF on behalf of the European Commission. The manual digitizing experiment was performed by the authors with additional contributions by Holger Frey and Raymond Le Bris.

Edited by: Reinhard Drews

References

Aizen, E. M., Aizen, V. B., Melack, J. M., Nakamura, T., and Ohta, T.: Precipitation and atmospheric circulation patterns at mid-latitudes of Asia, Int. J. Climatol., 21, 535–556, https://doi.org/10.1002/joc.626, 2001.

Aizen, V. B., Aizen, E. M., Melack, J. M., and Dozier, J.: Climatic and Hydrologic Changes in the Tien Shan, Central Asia, J. Climate, 10, 1393–1404, 1997.

Anderson, L. S. and Anderson, R. S.: Modeling debris-covered glaciers: response to steady debris deposition, The Cryosphere, 10, 1105–1124, https://doi.org/10.5194/tc-10-1105-2016, 2016.

Andreassen, L. M., Paul, F., Kääb, A., and Hausberg, J. E.: Landsat-derived glacier inventory for Jotunheimen, Norway, and deduced glacier changes since the 1930s, The Cryosphere, 2, 131–145, https://doi.org/10.5194/tc-2-131-2008, 2008.

Archer, D. R. and Fowler, H. J.: Spatial and temporal variations in precipitation in the Upper Indus Basin, global teleconnections and hydrological implications, Hydrol. Earth Syst. Sci., 8, 47–61, https://doi.org/10.5194/hess-8-47-2004, 2004.

Arendt, A., Bliss, A., Bolch, T., Cogley, J. G., Gardner, A. S., Hagen, J.-O., Hock, R., Huss, M., Kaser, G., Kienholz, C., Pfeffer, W. T., Moholdt, G., Paul, F., Radić, V., Andreassen, L. M., Bajracharya, S., Barrand, N. E., Beedle, M., Berthier, E., Bhambri, R., Brown, I., Burgess, E. W., Burgess, D., Cawkwell, F., Chinn, T., Copland, L., Davies, B., Angelis, H. de, Dolgova, E., Earl, L., Filbert, K., Forester, R., Fountain, A. G., Frey, H., Giffen, B., Glasser, N. F., Guo, W., Gurney, S. D., Hagg, W., Hall, D., Haritashya, U. K., Hartmann, G., Helm, C., Herreid, S., Howat, I., Kapustin, G., Khromova, T. E., König, M., Kohler, J., Kriegel, D., Kutuzov, S., Lavrentiev, I., Le Bris, R., Liu, S., Lund, J., Manley, W., Marti, R., Mayer, C., Miles, E. S., Li, X., Menounos, B., Mercer, A., Mölg, N., Mool, P., Nosenko, G., Negrete, A., Nuimura, T., Nuth, C., Pettersson, R., Racoviteanu, A., Ranzi, R., Rastner, P., Rau, F., Raup, B., Rich, J., Rott, H., Sakai, A., Schneider, C., Seliverstov, Y., Sharp, M. J., Sigurðsson, O., Stokes, C. R., Way, R. G., Wheate, R., Winsvold, S., Wolken, G., Wyatt, F., and Zheltihyna, N.: Randolph Glacier Inventory – A Dataset of Global Glacier Outlines: Version 5.0: GLIMS Technical Report, Global Land Ice Measurement from Space, Colorado, USA, Digital Media, https://doi.org/10.7265/N5-RGI-50 (last access: 13 March 2018), 2015.

Atwood, D. K., Meyer, F., and Arendt, A.: Using L-band SAR coherence to delineate glacier extent, Can. J. Remote Sens., 36, S186–S195, https://doi.org/10.5589/m10-014, 2010.

Bajracharya, S. R., Maharjan, S. B., Shrestha, F., Guo, W., Liu, S., Immerzeel, W., and Shrestha, B.: The glaciers of the Hindu Kush Himalayas: Current status and observed changes from the 1980s to 2010, Int. J. Water Resour. D., 31, 161–173, https://doi.org/10.1080/07900627.2015.1005731, 2015.

Barsch, D.: Rockglaciers: Indicators for the Present and Former Geoecology in High Mountain Environments, in: Springer Series in Physical Environment, Springer Berlin Heidelberg, Berlin, Heidelberg, 16, 331 pp., 1996.

Berthling, I.: Beyond confusion: Rock glaciers as cryo-conditioned landforms, Geomorphology, 131, 98–106, https://doi.org/10.1016/j.geomorph.2011.05.002, 2011.

Bhambri, R., Bolch, T., Kawishwar, P., Dobhal, D. P., Srivastava, D., and Pratap, B.: Heterogeneity in glacier response in the upper Shyok valley, northeast Karakoram, The Cryosphere, 7, 1385–1398, https://doi.org/10.5194/tc-7-1385-2013, 2013.

Bhambri, R., Hewitt, K., Kawishwar, P., and Pratap, B.: Surge-type and surge-modified glaciers in the Karakoram, Scientific

Reports, 7, 15391, https://doi.org/10.1038/s41598-017-15473-8, 2017.

Bishop, M. P., Shroder, J. F., Ali, G., Bush, A. B. G., Haritashya, U. K., Roohi, R., Sarikaya, M. A., and Weihs, B. J.: Remote Sensing of Glaciers in Afghanistan and Pakistan, in: Global Land Ice Measurements from Space, edited by: Kargel, J. S., Leonard, G. J., Bishop, M. P., Kääb, A., and Raup, B. H., Springer Berlin Heidelberg, Berlin, Heidelberg, 509–548, 2014.

Bliss, A., Hock, R., and Radić, V.: Global response of glacier runoff to twenty-first century climate change, J. Geophys. Res. Earth, 119, 717–730, https://doi.org/10.1002/2013JF002931, 2014.

Bodin, X., Rojas, F., and Brenning, A.: Status and evolution of the cryosphere in the Andes of Santiago (Chile, 33.5°S.), Geomorphology, 118, 453–464, https://doi.org/10.1016/j.geomorph.2010.02.016, 2010.

Böhner, J.: General climatic controls and topoclimatic variations in Central and High Asia, Boreas, 35, 279–295, https://doi.org/10.1111/j.1502-3885.2006.tb01158.x, 2006.

Bolch, T. and Gorbunov, A. P.: Characteristics and Origin of Rock Glaciers in Northern Tien Shan (Kazakhstan/Kyrgyzstan), Permafrost Periglac., 25, 320–332, https://doi.org/10.1002/ppp.1825, 2014.

Bolch, T. and Kamp, U.: Glacier mapping in high mountains using DEMs, Landsat and ASTER data, in: Proceedings of the 8th International Symposium on High Mountain Remote Sensing Cartography, La Paz, Bolivia, 21–27 March 2005, edited by: Kaufmann, V. and Sulzer, W., 2006.

Bolch, T., Buchroithner, M., Kunert, A., and Kamp, U.: Automated delineation of debris-covered glaciers based on ASTER data, in: GeoInformation in Europe, edited by: Gomarasca, A., Millpress, Rotterdam, 403–410, 2007.

Bolch, T., Menounos, B., and Wheate, R.: Landsat-based inventory of glaciers in western Canada, 1985-2005, Remote Sens. Environ., 114, 127–137, https://doi.org/10.1016/j.rse.2009.08.015, 2010.

Bolch, T., Kulkarni, A., Kaab, A., Huggel, C., Paul, F., Cogley, J. G., Frey, H., Kargel, J. S., Fujita, K., Scheel, M., Bajracharya, S., and Stoffel, M.: The state and fate of Himalayan glaciers, Science, 336, 310–314, https://doi.org/10.1126/science.1215828, 2012.

Bolch, T., Sandberg Sørensen, L., Simonsen, S. B., Mölg, N., Machguth, H., Rastner, P., and Paul, F.: Mass loss of Greenland's glaciers and ice caps 2003–2008 revealed from ICESat laser altimetry data, Geophys. Res. Lett., 40, 875–881, https://doi.org/10.1002/grl.50270, 2013.

Bolch, T., Pieczonka, T., Mukherjee, K., and Shea, J.: Brief communication: Glaciers in the Hunza catchment (Karakoram) have been nearly in balance since the 1970s, The Cryosphere, 11, 531–539, https://doi.org/10.5194/tc-11-531-2017, 2017.

Bookhagen, B. and Burbank, D. W.: Toward a complete Himalayan hydrological budget: Spatiotemporal distribution of snowmelt and rainfall and their impact on river discharge, J. Geophys. Res., 115, F03019, https://doi.org/10.1029/2009JF001426, 2010.

Braithwaite, R. J. and Müller, F.: On the parameterization of glacier equilibrium line altitude, in: Proceedings of the Workshop at Riederalp, Switzerland, 17–22 September 1978, IAHS-AISH Publ. No. 126, 263–271, 1980.

Braithwaite, R. J. and Raper, S.C.B.: Estimating equilibrium-line altitude (ELA) from glacier inventory data, Ann. Glaciol., 50, 127–132, https://doi.org/10.3189/172756410790595930, 2009.

Brock, B. W., Mihalcea, C., Kirkbride, M. P., Diolaiuti, G., Cutler, M. E. J., and Smiraglia, C.: Meteorology and surface energy fluxes in the 2005–2007 ablation seasons at the Miage debris-covered glacier, Mont Blanc Massif, Italian Alps, J. Geophys. Res., 115, D09106, https://doi.org/10.1029/2009JD013224, 2010.

Brun, F., Berthier, E., Wagnon, P., Kääb, A., and Treichler, D.: A spatially resolved estimate of High Mountain Asia glacier mass balances, 2000–2016, Nat. Geosci., 10, 668–673, https://doi.org/10.1038/NGEO2999, 2017.

Copland, L., Sylvestre, T., Bishop, M. P., Shroder, J. F., Seong, Y. B., Owen, L. A., Bush, A., and Kamp, U.: Expanded and Recently Increased Glacier Surging in the Karakoram, Arct. Antarct. Alp. Res., 43, 503–516, https://doi.org/10.1657/1938-4246-43.4.503, 2011.

Dehecq, A., Gourmelen, N., and Trouve, E.: Deriving large-scale glacier velocities from a complete satellite archive: Application to the Pamir–Karakoram–Himalaya, Remote Sens. Environ., 162, 55–66, https://doi.org/10.1016/j.rse.2015.01.031, 2015.

Dobreva, I., Bishop, M., and Bush, A.: Climate–Glacier Dynamics and Topographic Forcing in the Karakoram Himalaya: Concepts, Issues and Research Directions, Water, 9, 405, https://doi.org/10.3390/w9060405, 2017.

Falaschi, D., Bolch, T., Rastner, P., Lenzano, M. G., Lenzano, L., Lo Veccio, A., and Moragues, S.: Mass changes of alpine glaciers at the eastern margin of the Northern and Southern Patagonian Icefields between 2000 and 2012, J. Glaciol., 63, 258–272, https://doi.org/10.1017/jog.2016.136, 2017.

Frey, H. and Paul, F.: On the suitability of the SRTM DEM and ASTER GDEM for the compilation of topographic parameters in glacier inventories, Int. J. Appl. Earth Obs., 18, 480–490, https://doi.org/10.1016/j.jag.2011.09.020, 2012.

Frey, H., Paul, F., and Strozzi, T.: Compilation of a glacier inventory for the western Himalayas from satellite data: Methods, challenges, and results, Remote Sens. Environ., 124, 832–843, https://doi.org/10.1016/j.rse.2012.06.020, 2012.

Gardelle, J., Berthier, E., Arnaud, Y., and Kääb, A.: Region-wide glacier mass balances over the Pamir-Karakoram-Himalaya during 1999–2011, The Cryosphere, 7, 1263–1286, https://doi.org/10.5194/tc-7-1263-2013, 2013.

Gardner, A. S., Moholdt, G., Cogley, J. G., Wouters, B., Arendt, A. A., Wahr, J., Berthier, E., Hock, R., Pfeffer, W. T., Kaser, G., Ligtenberg, S. R. M., Bolch, T., Sharp, M. J., Hagen, J. O., van den Broeke, M. R., and Paul, F.: A reconciled estimate of glacier contributions to sea level rise: 2003 to 2009, Science, 340, 852–857, https://doi.org/10.1126/science.1234532, 2013.

Gorbunov, A. P. and Titkov, S. N.: Kamennye Gletchery Gor Srednej Azii (Rock glaciers of the Central Asian Mountains), Akademia Nauk SSSR, Irkutsk, 1989.

Granshaw, F. D. and Fountain, A. G.: Glacier change (1958–1998) in the North Cascades National Park Complex, Washington, USA, J. Glaciol., 52, 251–256, https://doi.org/10.3189/172756506781828782, 2006.

Guo, W., Liu, S., Xu, J., Wu, L., Shangguan, D., Yao, X., Wei, J., Bao, W., Yu, P., Liu, Q., and Jiang, Z.: The second Chinese glacier inventory: Data, methods and results, J. Glaciol., 61, 357–372, https://doi.org/10.3189/2015JoG14J209, 2015.

Haeberli, W. and Hoelzle, M.: Application of inventory data for estimating characteristics of and regional climate-change effects on mountain glaciers: A pilot study with the European Alps, Ann. Glaciol., 21, 206–212, https://doi.org/10.3189/S0260305500015834, 1995.

Haeberli, W., Hallet, B., Arenson, L., Elconin, R., Humlum, O., Kääb, A., Kaufmann, V., Ladanyi, B., Matsuoka, N., Springman, S., and Mühll, D. V.: Permafrost creep and rock glacier dynamics, Permafrost Periglac., 17, 189–214, https://doi.org/10.1002/ppp.561, 2006.

Haritashya, U. K., Bishop, M. P., Shroder, J. F., Bush, A. B. G., and Bulley, H. N. N.: Space-based assessment of glacier fluctuations in the Wakhan Pamir, Afghanistan, Climatic Change, 94, 5–18, https://doi.org/10.1007/s10584-009-9555-9, 2009.

Herreid, S., Pellicciotti, F., Ayala, A., Chesnokova, A., Kienholz, C., Shea, J., and Shrestha, A.: Satellite observations show no net change in the percentage of supraglacial debris-covered area in northern Pakistan from 1977 to 2014, J. Glaciol., 61, 524–536, https://doi.org/10.3189/2015JoG14J227, 2015.

Hewitt, K.: The Karakoram Anomaly? Glacier Expansion and the 'Elevation Effect' Karakoram Himalaya, Mt. Res. Dev., 25, 332–340, 2005.

Hewitt, K.: Glacier Change, Concentration, and Elevation Effects in the Karakoram Himalaya, Upper Indus Basin, Mt. Res. Dev., 31, 188–200, https://doi.org/10.1659/MRD-JOURNAL-D-11-00020.1, 2011.

Holzer, N., Golletz, T., Buchroithner, M., and Bolch, T.: Glacier Variations in the Trans Alai Massif and the Lake Karakul Catchment (Northeastern Pamir) Measured from Space, in: Climate Change, Glacier Response, and Vegetation Dynamics in the Himalaya, edited by: Singh, R. B., Schickhoff, U., and Mal, S., Springer International Publishing, Cham, 139–153, 2016.

Huss, M. and Hock, R.: A new model for global glacier change and sea-level rise, Front. Earth Sci., 3, 382, https://doi.org/10.3389/feart.2015.00054, 2015.

Immerzeel, W. W., van Beek, L. P. H., and Bierkens, M. F. P.: Climate change will affect the Asian water towers, Science, 328, 1382–1385, https://doi.org/10.1126/science.1183188, 2010.

Immerzeel, W. W., Wanders, N., Lutz, A. F., Shea, J. M., and Bierkens, M. F. P.: Reconciling high-altitude precipitation in the upper Indus basin with glacier mass balances and runoff, Hydrol. Earth Syst. Sci., 19, 4673–4687, https://doi.org/10.5194/hess-19-4673-2015, 2015.

Iturrizaga, L.: Trends in 20th century and recent glacier fluctuations in the Karakoram Mountains, Z. Geomorphol. Supp., 55, 205–231, https://doi.org/10.1127/0372-8854/2011/0055S3-0059, 2011.

Kääb, A., Berthier, E., Nuth, C., Gardelle, J., and Arnaud, Y.: Contrasting patterns of early twenty-first-century glacier mass change in the Himalayas, Nature, 488, 495–498, https://doi.org/10.1038/nature11324, 2012.

Kääb, A., Treichler, D., Nuth, C., and Berthier, E.: Brief Communication: Contending estimates of 2003–2008 glacier mass balance over the Pamir–Karakoram–Himalaya, The Cryosphere, 9, 557–564, https://doi.org/10.5194/tc-9-557-2015, 2015.

Kääb, A., Winsvold, S., Altena, B., Nuth, C., Nagler, T., and Wuite, J.: Glacier Remote Sensing Using Sentinel-2. Part I: Radiometric and Geometric Performance, and Application to Ice Velocity, Remote Sensing, 8, 598, https://doi.org/10.3390/rs8070598, 2016.

Khromova, T. E., Osipova, G. B., Tsvetkov, D. G., Dyurgerov, M. B., and Barry, R. G.: Changes in glacier extent in the eastern Pamir, Central Asia, determined from historical data and ASTER imagery, Remote Sens. Environ., 102, 24–32, https://doi.org/10.1016/j.rse.2006.01.019, 2006.

Kienholz, C., Hock, R., and Arendt, A. A.: A new semi-automatic approach for dividing glacier complexes into individual glaciers, J. Glaciol., 59, 925–937, https://doi.org/10.3189/2013JoG12J138, 2013.

Kienholz, C., Herreid, S., Rich, J. L., Arendt, A. A., Hock, R., and Burgess, E. W.: Derivation and analysis of a complete modern-date glacier inventory for Alaska and northwest Canada, J. Glaciol., 61, 403–420, https://doi.org/10.3189/2015JoG14J230, 2015.

Kirkbride, M. P. and Deline, P.: The formation of supraglacial debris covers by primary dispersal from transverse englacial debris bands, Earth Surf. Proc. Land., 38, 1779–1792, https://doi.org/10.1002/esp.3416, 2013.

Kotlyakov, V. M., Osipova, G. B., and Tsvetkov, D. G.: Monitoring surging glaciers of the Pamirs, central Asia, from space, Ann. Glaciol., 48, 125–134, https://doi.org/10.3189/172756408784700608, 2008.

Kraaijenbrink, P. D. A., Bierkens, M. F. P., Lutz, A. F., and Immerzeel, W. W.: Impact of a global temperature rise of 1.5 degrees Celsius on Asia's glaciers, Nature, 549, 257–260, https://doi.org/10.1038/nature23878, 2017.

Le Bris, R., Paul, F., Frey, H., and Bolch, T.: A new satellite-derived glacier inventory for western Alaska, Ann. Glaciol., 52, 135–143, https://doi.org/10.3189/172756411799096303, 2011.

Lee, C., Oh, J., Hong, C., and Youn, J.: Automated Generation of a Digital Elevation Model Over Steep Terrain in Antarctica From High-Resolution Satellite Imagery, IEEE T. Geosci. Remote, 53, 1186–1194, https://doi.org/10.1109/TGRS.2014.2335773, 2015.

Lin, H., Li, G., Cuo, L., Hooper, A., and Ye, Q.: A decreasing glacier mass balance gradient from the edge of the Upper Tarim Basin to the Karakoram during 2000–2014, Scientific Reports, 7, 6712, https://doi.org/10.1038/s41598-017-07133-8, 2017.

Lutz, A. F., Immerzeel, W. W., Shrestha, A. B., and Bierkens, M. F. P.: Consistent increase in High Asia's runoff due to increasing glacier melt and precipitation, Nat. Clim. Change, 4, 587–592, https://doi.org/10.1038/nclimate2237, 2014.

Maussion, F., Scherer, D., Mölg, T., Collier, E., Curio, J., and Finkelnburg, R.: Precipitation Seasonality and Variability over the Tibetan Plateau as Resolved by the High Asia Reanalysis*, J. Climate, 27, 1910–1927, https://doi.org/10.1175/JCLI-D-13-00282.1, 2014.

Minora, U., Bocchiola, D., D'Agata, C., Maragno, D., Mayer, C., Lambrecht, A., Vuillermoz, E., Senese, A., Compostella, C., Smiraglia, C., and Diolaiuti, G. A.: Glacier area stability in the Central Karakoram National Park (Pakistan) in 2001–2010, Prog. Phys. Geog., 40, 629–660, https://doi.org/10.1177/0309133316643926, 2016.

Mölg, N., Bolch, T., Rastner, P., Strozzi, T., and Paul, F.: Glacier inventory of Pamir and Karakoram, https://doi.org/10.1594/PANGAEA.894707, in review, 2018.

Monnier, S. and Kinnard, C.: Reconsidering the glacier to rock glacier transformation problem: New insights from the central Andes of Chile, Geomorphology, 238, 47–55, https://doi.org/10.1016/j.geomorph.2015.02.025, 2015.

Nicholson, L. and Benn, D. I.: Calculating ice melt beneath a debris layer using meteorological data, J. Glaciol., 52, 463–470, https://doi.org/10.3189/172756506781828584, 2006.

Nuimura, T., Sakai, A., Taniguchi, K., Nagai, H., Lamsal, D., Tsutaki, S., Kozawa, A., Hoshina, Y., Takenaka, S., Omiya, S., Tsunematsu, K., Tshering, P., and Fujita, K.: The GAMDAM glacier inventory: a quality-controlled inventory of Asian glaciers, The Cryosphere, 9, 849–864, https://doi.org/10.5194/tc-9-849-2015, 2015.

Paul, F.: Revealing glacier flow and surge dynamics from animated satellite image sequences: examples from the Karakoram, The Cryosphere, 9, 2201–2214, https://doi.org/10.5194/tc-9-2201-2015, 2015.

Paul, F. and Kääb, A.: Perspectives on the production of a glacier inventory from multispectral satellite data in Arctic Canada: Cumberland Peninsula, Baffin Island, Ann. Glaciol., 42, 59–66, https://doi.org/10.3189/172756405781813087, 2005.

Paul, F., Kääb, A., Maisch, M., Kellenberger, T., and Haeberli, W.: The new remote-sensing-derived Swiss glacier inventory: I. Methods, Ann. Glaciol., 34, 355–361, https://doi.org/10.3189/172756402781817941, 2002.

Paul, F., Kääb, A., and Haeberli, W.: Mapping of rock glaciers with optical satellite imagery, in: Permafrost: Extended Abstracts Reporting Current Research and new Information, edited by: Haeberli, W. and Brandová, D., International Conference on Permafrost, Zurich, Switzerland, 20.-25. July, Glaciology and Geomorphodynamics Group, Department of Geography, University of Zurich, 125–126, 2003.

Paul, F., Huggel, C., and Kääb, A.: Combining satellite multispectral image data and a digital elevation model for mapping debris-covered glaciers, Remote Sens. Environ., 89, 510–518, https://doi.org/10.1016/j.rse.2003.11.007, 2004.

Paul, F., Frey, H., and Le Bris, R.: A new glacier inventory for the European Alps from Landsat TM scenes of 2003: Challenges and results, Ann. Glaciol., 52, 144–152, https://doi.org/10.3189/172756411799096295, 2011.

Paul, F., Barrand, N. E., Baumann, S., Berthier, E., Bolch, T., Casey, K., Frey, H., Joshi, S. P., Konovalov, V., Le Bris, R., Mölg, N., Nosenko, G., Nuth, C., Pope, A., Racoviteanu, A., Rastner, P., Raup, B., Scharrer, K., Steffen, S., and Winsvold, S.: On the accuracy of glacier outlines derived from remote-sensing data, Ann. Glaciol., 54, 171–182, https://doi.org/:10.3189/2013AoG63A296, 2013.

Paul, F., Bolch, T., Briggs, K., Kääb, A., McMillan, M., McNabb, R., Nagler, T., Nuth, C., Rastner, P., Strozzi, T., and Wuite, J.: Error sources and guidelines for quality assessment of glacier area, elevation change, and velocity products derived from satellite data in the Glaciers_cci project, Remote Sens. Environ., 203, 256–275, https://doi.org/10.1016/j.rse.2017.08.038, 2017.

Pfeffer, W. T., Arendt, A. A., Bliss, A., Bolch, T., Cogley, J. G., Gardner, A. S., Hagen, J.-O., Hock, R., Kaser, G., Kienholz, C., Miles, E. S., Moholdt, G., Mölg, N., Paul, F., Radić, V., Rastner, P., Raup, B. H., Rich, J., and Sharp, M. J.: The Randolph Glacier Inventory: A globally complete inventory of glaciers, J. Glaciol., 60, 537–552, https://doi.org/10.3189/2014JoG13J176, 2014.

Quincey, D. J., Glasser, N. F., Cook, S. J., and Luckman, A.: Heterogeneity in Karakoram glacier surges, J. Geophys. Res.-Earth, 120, 1288–1300, https://doi.org/10.1002/2015JF003515, 2015.

Racoviteanu, A. and Williams, M. W.: Decision Tree and Texture Analysis for Mapping Debris-Covered Glaciers in the Kangchenjunga Area, Eastern Himalaya, Remote Sensing, 4, 3078–3109, https://doi.org/10.3390/rs4103078, 2012.

Ragettli, S., Pellicciotti, F., Immerzeel, W. W., Miles, E. S., Petersen, L., Heynen, M., Shea, J. M., Stumm, D., Joshi, S., and Shrestha, A.: Unraveling the hydrology of a Himalayan catchment through integration of high resolution in situ data and remote sensing with an advanced simulation model, Adv. Water Resour., 78, 94–111, https://doi.org/10.1016/j.advwatres.2015.01.013, 2015.

Ragettli, S., Bolch, T., and Pellicciotti, F.: Heterogeneous glacier thinning patterns over the last 40 years in Langtang Himal, Nepal, The Cryosphere, 10, 2075–2097, https://doi.org/10.5194/tc-10-2075-2016, 2016.

Rankl, M. and Braun, M.: Glacier elevation and mass changes over the central Karakoram region estimated from TanDEM-X and SRTM/X-SAR digital elevation models, Ann. Glaciol., 57, 273–281, https://doi.org/10.3189/2016AoG71A024, 2016.

Rankl, M., Kienholz, C., and Braun, M.: Glacier changes in the Karakoram region mapped by multimission satellite imagery, The Cryosphere, 8, 977–989, https://doi.org/10.5194/tc-8-977-2014, 2014.

Raper, S. C. B. and Braithwaite, R. J.: Glacier volume response time and its links to climate and topography based on a conceptual model of glacier hypsometry, The Cryosphere, 3, 183–194, https://doi.org/10.5194/tc-3-183-2009, 2009.

Rastner, P., Bolch, T., Notarnicola, C., and Paul, F.: A Comparison of Pixel- and Object-Based Glacier Classification With Optical Satellite Images, IEEE J. Sel. Top. Appl., 7, 853–862, https://doi.org/10.1109/JSTARS.2013.2274668, 2014.

Raup, B. and Khalsa, S. J. S.: GLIMS Analysis Tutorial, Global Land Ice Measurement from Space, available at: https://www.glims.org/MapsAndDocs/guides.html (last access: 5 October 2018), 2007.

RGI Consortium: Randolph Glacier Inventory – A Dataset of Global Glacier Outlines, Version 6.0, Technical Report, Global Land Ice Measurements from Space, Colorado, USA, Digital Media, https://doi.org/10.7265/N5-RGI-60 (last access: 31 March 2018), 2017.

Robson, B., Hölbling, D., Nuth, C., Stozzi, T., and Dahl, S.: Decadal Scale Changes in Glacier Area in the Hohe Tauern National Park (Austria) Determined by Object-Based Image Analysis, Remote Sensing, 8, 67, https://doi.org/10.3390/rs8010067, 2016.

Rowan, A. V., Egholm, D. L., Quincey, D. J., and Glasser, N. F.: Modelling the feedbacks between mass balance, ice flow and debris transport to predict the response to climate change of debris-covered glaciers in the Himalaya, Earth Planet. Sc. Lett., 430, 427–438, https://doi.org/10.1016/j.epsl.2015.09.004, 2015.

Sakai, A., Nuimura, T., Fujita, K., Takenaka, S., Nagai, H., and Lamsal, D.: Climate regime of Asian glaciers revealed by GAMDAM glacier inventory, The Cryosphere, 9, 865–880, https://doi.org/10.5194/tc-9-865-2015, 2015.

Sarıkaya, M. A., Bishop, M. P., Shroder, J. F., and Ali, G.: Remote-sensing assessment of glacier fluctuations in the Hindu Raj, Pakistan, Int. J. Remote Sens., 34, 3968–3985, https://doi.org/10.1080/01431161.2013.770580, 2013.

Satgé, F., Bonnet, M. P., Timouk, F., Calmant, S., Pillco, R., Molina, J., Lavado-Casimiro, W., Arsen, A., Crétaux,

J. F., and Garnier, J.: Accuracy assessment of SRTM v4 and ASTER GDEM v2 over the Altiplano watershed using ICESat/GLAS data, Int. J. Remote Sens., 36, 465–488, https://doi.org/10.1080/01431161.2014.999166, 2015.

Scherler, D., Bookhagen, B., and Strecker, M. R.: Hillslope-glacier coupling: The interplay of topography and glacial dynamics in High Asia, J. Geophys. Res., 116, F02019, https://doi.org/10.1029/2010JF001751, 2011a.

Scherler, D., Bookhagen, B., and Strecker, M. R.: Spatially variable response of Himalayan glaciers to climate change affected by debris cover, Nat. Geosci., 4, 156–159, https://doi.org/10.1038/NGEO1068, 2011b.

Seong, Y. B., Owen, L. A., Yi, C., and Finkel, R. C.: Quaternary glaciation of Muztag Ata and Kongur Shan: Evidence for glacier response to rapid climate changes throughout the Late Glacial and Holocene in westernmost Tibet, Geol. Soc. Am. Bull., 121, 348–365, https://doi.org/10.1130/B26339.1, 2009.

Shangguan, D., Liu, S., Ding, Y., Ding, L., Xiong, L., Cai, D., Li, G., Lu, A., Zhang, S., and Zhang, Y.: Monitoring the glacier changes in the Muztag Ata and Konggur mountains, east Pamirs, based on Chinese Glacier Inventory and recent satellite imagery, Ann. Glaciol., 43, 79–85, https://doi.org/10.3189/172756406781812393, 2006.

Shangguan, D., Liu, S., Ding, Y., Guo, W., XU, B., Xu, J., and Jiang, Z.: Characterizing the May 2015 Karayaylak Glacier surge in the eastern Pamir Plateau using remote sensing, J. Glaciol., 62, 944–953, https://doi.org/10.1017/jog.2016.81, 2016.

Shea, J. M., Immerzeel, W. W., Wagnon, P., Vincent, C., and Bajracharya, S.: Modelling glacier change in the Everest region, Nepal Himalaya, The Cryosphere, 9, 1105–1128, https://doi.org/10.5194/tc-9-1105-2015, 2015.

Shean, D. E., Alexandrov, O., Moratto, Z. M., Smith, B. E., Joughin, I. R., Porter, C., and Morin, P.: An automated, open-source pipeline for mass production of digital elevation models (DEMs) from very-high-resolution commercial stereo satellite imagery, ISPRS J. Photogramm., 116, 101–117, https://doi.org/10.1016/j.isprsjprs.2016.03.012, 2016.

Singh, P., Ramasastri, K. S., and Kumar, N.: Topographical Influence on Precipitation Distribution in Different Ranges of Western Himalayas, Nord. Hydrol., 26, 259–284, 1995.

Stokes, C. R., Popovnin, V., Aleynikov, A., Gurney, S. D., and Shahgedanova, M.: Recent glacier retreat in the Caucasus Mountains, Russia, and associated increase in supraglacial debris cover and supra-/proglacial lake development, Ann. Glaciol., 46, 195–203, https://doi.org/10.3189/172756407782871468, 2007.

Tran, T. A., Raghavan, V., Masumoto, S., Vinayaraj, P., and Yonezawa, G.: A geomorphology-based approach for digital elevation model fusion – case study in Danang city, Vietnam, Earth Surf. Dynam., 2, 403–417, https://doi.org/10.5194/esurf-2-403-2014, 2014.

United States Geological Survey: ASTER GDEM Version 2, available at: https://gdex.cr.usgs.gov/gdex/, last access: 1 January 2018a.

United States Geological Survey: SRTMGL30, available at: https://gdex.cr.usgs.gov/gdex/, last access: 1 January 2018b.

Vaughan, D. G., Comiso, J. C., Allison, I., Carrasco, J., Kaser, G., Kwok, R., Mote, P., Murray, T., Paul, F., Ren, J., Rignot, E., Solomina, O., Steffen, K., and Zhang, T.: Observations: Cryosphere, in: Climate Change 2013: Physical Science Basis. Contribution of Working Group I to the Fifth Assessment Report of the Intergovernmental Panel on Climate Change, edited by: Stocker, T. F., Qin, D., Plattner, G.-K., Tignor, M., Allen, S. K., Boschung, J., Nauels, A., Xia, Y., Bex, V., and Midgley, P. M., Cambridge University Press, Cambridge, United Kingdom and New York, NY, USA, 2013.

Wake, C. P.: Glaciochemical investigations as a tool for determining the spatial and seasonal variation of snow accumulation in the central Karakoram, northern Pakistan, Ann. Glaciol., 13, 279–284, 1989.

Wendt, A., Mayer, C., Lambrecht, A., and Floricioiu, D.: A Glacier Surge of Bivachny Glacier, Pamir Mountains, Observed by a Time Series of High-Resolution Digital Elevation Models and Glacier Velocities, Remote Sensing, 9, 388, https://doi.org/10.3390/rs9040388, 2017.

Winiger, M., Gumpert, M., and Yamout, H.: Karakorum-Hindukush-western Himalaya: Assessing high-altitude water resources, Hydrol. Process., 19, 2329–2338, https://doi.org/10.1002/hyp.5887, 2005.

Zech, R., Abramowski, U., Glaser, B., Sosin, P., Kubik, P. W., and Zech, W.: Late Quaternary glacial and climate history of the Pamir Mountains derived from cosmogenic 10Be exposure ages, Quaternary Res., 64, 212–220, https://doi.org/10.1016/j.yqres.2005.06.002, 2005.

13

Modulation of glacier ablation by tephra coverage from Eyjafjallajökull and Grímsvötn volcanoes, Iceland: an automated field experiment

Rebecca Möller[1,2], **Marco Möller**[3,4,1], **Peter A. Kukla**[2], **and Christoph Schneider**[4]

[1]Department of Geography, RWTH Aachen University, Aachen, Germany
[2]Geological Institute, Energy and Minerals Resources Group, RWTH Aachen University, Aachen, Germany
[3]Institute of Geography, University of Bremen, Bremen, Germany
[4]Geography Department, Humboldt-Universität zu Berlin, Berlin, Germany

Correspondence: Rebecca Möller (rebecca.moeller@geo.rwth-aachen.de)

Abstract. We report results from a field experiment investigating the influence of volcanic tephra coverage on glacier ablation. These influences are known to be significantly different from those of moraine debris on glaciers due to the contrasting grain size distribution and thermal conductivity. Thus far, the influences of tephra deposits on glacier ablation have rarely been studied. For the experiment, artificial plots of two different tephra types from Eyjafjallajökull and Grímsvötn volcanoes were installed on a snow-covered glacier surface of Vatnajökull ice cap, Iceland. Snow-surface lowering and atmospheric conditions were monitored in summer 2015 and compared to a tephra-free reference site. For each of the two volcanic tephra types, three plots of variable thickness (~ 1.5, ~ 8.5 and ~ 80 mm) were monitored. After limiting the records to a period of reliable measurements, a 50-day data set of hourly records was obtained, which can be downloaded from the Pangaea data repository (https://www.pangaea.de; doi:10.1594/PANGAEA.876656). The experiment shows a substantial increase in snow-surface lowering rates under the ~ 1.5 and ~ 8.5 mm tephra plots when compared to uncovered conditions. Under the thick tephra cover some insulating effects could be observed. These results are in contrast to other studies which depicted insulating effects for much thinner tephra coverage on bare-ice glacier surfaces. Differences between the influences of the two different petrological types of tephra exist but are negligible compared to the effect of tephra coverage overall.

1 Introduction

Deposits of sedimentary materials on the surface of glaciers are known to have significant influence on glacier melt as they alter the energy exchange processes at the surface (e.g., Nicholson and Benn, 2013; Mattson et al., 1993; Østrem, 1959). The thickness of the layer controls whether the dominant factor at the glacier surface is the decrease in albedo or the increase in thermal resistance (Möller et al., 2016). The former implies an increase in the energy gain to the glacier from solar radiation while the latter implies a decrease because of reduced heat conduction to the glacier surface. As a result, thin layers of supraglacial deposits lead to increased glacier melt, while thick layers imply decreased glacier melt or even insulation. With increasing layer thickness glacier melt peaks at the so-called effective thickness. With further increasing layer thickness, glacier melt decreases again and returns to the level of uncovered conditions at the so-called critical thickness. Beyond this thickness, glacier melt decreases further towards the limit of complete insulation (Adhikary et al., 1997).

The influence of tephra on glacier melt is usually parametrized using in situ data for calibration. However, most of the formulations developed thus far are designed to capture the effects of moraine debris deposits which are usually formed by layers with thicknesses on the order of meters or at least decimeters or centimeters. In recent years there have been numerous studies dealing with the relationship between debris thickness and resulting modification of ablation (e.g., Collier et al., 2015; Juen et al., 2014; Pratap et al., 2015; Rounce et al., 2015).

Volcanically active regions of the world in sub-polar and polar environments episodically experience the deposition of tephra on glacier surfaces after explosive volcanic eruptions. Volcanic tephra deposits show a wider range of depositional thicknesses than moraine debris, i.e., from sub-millimeter to meter scale. They also feature distinctly different thermal properties (Brock et al., 2007). The model formulation of Evatt et al. (2015) is valid for all thicknesses from dust to meter scale. However, dedicated studies dealing with the relationship between tephra thickness and the intensity of induced ablation change are remarkably less numerous than those dealing with moraine debris, even if supraglacial tephra deposits are known to significantly influence glacier surface processes and mass balance (e.g., Kirkbride and Dugmore, 2003; Möller et al., 2014; Nield et al., 2013). So far, only three recent studies have carried out a systematic, quantitative investigation of the influence of tephra deposits of varying thickness on glacier ablation (Dragosics et al., 2016; Juen et al., 2013; Möller et al., 2016). However, these studies were carried out on bare-ice surfaces and only rely on results obtained over short periods. The experiments covered periods of only 17 (Dragosics et al., 2016) or 13 days (Möller et al., 2016) of regular daily measurements. Moreover, the experiment of Dragosics et al. (2016) was carried out in an ex situ, non-local environment under controlled, partly laboratory-like conditions. The experiment of Juen et al. (2013) lasted for about 1 month, but ablation measurements were mostly carried out at an irregular frequency.

Here, we present data from automated, continuous measurements of meteorological conditions and snow-surface lowering under artificially installed plots of volcanic tephra of different type and thickness. The measurements were obtained from a field experiment which was carried out on Vatnajökull ice cap, Iceland, over the 2015 summer season. Snow-surface lowering rates under different thicknesses of tephra during days with and without precipitation are compared to illustrate the variability of snow-surface lowering with tephra thickness and the influence of different meteorological conditions. It has to be noted that our measurements are only a proxy for snow ablation, as snow density changes beneath the tephra plots (which also impact snow-surface lowering) were not quantified due to logistical limitations.

2 Field experiment

2.1 Study site

The field experiment was carried out at an elevation of ~ 970 m a.s.l. on Tungnaárjökull (64.3253° N, 18.0476° W), a glacier which is part of the western Vatnajökull ice cap,

Iceland (Fig. 1a). The site was situated on a slightly inclined surface, facing approximately west-southwest. It was characterized by wind-compacted snow coverage with a homogeneous depth of $\sim 2.7 \pm 0.2$ m throughout the site according to snow-depth probing. Layering of the snowpack was not well pronounced and snow density showed little variability over the vertical profile with an integrated mean of ~ 410 kg m^{-3}, which was obtained by stepwise measurements along a vertical profile.

2.2 Design and setup

The field experiment was designed to quantify the influence of volcanic tephra (with variable type and thickness) on snow-surface lowering and to relate the measured lowering to meteorological conditions. A set of six artificial plots of tephra coverage with a diameter of 0.7 m were installed at the study site. Three of these plots were made from tephra of Eyjafjallajökull volcano (EYV) and the other three from tephra of Grímsvötn volcano (GRV; Fig. 1a). Both types of tephra were spread out at thicknesses of ~ 1.5, ~ 8.5 and ~ 80 mm. This was done by weighing out tephra material according to its bulk density (1276 kg m^{-3} for EYV and 791 kg m^{-3} for GRV) as dispersal by thickness was not feasible at the millimeter scale. The three thicknesses approximately match the effective thickness (1.5 mm), the critical thickness (8.5 mm) and a thickness under which the dominance of insulation can be considered. These values were chosen according to results of a short, 13-day field experiment by Möller et al. (2016) carried out on bare glacier ice using tephra of GRV.

Contiguous to the tephra plots where snow-surface lowering was recorded, standard meteorological parameters were measured and recorded by an automatic weather station (AWS). The parameters include air temperature and relative humidity at two levels (initially 0.3 and 1.1 m above snow, but increasing according to snow-surface lowering), wind speed and direction (initially 2.1 m above snow), liquid precipitation and incoming and reflected shortwave radiation. For measuring snow-surface lowering at the tephra plots, an aluminum structure for sensor installation was mounted (Fig. 1b). Over each of the six plots ultrasonic height gauges measured snow-surface lowering at hourly intervals. In addition, sensors for surface temperature measurements were installed over the two ~ 80 mm plots. Table 1 gives an overview of all sensor and measurement specifications for both the AWS and the tephra plots. The snow-surface lowering measurement at the AWS provides a reference representing non-tephra covered conditions.

A camera system, taking photographs hourly, was setup to monitor and document the conditions of tephra plots and AWS. Unfortunately, it stopped working after a few days and we do not use these data here.

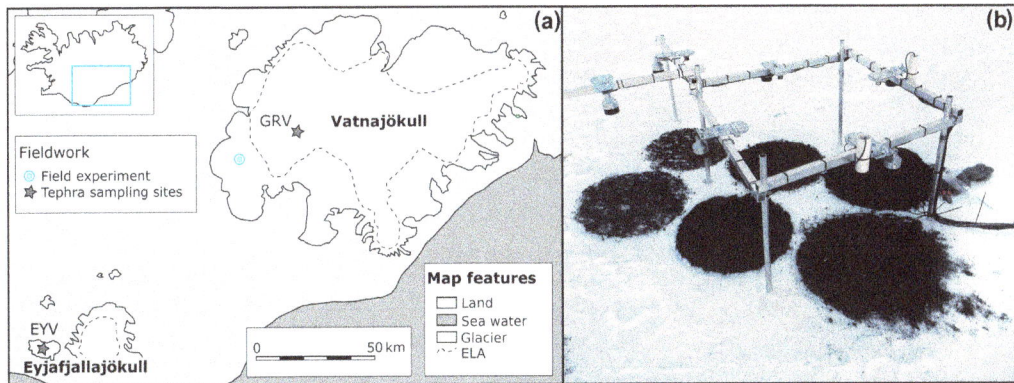

Figure 1. Overview of the field experiment. The locations of tephra sampling at the calderas of Eyjafjallajökull volcano (EYV) and Grímsvötn volcano (GRV) and the location of the field experiment are shown in **(a)**. The installation of the field experiment is shown in **(b)**. The three plots in the foreground are covered by EYV tephra and the three plots in the back by GRV tephra.

Table 1. Measured quantities at the field experiment installation and at the automatic weather station. For each variable the type of the sensor is given along with its uncertainty and the type of data aggregation over each 1 h record interval.

Variable	Sensor	Uncertainty	Aggregation
Air temperature	Vaisala HMP35C	± 0.4 K	Average
Relative humidity	Vaisala HMP35C	$\pm 3\,\%$	Average
Incoming SW radiation	Campbell Scientific CS300	$\pm 5\,\%$	Average
Reflected SW radiation	Campbell Scientific SP1110	$\pm 5\,\%$	Average
Rainfall	RM Young 52203	$\pm 2\,\%$	Total
Wind speed	RM Young 05103	$\pm 0.3\,\mathrm{m\,s^{-1}}$	Average
Wind direction	RM Young 05103	n.a.	Sample
Snow-surface lowering (reference)	Campbell Scientific SR50	± 1 cm	Sample
Surface temperature	Campbell Scientific IRTS-P	± 0.3 K	Average
Snow-surface lowering (tephra plots)	Campbell Scientific SR50A	± 1 cm	Sample

2.3 Tephra sampling

The tephra material was directly sampled at the calderas of EYV and GRV (Fig. 1a) in order to obtain pristine material. At EYV the tephra was acquired from inside the caldera (63.6314° N, 19.6373° W). This sampling was carried out on 7 May 2015. At GRV the tephra was collected at rocky outcrops near the southern caldera rim (64.4061° N, 17.2741° W). Here, sampling was done on 8 May 2015. At both locations, the tephra was taken from active geothermal areas.

2.4 Measurements and data preparation

The experiment started on 10 May 2015 and recorded hourly means and samples from the sensors described in Table 1 until 8 September 2015. Measurements stopped on 9 September 2015, when ablation was so advanced that the aluminum structure collapsed. During the collapse, the lowermost parts of the structure were still anchored inside the ice, but the center of mass of the overlying installation was probably too high above ground. The timing of the collapse was eas-

ily identifiable from abnormal radiation and distance measurements. For studying the influences of tephra coverage on snow-surface lowering, the records had to be narrowed down to a period without snow cover on top of the tephra. The selection of the suitable period is based on measured surface temperatures on the tephra packs of the two ~ 80 mm plots (Fig. 2).

Surface temperature is generally closely related to the intra-day cycles of air temperature and shortwave radiation. However, snow or ice surfaces cannot exceed 0 °C. This implies that surface temperatures which follow a regular above-zero intra-day cycle indicate a completely snow- or ice-free surface. In our field experiment, this is the case for the period after 15 June 2015 (Fig. 2). Up until this date, sub-zero surface temperatures prevail despite the presence of intra-day air temperature cycles which regularly exceed 0 °C. This indicates at least partly snow covered conditions on the surfaces of the tephra plots.

From 4 August 2015 onwards, the intra-day cycles of surface temperature start to become irregular. In addition, the periodic, substantially positive offsets of surface temperature over air temperature, which occurred consistently over

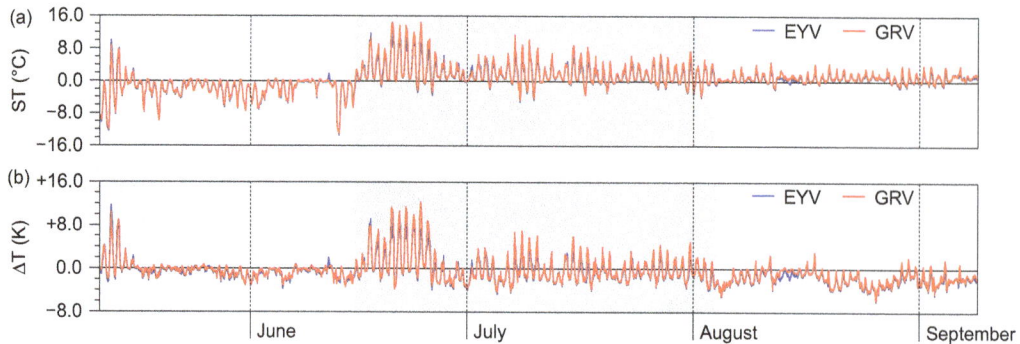

Figure 2. Records of measured hourly surface temperatures at the two ~ 80 mm tephra plots **(a)** and calculated differences between these surface temperatures and air temperatures measured at the automatic weather station **(b)** over 11 May to 8 September. The air temperatures are calculated as the mean of upper and lower air temperature sensor at the AWS. The types of tephra (EYV for Eyjafjallajökull volcano and GRV for Grímsvötn volcano) on which the surface temperatures were measured are indicated by color code. The grey shading in the center of the time series indicates the period considered in the final data set, i.e., 15 June to 3 August.

15 June to 3 August, were replaced by rather irregular, predominantly negative offsets (Fig. 2). This combination of observations suggests that the tephra packs started to disintegrate, providing space for snow or bare-ice outcrops which destroyed the homogeneous surfaces of the tephra plots. Over homogeneous, low-albedo tephra coverage, shortwave radiation adds considerably to the energy gain at the surface and thus drives surface temperatures far above the air temperature level. Over rather patchy tephra coverage with high-albedo bare-ice outcrops, the integrated energy gain due to absorbed shortwave radiation is much lower. In addition, the surface temperature of the outcrops is capped at 0 °C. The integrated surface temperature of the tephra plots might thus lie well below the air temperature level.

Based on these considerations, we limit the observations to the 50-day period covering 15 June to 3 August (Fig. 2). The final data set contains hourly averaged data for all meteorological parameters measured at the AWS (Fig. 3a). Moreover, it contains hourly data from all seven ultrasonic height gauges, i.e., from snow-surface lowering measurements at the six tephra plots and at the reference site at the AWS (Fig. 3b).

We compared the snow-surface lowering rates at the different plots. To facilitate this analysis, running 24 h differences, i.e., running daily snow-surface lowering rates, were calculated for the data of each of the seven sensors whenever valid measurements existed at all six tephra plots and at the reference site. This was undertaken in order to assure full comparability of the 24 h snow-surface lowering values. These running 24 h differences are also part of the published data set.

3 Results

Snow-surface lowering measurements over the chosen time period (15 June to 3 August 2015) reveal a loss of 2.25 m

of snow cover at the reference site and between 2.21 and 2.97 m at the tephra plots (Fig. 3b). During almost the entire period the study site showed snow coverage. Only for the plots with ~ 1.5 mm tephra coverage it cannot be ruled out that the snowpack beneath the plots disappeared just before the end of the study period. For the reference site, the snowpack completely disappeared during the second week of August according to the measured albedo values. The progressive snow-surface lowering led to an increasing measurement uncertainty towards the end of the study period because the sensors' footprints might have extended beyond the borders of the tephra plots and erosion of the tephra material might have destroyed the previously homogeneous dispersal across the plots. Nevertheless, the running daily snow-surface lowering rates, i.e., the slopes of the snow-surface lowering curves (Fig. 3b), show small variability with time, even if ephemeral increases sporadically occur at the end of June and during mid-July. The lowering rates at the six different plots become more similar over the second half of July, suggesting an incipient disintegration of the different tephra packs presumably due to erosion by meltwater.

Major disturbances occur in the snow-surface lowering curves of two of the GRV tephra plots (~ 8.5 and ~ 80 mm) in mid-July (Fig. 3b). On 14 July the measured distance at the ~ 8.5 mm GRV tephra plot increased by ~ 0.20 m, followed by an increase of ≥ 0.15 m at the ~ 80 mm GRV tephra plot on 16 July. These disturbances coincide with a major rain event (Fig. 3a). It can thus not be ruled out that partial destructions of the tephra plots and of the upper layers of the snowpack occurred at this date subsequently distorting the distance measurements at the six tephra plots.

The relationships between tephra thickness and running daily snow-surface lowering rates (Fig. 4) resemble the findings of previous studies dealing with bare-ice ablation (Kirkbride and Dugmore, 2003; Mattson et al., 1993; Möller et al., 2016). At the thin (~ 1.5 mm) tephra plots, snow-

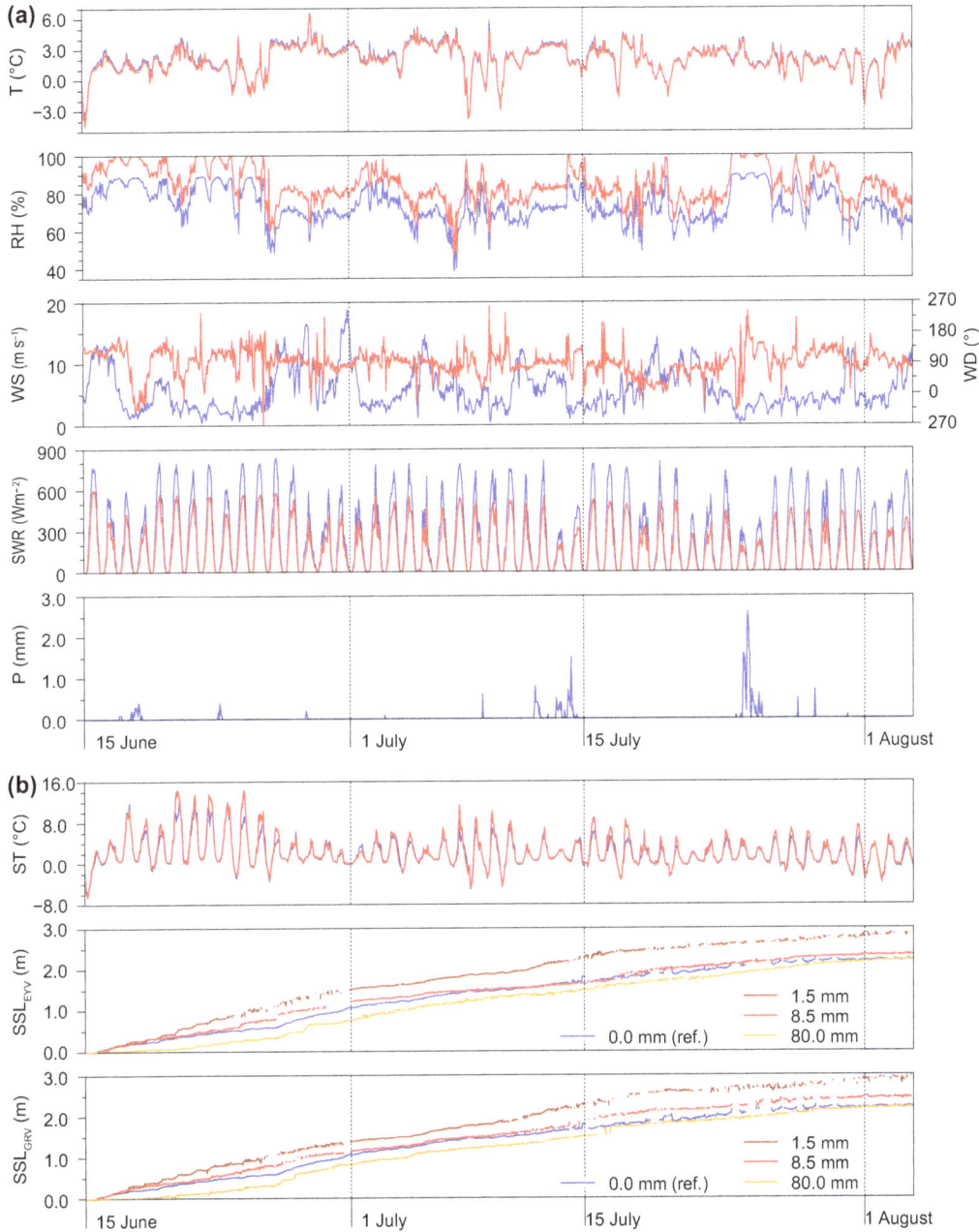

Figure 3. Hourly records of the measurements of all sensors installed at the automatic weather station are shown in **(a)**, and measurements of all sensors mounted at the field experiment installation are shown in **(b)**. Records are shown for 15 June to 3 August. For air temperature (T) and relative humidity (RH) the records of the upper sensor (blue line) are shown together with those of the lower sensor (red line). Wind speed (blue line) is shown together with wind direction (red line); note the different y axes here. Incoming shortwave radiation (SWR, blue line) is shown together with reflected shortwave radiation (red line). For precipitation (P), only the liquid fraction has been measured. Surface temperatures (ST) are shown for the ~ 80 mm plots of tephra from Eyjafjallajökull volcano (EYV, blue line) and from Grímsvötn volcano (GRV, red line). Cumulative snow-surface lowering (SSL) is shown over the different plots (indicated by color codes) of EYV tephra and GRV tephra.

surface lowering was substantially increased by a factor of 1.49 ± 0.88 (mean $\pm 1\sigma$ over time) under EYV tephra and by a factor of 1.51 ± 0.71 under GRV tephra. At the tephra plots geared to the critical thickness of the tephra (~ 8.5 mm),

snow-surface lowering was equal to uncovered conditions under EYV tephra (1.00 ± 0.61) and slightly increased under GRV tephra (1.17 ± 0.57). However, at the thick tephra plots (~ 80 mm) the observed snow-surface lowering did not

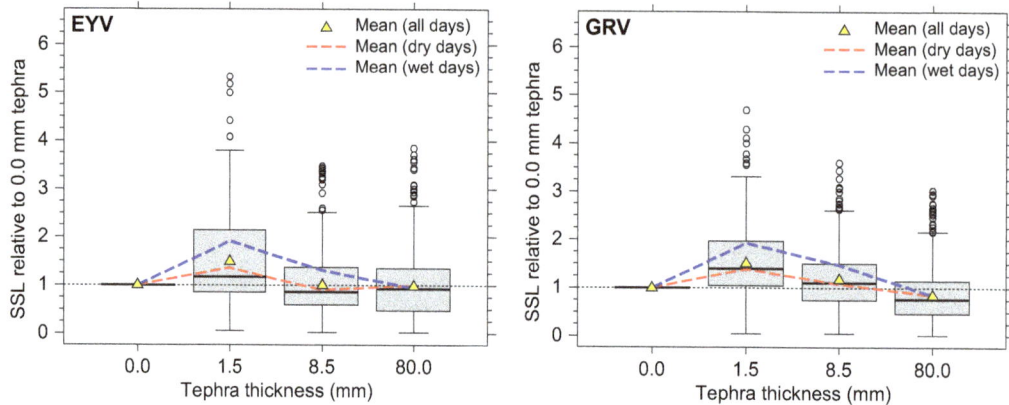

Figure 4. Running 24 h snow-surface lowering (SSL) rates at the different plots of tephra from Eyjafjallajökull volcano (EYV) and from Grímsvötn volcano (GRV) relative to the reduction rates measured at the non-tephra covered reference site. The box plots give an overview of the data spread across all running 24 h values recorded during the field experiment period (15 June to 3 August). Outliers are indicated as open circle symbols. Mean values over the entire field experiment period are indicated by yellow triangles, and the mean values over wet (precipitation > 0.1 mm) and dry (precipitation ≤ 0.1 mm) days are shown as color-coded line graphs.

match expectations drawn from previous bare-ice knowledge. Under EYV tephra, snow-surface lowering was close to uncovered conditions (0.98 ± 0.73) and under GRV tephra only a slight insulation effect was present (0.85 ± 0.59). The rather high standard deviations, however, suggest a considerable, misleading influence of sporadic, anomalously high and potentially erroneous values. Our assumption, which is supported by the distinctly more moderate medians of 0.93 (EYV) and 0.76 (GRV; Fig. 4), is of insulating conditions under both ∼ 80 mm tephra covers. Nevertheless, the high snow-surface lowering rates at the two sites with ∼ 80 mm tephra cover suggest substantially different snowpack behavior than bare glacier ice behavior under tephra coverage.

This unexpected and thus important finding cannot be explained in full detail here because of limitations in the experimental setup. One obvious explanation is the fact that pure snow ablation is masked by additional processes in the measurements conducted. Snow-surface lowering resulting from settling and compaction of the snowpack as well as from metamorphism on the snow-crystal level also definitely impact the measurements. Moreover, the rather small horizontal extent of the tephra plots probably permits lateral influences of weather conditions on the snowpack beneath the plots. Explanations beyond these influences cannot be given, because the pure, energy-balance-controlled ablation signal cannot be isolated from measured snow-surface lowering. It is thus recommended that future experiment setups at least account for snow density variations.

Distinct differences were observed between snow-surface lowering rates during periods with and without precipitation (Fig. 4). On wet days the increase in snow-surface lowering rates under the thin tephra covers compared to uncovered conditions is even more pronounced than it is on dry

days. This finding is in clear contrast to short-term measurements by Möller et al. (2016) on bare glacier ice. Their study shows that on wet days sub-tephra ice ablation rates are even decreased when compared to uncovered conditions. The increase in snow-surface lowering under the ∼ 8.5 mm tephra covers compared to uncovered conditions is also higher on wet days than on dry days. This implies that the critical thickness of wet tephra is generally higher than that of dry tephra. The strength of the small insulation effect at the thick ∼ 80 mm tephra plots is, however, independent of the allocation to dry or to wet days.

There were average summer meteorological conditions during the field experiment period (15 June to 3 August 2015; Fig. 3a). Air temperature mostly fluctuated between 0 and +4 °C (with few outliers) and showed a mean of +2.1 ± 1.4 °C (mean ± 1σ). Thereby, mean (± 1σ) air temperature gradients between lower and upper sensors amount to +0.20 ± 0.15 K m^{-1}. Daily albedo means decreased from ∼ 0.71 during the first week of the field experiment period to ∼ 0.58 during its last week. The associated daily mean of net shortwave radiation fluxes was 86.0 ± 22.4 W m^{-2}. The mostly undisturbed daily cycles of incoming shortwave radiation suggest little cloud coverage. Accordingly, total rainfall over the period sums up to only 40.2 mm. However, high wind speeds of 5.65 ± 3.34 m s^{-1} (mean ± 1σ) with peak wind periods reaching 12–19 m s^{-1} might have led to considerable undercatch of precipitation by the tipping-bucket rain gauge (Sugiura et al., 2006). The by far most frequently occurring wind directions (ENE to ESE) resemble the katabatic flow direction down the western slope of Vatnajökull.

Summary and outlook

A field experiment, studying the influences of different types of volcanic tephra on snow-surface lowering, was conducted on Vatnajökull ice cap, Iceland, in summer 2015. Two types of Icelandic tephra were compared, one from Eyjafjallajökull volcano and one from Grímsvötn volcano. Both tephras were sampled right before the start of the experiment at the calderas of the respective volcanoes. For the experiment, three different artificial plots of different thickness (~ 1.5, ~ 8.5 and ~ 80 mm) were installed from both tephras. Snow-surface lowering at all six tephra plots and at a tephra-free reference site was monitored automatically over the summer season jointly with surface temperature on the two ~ 80 mm tephra plots and concurrent atmospheric variables.

The experiment ran from mid-May to mid-September. Snow-surface lowering could be determined for 50 days (15 June to 3 August) at hourly resolution. The data set comprises records of air temperature and relative humidity at two levels, wind speed and direction, rainfall, incoming and reflected shortwave radiation and snow-surface lowering (in terms of distance from sensor to surface) over a non-tephra covered reference site and over the six tephra plots. Surface temperature was additionally measured at the two ~ 80 mm tephra plots. We presented a comparison of snow-surface lowering rates under the different tephra plots.

Snow-surface lowering showed substantial median increases at the two ~ 1.5 mm tephra plots (~ 17 % under Eyjafjallajökull tephra and ~ 40 % under Grímsvötn tephra). However, snow-surface lowering was also considerably increased at the ~ 8.5 mm Grímsvötn tephra plot (median of ~ 11 %), which contrasts with results of previous studies on bare-ice glacier surfaces. Insulation was small even under the thick ~ 80 mm plots (median reductions of ~ 7 % under Eyjafjallajökull tephra and ~ 24 % under Grímsvötn tephra). This also stands in contrast to earlier bare-ice results, where almost full insulation was found under comparably thick tephra covers. The increase in snow-surface lowering on days with rainfall under thinner tephra covers compared to uncovered conditions is markedly higher than on days without rainfall. This is in contrast to bare-ice conditions, where no ablation increase is present on rainfall days at all. This finding leaves room for further investigation. Influence of tephra type is small compared to the other factors.

For potential future experiments, the results and our experience in the field suggest that frequent snow profile analyses or at least snow density measurements over the experiment period are required to interpret the snow-surface lowering measurements obtained with regards to snow ablation. However, this is logistically challenging, as would be the suggested use of larger tephra plot diameters, which would bet-

ter prevent snow-surface lowering measurements from being influenced by lateral energy fluxes from the surface to the sub-tephra snowpack. Installing the six tephra plots with a diameter of 2.0 m instead of 0.7 m would have required the transport of over 320 kg of tephra (instead of ~ 115 kg) from the two sampling sites to the field experiment site.

In conclusion, the experiment delivers a data set which clearly illustrates that the influences of supraglacial tephra cover on glacier ablation are considerably different, depending on the surface of the glacier, i.e., snow or bare ice. To our knowledge, this data set is the first to continuously measure snow-surface lowering under different types and thicknesses of volcanic tephra. Together with the simultaneously acquired meteorological conditions, this data set allows for further in-depth study of the influence of weather conditions on sub-tephra snowmelt. Moreover, it can readily be included as a calibration or validation data set in broader studies on the influences of supraglacial particle cover on ablation.

Competing interests. The authors declare that they have no conflict of interest.

Acknowledgements. The field experiment was funded by grant no. SCHN680/6-1 and no. KU1476/5-1 of the German Research Foundation (DFG). We thank the Vatnajökull National Park administration for granting permission to carry out the experiment and the associated tephra sampling at Grímsvötn caldera. Helpful comments on the manuscript by Jan Lenaerts and Christoph Mayer are gratefully acknowledged.

Edited by: Reinhard Drews

References

Adhikary, S., Seko, K., Nakawo, M., Ageta, Y., and Miyazaki, N.: Effect of surface dust on snow melt, Bull. Glacier Res., 15, 85–92, 1997.

Brock, B., Rivera, A., Casassa, G., Bown, F., and Acuña, C.: The surface energy balance of an active ice-covered volcano: Villarrica Volcano, Southern Chile, Ann. Glaciol., 45, 104–114, https://doi.org/10.3189/172756407782282372, 2007.

Collier, E., Maussion, F., Nicholson, L. I., Mölg, T., Immerzeel, W. W., and Bush, A. B. G.: Impact of debris cover on glacier ablation and atmosphere-glacier feedbacks in the Karakoram, The Cryosphere, 9, 1617-1632, https://doi.org/10.5194/tc-9-1617-2015, 2015.

Dragosics, M., Meinander, O., Jónsdóttir, T., Dürig, T., De Leeuw, G., Pálsson, F., Dagsson-Waldhauserová, P., and Thorsteinsson, T.: Insulation effects of Icelandic dust and volcanic ash on snow and ice, Arab. J. Geosci., 9, 126, https://doi.org/10.1007/s12517-015-2224-6, 2016.

Evatt, G. W., Abrahams, D., Heil, M., Mayer, C., Kingslake, J., Mitchell, S. L., Fowler, A. C., and Clark, C. D.: Glacial melt under a porous debris layer, J. Glaciol., 61, 825–836, https://doi.org/10.3189/2015JoG14J235, 2015.

Juen, M., Mayer, C., Lambrecht, A., Wirbel, A., and Kueppers, U.: Thermal properties of a supraglacial debris layer with respect to lithology and grain size, Geogr. Ann. A, 95, 197–209, https://doi.org/10.1111/geoa.12011, 2013.

Juen, M., Mayer, C., Lambrecht, A., Han, H., and Liu, S.: Impact of varying debris cover thickness on ablation: a case study for Koxkar Glacier in the Tien Shan, The Cryosphere, 8, 377–386, https://doi.org/10.5194/tc-8-377-2014, 2014.

Kirkbride, M. P. and Dugmore, A. J.: Glaciological response to distal tephra fallout from the 1947 eruption of Hekla, south Iceland, J. Glaciol., 49, 420–428, https://doi.org/10.3189/172756503781830575, 2003.

Mattson, L. E., Gardner, J. S., and Young, G. J.: Ablation on debris covered glaciers: an example from the Rakhiot Glacier, Punjab, Himalaya, IAHS Redbooks, 218, 289–296, 1993.

Möller, R., Möller, M., Björnsson, H., Gudmundsson, S., Pálsson, F., Oddsson, B., Kukla, P. A., and Schneider, C.: MODIS-derived albedo changes of Vatnajökull (Iceland) due to tephra deposition from the 2004 Grimsvötn eruption, Int. J. Appl. Earth Obs. Geoinf., 26, 256–269, https://doi.org/10.1016/j.jag.2013.08.005, 2014.

Möller, R., Möller, M., Kukla, P. A., and Schneider, C.: Impact of supraglacial deposits of tephra from Grimsvötn volcano, Iceland, on glacier ablation, J. Glaciol., 62, 933–943, https://doi.org/10.1017/jog.2016.82, 2016.

Möller, R., Möller, M., Kukla, P. A., and Schneider, C.: Meteorological observations and ablation characteristics during the TIOGA experiment on Iceland in 2015, PANGAEA, https://doi.org/10.1594/PANGAEA.876656, 2017.

Nicholson, L. and Benn, D. I.: Properties of natural supraglacial debris in relation to modelling sub-debris ice ablation, Earth Surf. Proc. Land., 38, 490–501, https://doi.org/10.1002/esp.3299, 2013.

Nield, J. M., Chiverrell, R. C., Darby, S. E., Leyland, J., Vircavs, L. H., and Jacobs, B.: Complex spatial feedbacks of tephra redistribution, ice melt and surface roughness modulate ablation on tephra covered glaciers, Earth Surf. Proc. Landf., 38, 95–102, https://doi.org/10.1002/esp.3352, 2013.

Østrem, G.: Ice melting under a thin layer of moraine, and the existence of ice cores in moraine ridges, Geogr. Ann., 41, 228–230, 1959.

Pratap, B., Dobhal, D. P., Mehta, M., and Bhambri, R.: Influence of debris cover and altitude on glacier surface melting: a case study on Dokriani Glacier, central Himalaya, India, Ann. Glaciol., 56, 9–16, https://doi.org/10.3189/2015AoG70A971, 2015.

Rounce, D. R., Quincey, D. J., and McKinney, D. C.: Debris-covered glacier energy balance model for Imja-Lhotse Shar Glacier in the Everest region of Nepal, The Cryosphere, 9, 2295–2310, https://doi.org/10.5194/tc-9-2295-2015, 2015.

Sugiura, K., Ohata, T., and Yang, D.: Catch characteristics of precipitation gauges in high-latitude regions with high winds, J. Hydrometeorol., 7, 984–994, https://doi.org/10.1175/JHM542.1, 2006.

14

A complete glacier inventory of the Antarctic Peninsula based on Landsat 7 images from 2000 to 2002 and other preexisting data sets

Jacqueline Huber[1], Alison J. Cook[2,3], Frank Paul[1], and Michael Zemp[1]

[1]Department of Geography, University of Zürich–Irchel, Zürich, 8057, Switzerland
[2]Department of Geography, Swansea University, Swansea, SA2 SPP, UK
[3]Department of Geography, Durham University, Durham, DH1 3LE, UK

Correspondence to: Jacqueline Huber (jhuber@access.uzh.ch)

Abstract. The glaciers on the Antarctic Peninsula (AP) potentially make a large contribution to sea level rise. However, this contribution has been difficult to estimate since no complete glacier inventory (outlines, attributes, separation from the ice sheet) is available. This work fills the gap and presents a new glacier inventory of the AP north of 70° S, based on digitally combining preexisting data sets with geographic information system (GIS) techniques. Rock outcrops have been removed from the glacier basin outlines of Cook et al. (2014) by intersection with the latest layer of the Antarctic Digital Database (Burton-Johnson et al., 2016). Glacier-specific topographic parameters (e.g., mean elevation, slope and aspect) as well as hypsometry have been calculated from the DEM of Cook et al. (2012). We also assigned connectivity levels to all glaciers following the concept by Rastner et al. (2012). Moreover, the bedrock data set of Huss and Farinotti (2014) enabled us to add ice thickness and volume for each glacier.

The new inventory is available from the Global Land Ice Measurements from Space (GLIMS) database (doi:10.7265/N5V98602) and consists of 1589 glaciers covering an area of 95 273 km^2, slightly more than the 89 720 km^2 covered by glaciers surrounding the Greenland Ice Sheet. Hence, compared to the preexisting data set of Cook et al. (2014), this data set covers a smaller area and one glacier less due to the intersection with the rock outcrop data set. The total estimated ice volume is 34 590 km^3, of which one-third is below sea level. The hypsometric curve has a bimodal shape due to the unique topography of the AP, which consists mainly of ice caps with outlet glaciers. Most of the glacierized area is located at 200–500 m a.s.l., with a secondary maximum at 1500–1900 m. Approximately 63 % of the area is drained by marine-terminating glaciers, and ice-shelf tributary glaciers cover 35 % of the area. This combination indicates a high sensitivity of the glaciers to climate change for several reasons: (1) only slightly rising equilibrium-line altitudes would expose huge additional areas to ablation, (2) rising ocean temperatures increase melting of marine terminating glaciers, and (3) ice shelves have a buttressing effect on their feeding glaciers and their collapse would alter glacier dynamics and strongly enhance ice loss (Rott et al., 2011). The new inventory should facilitate modeling of the related effects using approaches tailored to glaciers for a more accurate determination of their future evolution and contribution to sea level rise.

1   Introduction

The ice masses of the Antarctic Peninsula (AP) potentially make a large contribution to sea level rise (SLR) since a large amount of water is stored in the ice and a high sensitivity to temperature increase has been reported (Hock et al., 2009). However, the glaciers on the AP were not separately taken into account for their individual sea level contribution in the Fifth Assessment Report of the IPCC (Vaughan et al., 2013) because a complete glacier inventory of the AP was not available at that time. As a result, only the ice masses of the surrounding islands were considered from the inventory compiled by Bliss et al. (2013). The freely available data sets for the AP were incomplete and of a varied nature (see Fig. 1), ranging from the World Glacier Inventory (WGI; WGMS and NSIDC, 2012), which provides extended parameters for most of the glaciers on the AP from the second half of the 20th century but without area information and only available as point data, to the vector data sets (two-dimensional outlines) from the Global Land Ice Measurements from Space (GLIMS; GLIMS and NSIDC, 2015) database and the Randolph Glacier Inventory (RGI; Arendt et al., 2015), which were spatially incomplete. Moreover, the spatial overlap of the WGI with the boundaries of individual glaciers in the RGI was limited (Fig. 1) so that an automated digital intersection (spatial join) for parameter transfer was not possible.

Conversely, for Graham Land, representing the part of the AP north of 70° S, several more specific data sets exist that could be combined for a full and coherent glacier inventory: a detailed 100 m resolution DEM was prepared by Cook et al. (2012); glacier catchment outlines based on this DEM and the Landsat Image Mosaic of Antarctica (LIMA; Bindschadler et al., 2008) were derived by Cook et al. (2014); a recently updated data set of rock outcrops for all of Antarctica is available from the Antarctic Digital Database (ADD; http://www.add.scar.org/home/add7); a modeled raster data set of bedrock topography is available from Huss and Farinotti (2014).

Here, we present the first comprehensive glacier inventory of the Antarctic Peninsula north of 70° S (Graham Land) and describe methods used to digitally combine the existing data sets. The final outline data set of the AP is supplemented with several glacier-specific parameters, such as topographic information and hypsometry, and thickness and volume information, as well as the earlier classification of glacier front characteristics. With these parameters we analyze similarities and differences with other glacierized regions, as well as glacier-specific contributions to sea level and climate sensitivities. For a clear handling by different modeling and remote sensing communities, each glacier is assigned one of three connectivity levels to the ice sheet (CL0 is no connection, CL1 is a weak connection and CL2 is a strong connection) following the approach introduced by Rastner et al. (2012) to separate the peripheral glaciers on Greenland from the ice sheet.

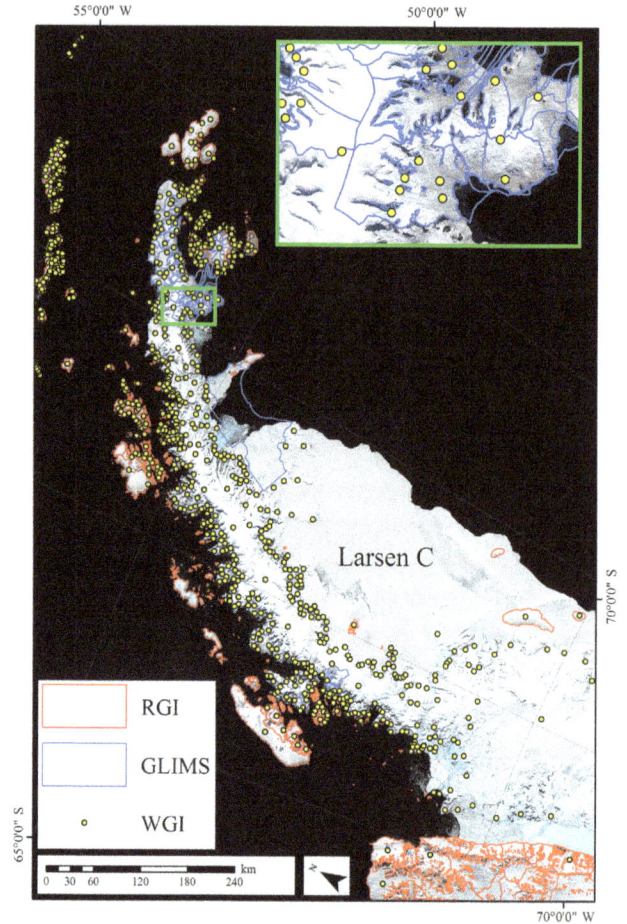

Figure 1. Landsat Image Mosaic of Antarctica (LIMA) overlaid by existing GLIMS and RGI glacier outlines and WGI glacier point locations for Graham Land on the AP. Inset map illustrating that the distribution of the WGI points does not enable assignation of points to individual glacier outlines.

2   Study region

The AP extends northwards of the mainland from approximately 75° S for more than 1500 km northeasterly to 63° S, and it is enclosed to the west by the Bellingshausen Sea and to the east by the Weddell Sea of the Southern Ocean. The part of the AP north of 70° S represents Graham Land and its peripheral islands, for which the glacier inventory is created. The South Shetland Islands are not regarded as being part of the AP and are therefore not included in the present inventory. The central part of the mainland is dominated by a narrow mountain chain with a mean height of 1500 m (maximum 3172 m) and an average width of 70 km. The unique topography, with an interior high-elevation plateau surrounded by steep slopes and flat valley bottoms results in distinct glacier types. In general, the highest regions are covered by ice caps, and much lower-lying valley glaciers are either connected to them and heavily crevassed in the steep regions, or they are entirely separated from them, uncovering several rock outcrops.

The AP has a polar-to-subpolar maritime climate, but the climatic and oceanographic regime varies across the AP, causing varying glacier dynamics (Arigony-Neto et al., 2014). The often polythermal glaciers experience a distinct melting period in austral summer, particularly the glaciers in the northern part of the AP. The special topographic characteristics of the AP make the flat, low-lying parts of its glaciers particularly vulnerable to climate change: for example, a small increase in temperature might cause large parts of their area to become ablation regions; most of them are marine-terminating glaciers that also experience melt from surrounding ocean waters (Cook et al., 2016), and many of them nourish ice shelves (Cook et al., 2014) that currently buttress them but can quickly disappear (Rott et al., 1996) causing rapid shrinkage of the related glaciers (Rott et al., 1996; Hulbe et al., 2008).

Since the early 1950s, significant atmospheric warming trends (Turner et al., 2009) and increasing ocean temperatures (Shepherd et al., 2003) have been observed across the AP. As a consequence, ice shelves are collapsing and glacier fronts are retreating (Pritchard and Vaughan, 2007; Davies et al., 2012; Cook et al., 2014, 2016). Conversely, knowledge about the mass balance of the glaciers of the AP is sparse (Rignot and Thomas, 2002), although a few studies exist that indicate a general mass loss (Helm et al., 2014; Kunz et al., 2012).

For the purpose of this study, the AP is additionally divided into four sectors (NW, NE, SW and SE) to reveal differences between climatically different regions of the AP. The division west–east is based on the main topographic divide, and north–south is based on the 66° S latitude.

3 Data sets

This section gives a short description of the preexisting data sets covering the AP (Graham Land) that are used for generating the glacier inventory. Table 1 summarizes their key characteristics, presenting their content, sources, access, references and application in this study. The following data sets are used:

1. the digital elevation model (DEM) by Cook et al. (2012);

2. the glacier catchment outlines by Cook et al. (2014);

3. the rock outcrop data set of Antarctica by Burton-Johnson et al. (2016);

4. the bedrock elevation grid by Huss and Farinotti (2014);

5. the Antarctic ice-sheet drainage divides by Zwally et al. (2012) and

6. the Landsat Image Mosaic of Antarctica (LIMA) by Bindschadler et al. (2008).

3.1 Digital elevation model

Cook et al. (2012) generated a 100 m resolution DEM of the AP (63–70° S), which is available from the National Snow and Ice Data Center (NSIDC; http://nsidc.org/data/NSIDC-0516) in the WGS84 Stereographic South Pole projection. This DEM is an improvement of the ASTER Global Digital Elevation Model (GDEM) product, which locally contained large errors and artifacts (see Cook et al., 2012). The accuracy of the DEM is in particular improved on gentle slopes of the high plateau region. However, they removed small anomalies, which has resulted in small inherent gaps along the coast, and some islands are missing (Cook et al., 2012). As a result, the DEM does not entirely cover the study region (approximately 1 % of the area is missing). This DEM has also been used by Cook et al. (2014) for the generation of catchment outlines (see next section) and is used in this study for the calculation of glacier-specific parameters (see Sect. 4.3) for the glacierized areas it covers.

3.2 Catchment outlines

Glacier inventories, such as those available in GLIMS or the RGI, require glaciers to be separated into individual entities (Paul et al., 2009). This can be accomplished by intersecting drainage divides derived from watershed analysis (e.g., Bolch et al., 2010; Kienholz et al., 2013) with outlines of glacier extents derived from semiautomated mapping techniques (e.g., Paul et al., 2002). Cook et al. (2014) automatically delineated glacier catchments of the AP in ArcGIS from ESRI by applying hydrological tools to the DEM described above (Fig. 2). They digitized the AP coastline and some islands in that data set based on images acquired by Landsat 7 between 2000 and 2002 for the LIMA (Bindschadler et al., 2008). Since the DEM misses some islands around the AP, mainly in the central western region, the drainage divide analysis is missing for these regions. Additionally, they used grounding lines from the Antarctic Surface Accumulation and Ice Discharge (ASAID) project data source (Bindschadler et al., 2011), modified in places with features visible on the LIMA to divide glaciers from ice shelves. Furthermore, the ice-velocity data set of Rignot et al. (2011) was considered by Cook et al. (2014) to manually verify and adjust the lateral boundaries of glaciers.

The resulting data set consists of 1590 glacier catchment outlines for the AP with an area of 96 982 km², covering the region between 63 and 70° S. Islands smaller than 0.5 km² and ice shelves are excluded. The data set provides a consistent time period of all basins and includes several parameters for each basin, such as location, time stamp, area, and a classification of glacier type, form and front. The definition of the parameters and category numbers conform to the

Table 1. Data sets used for the generation of the glacier inventory and a description of their properties.

	DEM	Glacier catchment outlines	Rock outcrops	Bedrock elevation grid	Antarctic ice-sheet drainage divides
Content	Elevation on a 100 m grid of the AP (Graham Land, 63–70° S)	Inventory of 1590 glacier basins of the AP (Graham Land, 63–70° S) on the mainland and surrounding islands	New rock outcrop data set for Antarctica	Bedrock data set for the AP (Graham Land, 63–70° S) on a 100 m grid	Drainage divides of the Antarctic ice sheet
Sources	ASTER Global Digital Elevation Model (GDEM)	DEM of Cook et al. (2012), LIMA (Bindschadler et al., 2008), grounding line based on the Antarctic Surface Accumulation and Ice Discharge (ASAID) project data source (Bindschadler et al., 2011)	Landsat 8 data	Simple ice-dynamic modeling with a variety of available data sets (surface mass balance, point ice thickness and ice flow velocity) (Haran et al., 2005)	GLAS/ICESat 500 m laser altimetry DEM (DiMarzio, 2007) Landsat Image Mosaic of Antarctica (LIMA; Bindschadler et al., 2008) and the MODIS Mosaic of Antarctica (Haran et al., 2005)
Reference	Cook et al. (2012)	Cook et al. (2014)	Burton-Johnson et al. (2016)	Huss and Farinotti (2014)	Zwally et al. (2012)
Access	http://nsidc.org/data/	http://add.scar.org/ (available only with a limited number of attributes)	http://add.scar.org/	Available online from the article's supplement (doi:10.5194/tc-8-1261-2014-supplement)	http://icesat4.gsfc.nasa.gov/cryo
Application in this study	Calculation of (a) glacier-specific topographic parameters (min, max, mean, median elevation, slope, aspect), (b) overall and glacier specific hypsometry, and (c) thickness grid combined with the bedrock elevation grid of Huss and Farinotti (2014)	Initial data set for the generation of glacier outlines	Used to remove the (ice-free) rock outcrops from the glacier catchment outlines to generate glacier outlines	Calculation of the thickness grid combined with the DEM of Cook et al (2012)	Separation of the glaciers form the ice sheet

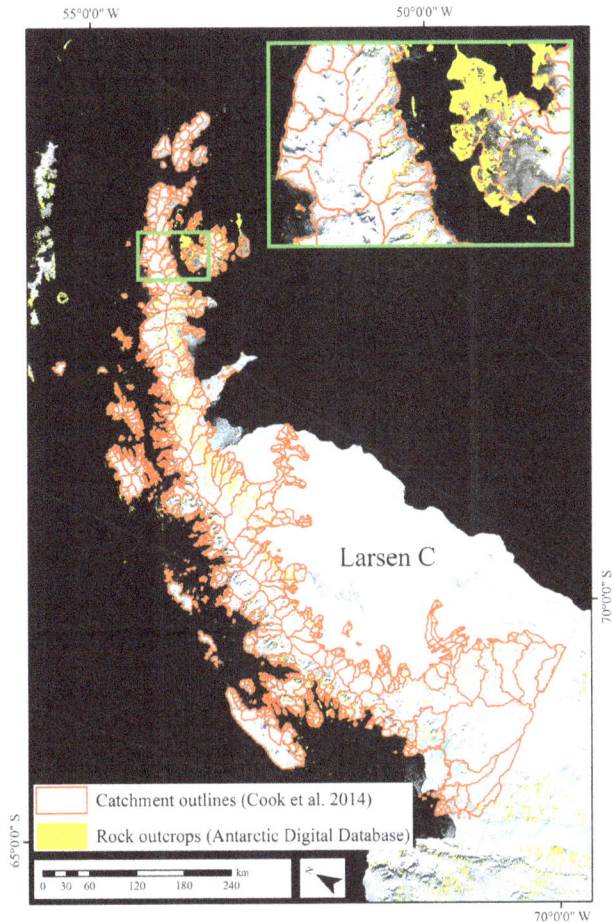

Figure 2. Glacier catchment outlines of Cook et al. (2014) and the newest rock outcrop data set from Burton-Johnson et al. (2016) overlaying the LIMA.

GLIMS classification system provided by the GLIMS Classification Manual (Rau et al., 2005) and based on the UNESCO (1970) guidelines as well as the Glossary of Glacier Mass Balance (Cogley et al., 2011). However, topographic parameters such as minimum, maximum, mean, and median elevation, or mean slope and aspect, are missing.

This catchment outline data set is available from the Scientific Committee on Antarctic Research (SCAR) ADD (http://add.scar.org/home/add7; ADD Consortium, 2012), but it does not include any of the glacier-specific attributes mentioned above aside from area and length. The data set with the complete information has not been published so far and has been generated and provided by A. Cook in the framework of this study in the WGS84 Stereographic South Pole projection. Whereas the catchment outlines provide a solid foundation for the generation of a glacier inventory, rock outcrops are part of the glacierized area and need to be removed (Raup and Khalsa, 2010).

3.3 Rock outcrops

The ADD website (www.add.scar.org) provides a detailed vector data set of rock outcrop boundaries in the WGS84 Stereographic South Pole projection that has recently been updated (see Burton-Johnson et al., 2016). A former rock outcrop data set, which has already been used (by Bliss et al. (2013) for instance) to create the inventory for the glaciers of the islands surrounding Antarctica, originated from a digitization of outcrops from different maps prepared in the 1990s at different scales and with variable accuracy. As a result, the data set has some major georeferencing inconsistencies, misclassifications and overestimations of the ice-free area of Antarctica (Burton-Johnson et al., 2016). The recently improved data set of exposed rock outcrops by Burton-Johnson et al. (2016) used here (Fig. 2), overcomes these issues and has a much better accuracy. It is based on a new automated method that identifies sunlit as well as shaded rock outcrops using multispectral classification of Landsat 8 satellite imagery. They manually removed incorrectly classified pixels (illuminated and shaded) such as snow, clouds and liquid water. The new data set reveals that 0.18 % of the total area of Antarctica is rock outcrops, which is approximately one-half of previous estimates (Burton-Johnson et al., 2016).

3.4 Bedrock elevation grid

Huss and Farinotti (2014) derived a new bedrock elevation grid with 100 m spatial resolution as well as the related ice thickness grid based on glacier surface topography and simple ice-dynamic modeling. Compared to the Bedmap2 data set by Fretwell et al. (2013) with a resolution of 1 km, the new version also captures the rugged subglacial topography in great detail. The narrow and deep subglacial valleys that are often below sea level are more accurately represented, allowing the modeling of even small-scale processes.

Their data set is available online from the article supplement (doi:10.5194/tc-8-1261-2014-supplement) on WGS84 Antarctic Polar Stereographic projection. Their data set already excluded the rock outcrops using the former version of the ADD (ADD Consortium, 2012). Since we have used the updated version of the rock outcrops data set for creating the glacier inventory, a new thickness grid is calculated (see Sect. 4.1).

3.5 Antarctic ice-sheet drainage divides

The Cryosphere Science Laboratory of NASA's Earth Sciences Divisions (Zwally et al., 2012) provides an Antarctic ice-sheet drainage divide data set developed by the Goddard Ice Altimetry Group from ICESat data based on the GLAS/ICESat 500 m laser altimetry DEM (DiMarzio, 2007). They used other sources, such as LIMA (Bindschadler et al., 2008) and the MODIS Mosaic of Antarctica (Haran et al., 2013), as a guide to refine the drainage divides. Ice-sheet drainage systems were delineated to identify regions that are broadly homogeneous regarding surface slope orientation relative to atmospheric advection and denoting the ice-sheet areas feeding large ice shelves. The AP is assigned to four different basins (drainage system ID numbers 24–27), with a relatively clear separation from the ice sheet along 70° S latitude (see Sect. 4.2).

4   Methods

The data generation workflow is roughly divided into four steps: (1) intersecting data sets, (2) defining connectivity levels, (3) calculating glacier-specific attributes (topographic parameters), including ice thickness and volume information, and (4) the calculation of the overall and glacier-specific hypsometry. All calculations are performed with various tools available in ESRI's ArcGIS version 10.2.2. All of the functionality is also available in other geographic information system (GIS) software packages. The four main steps are described in the following sections in more detail.

4.1   Intersecting data sets

When generating an inventory based on the semiautomated band ratio method (Paul et al., 2009), rock outcrops are automatically excluded from the glacier area. In this study the glacier catchment outlines are intersected with the latest vector data set of rock outcrop boundaries from the ADD (see Sect. 3.3). By removing the new rock outcrops from the catchment outlines of Cook et al. (2014), a mask of individual glaciers is generated, assuming that areas not identified since rock outcrops are ice covered. Apart from the rock outcrops, the data set of Cook et al. (2014) is generally in agreement with the procedures and GLIMS guidelines (Racoviteanu et al., 2009; Raup and Khalsa, 2010) for deriving glacier information.

To include glacier-specific ice thickness and volume information, the bedrock grid of Huss and Farinotti (2014) is subtracted from the DEM of Cook et al. (2012) and combined with the new glacier outlines. A grid with ice volume is then derived by multiplying the ice thickness grid with the cell area ($10\,000\,\mathrm{m}^2$).

4.2   Defining connectivity levels

Rastner et al. (2012) suggested that peripheral glaciers on Greenland with a strong dynamic connection to the Greenland Ice Sheet should be regarded as part of the ice sheet and assigned the connectivity level 2 (CL2). This is where glaciers have an extended connection to the ice sheet and the location of their drainage divide on the DEM is uncertain due to the low-sloping terrain. For the Antarctic ice-sheet drainage divides (see Sect. 3.5), basins south of 70° S are strongly connected to the West Antarctic ice sheet. Accordingly, they are assigned CL2 and are not included or further considered in the inventory presented here. The assignment

of CL1 (i.e., weak connectivity to ice sheet) to the glaciers on the mainland and north of 70° S is performed automatically within the GIS following the heritage rule introduced by Rastner et al. (2012), i.e., a glacier connected to a glacier assigned CL1 will also receive the attribute CL1. With this strategy, all glaciers on surrounding islands (i.e., those in the inventory from Bliss et al., 2013) are assigned the value CL0. Large glaciers that are theoretically separable but otherwise closely connected to the ice sheet (e.g., Pine Island and Thwaites) have the value CL2.

4.3   Glacier-specific topographic parameters, ice thickness and volume

All glacier-specific attributes (minimum; maximum; mean; and median elevation; mean slope, aspect, and thickness; total ice volume; and ice volume grounded below sea level) are calculated by combining the glacier outlines with the DEM, the ice thickness and volume grids using the zonal statistics tool in ArcGIS. This tool statistically summarizes the values of the underlying raster data sets (e.g., DEM, ice thickness) within specific zones with a unique ID (glacier outlines) and organizes the results into an attribute table. The table is joined with the attribute table of the glacier outlines data set based on a common and unique identifier in both tables (i.e., the glacier ID). All calculations are performed using the WGS84 South Pole Lambert Azimuthal Equal Area projection.

Since the bedrock and hence also thickness data sets are based, inter alia, on the DEM of Cook et al. (2012), they are not universally spatially congruent with the glacier outlines (i.e., the boundary limits differ between the outlines and the other data sets). Of the 1589 glacier outlines, the thickness and volume values could not be calculated for 50 glaciers of the inventory. Accordingly, the topographic parameters, thickness and volume values of the glaciers on the islands that are not completely covered by the ice thickness and bedrock data set do not represent values for complete glaciers. In addition, two glaciers are insufficiently covered by the $100\,\mathrm{m} \times 100\,\mathrm{m}$ pixel of the DEM. Hence these glaciers are not or insufficiently covered by the bedrock data set of Huss and Farinotti (2014). Hence, 1541 glaciers have topographic information and 1539 glaciers have thickness, volume and sea level equivalent (SLE) information, of which some only have partial ice thickness and volume information.

To estimate the volume grounded below sea level for each glacier, a grid representing the distribution of the volume grounded below sea level is calculated by extracting the areas of the bedrock grid with negative values (areas below sea level).

The SLE of the ice volume is calculated by assuming a mean ice density of $900\,\mathrm{kg\,m}^{-3}$ (not taking into account firn-air content) and dividing it by the ocean surface area ($3.625 \times 10^8\,\mathrm{km}^2$; Cogley, 2012), assuming all ice volume contributes to sea level if melted. This is not the case for the grounded ice

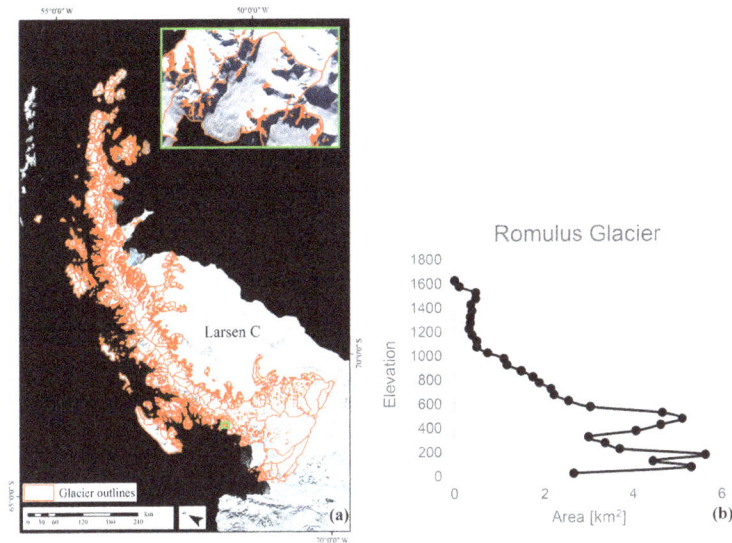

Figure 3. (a) Glacier outlines. Inset map showing Romulus Glacier referred to in Table 2 and **(b)** exemplifying the glacier-specific hypsometry.

below sea level, which has a negative (lowering) effect since this volume will be replaced by water with a higher density (Cogley et al., 2011). This effect has been considered in a second step in the SLE estimations presented. Other effects, such as the isostatic effect, the cooling and dilution effect on ocean waters by floating ice (Jenkins and Holland, 2007), are not taken into account here.

4.4 Glacier hypsometry

The distribution of the glacierized area with elevation (hypsometry) is calculated (a) for the entire AP in 100 m elevation bins, (b) for the four subregions also in 100 m bins and (c) for each individual glacier using 50 m elevation bins. The calculation is based on the DEM of Cook et al. (2012) that is converted to 100 m bins using the "reclassify" tool and the "extract by mask" tool for the respective subregions. Additionally, the hypsometry of the catchment outlines is calculated to determine the effect of removing rock outcrops from the hypsometry. For further comparisons we also calculated the hypsometry of the marine and ice-shelf-terminating glaciers and the hypsometry of the bedrock.

5 Results

5.1 Size distribution

The glacier inventory for the AP ranges from 63–70° S to 55–70° W and consists of 1589 glaciers covering an area of 95 273 km² (Fig. 3a) without rock outcrops, ice shelves and islands < 0.5 km². Hence, compared to the preexisting data set of Cook et al. (2014), this data set covers a smaller area and one glacier less due to the intersection with the

rock outcrop data set (we removed one glacier since all of its area was rock outcrop). The rock outcrops cover an area of 1709.4 km². The 619 glaciers located on islands (CL0) cover an area of 14 299 km², representing 15 % of the total glacierized area. The remaining 970 glaciers are located on the mainland (CL1), covering 80 974 km² and hence 85 % of the total area. Since the DEM is spatially not perfectly congruent with the glacier outlines, of the total 1589 glacier outlines, 48 outlines do not have any elevation information. As a result, the calculations including the DEM, the bedrock or the thickness data set are only applied to 1541 glaciers, of which some only have partial elevation information. In Table 2 all parameters of the attribute table are listed, including the corresponding values of an example glacier (for location see inset map in Fig. 3a). The hypsometry of each individual glacier, as exemplified in Fig. 3b, is stored and available separately in a csv file. Several parameters, such as primary classification, glacier form and front, and metadata about the satellite image, have been determined and provided by Cook et al. (2014), as defined for the GLIMS inventory. Others (i.e., connectivity levels, topographic parameters, ice thickness and volume) are the result of the calculations described in Sect. 4. The inventory is available for download from the GLIMS website: http://www.glims.org/maps/glims (doi:10.7265/N5V98602).

Regarding the connectivity levels, all glaciers on islands surrounding the AP are assigned CL0 (no connection) and the glaciers on the mainland are all assigned CL1 (weak connection). Even the glaciers at the very northern part of the AP have CL1 due to the applied topological heritage rule (a glacier connected to a glacier assigned CL1 also receives CL1). Since the glaciers further south are connected to the ice sheet, they are assigned CL2 (strong connection), are re-

Table 2. Glacier parameters in the attribute table of the inventory of the AP.

Name	Item	Glacier example	Description
Name	Name	Romulus Glacier	String, partially available
Satellite image date	SI_DATE	19.02.2001	Date of the satellite image used for digitizing
Year	SI_YEAR	2001	Year the outline is representing
Satellite image type	SI_TYPE	Landsat 7	Instrument name, e.g., Landsat 7
Satellite image ID	SI_ID	LE7220108000105050	Original ID of image
Coordinates	Lat, long	−68.391218, −66.82767	Decimal degree
Primary classification	Class	6 (mountain glacier)	See Cook et al. (2014)
Form	Form	2 (compound basin)	See Cook et al. (2014)
Front	Front	4 (calving)	See Cook et al. (2014)
Confidence	Confidence	1 Confident about all (class, form and front) classification types	See Cook et al. (2014)
Mainland/island	Mainl_Isl	1 (situated on mainland)	See Cook et al. (2014)
Area	Area	68.9 km^2	km^2
Connectivity level	CL	1 (weak connection)	See Sect. 4.2
Sector	Sector	SW	NW, NE, SW or SE
Minimum elevation	min_elev	4.6 m a.s.l.	m a.s.l.
Maximum elevation	max_elevation	1610.6 m a.s.l.	m a.s.l.
Mean elevation	mean_elev	466.5 m a.s.l.	m a.s.l.
Median elevation	med_elev	425.6 m a.s.l.	m a.s.l.
Mean aspect in degree	mean_asp_d	222°	°
Mean aspect nominal	mean_aspect	SW	Eight cardinal directions
Aspect sector	asp_sector	6	Clockwise numbering of the eight cardinal directions
Mean slope	mean_slope	13°	°
Total volume	tot_vol	13.4 km^3	km^3
Volume below sea level	vol_below	8.0 km^3	km^3
Mean thickness	mean_thick	191.4 m	m

garded as part of the ice sheet and hence are not included in the present data set.

Figure 4a portrays the percentages per size class in terms of number and area. The mean area (60.0 km^2) is considerably higher than the median area (8.2 km^2), reflecting the areal dominance of a few larger glaciers. Most of the glaciers can be found in the size classes 4–6 (1.0–50 km^2). These glaciers account for 77 % of the total number but only for 14 % of the total area. The glaciers larger than 100 km^2 cover the majority of the area (77 %) yet comprise only 11 % of the total number. With an area of 7018 km^2, Seller Glacier is the largest, accounting for 7 % of the total area and being twice as large as the second largest glacier (Mercator Ice Piedmont, 3499 km^2).

5.2 Topographic parameters

Figure 4b shows the distribution of glacier number and area as a percentage of the total for each aspect sector of the AP. The distribution is rather balanced and does not reveal any trends. Somewhat fewer glaciers and areas have aspects from south to southeast. The large value in area of the southwestern sector derives from the contribution of the largest glacier of the region (Seller Glacier).

Figure 5a and b present a scatter plot of area against mean and median and area against minimum and maximum elevation, revealing that mean, median and maximum elevation increase towards larger glaciers. Three glaciers have a maximum elevation above 3100 m a.s.l., being 300 m or more higher than all other glaciers. The highest elevation is in southern Graham Land with 3172 m. Many glaciers have a minimum elevation of (close to) zero m a.s.l. since most of them are marine terminating. The average mean elevation of the 1541 glaciers with elevation information is 409 m, and their median elevation is 317 m a.s.l. The spatial distribution of median elevation reveals an increase from the coast and islands (0–500 m a.s.l.) to the interior of the AP (up to about 1800 m a.s.l.). This can be seen in Fig. S1 in the Supplement.

When mean aspect is plotted against mean elevation (Fig. 6) there are also no significant trends. However, the highest mean elevation values are lower in the southeastern sector. The scatter plot of mean slope against area (Fig. S2) reveals the common dependence on glacier size, where mean slope decreases towards larger glaciers. Additionally, the scatter is smaller the larger the glacier, indicating that small glaciers exhibit a larger range of slope inclination.

The mean thickness of all 1539 glaciers involving thickness information is 130 m. The Eureka glacier, located in the south, has the largest mean thickness of all CL0 and

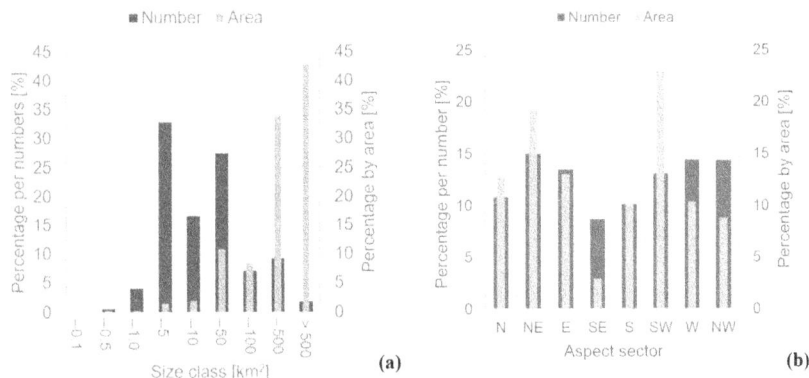

Figure 4. (a) Percentage of glacier count and area per size class (only upper boundary of each size class is given on the x axis) and **(b)** percentage of glacier count and area per aspect sector.

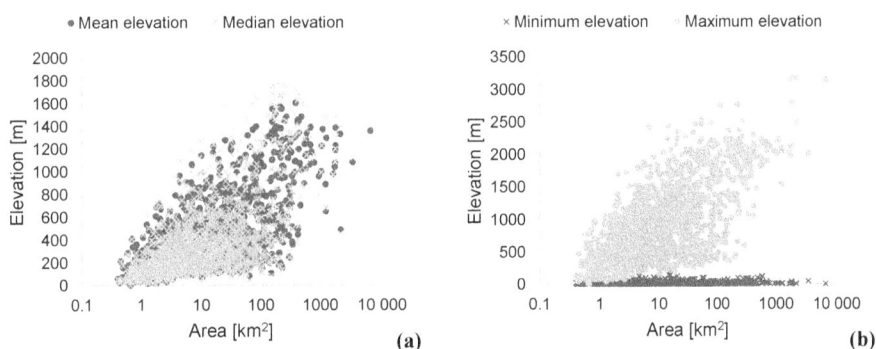

Figure 5. (a) Mean and median elevation vs. area and **(b)** minimum and maximum elevation vs. area of the 1541 glaciers, including elevation information.

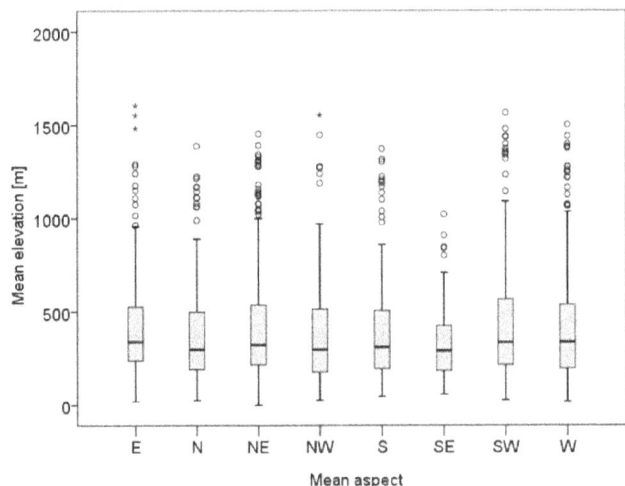

Figure 6. Mean glacier elevation vs. mean glacier aspect of 1541 glaciers. The top and bottom of the boxes indicate the 25th and 75th percentiles, respectively. The whiskers extend to 1.5 times the height of the box and to the minimum values.

CL1 glaciers with 851 m. The dependence of mean thickness on area and slope (indicating that the steeper or smaller the glacier, the thinner the ice) (Fig. 7a, b) is not surprising because ice thickness is modeled based on surface topography (Huss and Farinotti, 2012, 2014). However, low-sloping glaciers reveal a large range of mean thickness values. The large but low-sloping glaciers of the high plateau and those in the very south towards the Antarctic ice sheet form a cluster of glaciers with higher mean thicknesses. The many small glaciers along the coast are mostly thin. The mean thicknesses per sector and per mean aspect (Fig. S3) do not reveal any significant spatial patterns.

The total ice volume of the AP is 34 590 km^3. Since the volume is calculated based on the thickness data set, the volume distribution is basically a reflection of the thickness distribution. Table 3 lists the total volume per sector, revealing that most of the ice volume can be found in the southwestern and southeastern sectors (38.6 and 32 % of the total). This is not surprising because these two sectors make up 63 % of the total glacierized area. Regarding the glacier volume per glacier area for individual glaciers, the highest values are found for the large glaciers at the very south of the AP, adja-

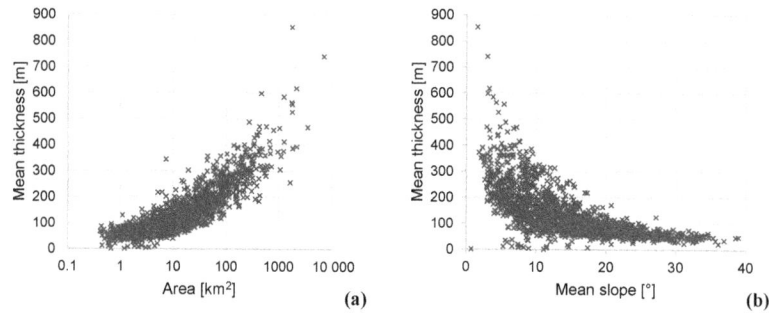

Figure 7. Scatter plot of the 1541 glaciers involving thickness information. **(a)** Mean thickness vs. area and **(b)** mean thickness vs. mean slope.

cent to the ice masses regarded as being a part of the Antarctic ice sheet.

Numerous, partly very pronounced, valleys lie below sea level, especially in the northeastern sector (the bedrock lying below sea level is visualized in Fig. S4). In total, approximately one-third of the total grounded ice volume is below sea level (Table 3), which has a negative effect on SLR (sea level lowering). About 50 % of the volume of the northeastern sector is grounded below sea level (Table 3). Although the negative effect on SLR is very small, this effect can now be better considered for future sea level estimations. Based on the results presented here, this results in a total estimated SLE of 54 mm (Table 3).

As mentioned before, the nominal glacier parameters primary classifications, glacier form and front, have been determined by and described in Cook et al. (2014). They further illustrate the number of glaciers within each classification and frontal type, which is therefore not repeated here.

5.3 Hypsometry

Figure 8a and b depict the glacier hypsometry (area–altitude distribution) for (a) the entire AP and for (b) each sector, revealing a bimodal shape of the hypsometry. Figure 8a additionally displays the hypsometry only for marine-terminating and ice-shelf-nourishing glaciers, as well as the hypsometry of the underlying bedrock. Exclusion of the rock outcrops, with a total area of 1709.4 km^2, does not change the general shape of the hypsometry. However, it slightly reduces the glacierized areas below 1500 m a.s.l., with a maximum areal reduction at 200–600 m and 1000–1200 m a.s.l. The hypsometry for marine-terminating and ice-shelf-nourishing glaciers confirms that most of the glacierized area is covered by these types. Additionally, these types extend over the entire elevation range. Accordingly, the bimodal shape of the curve does not arise from different glacier (types) at lower and higher elevations. Rather, it is determined by and reflects the topography of the AP: the low-sloping and low-lying coast regions covered by valley glaciers account for the maximum of the glacierized area between approximately 200 and 500 m a.s.l. The glacierized plateau region accounts for a secondary max-

imum at about 1500–1900 m a.s.l. The steep valley walls connecting the plateau with the coastal region result in the minimum at about 800–1400 m a.s.l. In addition, the hypsometry reveals that approximately 6000 km^2 of the 93 767 km^2 of glacierized area covered by the DEM is found in the lowest elevation band (0–100 m). These areas are in direct or in close contact with water or ice shelves

The hypsometry per AP sector (Fig. 8b; all excluding rock outcrops) reveals that in the two northern sectors both maxima of the hypsometric curve are less than those of the two southern sectors. The elevations of the maxima are about the same for NW, NE and SW, whereas both maxima of the SE sector are somewhat lower. The glacier cover per sector reflects the bedrock topography of each sector. The bedrock of the northern sectors has less area in the high plateau regions, and therefore most of the glacierized areas are at lower elevations. The southern sectors have a more dominant plateau region favoring more glacierized areas at higher elevations compared to the northern sectors. However, the northeastern sector has the largest fraction of glacierized area in the lowest 100 m and is therefore in direct or in close water or ice shelf contact.

5.4 Discussion

5.5 Source data

This study has presented a complete and now publicly available glacier inventory for the Antarctic Peninsula north of 70° S that has been compiled from the best and most recent preexisting data sets, complemented with information for individual glaciers that was not available before (topographic parameters, hypsography and ice thickness). To allow traceability of source data, we have not altered or corrected the available data sets despite some obvious shortcomings. For example, the DEM by Cook et al. (2012) does not cover all glaciers and covers several only partly, but we have not attempted to fill these missing regions with other source data (e.g., the ASTER GDEM). Consequently, the sample of glaciers with complete attribute information (1539) is reduced compared to the number of all glaciers in the study

Table 3. Glacier number, area, volume, volume grounded below sea level, the corresponding percentages and SLE per sector. For the estimation of SLE, see Sect. 4.3.

Sector	Count	Count with volume info	Area [km^2]	Count [%]	Area [%]	Volume [km^3]	Volume [%]	Volume$_{<0}$ [km^3]	Volume$_{<0}$ [%]	SLE [mm]
NW	704	679	17 218	44	18	4026	12	1093	27	7
NE	246	237	18 278	15	19	6133	18	2939	48	7
SW	378	362	31 130	24	33	13 365	39	4849	36	20
SE	261	261	28 647	17	30	11 065	32	2890	26	20
Total	1589	1539	95 273	100	100	34 590	100	11 771	100	54

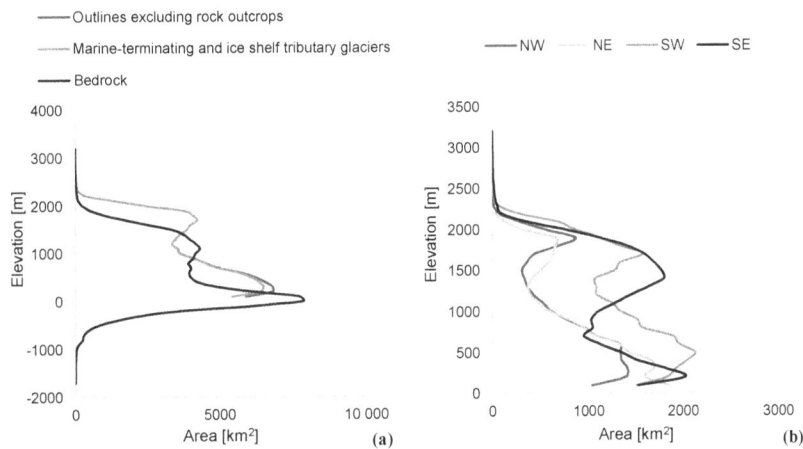

Figure 8. Glacier hypsometry of the total area covered by the DEM. **(a)** Total areal distribution excluding rock outcrops, areal distribution for marine-terminating and ice-shelf-nourishing glaciers and areal distribution of the underlying bedrock. **(b)** Areal distribution of the glacier cover per sector.

region (1589). The same applies for glaciers with a modeled ice thickness distribution. For 48 of these glaciers, the DEM information was incomplete and ice thickness was accordingly not modeled by Huss and Farinotti (2014). Similarly, for rock outcrops, although the reported accuracy is only $85 \pm 8\%$ and we could identify wrongly classified rock outcrops in comparison to LIMA, we used them as they are. This helps to also be consistent with other studies that will use the same data sets for their purposes. For the same reasons (consistency, traceability), we have also not corrected basin outlines or drainage divides using flow velocity fields derived from satellite sensors because this was also already been done by Cook et al. (2014) for the catchment outlines. Alterations here would also impact the already-existing detailed classification of glacier fronts and we think it is better not to change this at this stage. Overall, results are as good as the source data used and their errors or incompleteness fully propagate into the products we have created here. However, we do not expect any major changes in the glacier characteristics or our overall conclusions with such corrections being implemented. Conversely, addressing the shortcomings and improving the related data sets is certainly an issue to be considered for future work.

5.6   Comparison with other regions

In comparison with other recently compiled glacier inventories in regions of similar environmental conditions (mountainous coastal regions with maritime climate), such as Alaska (Kienholz et al., 2015), Greenland (Rastner et al., 2012) and Svalbard (Nuth et al., 2013), the AP has the largest glacierized area (95 273 km^2), closely followed by Greenland (89 720 km^2), Alaska (86 723 km^2) and with some distance Svalbard (33 775 km^2). The AP also has the largest absolute, although only the second largest relative, area covered by marine-terminating glaciers, which are expected to react very sensitively to small changes in climate and associated ocean temperature changes. The glacier number and area distributions in the corresponding studies of Alaska, Greenland, Svalbard and the AP reveal that a few larger glaciers contribute the most to the area in all regions. This dominance is also reflected in a median area, which is considerably smaller than the mean area. However, in Alaska, Greenland and Svalbard the number of small glaciers is distinctively higher, with maximum counts between 0.25 and 1 km^2. The glaciers of the AP do not exhibit this pattern, which confirms findings by Pfeffer et al. (2014) for glaciers in the RGI Antarctic and subantarctic regions. Only the glaciers on Svalbard have a favored northern aspect (Nuth et al., 2013), which is interpreted as evidence for the importance of solar radiation incidence for glacier distribution in this region (Evans and Cox, 2010).

The bimodal hypsometric curve for the glaciers on the AP (Fig. 8) is very important compared to the parabolic shape of the three other regions that have increasing area percentages towards their mid-elevation. Hence, the AP has most of its glacierized area at lower elevations (around 200–500 m), with a secondary peak at higher elevations (around 1500–1900 m). Since the hypsometry of a glacier is an indicator of its climatic sensitivity (Jiskoot et al., 2009), this comparison reveals that the future evolution of AP glaciers cannot be modeled with the same simplified approaches as glaciers in other regions (Raper et al., 2000) and that volume loss for a small rise in the equilibrium line altitude (ELA) might indeed be high (Hock et al., 2009). The aspect preference with poleward tendencies of glacier distribution that is common in other mountain ranges (Evans, 2006, 2007; Evans and Cox, 2005, 2010) could not be found for the AP because the entire AP is glacierized and most glaciers are marine terminating.

5.7 Uncertainties

5.7.1 Impacts on outlines and meta-information

A wide range of interconnected uncertainties impact the glacier outlines and the associated meta-information. Since we have taken data sets from the literature as they are, we restrict the analysis of uncertainties to the information reported in the related studies (see Sect. 5.7.2) and add here a more generalized description of the respective impacts. *Glacier outlines* are composed of (A) the outlines from LIMA, (B) the drainage divides from Cook et al. (2014) and (C) the rock outcrop data set by Burton-Johnson et al. (2016). Key factors influencing their accuracy are related to (A1) grounding line position (only for glaciers merging with ice shelves), (A2) accuracy of the digitizing, (B1) the accuracy of the DEM from Cook et al. (2012), (C1) correct mapping (yes or no) and (C2) positional accuracy of the rock outcrops. Whereas the impact of (A2) and (C2) on the derived glacier areas is small since deviations are generally normally distributed (i.e., they only impact precision), impacts of (A1) and (B1) on glacier area can be large. However, for (B1) the impact is mostly on the size class distribution of the glaciers since a shift of an internal drainage divide does not change the total area. Factor (C1) might have a larger impact on smaller glaciers (i.e., a missed rock outcrop can increase glacier area by 5 % or more), but for most of the larger glaciers the area overestimation will be less than 1 or 2 %. Therefore, the largest impact on glacier area comes from source (A1), albeit only for a subsample (264) of glaciers merging with ice shelves.

The accuracy of the meta-information provided with each glacier (topographic parameters, ice thickness) depends on (D) the DEM used to calculate them, (E) the bedrock data set by Huss and Farinotti (2012) and (F) the glacier outlines that provide the perimeter for the calculation. For these sources we can identify the following impacts: (D1) a glacier is not or only partly covered by DEM information, (D2) the parameter is more or less impacted by DEM accuracy, (E1) there is direct propagation of DEM accuracy (slope) into the ice thickness calculation, (E2) there is missing consideration of rock outcrops (C) in (E), and (F1) changes of the meta-information due to errors in the extent. For the latter (F1), one can expect under- or overestimation of mean slope in case of a grounding line being too extensive (resulting in more area with small slopes) or rock outcrops in steep terrain having been missed (resulting in more area with steep slopes). Positional uncertainty will impact all topographic parameters, but very likely not systematically (i.e., not resulting in a bias) since terrain differences should average out. Uncertainty source (E1) is highly variable and discussed by Huss and Farinotti (2012), and (E2) causes inconsistencies among the derived glacier volumes, but overall differences are likely small since they are not systematic. Glaciers not covered by DEM cells (D1) have simply no data, but for those partly covered, the existing DEM cells have been used for the calculation. Depending on the coverage, results might still be useful, but in general it might be better to also set them to no data to avoid misinterpretation. Finally, the impact of (D2) will vary with the parameter. These parameters calculated from individual cells (e.g., minimum or maximum elevation) will be more strongly influenced by DEM errors or artifacts than those based on aggregate numbers such as mean or median elevation (Frey and Paul, 2012). A quantitative assessment of the related impacts can only be performed once a better DEM is available for the region. So far, the manually corrected DEM from Cook et al. (2012) is likely the best data set available.

5.7.2 Input data uncertainties

The uncertainties for the input data sets used are given as follows. The positional accuracy of the grounding line varies strongly with the nature of the boundary and is given as ±502 m for the outlet glacier boundaries merging with ice shelves (Bindschadler et al., 2011). For the outline positions of other glaciers, an uncertainty of ±2 pixel (30 m) is assumed. The classification of rock outcrops is based on an automated but manually checked classification. The mean value for correct pixel identification is given as 85 ± 8 % (Burton-Johnson et al., 2016). The application of the DEM causes uncertainties in drainage divides and topographic parameters. According to Cook et al. (2012), the accuracy of the DEM is < 200 m horizontally and about ±25 m vertically, but it varies regionally. Large shifts of the outlines in flat terrain are thus possible, causing the highly variable impacts on glacier area as described before. The impact of DEM uncertainties on topographic parameters is higher for smaller glaciers and those depending on single-cell values. Given our experiences with other DEMs, we estimate the uncertainty to be ± 50 m for all elevations and $\pm 5°$ for mean slope and aspect. For the regionally varying uncertainty of thickness and the corresponding uncertainty in volume and SLE, we refer to the detailed estimates of Huss and Farinotti (2012).

5.7.3 Further comments on the input data sets

The DEM of Cook et al. (2012) currently provides the highest resolution and quality for the area of the AP. However, the DEM only covers $93\,250$ km^2 and hence 98.4 % of the total glacierized area. In consequence, the calculation of the topographic parameters (mean, median, minimum and maximum elevation, slope, aspect) was not possible for 48 glaciers, representing an area of 1044 km^2, or 3 % (1 %) of the total number (area). For example, the region of Renaud and Biscoe islands at the midwestern coast of the AP does not have any elevation information. As some glaciers are only partially covered by the DEM, their parameters are likely based on a nonrepresentative part of the glacier. Moreover, glacier hypsometry is calculated based on the DEM and represents only the area covered by the DEM. Since the ice thickness and bedrock data set of Huss and Farinotti (2014) is also based on the DEM of Cook et al. (2012), mean thickness and volume could not be calculated for 50 glaciers. We suggest adding the now missing topographic information as soon as these glaciers are covered by a DEM of appropriate quality (e.g., the forthcoming TanDEM-X DEM). This new DEM might also then be used to recalculate drainage divides, grounding lines, glacier extents and ice thickness distribution considering the improved rock outcrop data set. This would also help to overcome the current inconsistencies among the applied data sets.

The new rock outcrop data set already has a higher accuracy and is more consistent than the former data set provided by the ADD. However, as mentioned above, some areas are still misclassified and an in-depth check and correction for the glaciers on the AP would help further improve the new inventory. Conversely, manual correction of these errors for the entire region would remove the traceability to the source data sets and we decided to maintain it for this first version.

To divide the glaciers from ice shelves, Cook et al. (2014) used the grounding line based on the ASAID project data source (Bindschadler et al., 2011), modified in places with features visible in the LIMA. Because the definition of the location of the grounding line significantly influences the extent of a glacier flowing into an ice shelf, grounding line positions obtained from new and forthcoming techniques will also alter glacier extent. However, this is then more a matter of definition rather than uncertainty.

5.8 Assignment of connectivity levels

In the south, the assignment of connectivity levels corresponds with the Antarctic ice-sheet drainage divides from the Cryosphere Science Laboratory of NASA's Earth Sciences Divisions (Zwally et al., 2012), which has assigned all unconnected glaciers (on islands) a CL0 and all glaciers on the AP CL1, following the suggestion by Rastner et al. (2012) for peripheral glaciers on Greenland. It is certainly the simplest possibility for such an assignment, but we think it is nevertheless sensible and fulfils its purpose. Consistency with earlier applications (e.g., all glaciers in the inventory by Bliss et al., 2013 have CL0) and transparency of the method are further benefits. It also allows the glacier and ice-sheet measuring and modeling communities to perform their work with their respective methods and determine, for example, past or future mass loss and/or sea level contributions independently. This would allow a cross check of methods for individual glaciers that are not resolved (such as results from gravimetry and glacier models) and possibly also explain remaining differences between methods (Shepherd et al., 2012; Briggs et al., 2017). The problem of double counting the contributions can also be avoided.

5.9 Specific characteristics of the AP glaciers

The ELA for a balanced budget (ELA$_0$) of land-terminating glaciers can be well approximated from topographic indices such as the mean, median or midpoint elevation (e.g., Braithwaite and Raper, 2009). The ELA is also a good proxy for precipitation (Ohmura, 1992; Oerlemans, 2005) and useful for modeling the effect of rising temperatures on future glacier extent (Zemp et al., 2006, 2007; Paul et al., 2007; Cogley et al., 2011;). However, since the glaciers of the AP are mainly marine-terminating glaciers (Vaughan et al., 2013), the lower limits of these glaciers are predefined and the ELA variability is largely determined by the variability of the topography (i.e., its maximum elevation). The increasing median elevation towards the interior (Fig. S1) does not result from decreasing precipitation towards the interior but is a consequence of glacier hypsometry and depends on whether a glacier reaches sea level or not.

The bimodal shape of the hypsometry, revealing that half of the glacierized areas are situated below 800 m a.s.l., as well as the high areal fraction of marine-terminating glaciers, indicates a high sensitivity of the AP glaciers to rising air and water temperatures (Hock et al., 2009). Due to their special hypsometry, their sensitivity is likely higher than for glaciers in Alaska, Greenland or Svalbard since these have a smaller fraction of marine-terminating glaciers and a smaller share of area at very low elevations.

The total ice volume and the volume below sea level are necessary for accurate estimations of the sea level contribution. At 54 mm the AP's glaciers have a higher contribution potential than the glaciers of Alaska (45 mm), Central Asia

(10 mm), the Greenland periphery (38 mm), the Russian Arctic (31 mm) or Svalbard (20 mm) (Huss and Hock, 2015). In total, global glaciers have a potential SLR of approximately 374 mm (Huss and Hock, 2015) to 500 mm (Huss and Farinotti, 2012; Vaughan et al., 2013), which is still significant for low-lying coastal regions (Paul, 2011; Marzeion and Levermann, 2014). Compared to the Antarctic ice sheet, with a SLE of 58.3 m (Vaughan et al., 2013), the SLE of the AP seems negligible. However, regarding the high sensitivity and much shorter response times of these glaciers to climate change, they are expected to be major contributors to SLR in the next decades (Hock et al., 2009). As the contribution of the AP's glaciers has not yet been fully considered in most studies, the new inventory can now be used to model their evolution explicitly with the current best approaches (e.g., Huss and Hock, 2015).

The results presented here allow a rough approximation of the consequences of ongoing climate change for the AP: with respect to the hypsometry, the lowest 800 m and hence 50 % of the glacierized area is prone to rising ablation and mass loss, causing a sea level contribution of roughly 50 % of the total AP SLE (27 mm). Regarding the glacier termini, about 30 % of the glacierized areas flow into the Larsen C ice shelf. Collapse of this ice shelf (similar to Larsen A and B), which may happen soon due to a growing rift (Jansen et al., 2015), would likely cause rapid dynamic thinning of its tributary glaciers (e.g., Rott et al., 2011) due to debuttressing. About 15 % of the SLE (9 mm) from the lowest 800 m is attached to the Larsen C ice shelf.

6   Conclusions

The compilation of a glacier inventory of the AP (63–70° S, Graham Land), consisting of glacier outlines accompanied by glacier-specific parameters, was achieved by combining already existing data sets with GIS techniques. The exclusion of rock outcrops by using the latest corresponding data set of the ADD (Burton-Johnson et al., 2016) from the glacier catchment outlines of Cook et al. (2014) resulted in 1589 glacier outlines (excluding ice shelves and islands < 0.5 km^2), covering an area of 95 273 km^2. Combining the outlines with the DEM of Cook et al. (2012) enabled us to derive several topographic parameters for each glacier. By applying the bedrock data set of Huss and Farinotti (2014), volume and mean thickness information was calculated for each glacier.

Connectivity levels with the ice sheet were assigned to all glaciers following Rastner et al. (2012) to facilitate observations and modeling by different groups. We started with a simple and transparent rule: glaciers south of 70° S (Palmer Land) are assigned CL2 and are regarded as being part of the ice sheet, while all glaciers north of it and on the AP are assigned CL1 and all glaciers on surrounding islands are assigned CL0. The resulting inventory and its quality are largely influenced by the availability and accessibility of accurate auxiliary data sets. For instance, the DEM does only cover 98.4 % of the glacierized area. Hence, for 50 glaciers the topographic parameters, thickness and volume information are missing. For other glaciers, the values are not representative for the entire glacier because smaller parts have no DEM information. Future improved DEMs might help completely cover these glaciers.

Since GLIMS now provides the complete glacier outlines data set of the AP (see glims.org), a significant gap in the global glacier inventory has been closed and a major contribution for forthcoming regional and global glaciological investigations can be made. Furthermore, the new inventory demonstrates the potential for improving knowledge about glacier characteristics, sensitivities and similarities and differences to glaciers in other regions. With the full inventory now freely available, approaches to improving, extending and further investigating the glaciers of the AP are strongly encouraged.

Author contributions. Jacqueline Huber compiled and analyzed this data set in the framework of a master's thesis and prepared the manuscript. Frank Paul and Michael Zemp supervised the thesis and helped with the preparation of the manuscript. Alison J. Cook generated the glacier catchment data set and provided this data set, including information about the generation process.

Competing interests. The authors declare that they have no conflict of interest.

Acknowledgements. Jacqueline Huber and Michael Zemp acknowledge financial support by the Swiss GCOS at the Federal Office of Meteorology and Climatology MeteoSwiss. The work of Frank Paul is funded by the ESA project Glaciers_cci (4000109873/14/I–NB). We are grateful to the LIMA Project, the Antarctic Digital Database and the National Snow and Ice Data Center and The Cryosphere Science Laboratory of NASA's Earth Sciences Divisions for free download of their data, allowing us to realize this study. We would also like to thank Matthias Huss and Daniel Farinotti for making their ice thickness data available as well as for their practical input. The inventory and the study benefited greatly from their bedrock data set. Finally, we would like to thank R. Drews, B. Marzeion and the anonymous referee for the constructive suggestions for improving the quality of this paper.

Edited by: R. Drews

References

ADD Consortium: Antarctic Digital Database, Version 6.0, available at: http://add.scar.org/home/add6, last access: 4 May 2015, 2012.

Arendt, A., Bolch, T., Cogley, J. G., Gardner, A., Hagen, J.-O., Hock, R., Kaser, G., Pfeffer, W. T., Moholdt, G., Paul, F., Radić, V., Andreassen, M., Bajracharya, S., Beedle, M. J., Berthier, E., Bhambri, R., Bliss, A., Brown, I., Burgess, D. O., Burgess, E. W., Cawkwell, F., Chinn, T., Copland, L., Davies, B., de Angelis, H., Dolgova, H., Filbert, K., Forester, R., Fountain, A., Frey, H., Giffen, B., Glasser, N., Gurney, S., Hagg, W., Hall, D., Haritashya, U. K., Hartmann, G., Helm, C., Herreid, S., Howat, I. M., Kapustin, G., Khromova, T., Kienholz, C., Koenig, M., Kohler, J., Kriegel, D., Kutuzov, S., Lavrentiev, I., Le Bris, R., Lund, J., Manley, W. F., Mayer, C., Miles, E. S., Li, X., Menounos, B., Mercer, A., Mölg, N., Mool, P., Nosenko, G., Negrete, A., Nuth, C., Pettersson, R., Racoviteanu, A., Ranzi, R., Rastner, P., Rau, F., Raup, B. H., Rich, J., Rott, H., Schneider, C., Seliverstov, Y., Sharp, M. J., Sigurosson, O., Stokes, C. R., Wheate, R., Wolken, G. J., Wyatt, F., and Zheltyhina, N.: Randolph Glacier Inventory – A Dataset of Global Glacier Outlines: Version 5.0, Global Land Ice Measurements from Space, Boulder CO, USA, 2015.

Arigony-Neto, J., Skvarca, P., Marinsek, S., Braun, M., Humbert, A., Júnior, C. W. M., and Jaña, R.: Monitoring Glacier Changes on the Antarctic Peninsula, in: Global Land Ice Measurements from Space, edite by: Kargel, J. S., Leonard, G. J., Bishop, M. P., Kääb, A., and Raup, B. H., Springer Berlin Heidelberg, Berlin, Heidelberg, 717–741, doi:10.1007/978-3-540-79818-7_30, 2014.

Bindschadler, R., Vornberger, P., Fleming, A. H., Fox, A. J., Mullins, J., Binnie, D., Paulsen, S. J., Granneman, B., and Gorodetzky, D.: The Landsat Image Mosaic of Antarctica, Remote Sens. Environ., 112, 4214–4226, doi:10.1016/j.rse.2008.07.006, 2008.

Bindschadler, R., Choi, H., Wichlacz, A., Bingham, R., Bohlander, J., Brunt, K., Corr, H., Drews, R., Fricker, H., Hall, M., Hindmarsh, R., Kohler, J., Padman, L., Rack, W., Rotschky, G., Urbini, S., Vornberger, P., and Young, N.: Getting around Antarctica: new high-resolution mappings of the grounded and freely-floating boundaries of the Antarctic ice sheet created for the International Polar Year, The Cryosphere, 5, 569–588, doi:10.5194/tc-5-569-2011, 2011.

Bliss, A., Hock, R., and Cogley, J. G.: A new inventory of mountain glaciers and ice caps for the Antarctic periphery, Ann. Glaciol., 54, 191–199, doi:10.3189/2013AoG63A377, 2013.

Bolch, T., Menounos, B., and Wheate, R.: Landsat-based inventory of glaciers in western Canada, 1985–2005, Remote Sens. Environ., 114, 127–137, doi:10.1016/j.rse.2009.08.015, 2010.

Braithwaite, R. J. and Raper, S.: Estimating equilibrium-line altitude (ELA) from glacier inventory data, Ann. Glaciol., 50, 127–132, doi:10.3189/172756410790595930, 2009.

Briggs, K. H., Shepherd, A., Muir, A., Gilbert, L., McMillan, M., Paul, F., and Bolch, T.: Sustained high rates of mass loss from Greenland's glaciers and ice caps, Geophys. Res. Lett., submitted, 2017.

Burton-Johnson, A., Black, M., Fretwell, P. T., and Kaluza-Gilbert, J.: An automated methodology for differentiating rock from snow, clouds and sea in Antarctica from Landsat 8 imagery: a new rock outcrop map and area estimation for the entire Antarctic continent, The Cryosphere, 10, 1665–1677, doi:10.5194/tc-10-1665-2016, 2016.

Cogley, J. G.: Area of the Ocean, Mar. Geod., 35, 379–388, doi:10.1080/01490419.2012.709476, 2012.

Cogley, J. G., Hock, R., Rasmussen, L. A., Arendt, A. A., Bauder, A., Braithwaite, R. J., Jansson, P., Kaser, G., Möller, M., Nicholson, L., and Zemp, M.: Glossary of glacier mass balance and related terms, Technical documents in hydrology, IHP-VII, No. 86, UNESCO, Paris, France, vi, 114, doi:10.1657/1938-4246-44.2.256b, 2011.

Cook, A. J., Murray, T., Luckman, A., Vaughan, D. G., and Barrand, N. E.: A new 100-m Digital Elevation Model of the Antarctic Peninsula derived from ASTER Global DEM: methods and accuracy assessment, Earth Syst. Sci. Data, 4, 129–142, doi:10.5194/essd-4-129-2012, 2012.

Cook, A. J., Vaughan, D. G., Luckman, A. J., and Murray, T.: A new Antarctic Peninsula glacier basin inventory and observed area changes since the 1940s, Antarct. Sci., 26, 614–624, doi:10.1017/S0954102014000200, 2014.

Cook, A. J., Holland, P. R., Meredith, M. P., Murray, T., Luckman, A., and Vaughan, D. G.: Ocean forcing of glacier retreat in the western Antarctic Peninsula, Science, 353, 283–286, doi:10.1126/science.aae0017, 2016.

Davies, B. J., Hambrey, M. J., Smellie, J. L., Carrivick, J. L., and Glasser, N. F.: Antarctic Peninsula Ice Sheet evolution during the Cenozoic Era, Quaternary Sci. Rev., 31, 30–66, doi:10.1016/j.quascirev.2011.10.012, 2012.

DiMarzio, J. P.: GLAS/ICESat 500 m Laser Altimetry Digital Elevation Model of Antarctica, 1st Edn., National Snow and Ice Data Center (NSIDC), Boulder, Colorado USA, doi:10.5067/K2IMI0L24BRJ, 2007.

Evans, I. S.: Local aspect asymmetry of mountain glaciation: A global survey of consistency of favoured directions for glacier numbers and altitudes, Geomorphology, 73, 166–184, doi:10.1016/j.geomorph.2005.07.009, 2006.

Evans, I. S.: Glacier distribution and direction in the Arctic: The unusual nature of Svalbard, Landform Analysis, 5, 21–24, 2007.

Evans, I. S. and Cox, N. J.: Global variations of local asymmetry in glacier altitude: Separation of north–south and east–west components, J. Glaciol., 51, 469–482, doi:10.3189/172756505781829205, 2005.

Evans, I. S. and Cox, N. J.: Climatogenic north–south asymmetry of local glaciers in Spitsbergen and other parts of the Arctic, Ann. Glaciol., 51, 16–22, doi:10.3189/172756410791392682, 2010.

Fretwell, P., Pritchard, H. D., Vaughan, D. G., Bamber, J. L., Barrand, N. E., Bell, R., Bianchi, C., Bingham, R. G., Blankenship, D. D., Casassa, G., Catania, G., Callens, D., Conway, H., Cook, A. J., Corr, H. F. J., Damaske, D., Damm, V., Ferraccioli, F., Forsberg, R., Fujita, S., Gim, Y., Gogineni, P., Griggs, J. A., Hindmarsh, R. C. A., Holmlund, P., Holt, J. W., Jacobel, R. W., Jenkins, A., Jokat, W., Jordan, T., King, E. C., Kohler, J., Krabill, W., Riger-Kusk, M., Langley, K. A., Leitchenkov, G., Leuschen, C., Luyendyk, B. P., Matsuoka, K., Mouginot, J., Nitsche, F. O., Nogi, Y., Nost, O. A., Popov, S. V., Rignot, E., Rippin, D. M., Rivera, A., Roberts, J., Ross, N., Siegert, M. J., Smith, A. M., Steinhage, D., Studinger, M., Sun, B., Tinto, B. K., Welch, B. C., Wilson, D., Young, D. A., Xiangbin, C., and Zirizzotti, A.: Bedmap2: improved ice bed, surface and thickness datasets for Antarctica, The Cryosphere, 7, 375–393, doi:10.5194/tc-7-375-2013, 2013.

Frey, H. and Paul, F.: On the suitability of the SRTM DEM and ASTER GDEM for the compilation of topographic parameters in glacier inventories, Int. J. Appl. Earth Obs., 18, 480–490, doi:10.1016/j.jag.2011.09.020, 2012.

GLIMS and NSIDC: Global Land Ice Measurements from Space glacier database, Compiled and made available by the international GLIMS community and the National Snow and Ice Data Center, Boulder CO, USA, doi:10.7265/N5V98602, 2005, updated 2016.

Haran, T., Bohlander, J., Scambos, T., Painter, T., and Fahnestock, M.: MODIS Mosaic of Antarctica 2003–2004 (MOA2004), NSIDC: National Snow and Ice Data Center, Boulder Colorado, USA, doi:10.7265/N5ZK5DM5, 2005, updated 2013.

Helm, V., Humbert, A., and Miller, H.: Elevation and elevation change of Greenland and Antarctica derived from CryoSat-2, The Cryosphere, 8, 1539-1559, doi:10.5194/tc-8-1539-2014, 2014.

Hock, R., de Woul, M., Radić, V., and Dyurgerov, M.: Mountain glaciers and ice caps around Antarctica make a large sea-level rise contribution, Geophys. Res. Lett., 36, L07501, doi:10.1029/2008GL037020, 2009.

Hulbe, C. L., Scambos, T. A., Youngberg, T., and Lamb, A. K.: Patterns of glacier response to disintegration of the Larsen B ice shelf, Antarctic Peninsula, Global Planet. Change, 63, 1–8, doi:10.1016/j.gloplacha.2008.04.001, 2008.

Huss, M. and Farinotti, D.: Distributed ice thickness and volume of all glaciers around the globe, J. Geophys. Res., 117, F04010, doi:10.1029/2012JF002523, 2012.

Huss, M. and Farinotti, D.: A high-resolution bedrock map for the Antarctic Peninsula, The Cryosphere, 8, 1261–1273, doi:10.5194/tc-8-1261-2014, 2014.

Huss, M. and Hock, R.: A new model for global glacier change and sea-level rise, Front. Earth Sci., 3, 382, doi:10.3389/feart.2015.00054, 2015.

Jansen, D., Luckman, A. J., Cook, A., Bevan, S., Kulessa, B., Hubbard, B., and Holland, P. R.: Brief Communication: Newly developing rift in Larsen C Ice Shelf presents significant risk to stability, The Cryosphere, 9, 1223–1227, doi:10.5194/tc-9-1223-2015, 2015.

Jenkins, A. and Holland, D.: Melting of floating ice and sea level rise, Geophys. Res. Lett., 34, L16609, doi:10.1029/2007GL030784, 2007.

Jiskoot, H., Curran, C. J., Tessler, D. L., and Shenton, L. R.: Changes in Clemenceau Icefield and Chaba Group glaciers, Canada, related to hypsometry, tributary detachment, length–slope and area–aspect relations, Ann. Glaciol., 50, 133–143, doi:10.3189/172756410790595796, 2009.

Kienholz, C., Herreid, S., Rich, J. L., Arendt, A. A., Hock, R., and Burgess, E. W.: Derivation and analysis of a complete modern-date glacier inventory for Alaska and northwest Canada, J. Glaciol., 61, 403–420, doi:10.3189/2015JoG14J230, 2015.

Kienholz, C., Hock, R., and Arendt, A. A.: A new semi-automatic approach for dividing glacier complexes into individual glaciers, J. Glaciol., 59, 925–937, doi:10.3189/2013JoG12J138, 2013.

Kunz, M., King, M. A., Mills, J. P., Miller, P. E., Fox, A. J., Vaughan, D. G., and Marsh, S. H.: Multi-decadal glacier surface lowering in the Antarctic Peninsula, Geophys. Res. Lett., 39, L19502, doi:10.1029/2012GL052823, 2012.

Marzeion, B. and Levermann, A.: Loss of cultural world heritage and currently inhabited places to sea-level rise, Environ. Res. Lett., 9, 034001, doi:10.1088/1748-9326/9/3/034001, 2014.

Nuth, C., Kohler, J., König, M., von Deschwanden, A., Hagen, J. O., Kääb, A., Moholdt, G., and Pettersson, R.: Decadal changes from a multi-temporal glacier inventory of Svalbard, The Cryosphere, 7, 1603–1621, doi:10.5194/tc-7-1603-2013, 2013.

Oerlemans, J.: Extracting a climate signal from 169 glacier records, Science, 308, 675–677, doi:10.1126/science.1107046, 2005.

Ohmura, A.: Energy and mass balance during the melt season at the equilibrium line altitude, Paakitsoq, Greenland Ice Sheet (69°34′25.3″ North, 49°17′44.1″ West, 1155 M A.S.L.), ETH Greenland Expedition Progress Report No. 2, Dept of Geography, ETH Zürich, 94 pp., 1992.

Paul, F.: Sea-level rise: Melting glaciers and ice caps, Nat. Geosci., 4, 71–72, doi:10.1038/ngeo1074, 2011.

Paul, F., Kääb, A., Maisch, M., Kellenberger, T., and Haeberli, W.: The new remote-sensing-derived Swiss glacier inventory: I. Methods, Ann. Glaciol., 34, 355–361, doi:10.3189/172756402781817941, 2002.

Paul, F., Maisch, M., Rothenbühler, C., Hoelzle, M., and Haeberli, W.: Calculation and visualisation of future glacier extent in the Swiss Alps by means of hypsographic modelling, Global Planet. Change, 55, 343–357, doi:10.1016/j.gloplacha.2006.08.003, 2007.

Paul, F., Barry, R. G., Cogley, J. G., Frey, H., Haeberli, W., Ohmura, A., Ommanney, C., Raup, B., Rivera, A., and Zemp, M.: Recommendations for the compilation of glacier inventory data from digital sources, Ann. Glaciol., 50, 119–126, doi:10.3189/172756410790595778, 2009.

Pfeffer, W. T., Arendt, A. A., Bliss, A., Bolch, T., Cogley, J. G., Gardner, A. S., Hagen, J.-O., Hock, R., Kaser, G., Kienholz, C., Miles, E. S., Moholdt, G., Mölg, N., Paul, F., Radić, V., Rastner, P., Raup, B. H., Rich, J., and Sharp, M. J.: The Randolph Glacier Inventory: a globally complete inventory of glaciers, J. Glaciol., 60, 537–552, doi:10.3189/2014JoG13J176, 2014.

Pritchard, H. D. and Vaughan, D. G.: Widespread acceleration of tidewater glaciers on the Antarctic Peninsula, J. Geophys. Res., 112, F03S29, doi:10.1029/2006JF000597, 2007.

Racoviteanu, A. E., Paul, F., Raup, B., Khalsa, S. J. S., and Armstrong, R.: Challenges and recommendations in mapping of glacier parameters from space: results of the 2008 Global Land Ice Measurements from Space (GLIMS) workshop, Boulder, Colorado, USA, Ann. Glaciol., 50, 53–69, doi:10.3189/172756410790595804, 2009.

Raper, S. C. B., Brown, O., and Braithwaite, R. J.: A geometric glacier model for sea-level change calculations, J. Glaciol., 46, 357–368, doi:10.3189/172756500781833034, 2000.

Rastner, P., Bolch, T., Mölg, N., Machguth, H., Le Bris, R., and Paul, F.: The first complete inventory of the local glaciers and ice caps on Greenland, The Cryosphere, 6, 1483–1495, doi:10.5194/tc-6-1483-2012, 2012.

Rau, F., Mauz, F., Vogt, S., Khalsa, S. J. S., and Raup, B.: Illustrated GLIMS glacier classification manual: glacier classification guidance for the GLIMS inventory, Version 1, available at: www.glims.org/MapsAndDocs/guides.html, last acccess: 7 January 2016, 2005.

Raup, B. and Khalsa, S. J. S.: GLIMS Analysis Tutorial, available at: www.glims.org/MapsAndDocs/guides.html, last acccess: 7 January 2016, 2010.

Rignot, E. and Thomas, R. H.: Mass balance of polar ice sheets, Science, 297, 1502–1506, doi:10.1126/science.1073888, 2002.

Rignot, E., Mouginot, J., and Scheuchl, B.: Antarctic grounding line mapping from differential satellite radar interferometry, Geophys. Res. Lett., 38, L10504, doi:10.1029/2011GL047109, 2011.

Rott, H., Skvarca, P., and Nagler, T.: Rapid Collapse of Northern Larsen Ice Shelf, Antarctica, Science, 271, 788–792, doi:10.1126/science.271.5250.788, 1996.

Rott, H., Müller, F., Nagler, T., and Floricioiu, D.: The imbalance of glaciers after disintegration of Larsen-B ice shelf, Antarctic Peninsula, The Cryosphere, 5, 125–134, doi:10.5194/tc-5-125-2011, 2011.

Shepherd, A., Wingham, D., Payne, T., and Skvarca, P.: Larsen ice shelf has progressively thinned, Science, 302, 856–859, doi:10.1126/science.1089768, 2003.

Shepherd, A., Ivins, E. R., A, G., Barletta, V. R., Bentley, M. J., Bettadpur, S., Briggs, K. H., Bromwich, D. H., Forsberg, R., Galin, N., Horwath, M., Jacobs, S., Joughin, I., King, M. A., Lenaerts, J. T. M., Li, J., Ligtenberg, S. R. M., Luckman, A., Luthcke, S. B., McMillan, M., Meister, R., Milne, G., Mouginot, J., Muir, A., Nicolas, J. P., Paden, J., Payne, A. J., Pritchard, H., Rignot, E., Rott, H., Sorensen, L. S., Scambos, T. A., Scheuchl, B., Schrama, E. J. O., Smith, B., Sundal, A. V., van Angelen, J. H., van de Berg, W. J., van den Broeke, M. R., Vaughan, D. G., Velicogna, I., Wahr, J., Whitehouse, P. L., Wingham, D. J., Yi, D., Young, D., and Zwally, H. J.: A Reconciled Estimate of Ice-Sheet Mass Bal-

ance, Science, 338, 1183–1189, doi:10.1126/science.1228102, 2012.

Turner, J., Bindschadler, R., Convey, P., DiPrisco, G., Fahrbach, E., Gutt, J., Hodgson, D., Mayewski, P., and Summerhayes, C.: Antarctic Climate Change and the Environment. A contribution to the International Polar Year 2007–2008, Scientific Committee on Antarctic Research, Cambridge, 526 pp., 2009.

UNESCO: Combined Heat, Ice and Water Balances at Selected Glacier Basins – A contribution to the International Hydrological Decade – A guide for compilation and assemblage of data for glacier mass balance measurements, UNESCO Technical papers in hydrology, 5, 35 pp., 1970.

Vaughan, D. G., Comiso, J. C., Allison, I., Carrasco, J. Kaser, G., Kwok, R., Mote, P., Murray, T., Paul, F., Ren, J., Rignot, E., Solomina, O., Steffen, K., and Zhang, T.: Observations: Cryosphere, in: Climate Change 2013: The Physical Science Basis, Contribution of Working Group I to the Fifth Assessment Report of the Intergovernmental Panel on Climate Change, edited by: Stocker, T. F., Qin, D., Plattner, G.-K., Tignor, M., Allen, S. K., Boschung, J., Nauels, A., Xia, Y., Bex, V., and Midgley, P. M., Cambridge University Press, Cambridge, United Kingdom and New York, NY, USA, 2013.

WGMS and NSIDC: World Glacier Inventory, Compiled and made available by the World Glacier Monitoring Service, Zurich, Switzerland, and the National Snow and Ice Data Center, Boulder CO, USA, doi:10.7265/N5/NSIDC-WGI-2012-02, 1989, updated 2012.

Zemp, M., Haeberli, W., Hoelzle, M., and Paul, F.: Alpine glaciers to disappear within decades?, Geophys. Res. Lett., 33, L13504, doi:10.1029/2006GL026319, 2006.

Zemp, M., Hoelzle, M., and Haeberli, W.: Distributed modelling of the regional climatic equilibrium line altitude of glaciers in the European Alps, Global Planet. Change, 56, 83–100, doi:10.1016/j.gloplacha.2006.07.002, 2007.

Zwally, H. J., Giovinetto, M. B., Beckley, M. A., and Saba, J. L.: Antarctic and Greenland Drainage Systems, available at: http://icesat4.gsfc.nasa.gov/cryo_data/ant_grn_drainage_systems.php, last access: 7 January 2016, 2012.

PERMISSIONS

The contributors of this book come from diverse backgrounds, making this book a truly international effort. This book will bring forth new frontiers with its revolutionizing research information and detailed analysis of the nascent developments around the world.

We would like to thank all the contributing authors for lending their expertise to make the book truly unique. They have played a crucial role in the development of this book. Without their invaluable contributions this book wouldn't have been possible. They have made vital efforts to compile up to date information on the varied aspects of this subject to make this book a valuable addition to the collection of many professionals and students.

This book was conceptualized with the vision of imparting up-to-date information and advanced data in this field. To ensure the same, a matchless editorial board was set up. Every individual on the board went through rigorous rounds of assessment to prove their worth. After which they invested a large part of their time researching and compiling the most relevant data for our readers.

The editorial board has been involved in producing this book since its inception. They have spent rigorous hours researching and exploring the diverse topics which have resulted in the successful publishing of this book. They have passed on their knowledge of decades through this book. To expedite this challenging task, the publisher supported the team at every step. A small team of assistant editors was also appointed to further simplify the editing procedure and attain best results for the readers.

Apart from the editorial board, the designing team has also invested a significant amount of their time in understanding the subject and creating the most relevant covers. They scrutinized every image to scout for the most suitable representation of the subject and create an appropriate cover for the book.

The publishing team has been an ardent support to the editorial, designing and production team. Their endless efforts to recruit the best for this project, has resulted in the accomplishment of this book. They are a veteran in the field of academics and their pool of knowledge is as vast as their experience in printing. Their expertise and guidance has proved useful at every step. Their uncompromising quality standards have made this book an exceptional effort. Their encouragement from time to time has been an inspiration for everyone.

The publisher and the editorial board hope that this book will prove to be a valuable piece of knowledge for researchers, students, practitioners and scholars across the globe.

LIST OF CONTRIBUTORS

Adam Treverrow and Tim H. Jacka
Antarctic Climate and Ecosystems Cooperative Research Centre, University of Tasmania, Hobart 7004, Australia

Li Jun
SGT Inc., NASA Goddard Space Flight Center, Greenbelt, MD, USA

Anny Cazenave
LEGOS, France, and ISSI, Switzerland

Benoit Meyssignac
LEGOS, France

Michael Ablain
CLS, France

Magdalena Balmaseda
ECMWF, UK

Jonathan Bamber
U. Bristol, UK

Kurt Lambeck
ANU, Australia, and ISSI, Switzerland

Felix Landerer
JPL/Caltech, USA

Paul Leclercq
UIO, Norway

Benoit Legresy
CSIRO, Australia

Eric Leuliette
NOAA, USA

David Wiese
JPL/Caltech, USA

Susan Wijffels
CSIRO, Australia

Richard Westaway
U. Bristol, UK

Guy Woppelmann
U. La Rochelle, France

Bert Wouters
U. Utrecht, The Netherlands

Francisco Machío
Escuela Superior de Ingeniería y Tecnología, Universidad Internacional de La Rioja (UNIR), Calle Almansa, 101, 28040 Madrid, Spain

Ricardo Rodríguez-Cielos
Departamento de Señales, Sistemas y Radiocomunicaciones, ETSI de Telecomunicación, Universidad Politécnica de Madrid, Av. Complutense, 30, 20040 Madrid, Spain

Francisco Navarro, Javier Lapazaran and Jaime Otero
Departamento de Matemática Aplicada a las Tecnologías de la Información y las Comunicaciones, ETSI de Telecomunicación, Universidad Politécnica de Madrid, Av. Complutense, 30, 20040 Madrid, Spain

Ulrich Strasser, Thomas Marke, Rudolf Sailer and Johann Stötter
Department of Geography, University of Innsbruck, Innsbruck, 6020, Austria

Irmgard Juen, Michael Kuhn, Fabien Maussion, Lindsey Nicholson and Georg Kaser
Department of Atmospheric and Cryospheric Sciences, University of Innsbruck, Innsbruck, 6020, Austria

Ludwig Braun, Heidi Escher-Vetter and Christoph Mayer
Geodesy and Glaciology, Bavarian Academy of Sciences and Humanities, Munich, 80539, Germany

Klaus Niedertscheider
Hydrographic Service of Tyrol, Innsbruck, 6020, Austria

Markus Weber
Photogrammetry and Remote Sensing, Technical University of Munich, Munich, 80333, Germany

David Mennekes and Markus Weiler
Chair of Hydrology, University of Freiburg, 79098, Freiburg, Germany

Jan Seibert
Hydrology and Climate Unit, Department of Geography, University of Zurich, 8057, Zurich, Switzerland

Daphné Freudiger
Chair of Hydrology, University of Freiburg, 79098, Freiburg, Germany
Hydrology and Climate Unit, Department of Geography, University of Zurich, 8057, Zurich, Switzerland

Janin Schaffer, Ralph Timmermann, Jan Erik Arndt and Daniel Steinhage
Alfred Wegener Institute, Helmholtz Centre for Polar and Marine Research, Bremerhaven, Germany

Steen Savstrup Kristensen
DTU Technical University of Denmark, 2800 Lyngby, Denmark

Christoph Mayer
Bavarian Academy of Sciences and Humanities, Commission for Geodesy and Glaciology, Munich, Germany

Mathieu Morlighem
University of California, Irvine, Department of Earth System Science, Croul Hall, Irvine, California 92697-3100, USA

Christophe Genthon, Claudio Durán Alarcón and Brice Boudevillain
Univ. Grenoble Alpes, CNRS, IRD, Grenoble INP, IGE, 38000 Grenoble, France

Alexis Berne and Christophe Praz
Environmental Remote Sensing Laboratory, Environmental Engineering Institute, School of Architecture, Civil and Environmental Engineering, École Polytechnique Fédérale de Lausanne, 1015 Lausanne, Switzerland

Jacopo Grazioli
Federal Office of Meteorology and Climatology, MeteoSwiss, Locarno-Monti, Switzerland

Martin J. Siegert
Grantham Institute and Department of Earth Science and Engineering, Imperial College London, South Kensington, London, UK

Hafeez Jeofry
Grantham Institute and Department of Earth Science and Engineering, Imperial College London, South Kensington, London, UK
School of Marine Science and Environment, Universiti Malaysia Terengganu, Kuala Terengganu, Terengganu, Malaysia

Neil Ross
School of Geography, Politics and Sociology, Newcastle University, Claremont Road, Newcastle Upon Tyne, UK

Hugh F. J. Corr
British Antarctic Survey, Natural Environment Research Council, Cambridge, UK

Jilu Li
Center for the Remote Sensing of Ice Sheets, University of Kansas, Lawrence, Kansas, USA

Mathieu Morlighem
Department of Earth System Science, University of California, Irvine, Irvine, California, USA

Prasad Gogineni
Department of Electrical and Computer Engineering, The University of Alabama, Tuscaloosa, Alabama 35487, USA

Katrin Lindbäck, Jack Kohler, Alexandra Messerli and Kenichi Matsuoka
Norwegian Polar Insitute, Framsentret, Postboks 6606, Langnes, 9296 Tromsø, Norway

Rickard Pettersson and Dorothée Vallot
Department of Earth Sciences, Uppsala University, Villavägen 16, 752 36 Uppsala, Sweden

Christopher Nuth
University of Oslo, Postboks 1047 Blindern, 0316 Oslo, Norway

Kirsty Langley
Asiaq Greenland Survey, Postboks 1003, 3900 Nuuk, Greenland

Ola Brandt
Norwegian Coastal Administration, Kystveien 30, 4841 Arendal, Norway

Esteban Alonso-González, Alba Sanmiguel-Vallelado, Francisco Navarro-Serrano and Juan Ignacio López-Moreno
Pyrenean Institute of Ecology, CSIC, Zaragoza, Spain

Jesús Revuelto
Pyrenean Institute of Ecology, CSIC, Zaragoza, Spain
Météo-France – CNRS, CNRM (UMR3589), Centre d'Etudes de la Neige, Grenoble, France

Cesar Azorin-Molina
Pyrenean Institute of Ecology, CSIC, Zaragoza, Spain
Regional Climate Group, Department of Earth Sciences, University of Gothenburg, Gothenburg, Sweden

Ibai Rico
Pyrenean Institute of Ecology, CSIC, Zaragoza, Spain
University of the Basque Country. Department of Geography, Prehistory and Archaeology, Vitoria, Spain

Ricardo Rodríguez Cielos
Departamento de Señales, Sistemas y Radiocomunicaciones, ETSI de Telecomunicación, Universidad Politécnica de Madrid, Madrid, Spain

Julián Aguirre de Mata and Andrés Díez Galilea
Departamento de Ingeniería Topográfica y Cartografía, ETSI en Topografía, Geodesia y Cartografía, Universidad Politécnica de Madrid, Madrid, Spain

Marina Álvarez Alonso
Departamento de Lenguajes y Sistemas Informáticos e Ingeniería de Software, ETS de Ingenieros Informáticos, Universidad Politécnica de Madrid, Madrid, Spain

Pedro Rodríguez Cielos
Departamento de Matemática Aplicada, ETSI de Telecomunicación, Universidad de Málaga, Málaga, Spain

Francisco Navarro Valero
Departamento de Matemática Aplicada a las Tecnologías de la Información y las Comunicaciones, ETSI de Telecomunicación, Universidad Politécnica de Madrid, Madrid, Spain

Nico Mölg, Tobias Bolch, Philipp Rastner and Frank Paul
Department of Geography, University of Zurich, Winterthurerstr. 190, 8057 Zurich, Switzerland, Switzerland

Tazio Strozzi
Gamma Remote Sensing, Worbstr. 225, 3073 Gümligen, Switzerland

Rebecca Möller
Department of Geography, RWTH Aachen University, Aachen, Germany
Geological Institute, Energy and Minerals Resources Group, RWTH Aachen University, Aachen, Germany

Peter A. Kukla
Geological Institute, Energy and Minerals Resources Group, RWTH Aachen University, Aachen, Germany

Marco Möller
Department of Geography, RWTH Aachen University, Aachen, Germany
Institute of Geography, University of Bremen, Bremen, Germany
Geography Department, Humboldt-Universität zu Berlin, Berlin, Germany

Christoph Schneider
Geography Department, Humboldt-Universität zu Berlin, Berlin, Germany

Jacqueline Huber, Frank Paul and Michael Zemp
Department of Geography, University of Zürich–Irchel, Zürich, 8057, Switzerland

Alison J. Cook
Department of Geography, Swansea University, Swansea, SA2 SPP, UK
Department of Geography, Durham University, Durham, DH1 3LE, UK

Index

www.ingramcontent.com/pod-product-compliance
Lightning Source LLC
Chambersburg PA
CBHW082051190326
41458CB00010B/3505